Design and Analysis of Long-term Ecological Monitoring Studies

To provide useful and meaningful information, long-term ecological programs need to implement solid and efficient statistical approaches for collecting and analyzing data. This volume provides rigorous guidance on quantitative issues in monitoring, with contributions from world experts in the field. These experts have extensive experience in teaching fundamental and advanced ideas and methods to natural resource managers, scientists, and students.

The chapters present a range of tools and approaches, including detailed coverage of variance component estimation and quantitative selection among alternative designs; spatially balanced sampling; sampling strategies integrating design- and model-based approaches; and advanced analytical approaches such as hierarchical and structural equation modeling. Making these tools more accessible to ecologists and other monitoring practitioners across numerous disciplines, this is a valuable resource for any professional whose work deals with ecological monitoring.

Supplementary example software code is available at www.cambridge.org/9780521191548.

Robert A. Gitzen is a Research Scientist at the _____ _ources, University of Missouri, Columbia. He has worked closel_ _ational Park Service to develop monitoring studies in the northern Grea_ _ _ _ has conducted numerous research studies focused on quantitative methods and wi_ _ e conservation.

Joshua J. Millspaugh is Professor and Pauline O'Connor Distinguished Professor of Wildlife Management at the University of Missouri, Columbia. He has written and edited three previous books on quantitative methods in ecology, received state and national awards for teaching, and serves frequently on scientific panels addressing pressing conservation issues.

Andrew B. Cooper is Associate Professor in the School of Resource and Environmental Management, Simon Fraser University, British Columbia. A quantitative ecologist, he has worked extensively with federal, state/provincial, and regional fish and wildlife management agencies as well as with a number of environmental conservation organizations in the USA and Canada.

Daniel S. Licht is Regional Wildlife Biologist for the Midwest Region of the US National Park Service. Having worked on wildlife issues in many parts of the USA, his experience includes wildlife and habitat management and restoration, inventory and monitoring, research, and program administration.

Design and Analysis of Long-term Ecological Monitoring Studies

Edited by

ROBERT A. GITZEN

School of Natural Resources, University of Missouri, Columbia, USA

JOSHUA J. MILLSPAUGH

School of Natural Resources, University of Missouri, Columbia, USA

ANDREW B. COOPER

School of Resource and Environmental Management, Simon Fraser University, British Columbia, Canada

DANIEL S. LICHT

Midwest Region, National Park Service, Rapid City, South Dakota, USA

CAMBRIDGE
UNIVERSITY PRESS

CAMBRIDGE UNIVERSITY PRESS
Cambridge, New York, Melbourne, Madrid, Cape Town,
Singapore, São Paulo, Delhi, Mexico City

Cambridge University Press
The Edinburgh Building, Cambridge CB2 8RU, UK

Published in the United States of America by Cambridge University Press, New York

www.cambridge.org
Information on this title: www.cambridge.org/9780521191548

First published 2012

Printed in the United Kingdom at the University Press, Cambridge

A catalog record for this publication is available from the British Library

Library of Congress Cataloging in Publication data
Design and analysis of long-term ecological monitoring studies / edited by Robert A. Gitzen . . . [et al.].
    p.   cm.
Includes bibliographical references and index.
ISBN 978-0-521-19154-8 (hbk.) – ISBN 978-0-521-13929-8 (pbk.)
1. Environmental monitoring.   2. Ecology – Statistical methods.   I. Gitzen, Robert A., 1968–
QH541.15.M64D46   2012
363.17′63 – dc23   2012002684

ISBN 978-0-521-19154-8 Hardback
ISBN 978-0-521-13929-8 Paperback

Additional resources for this publication at
www.cambridge.org/9780521191548

Contents

See color plate section between pp. 360 and 361.

List of contributors

James K. Agee
School of Environmental and Forest Sciences, University of Washington, Seattle, Washington, USA.

Jonathan Bart
US Geological Survey, Forest and Rangeland Ecosystem Science Center, Snake River Field Station, Boise, Idaho, USA.

Robert E. Bennetts
National Park Service, Southern Plains Inventory and Monitoring Network, Des Moines, New Mexico, USA.

Hawthorne L. Beyer
Department of Ecology & Evolutionary Biology, University of Toronto, Toronto, Ontario, Canada.

Kenneth A. Bollen
Department of Sociology, University of North Carolina, Chapel Hill, North Carolina, USA.

Shawn L. Carter
US Geological Survey, National Climate Change & Wildlife Science Center, Reston, Virginia, USA.

Sarah J. Converse
US Geological Survey, Patuxent Wildlife Research Center, Laurel, Maryland, USA.

Steven G. Fancy
National Park Service, Inventory and Monitoring Division, Fort Collins, Colorado, USA.

Daniel Fink
Lab of Ornithology, Cornell University, Ithaca, New York, USA.

Richard A. Fuller
The Ecology Centre, School of Biological Sciences, The University of Queensland, St Lucia, Queensland, Australia.

Steven L. Garman
US Geological Survey, Rocky Mountain Geographic Science Center, Denver, Colorado, USA.

Robert A. Gitzen
Department of Fisheries and Wildlife Sciences, School of Natural Resources, University of Missouri, Columbia, Missouri, USA.

James B. Grace
US Geological Survey, National Wetlands Research Center, Lafayette, Louisiana, USA.

Brian R. Gray
US Geological Survey, Upper Midwest Environmental Sciences Center, La Crosse, Wisconsin, USA.

Wesley M. Hochachka
Lab of Ornithology, Cornell University, Ithaca, New York, USA.

Mevin B. Hooten
Department of Mathematics and Statistics, Utah State University, Logan, Utah, USA. (Current Address: US Geological Survey, Colorado Cooperative Fish and Wildlife Research Unit, Colorado State University, Fort Collins, Colorado, USA.)

Darren J. Johnson
IAP World Services, Inc., at US Geological Survey, National Wetlands Research Center, Lafayette, Louisiana, USA.

Douglas H. Johnson
US Geological Survey, Northern Prairie Wildlife Research Center, St. Paul, Minnesota, USA.

Liana N. Joseph
WCS Institute, Wildlife Conservation Society, The Bronx, New York, New York, USA and The Ecology Centre, School of Biological Sciences, The University of Queensland, St Lucia, Queensland, Australia.

Jon E. Keeley
US Geological Survey, Western Ecological Research Center, Sequoia and Kings Canyon Field Station, Three Rivers, California, USA.

William L. Kendall
US Geological Survey, Patuxent Wildlife Research Center, Laurel, Maryland, USA. (Current address: US Geological Survey, Colorado Cooperative Fish and Wildlife Research Unit, Colorado State University, Fort Collins, Colorado, USA.)

Thomas M. Kincaid
US Environmental Protection Agency, Western Ecology Division, Corvallis, Oregon, USA.

Yuancai Lei
Institute of Forest Resource Information Techniques, Chinese Academy of Forestry, Beijing, PR China.

Todd R. Lookingbill
Department of Geography and the Environment, University of Richmond, Richmond, Virginia, USA.

Darryl I. MacKenzie
Proteus Wildlife Research Consultants, Dunedin, New Zealand.

Daniel J. Manier
Arctic Slope Regional Corporation Management Services, contracted to: US Geological Survey, Fort Collins Science Center, Fort Collins, Colorado, USA.

Trent McDonald
Western EcoSystems Technology, Inc., Cheyenne, Wyoming, USA.

Joshua J. Millspaugh
Department of Fisheries and Wildlife Sciences, School of Natural Resources, University of Missouri, Columbia, Missouri, USA.

Clinton T. Moore
US Geological Survey, Patuxent Wildlife Research Center, Warnell School of Forestry and Natural Resources, University of Georgia, Athens, Georgia, USA. (Current address: US Geological Survey, Georgia Cooperative Fish and Wildlife Research Unit, Warnell School of Forestry and Natural Resources, University of Georgia, Athens, Georgia, USA.)

Anthony R. Olsen
US Environmental Protection Agency, Western Ecology Division, Corvallis, Oregon, USA.

Quinn Payton
US Environmental Protection Agency, Western Ecology Division, Corvallis, Oregon, USA.

Hugh P. Possingham
The Ecology Centre, School of Biological Sciences, The University of Queensland, St Lucia, Queensland, Australia.

Song S. Qian
Nicholas School of the Environment, Duke University, Durham, North Carolina, USA. (Current address: Center for Ecological Sciences, Tetra Tech, Inc., Research Triangle Park, North Carolina, USA.)

Joel H. Reynolds
US Fish and Wildlife Service, National Wildlife Refuge System, Division of Realty & Natural Resources, Anchorage, Alaska, USA. (Current address: Western Alaska Landscape Conservation Cooperative, Anchorage, Alaska, USA.)

Beth E. Ross
Department of Wildland Resources, Utah State University, Logan, Utah, USA.

J. Andrew Royle
US Geological Survey, Patuxent Wildlife Research Center, Laurel, Maryland, USA.

John Paul Schmit
National Park Service – Center for Urban Ecology, National Capital Region Inventory and Monitoring Program, Washington, DC, USA.

E. William Schweiger
National Park Service, Rocky Mountain Inventory and Monitoring Network, Fort Collins, Colorado, USA.

John R. Skalski
School of Aquatic & Fishery Sciences, University of Washington, Seattle, Washington, USA.

David R. Smith
US Geological Survey, Kearneysville, West Virginia, USA.

Michael B. Soma
US Geological Survey, Northern Prairie Wildlife Research Center, Minneapolis, Minnesota, USA.

N. Scott Urquhart
Retired, Department of Statistics, Colorado State University, Fort Collins, Colorado, USA.

Christopher A. Walter
The University of Montana, Missoula, Montana, USA.

Christopher K. Wikle
Department of Statistics, University of Missouri, Columbia, Missouri, USA.

John A. Young
US Geological Survey, Kearneysville, West Virginia, USA.

Benjamin Zuckerberg
Lab of Ornithology, Cornell University, Ithaca, New York, USA. (Current address: Department of Forest and Wildlife Ecology, University of Wisconsin – Madison, Madison, Wisconsin, USA.)

Foreword: Ecology, management, and monitoring

Ecological monitoring has a somewhat checkered history, in part because it has been a fuzzy concept. It has ranged from measuring a single variable at a single location over time to measuring multiple variables at a national scale, sometimes with quantitatively defensible designs and sometimes not. Most currently accepted definitions include measuring in some *convincing* way (and that adjective is critical) some aspect of ecological composition, structure, or function over time. In the past, managers have received more rewards from doing things than understanding the effectiveness of what they did. Funding for ecological monitoring, therefore, has been variable and usually has been the first victim of budget cuts. The scientific community has also been partly to blame. Where monitoring was employed, sometimes the wrong thing was measured, or measured in the wrong way. Monitoring was often designed with little attention to a conceptual framework, and seldom had a firm tie to policy and decision making.

In part due to this history, research and monitoring have been viewed as two distinct and separable activities. The former was seen as controlled, largely experimental, and rigorous, while the latter was seen as uncontrolled, largely observational, and inexact. This distinction has thankfully faded over time, and now research and monitoring are seen more as a continuum (Busch and Trexler 2003). Research is generally stronger at determining cause and effect. Monitoring contributes a spatial and temporal depth that is less commonly seen in research, and both are capable of testing hypotheses.

Ecological monitoring is as complex as the ecosystems it is intended to measure. Those ecosystems are in a constant state of flux. The only constant is change, and that change can be linear or nonlinear. Natural variation can make analysis difficult. Some populations will be naturally cyclic on an annual scale (salmon), or in the case of migratory animal populations, seasonally variable. Monitoring needs to acknowledge the following features of ecological systems: the significance of natural processes, the dynamic nature of ecological systems, the uncertainty and inherent variability of those systems, and the importance of cumulative effects (Dale *et al.* 1999).

How the system works, identified by conceptual models, is an important precursor to development of a monitoring plan. Evaluating ecosystem linkages helps to define the right things to measure, and may help in explaining the source of the ecological change being measured. Models help to justify and explain why a particular resource or species is important to measure.

Well-defined monitoring plans can help define ecological change in a reliable way. However, change alone is an insufficient metric for a monitoring plan. There must be thresholds of change that trigger action by managers, and managers need to be involved in defining those thresholds. This implicitly incorporates values in addressing ecological change as good, bad, or ugly. Consider a resource being monitored where, over time, dissimilarity among sites increases, remains stable, or decreases. Those data, independent of incorporating values, are not very meaningful without a management context. Increasing dissimilarity among sites may indicate increasing species richness, perhaps not desirable if invasive species are the cause. Decreasing dissimilarity among sites may indicate homogenization, perhaps desirable if disturbed sites are recovering to the levels of pristine sites, or perhaps undesirable if generalists are gaining at the expense of less common species. Management objectives provide an interpretive context for change. At some point, sufficient change occurs that management actions are needed, triggered not only by statistically significant change, but also its relevance to management objectives.

Ecological monitoring is an essential part of the process known as adaptive management (Holling 1978). Effective monitoring may show a need for new ecosystem assessments, new decisions concerning resource allocation or priorities for action, or changes in implementing actions. Substituting space for time (spatially separated sites where treatment occurred in the past) can provide short-term insights into the temporal change that might be expected over time, but is criticized due to assumptions of site uniformity before treatment/disturbance, uniformity of conditions following treatment/disturbance, and often subjective sampling designs. Ecological monitoring demands a patient approach, as most change in "noisy" ecosystems requires multiple years of data to provide reliable inference. Some monitoring protocols may be able to show change over a short time, but others may require a decade or two to show, for example, a 25% population change with 80% power at $\alpha = 0.10$. Higher sampling intensities may shorten that time, but with associated higher costs.

As scale increases from point to landscape, the issue of spatial balance of sampling emerges. Many monitoring plans are now at these larger ecoregion scales, so it is important to address sampling design as much as measurement methodology. Newer spatially balanced probability designs, such as the Generalized Random Tessellation Stratified Design (GRTS; Stevens and Olsen 2004) help in allowing unbiased inference from monitoring at these larger scales.

Monitoring programs still face many challenges. First, some monitoring plans are considered complete when they are not. In a recent survey of plant and animal population monitoring programs, only half the programs had determined in advance what statistical methods would be employed in evaluating change, and only 20% of the programs included issues of statistical power, with little difference between older (> 20 years) and recent (< 6 years) programs (Marsh and Trenham 2008). These data suggest that a majority of current monitoring plans (at least for plants and animals) are likely inept, due either to ignorance of statistics or a lack of ability to incorporate quantitative techniques. Those with excellent design and analysis will be of the highest utility to managers because statements of change (or lack of change) will be convincing. Second, stable funding of long-term monitoring is a real concern. Most funding institutions are public, and with that comes the inevitable cycle of periodic budget cuts associated with new laws, recessions, and the like. Ecological monitoring programs are in direct competition with social programs, and few have developed adequate rationales to compete with those programs in times of shrinking budgets. Not only is the scale of monitoring restricted when the budgets are cut, but the sample design itself may be compromised if only a portion of the monitoring can be funded.

These problems can be at least partly alleviated with defensible monitoring approaches. Rigorous design, implementation, and analysis for ecological monitoring are the wave of the future, destined to have the power of a tsunami. And the earthquake that spawns that tsunami is contained within the chapters of this book.

Dr. James K. Agee
Emeritus Professor of Forest Ecology
School of Environmental and Forest Sciences
University of Washington
Seattle, Washington, USA

Preface

Environmental monitoring is of fundamental importance to natural resource managers, scientists, and human society in general – consider the inarguable importance of quantifying changes in climate, air and water quality, surface and ground water dynamics, and similar attributes. However, monitoring studies also have the potential to be a significant waste of time and money (see, for example, discussions by Legg and Nagy 2006). To have value, a monitoring program needs to produce information of sufficient accuracy relevant to a clearly defined purpose, and to do so cost-effectively. Yet, even in the short term, natural populations and systems are inherently variable and usually difficult to study. Adding in a multi-year (usually multi-decade) focus creates many additional challenges and scales of uncertainty – and increases the potential amount of time and money wasted if these challenges are not adequately addressed. Many monitoring efforts have failed or will fail due to poorly defined objectives and inadequate designs (Yoccoz *et al.* 2001, Noon 2003, Legg and Nagy 2006, Lindenmayer and Likens 2010a). Yet, statisticians and ecologists have developed, and continue to develop, a rich body of knowledge and practical methods for addressing these challenges, and have applied these methods successfully at a variety of scales for a diversity of attributes. *Our goal for this volume is to help make some key components of this knowledge base, as well as new extensions, readily available and accessible to quantitative and applied natural resource scientists and managers, program managers, students, and consulting biometricians involved with environmental monitoring worldwide.*

We have simple motivations for producing this volume. As a result of three of the four editors' experience working or collaborating with the US National Park Service (NPS) Inventory and Monitoring Program, it became apparent that sampling tools and expert guidance not available in published texts were being regularly applied in the development of monitoring programs by or in partnership with the US Environmental Protection Agency (USEPA), US Geological Survey (USGS), NPS, and other entities. More generally, throughout the monitoring world, there were many pools of expertise and situations where existing analytical tools were being applied and extended for use in monitoring. However, much of this relevant expertise was not easily accessible to the broad audience of ecologists involved with monitoring. Information on a diversity of approaches, tools, and current developments was scattered widely in sources such as statistical journals, other texts not focused on monitoring, unpublished sources, and agency web pages, as well as in the collective professional experience of many biometricians and quantitative ecologists. We wanted to make this information and expertise more readily available to help all environmental monitoring programs increase their effectiveness and to stimulate further extensions of existing methods. Through this edited volume, we also wanted to help provide readers with diverse views and expert guidance directly from many world experts who have developed and guided the implementation of quantitative methodology in ecological monitoring.

Scope

This volume is intended to offer broad guidance on defining objectives for monitoring and on developing a survey design and analytical approach to meet these objectives. It is organized into five sections. The first section gives perspectives about defining the purpose for a monitoring program, an overview of important quantitative issues, and a review of the necessary statistical background for readers with relatively basic statistical training and knowledge. The second section focuses on critical issues and tools for designing monitoring programs – including probability sampling designs; planning the temporal component of monitoring; and estimating sources of variability that will affect the quality of information produced by the study. The third section focuses on a wide variety of methods and perspectives for analyzing monitoring data. Chapters in the fourth section focus on additional topics related to developing a monitoring program, planning a survey design, and analyzing data; this section focuses on advanced and specialized issues (e.g. hierarchical modeling, planning demographic monitoring of populations) and applications. The fifth, concluding, section illustrates how quantitative issues and other aspects of a monitoring program can be integrated, and how decisions about monitoring can be evaluated in the context of broader conservation and management goals of an organization.

This volume is intended to highlight general challenges in monitoring design and analysis, to demonstrate principles and some widely useful methods for addressing these challenges, and to provide perspectives from a diverse group of experts. Our goal is not to provide an exhaustive source of statistical methods or decision keys leading readers through the realm of design options to the specific approach they should use in their situation. Our expectation is that readers will use this volume in conjunction with general sampling texts, literature on monitoring, and other specialized references – and particularly in collaboration with statisticians with experience in monitoring. Similarly, chapter authors provide equations and software commands to help readers understand general methods, and steer readers to more specialized references for more detailed guidance.

Our intent was that every chapter in this volume would be relevant across a wide range of environmental disciplines. In terms of specific examples and scenarios discussed, there is a moderate bias towards wildlife-focused monitoring given the backgrounds of the editors, but the quantitative issues and tools discussed are of general importance. Conversely, some tools and approaches that have been a regular component of some large-scale monitoring programs, particularly programs coordinated by the USEPA, for >10 years (e.g. spatially balanced sampling) have become a standard part of the survey-design toolbox for other programs and disciplines only in the last few years. Therefore, while examples presented in chapters obviously are discipline-specific, chapters focus on concepts and methods of general importance to any environmental discipline. More generally, the volume offers a diversity of perspectives about how the experts involved see the world of environmental monitoring, and even how they define "monitoring".

Features

Technical accessibility

Besides relevance across disciplines, we also wanted this volume to have broad relevance to readers with varying levels of technical expertise and quantitative interests. We sought this breadth both at the level of the entire volume and within individual chapters. Collectively, chapters in the first and last sections ("Overview" and "Conclusion") focus less on quantitative details and more on large picture issues relevant to all readers. The middle sections focus on design and analysis, with most chapters focusing partly on technical details and considerations. All readers with some coursework or basic knowledge of applied environmental statistics should benefit from the general discussion in these chapters. In some chapters, ecologists with advanced backgrounds in statistics and some biometricians will benefit most from the technical details. In these latter cases, the authors remain focused on applied and practical issues, and on explaining methods at a level more accessible to ecologists compared to discussions in statistical journals. Most chapters include either multiple small real-world or realistic examples, or an extended case study, to make clear the relevance of the topics being discussed and to help readers better understand the application of methods discussed.

Chapter features

In addition, each chapter was structured to ensure its practical value to all readers, through the following features:

Introduction and summary. Each chapter has a general introduction and summary to document the context for the specific topics discussed in the chapter and to reiterate key messages.

Take-home messages. In the long term, time and money spent on sound statistical planning and investing in partnership with statistical experts pays off. However, time and money first need to be spent. Such investments require decisions by program managers and administrators who certainly want and need to understand the big picture, but possibly not all the technical details. Therefore, each chapter in this volume includes a "Take-home Messages for Program Managers" call-out box, to give readers the "big picture" in a less technical fashion and to help program managers see the relevance of the chapter to their program.

Common problems and difficult gray areas. The authors involved in this volume have extensive practical experience with monitoring programs, and understand possible common difficulties and sticking points related to their chapter. Therefore, chapters each have a "Common Challenges" call-out box in which authors briefly emphasize some common issues that readers likely will need to consider further as they apply what they learned.

Future research and development. To help define the limits of current knowledge or available tools and outline potential high-priority research/development needs, authors provide a *Future Research and Development* section near the end of each chapter. Moreover, the above-mentioned "Common Challenges" also often point out issues in need of further research and development of methodology.

Software applications

Software commands provided in chapters and online supplements are intended to point readers towards available tools (most commonly in R, SAS, or WinBUGS), help demonstrate the application of design and analytical approaches, and help some readers build on their existing knowledge to begin applying these approaches in their situations. In several cases, authors provide software code, background information, data sets used for example, and even analytical output as online supplements (www.cambridge.org/9780521191548). Authors usually steer readers to more extensive instruction and demonstration of software applications available from user's guides, web sites, specialized training, and other texts. (Throughout the volume, references to proprietary software by authors who are US government employees do not imply endorsement by the US government.)

Acknowledgments
We are highly grateful to all authors involved in this volume, for eagerly contributing their expertise and being very responsive to questions and requests. Doug Johnson, Tony Olsen, and Tom Kincaid deserve special mention for contributing more than one chapter. We thank authors for graciously allowing us, during final editing, to insert cross-references to other chapters to increase cohesiveness; any errors in such links are the responsibility of the editors. Peer review is critical to maximizing the quality and utility of this sort of edited volume, and we are very grateful to peer reviewers of each chapter, several of whom reviewed more than one chapter each. A complete list is provided of the reviewers from whom we obtained comments. Rob Bennetts and Steve Garman deserve special thanks for contributing chapters and serving as peer reviewers. Glen Sargeant, John Gross, and Tom Philippi provided helpful comments during development of the initial outline for this volume. Support and encouragement for development of this volume was provided by the NPS Inventory and Monitoring Program, Northern Great Plains Network (NGPN), and we are grateful to NGPN staff (particularly Kara Paintner-Green, Marcia Wilson, other I&M personnel, and park staff) and collaborators (particularly Amy Symstad of USGS). We thank the University of Missouri, School of Natural Resources and Department of Fisheries and Wildlife Sciences for their support in

completing this volume. Dr. Jack Jones, Department Chair, and Dr. Mark Ryan, Director of the School of Natural Resources, deserve special recognition for their encouragement and support of this project. We also thank the NPS Midwest Region and the School of Resource and Environmental Management, Simon Fraser University, for additional support. Finally, we are grateful to Abigail Jones at Cambridge University Press for her careful work and support in producing this volume.

Robert A. Gitzen
Joshua J. Millspaugh
Andrew B. Cooper
Daniel S. Licht

Acknowledgments

We acknowledge the following people for their help in reviewing the manuscript:

Michael Adams, US Geological Survey, Forest and Rangeland Ecosystem Science Center, Corvallis, Oregon, USA.

James Agee, School of Environmental and Forest Sciences, University of Washington, Seattle, Washington, USA.

Erik Beever, US Geological Survey, Alaska Science Center, Anchorage, Alaska, USA.

Robert Bennetts, National Park Service, Southern Plains Inventory and Monitoring Network, Des Moines, New Mexico, USA.

Michael Bevers, US Forest Service, Rocky Mountain Research Station, Fort Collins, Colorado, USA.

William Block, US Forest Service, Rocky Mountain Research Station, Flagstaff, Arizona, USA.

G. Scott Boomer, US Fish & Wildlife Service, Division of Migratory Bird Management, Laurel, Maryland, USA.

Jennifer Brown, Department of Mathematics and Statistics, University of Canterbury, Christchurch, New Zealand.

Jacob Carstensen, National Environmental Research Institute, Department of Marine Ecology, Roskilde, Denmark.

Duane Diefenbach, Pennsylvania Cooperative Fish & Wildlife Research Unit, Pennsylvania State University, University Park, Pennsylvania, USA.

David Elston, Biomathematics & Statistics Scotland, The James Hutton Institute, Craigiebuckler, Aberdeen, Scotland, UK.

E. David Ford, School of Environmental and Forest Sciences, University of Washington, Seattle, Washington, USA.

Steven Garman, US Geological Survey, Rocky Mountain Geographic Science Center, Denver, Colorado, USA.

John Gross, National Park Service, Inventory and Monitoring Program, Fort Collins, Colorado, USA.

Jonathan Haufler, Ecosystem Management Research Institute, Seeley Lake, Montana, USA.

Jean-Yves Humbert, Institute of Ecology and Evolution, University of Bern, Bern, Switzerland.

Kathryn Irvine, Department of Mathematical Sciences, Montana State University, Bozeman, Montana, USA.

Barbara Keller, Department of Fisheries and Wildlife Sciences, School of Natural Resources, University of Missouri, Columbia, Missouri, USA.

David P. Larsen, Pacific States Marine Fisheries Commission, c/o US Ecological Protection Agency, Western Ecology Division, Corvallis, Oregon, USA.

David Larsen, Department of Forestry, School of Natural Resources, University of Missouri, Columbia, Missouri, USA.

Michael Larson, Minnesota Department of Natural Resources, Forest Wildlife Populations and Research Group, Grand Rapids, Minnesota, USA.

John Lehmkuhl, US Forest Service, Pacific Northwest Research Station, Wenatchee Washington, USA.

Lyman McDonald, Western EcoSystems Technology, Inc., Laramie, Wyoming, USA.

Donald McKenzie, US Forest Service, Pacific Wildland Fire Sciences Lab, Seattle, Washington, USA.

Lloyd Morrison, US National Park Service, Inventory and Monitoring Program, Heartland Network/Department of Biology, Missouri State University, Springfield, Missouri, USA.

Vito Muggeo, Dipartimento Scienze Statistiche e Matematiche 'Vianelli', Università di Palermo, Palermo, Italy.

Barry Noon, Department of Fish, Wildlife, and Conservation Biology, Colorado State University, Fort Collins, Colorado, USA.

Craig Paukert, US Geological Survey, Missouri Cooperative Fish and Wildlife Research Unit, University of Missouri, Columbia, Missouri, USA.

Eric Rexstad, Research Unit for Wildlife Population Assessment, Centre for Research into Ecological and Environmental Modelling, University of St Andrews, St Andrews, Scotland, UK.

Christopher Rota, Department of Fisheries and Wildlife Sciences, School of Natural Resources, University of Missouri, Columbia, Missouri, USA.

Steve Sheriff, Missouri Department of Conservation, Resource Science Division, Columbia, Missouri, USA.

Stephen Stehman, Department of Forest and Natural Resources Management, SUNY College of Environmental Science and Forestry, Syracuse, New York, USA.

Amy Symstad, US Geological Survey, Northern Prairie Wildlife Research Center, Hot Springs, South Dakota, USA.

Frank R. Thompson, III, US Forest Service, Northern Research Station, Columbia, Missouri, USA.

William Thompson, National Park Service, Inventory and Monitoring Program, Anchorage, Alaska, USA.

William Venables, CSIRO/CMIS Cleveland Laboratories, Cleveland, Queensland, Australia.

Sanford Weisberg, School of Statistics, University of Minnesota, Minneapolis, Minnesota, USA.

Abbreviations

ACS	adaptive cluster sampling
AIC	Akaike's information criterion
ALT	autoregressive latent trajectory model
ARCL	autoregressive cross-lagged model
ARM	Adaptive Resource Management
ASH	average shifted histogram
ASTER	Advanced Spaceborne Thermal Emission and Reflection Radiometer
BACI	Before–After, Control–Impact
BBS	Breeding Bird Survey
CAR	conditional autoregressive structure
CART	classification and regression tree
CDF	cumulative distribution function
CJS	Cormack–Jolly–Seber
CMR	capture–mark–recapture
CvM	Cramér–von Mises
CWS	Canadian Wildlife Service
DEM	Digital Elevation Model
DIC	deviance information criterion
EMAP	Environmental Monitoring and Assessment Program
EPA	(US) Environmental Protection Agency
FIA	Forest Inventory and Analysis (Program)
FIML	full-information maximum likelihood
GAM	generalized additive model
GAMM	generalized additive mixed model
GCG	Gene Conservation Groups
GLM	generalized linear model
GLMM	generalized linear mixed model
GRS	general random sample
GRTS	Generalized Random Tessellation Stratified design
GUIDOS	Graphical User Interface for the Description of image Objects and their Shapes
IBI	index of biotic integrity
ICAR	Intrinsic-CAR
ICC	intracluster or intraclass correlation coefficient

IDEs	integro-difference equations
INLA	integrated nested Laplace approximations
KS	Kolmogorov–Smirnov
LISREL	linear structural relations
LTM	Latent Trajectory Model
MAPS	Monitoring Avian Productivity and Survivorship (Program)
MAR	missing at random
MCAR	missing completely at random
MCMC	Markov chain Monte Carlo
ML	maximum likelihood
MLE	ML estimator
MMDM	mean maximum distance moved
MNAR	missing not completely at random
MSE	mean squared error
NARS	National Aquatic Resource Surveys
NASS	National Agricultural Statistics Service
NES	National Eutrophication Survey
NGPN	Northern Great Plains Network
NPS	National Park Service
PAH	polycyclic aromatic hydrocarbons
PDEs	partial differential equations
PPS	probability proportional to size
QA	quality assurance
REML	restricted (residual) maximum likelihood
RRMSE	relative root mean square error
s.m.a.r.t.	*specific*, *measurable*, *achievable*, *results*-oriented, *time*-frame
SEM	structural equation modeling
SRS	simple random sampling
TCI	topographic convergence index
USFWS	US Fish and Wildlife Service
USGS	US Geological Survey
VPC	variance partition coefficients

Section I

Overview

1 Ecological monitoring

The heart of the matter

Robert A. Gitzen and Joshua J. Millspaugh

Introduction

Most environmental scientists and natural resource managers view long-term monitoring as a good thing. To fulfill their mandates and meet their goals, natural resource management and conservation organizations need information on the current status and patterns of change in high-priority resources, critical ecological processes, and stressors of high concern. To inform policy makers and the general public as well as management organizations, government agencies may have mandates for assessing the condition of selected natural resources in perpetuity because of the special importance of these resources to society and because of the possibility of both suspected and unforeseen changes to these resources. More directly, these organizations need information for assessing whether current management is effectively maintaining the state of managed systems and populations within the desired range of conditions, and for reducing uncertainty affecting management performance.

Moreover, there is incredibly high scientific value in many long-term data sets collected for management and conservation purposes, as well as data collected as part of long-term ecological research programs. Natural systems usually are characterized by complex temporal dynamics and interactions often not evident from a collection of short-term research studies. Data from long-term monitoring can be invaluable for empirical examination of hypotheses about spatial and temporal dynamics – for example, about community dynamics, population growth and density dependence, and influences of infrequent extreme climatic events. With the possibility of rapid and major changes to the Earth's climate, there is unprecedented recognition of the importance of existing long-term data sets and demand for new monitoring studies (e.g. see Chapter 22).

In light of these multiple potential uses of information from monitoring, countless monitoring programs have been implemented or are being developed by national, regional, and local agencies (see numerous examples in subsequent chapters); private entities such as forest management companies; conservation groups and other non-government organizations and citizen groups (e.g. Curtin 2002, Topp-Jørgensen et al. 2005, Fernandez-Gimenez et al. 2008; www.monitoringmatters.org; Chapter 21); and individuals (e.g. Bradley et al. 1999). There are few if any organizations in the field of

Design and Analysis of Long-term Ecological Monitoring Studies, ed. R.A. Gitzen, J.J. Millspaugh, A.B. Cooper, and D.S. Licht. Published by Cambridge University Press. © Cambridge University Press 2012.

natural resources that do not profess an interest in long-term monitoring, and most are planning or conducting monitoring in some form.

Although interest in monitoring is not new, recently there has been an explosion of discussion in the scientific literature about its design and analysis. Much of this discussion has focused on (i) the purpose and role of monitoring, including discussion of objectively assessing costs: benefits of monitoring in relation to other uses of funding (Nichols and Williams 2006, Hauser *et al.* 2006, Lindenmayer and Likens 2010b, McDonald-Madden *et al.* 2010, Wintle *et al.* 2010); and (ii) statistical aspects of monitoring. The latter focus includes guidance and case studies discussing how monitoring can be designed properly as well as continued development and application of new approaches for design and analysis (e.g. Legg and Nagy 2006, Field *et al.* 2007, Chandler and Scott 2011, MacKenzie *et al.* 2011, Reynolds *et al.* 2011). Availability of broad qualitative and quantitative guidance about how to do things "right" in monitoring continues to increase rapidly (e.g. Thompson *et al.* 1998, Elzinga *et al.* 2001, de Gruijter *et al.* 2006, Lindenmayer and Likens 2010a, McComb *et al.* 2010). Yet, despite this high interest in monitoring and increased understanding of the importance of design and analytical considerations, there have been no broad syntheses addressing these key methodological issues. In particular, there has been a lack of comprehensive guidance providing diverse perspectives and recommendations from monitoring experts, incorporating some design and analytical approaches that are or are rapidly becoming standard tools in ecological monitoring (e.g. spatially balanced sampling, hierarchical modeling, structural equation modeling, modeling threshold changes; see subsequent chapters), and providing advanced and extensive coverage of both design and analysis of complex monitoring studies.

This volume focuses on key questions and some important methods related to the design and analysis of long-term ecological monitoring studies. The choices of survey designs and analytical approaches for monitoring are questions with inherent statistical aspects. However, choices among design and analysis alternatives are not made in a statistical vacuum; they are intricately tied to the purpose and objectives of monitoring. Therefore, from the standpoint of this volume, "design" of monitoring includes not only specification of how, where, and when to collect data, but also specification of the specific objectives for monitoring, which in turn is driven by the general rationale for monitoring. Another reason why such a broad definition is appropriate for this quantitative text is because decisions about why and what we should monitor – and even whether to start or continue monitoring – increasingly are being placed in a quantitative framework (e.g. Wintle *et al.* 2010; Chapter 23).

In this chapter, we provide our perspectives on the context for this volume, and discuss challenges and recommendations related to the design and analysis of monitoring. We start by briefly emphasizing the importance and role of quantitative considerations in monitoring (Box 1.1); these issues are discussed further in Chapters 2 and 22, and emphasized, implicitly or explicitly, by every other chapter. We next discuss some potential reasons why inadequate attention to qualitative and quantitative design issues has been reported to be such a common problem in monitoring programs, and provide our perspectives on current risks from these causal factors. We then highlight several indicators

> **Box 1.1** Take-home messages for program managers
>
> Quantitative issues are not the only critical aspects of an environmental monitoring program. However, if monitoring is intended to produce useful and conclusive information, none of the other components of the program can compensate for inadequate attention to design and analytical issues. Long-term monitoring involves complex statistical and subject-matter considerations. An organization can either invest in the expertise needed to carefully address these issues during design and analysis (see Chapter 2), or it can take short cuts and rely on blind luck in the hope that money spent on monitoring produces something useful, eventually. We recommend the former.
>
> An important issue during development of a monitoring study is the identification of specific quantitative sampling objectives (desired statistical power, precision) related to primary variables or processes of interest, and of specific high-priority statistical analyses that are planned once monitoring is operational. These are essential inputs into quantitative study design. For example, a power or sample size analysis has to be based on a specific analytical method expected to be appropriate for analyzing the monitoring data. The fact that these features have been defined, and the process by which they are identified, are also critical indicators of whether a program knows why it is planning to conduct monitoring. It is not a good sign if they are selected in an offhand fashion. If a study is intended to be relevant to natural resource managers, it is worrisome if managers do not take the lead or collaborate closely in specifying these quantitative objectives. This issue should be addressed in the same way as all quantitative issues in monitoring: with integrated consideration of statistical, ecological, and practical aspects of the problem, based on clearly defined reasons for monitoring.

of whether a monitoring program is thoroughly addressing quantitative issues. Finally, we suggest selected general steps for continued progress in ensuring high quantitative standards for monitoring.

The role of quantitative considerations in monitoring studies

For a monitoring program to succeed, quantitative issues must be addressed adequately unless the program is focused primarily on education, making monitoring clients feel good in the short term, or allowing an organization to check a box indicating that "monitoring" occurred. The critical importance of design and analytical issues is the rationale for this volume. It may seem self-evident to many readers, yet many reviews suggest that inadequate attention to quantitative issues is one of the common causes for failure or ineffectiveness of monitoring programs (Noon 2003, Legg and Nagy 2006, Lindenmayer and Likens 2010a, and multiple other reviews).

If a major purpose of a monitoring study is to produce useful and relevant information for managers, policy makers and scientists, quantitative issues have to be addressed adequately (Box 1.2). The meaning of "useful and relevant" must be defined for each individual study in the process of specifying clear objectives. To be useful, information also needs to be timely, based on sufficiently accurate data and analyses, and able to stand up to objective peer scrutiny as well as agenda-driven criticisms. Adequate consideration of quantitative issues includes assessing whether reliable data can be produced with feasible survey effort, determining how to allocate available effort into an effective, efficient, and defensible survey design, and determing how to produce useful information from the resulting data; see Chapter 2 for further discussion.

Box 1.2 Common challenges: how low can you go?

In practice, the selection of meaningful quantitative objectives usually involves some degree of iteration: defining a target level of information quality, conducting an examination that often shows there is no realistic way of obtaining this quality, and lowering expectations, such as accepting a lower expected precision or restricting the target population to focus available resources more effectively (Chapter 10). Also, in practice, there is a strong temptation to take the following approach: set the budget for a particular study, determine what level of power or precision can be obtained with the feasible effort, and set this level as the sampling objective. The problem is that this requires no thought about what quality of information the study needs to provide to be successful. The first, iterative approach of progressively lowering expectations can also simply end up at the same point as the second approach – "we don't know what we want, but we do know what we can afford" – despite the best of initial intentions. Money is critical – but shouldn't there be some safeguards against simply spending money because it is available?

Because of this, we recommend defining "quality" and "usefulness" by also defining "worthlessness". In addition to setting the desired level of power or precision needed by the program, also define some hard-line threshold or "floor" for minimum acceptable expected performance. Prior to running any quantitative examination, what is the highest reasonable threshold below which you would say, with confidence, that "if this is the best we can do, we should not even bother collecting data"? This is a useful but uncomfortable question. If you can't define such a threshold, then you essentially are assuming one of the following: (a) if we collect data, we have faith that some useful information will materialize; (b) our primary focus is not on producing useful information; or (c) any data are better than no data. The first two possibilities relate to organizational priorities and efficiency, and are not inherently wrong or right. "Any data are better than no data" is the most worrisome. From an exploratory standpoint, there is value in what may turn out to be a 15-year pilot study. However, for addressing specific objectives, too little information or low quality information can produce the wrong conclusion and perhaps a wasted or harmful management response, or lack of response (Chapters 2, 18).

A sticky issue is that terms such as "suitable accuracy", "adequately", and "defensibility" are partially linked to subjective professional judgment. Accuracy should be assessed in reference to quantified sampling objectives, but at some level professional judgment and consensus usually are used to set desired precision, power, and other criteria. The closest we may come to a priori "objective" sampling objectives may be the precision benchmarks of Robson and Regier (1964) in an ecological context and perhaps the effect sizes of Cohen (1988) in a human-behavioral context, and these authors offer such standards simply as helpful but arbitrary conventions. Yet, professional judgments about desired accuracy are necessary, appropriate, and well-informed if based on hard thought about some intended uses of the monitoring data. Adequate consideration of quantitative issues does not rule out the need for some professional judgments, but rather leads to reduced unnecessary use of subjective judgments and greater use of structured scientific thinking, numerical evidence, and available statistical guidance. Later in the chapter we suggest some general criteria for assessing whether a monitoring study is likely to have adequately addressed quantitative issues.

Our emphasis in this chapter is on quantitative issues, but by no means are we implying that quantitative issues on their own determine the success or failure of a program, or that other factors have not been just as important in driving many past programs to failure (Noon 2003, Lindenmayer and Likens 2010a). Chapter 22 provides a broad context for this volume by discussing other critical factors which must be addressed in addition to those related to statistical planning and implementation, and illustrating how these factors have been addressed in the US National Park Service's long-term monitoring program. For example, monitoring programs must have efficient administration, strong support by intended clients of the program, effective data management, and reporting that is regular, timely, and useful, such that the program is widely viewed as productive and relevant (Elzinga *et al.* 2001, Fancy *et al.* 2009, Lindenmayer and Likens 2010a, McComb *et al.* 2010; see also Chapter 3). Ultimately, programs must have money; high-quality, productive, well-supported programs still face the budget axe. Programs try to maximize their survival probability by simultaneously and successfully meeting all of these requirements, quantitative and qualitative.

Flawed designs: their causes and avoidance

As noted above, monitoring programs have been highly susceptible to inefficiency and failure as a result of poor designs. If this problem was over-stated, we doubt that monitoring experts and biometricians would feel the need to continually emphasize some basic design issues, such as the critical importance of probability sampling in surveys when inference about a target population of interest, such as a study area, is to be based on measurement of only a subset of the population (e.g. Anderson 2001; Chapters 2, 5, 6, 18, 22). Why have inadequate designs been too frequently developed, particularly when simultaneously there have been many successful "status and change" surveys with rigorous sampling designs in place, at national and even cross-boundary scales (e.g. see examples in subsequent chapters and in McComb *et al.* 2010)? We suggest the problems can be traced at least partially to inadequate knowledge and attitudes of scientists and managers involved in monitoring.

Frequent use of non-defensible survey designs in ecology (Anderson 2001) suggests that many scientists and managers either do not understand or have tended to gloss over the importance of carefully, sometimes painfully developing rigorous survey designs to meet defined quantitative objectives. Interest in monitoring has long been high, but this often has been paired with limited knowledge and experience in carefully designing surveys. In many ecological disciplines, mandatory statistical training for students typically has focused primarily on a cookbook of standard analyses, with some coverage of experimental designs, often little coverage of survey design (except perhaps in forestry programs), and even less emphasis on conceptual thinking about objectives, important sources of bias and variance, and integrated choice of design and analytical approaches. Arguably, an enthusiasm to get started on monitoring combined with insufficient backgrounds in study designs often led to ineffective efforts, especially prior to the availability of rigorous guidance accessible to field personnel that emphasized broader design issues (e.g. Wilson *et al.* 1996, Elzinga *et al.* 1998, 2001, Thompson *et al.* 1998) as much as choice of measurement techniques. For example, in an important reference for habitat monitoring (Cooperrider *et al.* 1986), sample selection received very little attention, perhaps under the assumption that most field biologists and managers already had a firm understanding of the importance of proper sampling. This does not seem to have been a valid assumption.

Engagement of academic researchers to help design monitoring often probably has helped avoid some monitoring pitfalls, but not all. The default approach of such researchers when dealing with observational studies is to set up comparisons that mirror standard experimental designs *sans* randomization. In ecological disciplines, such comparative studies are an essential complement to controlled experiments, although proper use of these studies requires explicitly acknowledging limits to the inference possible (Shaffer and Johnson 2008). On the one hand, the researcher's natural tendency to incorporate careful comparisons into monitoring has important advantages. When monitoring is viewed as data collection free of specific scientific questions and hypotheses other than about presence/absence of a population-wide trend, the potential management and scientific value of the study is short-changed (Noon 2003, Lindenmayer and Likens 2010a).

However, a monitoring study usually is not just a research study with a multi-year component. Often the goals of surveys, such as in monitoring programs, include estimation of population-wide parameters as well as comparison of subpopulations of interest and examination of relationships (Cochran 1977). However, many ecological researchers either did not have much experience viewing observational studies with a survey-sampling perspective rather than a purely quasi-experimental design perspective, or perhaps viewed their role partly as shifting descriptive surveys into the realm of research. Either way, this probably has contributed to the traditional overemphasis of null hypothesis testing vs. parameter estimation and other approaches to inference in ecological studies (Johnson 1999, Burnham and Anderson 2002). Conversely, unless framed as research comparisons, observational studies and surveys, including monitoring, often have been seen simply as the purview of management, not science (Noon 2003, Lindenmayer and Likens 2010b). The view that monitoring is distinct from "science" may have led both ecological scientists and, ironically, managers to have fairly low standards for the quality of data expected from monitoring. In addition, the attitude that "we are simply trying to detect changes" often has led practitioners to move forward with poorly defined objectives, a bane of many programs.

For comparative studies viewed as non-randomized experiments, judgment or convenience sampling (Chapters 2, 5) of "representative" sites in different groups being compared has been widespread, rather than random selection from a pool of available sites. Such an approach, problematic enough for research comparisons, is simply a fatal flaw for monitoring studies seeking to make population-wide inferences based on a sample of sites. However, in monitoring programs these types of critical flaws often have been seen as necessary concessions to the reality of working "in the real world."

In truth, time and money always are limited. Perhaps we are underestimating the effect of financial limitations (Noon 2003). When budget cuts doom a monitoring program, this may have nothing to do with its statistical design. In other situations, financial limitations may be the proximate reason for inadequate design, but underlying attitudes and philosophies are part of the problem. For example, although money always has limited sample sizes and access to expertise, this often was and is combined with attitudes that "whatever we can afford has to be good enough" and "getting some data is always better than having no data." Later in the chapter we will discuss the importance of avoiding these dangerous attitudes, but their prevalence is demonstrated by the relatively low use of quantitative evaluations (power/precision examinations) by many past monitoring efforts to justify their designs (Legg and Nagy 2006, Marsh and Trenham 2008). To some extent, this low use is fairly sensible, if we simplistically assume for a moment that the only goal of these examinations is to assess whether there is any chance of meeting quantitative objectives with available effort, and less naively assume that the planned design was going to be implemented with only minor tweaks regardless of the results of the examination. Also, such examinations are based on specific intended uses of monitoring data. When the first step in quantitative design examinations indicates that that objectives have not yet been specified clearly enough to support meaningful examinations, there is a strong temptation simply to skip that part of the process. A rushed, non-careful design phase is also likely when the impact of quantitative considerations on the quality of information produced by monitoring is underestimated.

We have argued that frequent monitoring-design problems are often caused by insufficient knowledge and insufficiently careful approaches to monitoring design. These factors often lead ecologists (students, scientists, and managers) to underestimate their need for additional input from statistical experts (Millspaugh and Gitzen 2010), in monitoring as well as other contexts. Typically, ecological statisticians have been most commonly viewed as quantitative plumbers, to be consulted after data are collected if the analysis becomes clogged. Still, even when monitoring practitioners recognized their need for additional quantitative help, they may not have had feasible access to such experts, a deficiency needing continued attention. Overall, if suitable statisticians or other personnel with suitable training and expertise weren't involved from the beginning, there is a good chance that a monitoring study would have an inadequate design that either doomed it to failure, led it to produce information of much lower value than could have been obtained, or, in the worst-case scenario, led it to survive and produce misleading data. Are those days in the past?

What has and hasn't changed?

Progress

Although historically it was mainly a few regional- or national-scale programs that had high standards regarding quantitative aspects of monitoring, high quantitative standards are becoming the norm at many levels of monitoring. Partly this may be an outcome of learning by experience, given frequent commentaries on the prevalence and problems of inadequate designs. Moreover, there is greater awareness of the success of well-established programs with probability survey designs that have an ongoing history of supporting flexible inference at local to national scales (e.g. Nusser *et al.* 1998, Bechtold and Patterson 2005). Hopefully, the number of good examples is on its way to exceeding the number of bad examples. For example, in the USA, ecologists with the National Park Service (NPS) and their collaborators from other agencies and organizations have implemented or are in the process of designing hundreds of monitoring protocols with clear objectives and defensible survey designs (Chapters 10, 16, 22), even for very small target populations (e.g. vegetation monitoring at Fort Union Trading Post National Historic Site in North Dakota/Montana, USA, uses spatially balanced sampling from a 133-ha sample frame; Symstad *et al.* 2011). This type of national-scale program with a local-scale focus broadly expands awareness of available methods, and pulls in many professional ecologists and statisticians to further develop and apply diverse design and analysis approaches. It also leads many managers and biologists to think about the importance and requirements of sound statistical design. Managers and environmental scientists who learn about and successfully implement sound designs motivate and raise the bar for other programs, and build more widespread availability of quantitatively skilled people with experience in monitoring design and analysis.

There also has been a rapid decrease in the tendency to view monitoring as less interesting, and requiring less rigor, than "real science." This may be driven partly by the blurring of any past divisions between basic ecology and highly applied, management/conservation-focused science; an increased focus on estimation rather than rote hypothesis testing in ecological science (e.g. Burnham and Anderson 2002); and by increased realization about the potential for monitoring programs – and management/policy-focused data collection in general – to provide data highly useful for addressing ecological questions. As a result, there is broader recognition of the need for strong scientific partnerships in the design and analysis of monitoring (e.g. Lindenmayer and Likens 2010a; Chapter 22), and in turn ecological scientists are highly interested in the why and how of monitoring (e.g. Nichols and Williams 2006, Lindenmayer and Likens 2010a, b). In addition to being able to help monitoring programs develop strong objectives and clear questions, an increasing number of ecological research scientists likely have developed expertise in, or at least solid familiarity with, survey-design methodology as well as experimental design. Comprehensive, often advanced guidance on design and analysis of monitoring surveys continues to become more available (e.g. Elzinga *et al.* 2001, de Gruijter *et al.* 2006, McComb *et al.* 2010, Chandler and Scott 2011) and quantitative-ecology texts with more specialized focus typically directly address or are highly relevant to monitoring contexts (e.g. Thompson *et al.* 1998, Buckland *et al.* 2004, Skalski *et al.* 2005, MacKenzie *et al.* 2006, Royle and Dorazio 2008). Overall, there has been a concurrent increase in expectations for what constitutes quantitative fluency by ecologists in general (e.g. Ellison and Dennis 2010), which likely also has raised expectations for quantitative aspects of monitoring programs.

For biometricians, the spatial and temporal dimensions of monitoring have always provided a rich realm of interesting research-methodology questions with very practical implications for existing and future programs. However, in some cases, the rates at which new techniques become widely available are almost breathtaking. As an example, consider occupancy modeling, focused (in its most typical application) on species presence/absence and occurrence dynamics (Chapter 18). Relatively speaking, almost as soon as the general likelihood framework for occupancy estimation was formalized, it was extended to monitoring contexts (e.g. see MacKenzie *et al.* 2006 and Chapter 18). Software for implementing occupancy estimation also was rapidly made available to ecologists. Increasingly, few monitoring programs focused on occurrence of species with imperfect detectability would even propose a monitoring design that does not take advantage of developments in this field that are only a few years old, and few reviewers of proposed study designs or subsequent results would be tolerant of those that ignored these developments.

Given this increased interest, scrutiny, and awareness, we expect few if any new individual monitoring studies focused at state/regional or larger scales will move forward without detailed conceptual understanding of the monitoring context, solid objectives, and carefully planned survey designs that thoroughly address major sources of bias and variability (e.g. biases due to incomplete detection of individuals and species present on measured plots).

Opportunities for continued progress

However, these aren't all happy days. One messy issue is the potential for an increasing divide between the "haves" and the "have nots", in terms of advanced quantitative sophistication and resources (e.g. see discussion by Dennis and Ellison 2010, Ellison and Dennis 2010, and Millspaugh and Gitzen 2010). To be sure, an ever-increasing number of students, scientists, and managers who certainly would not consider themselves to be "statisticians" have advanced quantitative skills with the ability, for example, to properly develop complex spatial designs or implement Bayesian hierarchical modeling. The term "quantitative ecologist" no longer applies only to highly advanced theoreticians, but encompasses a broader range of ecologists, including many serving as a bridge between statistical experts and other scientists and managers, and many who are routinely extending and tailoring statistical methodology as part of their primary focus on scientific and applied management questions.

At the same time, it is easy to overestimate the percentage of current or future ecological practitioners whose jobs will require the quantitative interest and statistical-software skills needed even to mis-apply advanced methods. We are not arguing that this should be the case – it simply is the reality. For example, at least in the USA, the skills and interests of the average undergraduate student in a natural resource curriculum do not focus on quantitative ecology. Quantitative training for graduate students varies widely depending on the student, faculty supervisor, and program. Yet, these students ultimately will become the professionals who are the primary clients for much monitoring information and who often will be sincerely eager to initiate monitoring. Despite having little experience (in some cases) in the conceptual and quantitative aspects of study planning, they may have primary oversight for monitoring design, particularly at small to medium spatial scales. If they do not see the importance of quantitative considerations in monitoring or do not have access to the direct help they may want, they will still forge ahead to the best of their abilities. These ecological practitioners need to be considered in the effort towards continued overall increases in the quality of monitoring. This is one of the reasons this volume provides advice aimed at such ecologists in every chapter.

It also needs to be recognized that a primary limitation often isn't technical expertise but rather attitudes and attention to general principles – and this issue may be just as relevant for practitioners with stronger quantitative backgrounds. Perhaps partly based on the way statistics normally has been taught to them and partly because of a strong motivation to get tangible results quickly, ecologists have a tendency to focus on the mechanics of statistical tools, and "can I do this?" rather than "should I do this?" or "why am I doing this?". For example, let's think again about occupancy modeling. There is a tendency in some practitioners to see the primary question about the appropriateness of this method as being whether or not an existing data set can somehow be subdivided into a collection of 0's and 1's for each of multiple study sites. Probably partly in recognition of this temptation, experts have very explicitly emphasized the critical importance of underlying assumptions and proper study design (e.g. Chapter 18).

Consider also the example of statistical power analysis in monitoring. If examination of expected design performance in relation to specific objectives is becoming more routine, this is a step in the right direction (e.g. Legg and Nagy 2006). It is important to have more widespread availability of software tools (e.g. Gibbs and Ene 2010) for such quantitative examinations, particularly when dealing with complex panel designs, non-normal data types, and other complicated situations (see Chapters 7–10). That said, once the quantitative framework has been developed and the appropriate tools made available, the basic mechanics of power examinations often are straightforward, or can be addressed with fairly simple simulations easily programmed in packages such as R (R Development Core Team 2011). A cynic might argue there is sometimes a tendency to see quantitative design in monitoring as simply sprinkling a few trend-analysis power curves in suitable places to satisfy superficial review. Admittedly this is one situation in which any effort may be better than nothing (Chapter 8); even rapid power explorations can help elucidate trade-offs among alternative designs and levels of effort. However, the primary value and challenge of quantitative examinations are not driven by mechanics. These examinations require a mix of expertise and perspectives, and are best addressed – like all aspects of monitoring design – by a team incorporating experts in statistics, ecology, and the management context (Chapter 2). The non-mechanical issues involved are highly technical (Chapters 8, 9), conceptual, and practical: what quality of information do we need from monitoring data, what is the best benchmark of quality (power for trend, power to assess whether a defined threshold has been crossed, precision, value in updating model weights as in Chapter 4), what analysis will we use to turn data into conclusions, what are the sources of significant variance and bias affecting the study, and how do we estimate and address these components? The design of monitoring programs will continue to improve significantly if a focus on power analyses matures into broader incorporation of such careful, integrative thinking in quantitative study design.

Some indicators of a monitoring study that has adequately addressed quantitative issues

We have stressed the general importance of quantitative considerations in ecological monitoring. We have argued that, although there are a rapidly increasing number of well-designed monitoring studies, throughout the monitoring world as a whole there is a widely recognized need for continued progress in adequately addressing fundamental quantitative issues (e.g. Reynolds *et al.* 2011). Such progress will be based partly on continued general developments in methodology and training, and partly on the collective effort of individual monitoring programs each striving to address concrete quantitative considerations more thoroughly.

In this section, we discuss selected specific indicators that may help monitoring practitioners assess whether their specific monitoring study is adequately addressing important quantitative issues. This set of indicators is not intended to be novel (given that it is based on fundamental considerations) or comprehensive; Chapter 2 and an increasing number of other references provide more comprehensive guidance (e.g. Elzinga *et al.* 2001, de Gruijter *et al.* 2006, McComb *et al.* 2010). Most of these indicators are recognized as important standard components of any monitoring protocol (Oakley *et al.* 2003); others should be.

(i) An explicit summary of the past and ongoing roles, and plans for future involvement, of quantitative experts involved in addressing these components

This may seem pedantic, but the design and analysis of monitoring involves complex quantitative issues (e.g. Chapters 8, 9) which should be addressed with a comprehensive, "whole-problem" approach (Chapter 2). "Good design is an inherently statistical process" (Lindenmayer and Likens 2010a:19). A study intended to extend for multiple years or decades should think strategically about how it will address these challenges, and should be explicit about whether quantitative experts have been involved, and what roles they have played. By extension, any monitoring study intending to be management relevant and offer scientific value should be equally clear about the role of managers and scientists in design and analysis.

(ii) Specific monitoring objectives and clear definitions of the target population of interest and the sample frame (in time and space)

We refer readers to Chapters 2, 3, 5, 22, and others for more detailed discussion of these critical issues. Quantitative sampling objectives may (Chapter 2) or may not be seen as a standard component of general monitoring objective statements. Frequently there may be a statement of specific overall monitoring objectives for a protocol focusing on many variables, plus further statements of specific sampling objectives (NPS 2008a; Chapter 22). We focus separately on sampling objectives, but this organization is simply to facilitate discussion. What matters is that all aspects of "s.m.a.r.t." (Chapter 2) objectives are clearly defined and stated. Collectively, the qualitative and quantitative objectives for monitoring are all-important; many of the criteria discussed in this section are motivated by their value in indicating whether a study has well-defined objectives.

(iii) Quantified sampling objectives regarding focal parameters of interest, with clear practical justification for how these were selected

(iv) Tied to these objectives, specific description of selected high-priority analyses that are expected to be routinely conducted

These two criteria are essential and powerful indicators of whether monitoring practitioners have a clear idea of why they are monitoring. See Chapters 2 and 8 for directly relevant discussion; Chapter 10 provides an example of a carefully selected sampling objective. Power-for-trend-detection is often a focus, but inclusion of a visually pleasing curve showing power vs. time should not be seen as the defining test of whether quantitative aspects of a study have been addressed adequately. Examining power-for-trend-detection can offer important general insights about effective survey designs (e.g. Chapter 7). In any specific situation, focusing on power-for-trend may (see examples in Chapters 2 and 10) or may not (see comments in Chapter 18) be an indicator of careful thought about specific objectives. Other metrics may be more relevant to managers, such as power to detect a change in some population summary between two sampling periods (e.g. Chapter 14), or desired precision of estimates of status for the population and/or for strata or other subpopulations of interest. Metrics may be related to predictive accuracy or ability to correctly select from a set of alternative statistical models tied to subject-matter scientific hypotheses (Burnham and Anderson 2002; Chapter 15).

Clearly, in most broad-scale long-term monitoring studies, there will be numerous unintended uses of data, unexpected changes in objectives or questions of interest, and the potential for new analytical methods. Monitoring programs have to count on, and plan for, the unexpected. However, if a monitoring study has not identified at least a few high-priority planned specific analyses and defined quantitative benchmarks related to the quality of key information desired from the study, then it is not ready for implementation except as an extended pilot study.

The same is true if quantitative standards are purely an outcome of, rather than an input to, power and sample-size analyses. Costs are always limiting and critical; it would be foolish to argue otherwise. However, often quantitative examinations are simply a quantification of what magnitude of trend can be detected or level of precision obtained given the available funding. Such examinations have high value in assessing survey design trade-offs, and give program managers and peer reviewers some basis for assessing what level of accuracy might be expected in information produced by the study. But in general, in any individual study, the approach of plotting a generic power curve, then letting this curve dictate the sampling objective, is a sign of a vague purpose for monitoring, and usually a sign of shallow quantitative thinking in the design of monitoring (Box 1.2). This approach is particularly problematic when a monitoring program is intended to provide useful information to managers, but managers do not play a major role in defining what information is needed and at what level of accuracy.

(v) Careful conceptual identification of major relevant sources of variance and bias for the study at hand

Chapters 2, 3, 8, and 11 explicitly discuss the importance of this issue, and it is either directly or implicitly emphasized in Chapters 7, 9, 11, 18, and most others. This importance is the reason why this volume contains extensive coverage of variance component identification and estimation. It is a critical aspect of study design, analysis, and interpretation. It requires a careful integration of both statistical and ecological knowledge and experience (Chapter 8). Ecological practitioners bring critical subject-matter knowledge for identifying sources of inaccuracy and helping prioritize which are of practical relevance. However, many ecologists have not been trained to think in this manner. For example, applied statistics courses for ecologists typically expose them to a single generic σ^2, but even in the simplest study-design situations, this needs to be partitioned correctly into multiple components to be useful in making choices among alternative survey designs and levels of effort (e.g. Urquhart *et al.* 1998, Kincaid *et al.* 2004; Chapters 7, 8, 9).

Monitoring experts strongly emphasize the value of conceptual models – which may take on a variety of formats: tabular, graphical, narrative, and quantitative (Gross 2003) – for summarizing relevant variables and relationships as part of determining objectives, identifying supplemental measurements to help increase the precision and explanatory power of monitoring, and guiding analyses (e.g. Noon 2003, Lindenmayer and Likens 2010a; Chapters 2, 22). This same conceptual-modeling strategy is highly useful for helping ensure broad and systematic thinking about current and future sources of inaccuracy. In fact, these strategies are not separate; conceptual ecological models based on hard thought will (and often are intended to) simultaneously help define the variance structure of the resource of interest. For example, ecological dynamics largely define the temporal variance structure – helping to identify, for example, an "index period" each year during which the responses of interest are most stable (Chapter 7) and may offer the most meaningful insights into ecological status. As another example, ecological thinking is a necessary starting point for considering the relative strength of population-wide year-to-year variation; e.g. are the resources of interest likely to show high or low sensitivity to park- or region-wide year-to-year random fluctuations in weather?

In addition, this conceptual thinking must encompass aspects of the observation process – sources of sampling and measurement errors, as well as those that could be produced by incomplete implementation of the survey design, etc. In turn, this requires conceptual thought about the interaction between ecological and observation factors. For example, in demographic monitoring, consideration of potential bias and proper interpretation of estimates requires direct consideration of the scale and magnitude of spatial, temporal, and behavioral factors (Chapter 18). Similarly, is there recognized potential for future changes in the population's variance structure (Chapter 10), or for future reduction in the spatial and temporal overlap between the sampled and target populations of interest (Chapters 2, 3, 11)?

(vi) Quantitative estimates of variance components and measures of the uncertainty of these estimates

See Chapters 7–10 for fuller discussion. Uncertainty of these estimates is of interest because it translates into uncertainty in estimates of expected power, precision, or other metrics (Chapter 8). If there is high uncertainty in a variance component estimate, then there may be a high probability of much lower power or precision than that calculated using the variance point estimates (Sims *et al.* 2007). Identification and estimation of variance components often requires careful consultation with a statistician, and quantifying uncertainty in these estimates involves additional technical complexities; for example, due to the potentially highly skewed sampling distribution of variance estimators (e.g. Papritz *et al.* 2011; D. M. Bates, *lme4: Mixed effects modeling with R*, in preparation; Chapters 8, 9). In practice, lack of routine presentation of uncertainty estimates for variance components – both in terms of estimation uncertainty regarding their current magnitude and concerning potential future changes in their magnitude (Chapter 10) – should be seen as an area for future growth, considering the pressing need for more routinely estimating and using variance components for study design (Chapter 8).

(vii) A specific plan for addressing expected sources of bias or at worst periodically assessing their current magnitude, or very strong justification for ignoring them

See Chapters 2, 3, and 11. Examples in Chapter 11 document a reality: it is extremely messy, expensive, and frustrating to try to explain, estimate, or analyze away the effects of plausible sources of bias. Particularly in a monitoring context, well-planned survey designs do not guarantee that unexpected sources of bias won't develop. However, scientifically and strategically, simply ignoring reasonably plausible sources of bias at any stage is simply a bad idea if the monitoring study is attempting to produce conclusive information. At the design phase of new long-term studies, the objectives and intended use of data determine what degree of bias and imprecision are acceptable, or conversely, to what extent the study should go to account for sources of inaccuracy. However, this implies careful thought. For example, it would be idiotic to take a "hope for the best" approach to addressing widely discussed, plausible sources of bias that might make resulting information worthless *relevant to its intended use*, and that almost certainly will cause major headaches for the monitoring program when the time comes to analyze data and attempt to defend the results.

(viii) Quantitative assessments of expected performance of candidate survey designs relative to primary sampling objectives

See Chapters 2 and 7–10 for fuller discussion, as well as previous comments in this section. We emphasize again that such an examination is based on one (or a set of alternative) *specific* analytical approaches (Chapter 8) – i.e. for how data will be analyzed once collected – and survey-design alternatives. For example, even if one just wants to blindly incorporate a nice-looking generic power curve without any real thought, one faces the problem that the plot has to be based on a specific design and survey approach.

Beyond this, it is important to recognize that the limits of such formal examinations do not negate their value. For example, to accurately predict what level of power we can expect to detect a specified trend, we need an accurate estimate of what population-wide year-to-year variability will be over the next 1–2 decades. If we lack confidence in our omniscience, we may be tempted to see little value for power analyses except for examining general performance of alternative designs. We argue for the value of planning as carefully as possible within the limits of current knowledge, and learning from new information as it is collected.

(ix) Justification for choice of design and very clear documentation of the design

What was the overall role of quantitative criteria, budget considerations, logistical constraints, and other influences (e.g. professional judgment) in determining the selected design and level of effort? Transparently documenting the decision factors helps a monitoring program think clearly and explicitly about the process, helps ensure that the role of professional judgment has been appropriate, and may be invaluable for those involved with the study in the future for understanding its history.

The latter, "documentation" rationale is especially important when it comes to recording specifics about the sampling design. Most of these details are part of the documented sample frame [see Indicator (i)] and are a standard component of a suitable monitoring protocol (Oakley *et al.* 2003). Some of this basic information includes:

- Spatial boundaries of the sample frame and careful description of temporal dimensions.
- For frames with a finite set of sample units, attributes for each unit such as indicators of strata and/or subpopulation membership if this was considered in the sample selection and specified inclusion probabilities for every sample unit.
- Criteria for rejecting selected sample units in the field (e.g. due to frame errors that incorrectly defined a sample unit as being part of the target population, such as a vegetation monitoring sample unit that actually is in the middle of an unmapped permanent lake).
- The ordered list of the selected sampled units for spatially balanced sampling (Chapter 6), and an ultimate record of sites on this list that were initially selected but rejected appropriately.

Such information is critical for future design-based analyses, and for any future adjustment to, or re-assessments of the sampling design.

(x) Identification of areas of flexibility /inflexibility for change

A monitoring study that survives more than a few years will face potential changes in primary objectives, supplemental questions of interest, funding, and methods (Chapters 2, 3, 20). An active approach is also needed that iteratively learns from monitoring information as it is produced (Lindenmayer and Likens 2009; Chapter 2). For example, survey designs often must be selected with only coarse estimates of some variance components, but these estimates can be improved as data are collected – which may indicate a need for altering the number of sites or the revisit plan.

Given these realities, it is important to think explicitly about how much flexibility and extensibility is desired in the survey design, and how this will be achieved. For example, forms of spatially balanced Generalized Random Tessellation Stratified (GRTS) designs as well as basic simple random samples can be upscaled or downscaled as funding dictates; for the GRTS design, this needs to be addressed at the design phase (Chapters 5, 6). Such an adjustable sample may also be useful as a "master sample" for other data-collection efforts as funding and objectives permit. It is a healthy sign when a monitoring study documents explicit consideration of such future scenarios.

(xi) Ongoing records of study implementation and data on necessary covariates and subpopulations of interest

In addition to the primary measurements of interest, a monitoring program will require supplemental information such as covariates (e.g. weather for yearly variability) and management-related data for defining subpopulations (e.g. areas that have or have not been treated for invasive species within the last 10 years). As important supplemental information is identified – and hopefully this has been done as part of the design phase – clearly a program needs to ensure that this information is routinely collected. Additional information such as deviations from the protocol, survey effort at a site each year (Chapter 21), and records and reasons for missed visits and other missing data (Chapters 13, 15) likely will be important during analyses.

(xii) Explicit specification of how major assumptions related to the methodology and about potential sources of bias were addressed during data analysis

Essentially, this is an opportunity for revisiting how design assumptions have been addressed, documenting unforeseen sources of inaccuracy, formally considering statistical assumptions of analytical methods, and being explicit about which problematic factors (e.g. sources of bias) have been addressed, and which have been wished away. In analyses focused on modeling rather than design-based estimation, explicit definition of a priori and exploratory aspects of the analysis also are simply good scientific practice (Burnham and Anderson 2002).

Continued general needs regarding quantitative issues in monitoring

For any specific monitoring study, the indicators above suggest how well the program is addressing quantitative issues. Ultimately, the motivation to address these issues must come from within the monitoring program, a motivation which depends on the importance placed on these issues by monitoring practitioners. In turn, though, this is affected partly by how ecologists have learned to think about fundamental quantitative principles and conceptual study-design issues. Moreover, it is the practical needs of monitoring programs that drive many continued statistical developments in this field (e.g. see comments in Chapter 18), as demonstrated throughout this volume. In light of these factors, we identify several general "needs" to help monitoring continue to improve its effectiveness.

(i) The field of monitoring benefits greatly from continued increases in the number of ecologists with advanced quantitative skills. Much of the content of this volume targets these ecologists. However, it will also benefit from continued increases in the number of practitioners with solid training in fundamental concepts, and who apply this training properly as a matter of habit. A basis in fundamental concepts ensures ecologists and statisticians speak the same language. For example, earlier in the chapter, we hypothesized that lack of familiarity and experience with survey-sampling methodology has been problematic in monitoring. Experimental design is a critical statistical and scientific topic, but survey design is equally important (e.g. Johnson *et al.* 2001). The underlying fundamental issues of bias and variance are the same. However, for whatever reason, there are more than a few natural resources practitioners who might not consider designing an experiment without randomization, but who simultaneously see probability sampling as a nuisance rather than a vital tool for bias reduction. Terminology also matters; referring to any statistical design as "experimental design" tends to de-emphasize the needs and limits of non-experimental studies. In terms of more advanced skill sets, there clearly is a recognized need for continued training of quantitative ecologists in selected analytical approaches, such as general hierarchical modeling and structural equation modeling (Chapters 12, 15, 19, 20). Such approaches depend on careful understanding and specification of ecological and observational dynamics in the form of statistical models. In addition, there are large numbers of monitoring surveys being developed or implemented with complex spatial and/or temporal survey designs. There are far more quantitative ecologists with advanced skills in statistical modeling than in design-based analyses of status and change in complex multi-year designs; this may suggest a need for more training resources and a realistic expectation of the statistical help needed. It should be recognized that many commonly encountered quantitative challanges in monitoring involve areas of active research. For example, several experts in this volume point out areas of active research relevant to design-based analyses in monitoring (Chapters 5–7, 14, 17).

(ii) Increased recognition of the role of qualified statisticians as team members at all stages of the monitoring program, and organizational recognition of the need for this collaboration. (Chapter 2 discusses this need at length.) This is a difficult problem. Arguably those who view quantitative expertise as something needed only at key sticking points in the design or analysis stage are also those who would most benefit from additional help in systematically thinking through all stages of the monitoring process. If a monitoring practitioner without advanced relevant quantitative expertise firmly understands the importance of statistical considerations in ecological studies, we expect this will lead to increased and more effective engagement of quantitative experts. At the same time, the demand for statistical help in design and analysis of monitoring data often is much greater than the affordable availability of statistical experts. An increasing pool of statisticians and quantitative ecologists with monitoring expertise will help in this regard.

(iii) Increased understanding of the link between objectives – including intended analyses and use of the data – and all other quantitative considerations. This is both a fundamental scientific concept and a key part of effective collaborations between ecologists/managers and statisticians. There are few absolutes in monitoring that are independent of the objective or the ecological context. For example, there is nothing inherently wrong about monitoring specific purposely selected high-priority sites; the problem comes in trying to claim that these sites are of broader representative value (Chapters 5, 18). Ineffective collaborations result if the ecologist thinks a method can be selected without reference to specific objectives, or if either the ecologist or the statistician thinks it is primarily the latter party's role to specify the objectives, qualitatively or quantitatively.

Future research and development

This volume contains a diversity of perspectives about appropriate uses of monitoring, and of the funding allocated to monitoring. Under the assumption that underlying management and scientific goals dictate the proper use of monitoring, it is clear that there is a definite need for many forms of monitoring. Agencies may have a mandated or assumed responsibility for tracking the general condition of specific resources or areas apart from any well-defined, re-occurring decision-making process (e.g. Chapters 3, 7, 22). For example, society and policy-makers need ongoing monitoring of climate, water quality, and land use at local to global scales, with flexible designs that allow a variety of questions to be examined at a variety of scales. Conversely, there is a strong rationale for increasing the efficiency and value of monitoring in the context of Adaptive Resource Management (ARM) and scientific learning (Nichols and Williams 2006, Lindenmayer and Likens 2010a; Chapter 4). Regardless of whether it is defined as monitoring or not, managers also need to address additional high-priority management questions about specific management actions or one-time decisions.

In other words, there is very clear desire to satisfy multiple broad purposes through monitoring studies (Chapter 22). Authors in this volume discuss (Chapters 3, 4) or illustrate (Chapter 16) some strategies for such multi-purpose monitoring, in which a monitoring framework focusing on one purpose is supplemented with additional data collection relevant to other objectives. This is a high priority for continued work. Ecologists will benefit as more examples become available of applying existing methods to meet multiple purposes. Some highly useful approaches may combine probability and model-assisted/model-based sampling (Chapters 16, 17), make greater use of domain and small-area estimation to answer more focused questions (Rao 2003, de Gruijter *et al.* 2006), or combine permanent/static monitoring sites with a dynamic monitoring network to address shorter-term questions (Hooten *et al.* 2009a; Chapter 16).

Related to this issue is the need for more effective use of routinely collected management data sets for monitoring (Skalski *et al.* 2005). State and federal agencies collect large amounts of data annually to monitor the status and trends of harvested wildlife populations and other resources. These data fill file cabinets and lead to summaries that are included in annual reports. But otherwise the data are often viewed simply as "management data" without any real appreciation of their potential value. This "collect and neglect" syndrome is unfortunate because there is value, particularly if combined with additional, rigorously collected data. Often, these data are broad-scale and extensive, relatively inexpensive to obtain, collected over many years, and serve as the only basis for understanding processes and dynamics at large spatial scales. When inference is expected at these large spatial scales, we view it short-sighted to avoid data collected annually by management agencies. Skalski *et al.* (2005) and Broms *et al.* (2010) provide examples of combining management data with other data (e.g. age-at-harvest and mark–recapture or radiotelemetry data) for more powerful insights. The wave of the future in broad-scale monitoring of harvested wildlife populations, for example, should combine extensive data with intensive studies. More generally, there is high potential utility in developing and applying flexible approaches for quantitative integration of data from management-focused studies, general monitoring and inventories, and research.

Summary

A monitoring program focused on producing useful information needs to carefully integrate quantitative expertise, general scientific thinking, and subject-matter expertise about the resource being monitored and its management. Ecology is becoming an increasingly quantitative field, and statisticians continue to develop novel and practical approaches for monitoring design and analysis. However, increasing the overall success of monitoring programs in addressing quantitative considerations does not depend only, or perhaps primarily, on those with advanced quantitative backgrounds. Advanced tools and training for some ecologists need to be balanced with continued improvements in training of all natural resources students in fundamental statistical concepts. There needs to be a greater understanding of the value of partnerships with appropriate statisticians. This also involves understanding that core quantitative issues in monitoring cannot be addressed by simply finding the right equation or entering the proper software commands. For example, ecologists often are not used to thinking conceptually about sources of bias and variability, but defining these requires both statistical experience and familiarity with the resource to be monitored. Even more critically, managers and scientists must understand that they play the lead roles in developing precise objectives for monitoring and planning how data will be used once it is collected.

Acknowledgments

For their help in the development of perspectives expressed in this volume, we thank many personnel of the National Park Service Inventory and Monitoring Program. We owe special thanks to monitoring personnel and park managers of the Northern Great Plains Inventory and Monitoring Program including Kara Paintner-Green, Dan Licht, Marcia Wilson, and others, as well as Amy Symstad and other scientific collaborators with this program. For their role in increasing our knowledge and stimulating our thoughts about the design and analysis of monitoring, we thank authors in this volume, peer reviewers of chapters, and others who shared suggestions or knowledge about monitoring. However, the opinions and recommendations provided in this chapter are our responsibility alone.

2 An overview of statistical considerations in long-term monitoring

Joel H. Reynolds

Management decisions all too often are based on data of unknown reliability – that is, from research conducted using biased methods, low sample sizes, and inappropriate analyses.... The general dearth of monitoring, validation, and adaptive management research forces a vicious cycle. This does not need to – and should not – persist.

(Morrison *et al.* 2006: xix)

Introduction

Developing a monitoring program requires a chain of decisions, many complex, about a variety of issues. These decisions deal with questions ranging from policy and management ("What information is required to decide among the feasible actions?") to the science of the system under study ("What is the best indicator of the ecological process of interest, and what is the best measurement protocol for that attribute?") to purely statistical ("How much effort will it take to estimate a parameter with a precision of $\pm 10\%$ of its true value?"). This variety of challenging questions implicitly encourages compartmentalization and isolation in decision making. However, this compartmentalized approach can lead to suboptimal and even counterproductive decisions. Instead, a "whole-problem" perspective helps us see that all of the necessary decisions interact to influence the quality and cost of the monitoring program's ultimate product, information (Box 2.1).

A whole-problem perspective reveals that statistical considerations permeate the monitoring process. Information quality, a key measure of a monitoring program's effectiveness, inherently entails consideration of the *statistical concepts* of bias and uncertainty (Box 2.2). Bias and uncertainty are controlled and accounted for through *statistical methods* of survey design, measurement, and analysis. Most importantly, effectively incorporating these statistical methods into the broader monitoring program requires *statistical thinking* to see how each source of bias and variability ultimately impacts information quality and cost and to identify where to allocate resources among the many program components to most efficiently improve information quality (Hooke 1980, Snee 1990, Chance 2002, Hoerl and Snee 2010).

Design and Analysis of Long-term Ecological Monitoring Studies, ed. R.A. Gitzen, J.J. Millspaugh, A.B. Cooper, and D.S. Licht. Published by Cambridge University Press. © Cambridge University Press 2012.

Box 2.1 Take-home messages for program managers

Monitoring is a process to produce information repeatedly through time. To be successful, a monitoring program must be *effective* (produce information of the necessary quality), *efficient* (produce it for low relative cost), and *feasible* (produce it with the available resources). Meeting all three criteria requires both *statistical thinking*, to understand and control the interplay of the many program decisions underlying the information cost and quality (e.g. bias and uncertainty), and *thorough planning*, to identify the combination of decision choices that satisfies all three criteria. Without sufficient statistical thinking and thorough planning, a monitoring program may make developmental decisions (e.g. regarding objectives, survey designs, or intended statistical analyses) with major flaws that may not be recognized and resolved until monitoring has been implemented for many years. Such a program risks delaying recognition of system changes, reducing management response options, slowing learning about the system, and wasting resources. In the long run, there is the risk of severe impacts to the system that might have been averted, and the loss of institutional support for the monitoring program.

Monitoring involves myriad decisions ranging from policy and management to the science of the system under study to applied statistics (Fig. 2.1b). Experienced applied statisticians provide "whole-problem" thinking and can recognize the implications of, say, a study's objectives on the analysis, design, and measurements choices, and vice versa. They can provide invaluable critique of the feasibility and overall efficiency of alternative choices. Bringing applied statisticians on to the monitoring program planning team from the start, along with representatives of the decision makers and field scientists, can help prevent problems in the design, data collection, and analysis phases and identify efficient allocation of effort (e.g. staff time, equipment, money) among the many program components (e.g. sampling, measuring, modeling) so that monitoring cost-effectively provides information of sufficient quality to meet the program's objectives. The cost of thorough planning is generally negligible relative to the overall costs of implementing and maintaining a monitoring program for even just a brief period. Such planning also helps garner institutional support and can be used to assess the potential effects of declining budgetary support.

Statistical thinking can help a monitoring program produce high-quality information by bringing attention to the linkages among the component decisions. For example, "What are the implications of deciding to monitor this attribute over this time period at this spatial scale, in terms of both methods and required effort to achieve a particular level of information quality?" Restricting the role of statistical advice to isolated choices of specific tools ("use logistic regression rather than a chi-square test") encourages isolated decision making. Ignoring the linkages during program development just delays discovering them until analysis, when the effects of flawed decisions are finally revealed.

Box 2.2 What are bias and uncertainty?[1]

Ecological monitoring and other field studies usually seek to make inference from a sample (i.e. a subset selected for measurement) to the entire population of interest. The information from a sample, call it the "evidence", usually will differ with each possible sample that could have been selected, generating uncertainty. The distribution of all possible values that could have resulted from the specific sampling design and sample effort is the *sampling distribution*. The sampling distribution has three characteristic features: its center, the mean of all possible sample values; its scale, a summary of the variation among the possible values; and its relative shape (e.g. symmetric or skewed).

  Bias is the difference between the sampling distribution's mean and the true value from the population of interest. The uncertainty due to the sampling process is embodied in the sampling distribution's scale and relative shape. The more tightly clustered the sampling distribution around its center, the smaller its scale and the less uncertainty associated with the evidence. Scale is commonly summarized in terms of the standard error. While the formula for estimating the standard error depends on the sampling method and population parameter being estimated, the standard error itself provides a universal yardstick for summarizing the sampling distribution's scale, allowing sampling uncertainties to be compared. Other sources of uncertainty are discussed in the text.

  Paying attention to linkages before engaging in data collection requires thorough planning. This planning should be conducted by a team that includes, from the beginning, representatives of the natural resource managers and decision makers, the field scientists, and an expert in statistical thinking in the program's field of application, i.e. an experienced applied statistician[2] (Hahn and Meeker 1993, Stevens *et al.* 2007, Lindenmayer and Likens 2009, Hoerl and Snee 2010). Some readers may have a tendency to see a narrower role for the applied statistician compared to the resource manager or field scientist. However, every decision made during planning and implementation of a monitoring program must be considered from a statistical perspective to assess the decision's effect on the quality of the final information product and on the program's efficiency. Moreover, the applied statistician brings a passion for methods of learning about the world to match the field scientists' passion for answering specific questions about specific processes or populations.

---

[1] Adopted from Lurman *et al.* (2010).

[2] Like biology or ecology, statistics is a large field. It includes both experts trained in theory and those trained in application, with as many fields of application as there are areas of inquiry (e.g. industrial statistics, demographics, medical research, wildlife studies, etc.). For developing a monitoring program, one should search for an applied statistician specializing in monitoring and sampling designs specific to the program at hand. The applied statistician experienced in the relevant field of application will already understand many of that field's methodological issues, pitfalls, and constraints as well as statistical methods for addressing them proactively, through the survey design, or retroactively, through the analysis.

What are the costs of not planning?

Thorough planning, including statistical thinking, is essential to the success of any monitoring effort (Lindenmayer and Likens 2009, 2010b). Without it, one is more likely to make flawed decisions. The negative effects of flawed decisions may not be recognized and resolved for a long time. The often large natural variability of the ecosystem feature of interest combined with typically modest program staff and funds result in a need for long periods of sustained commitment before signals of change are expected to be detected. Meanwhile, the poorer-quality information, relative to what could have been achieved, further delays recognition of system changes. This slows the rate of learning about the system, reduces management's ability to respond to changes, or even may lead to misguided management actions resulting from poor-quality information (Fairweather 1991, Taylor and Gerrodette 1993, Gibbs *et al.* 1999, Reid 2001, Field *et al.* 2005, Legg and Nagy 2006, Taylor *et al.* 2007). In conservation management, failing to detect changes in time to prevent potentially irreversible shifts, declines, or losses can have severe impacts – on the ecosystem structure and function, on program "clients" (e.g. subsistence, sport, and commercial users), and on institutional support. These dangers of ineffectiveness should be reason enough to implement statistical thinking and thorough planning from the outset.

Because poor-quality information reduces management effectiveness and efficiency in addressing current issues, there are fewer resources for other strategic management activities, compounding the shift to a more crises-driven management mode. In the long run, this can result in severe organizational costs from reduced learning about the system and from lost opportunity to acquire high-quality current information for interpreting future system changes and predicting future responses (Nichols and Williams 2006, Lyons *et al.* 2008). In turn, these impacts contribute to a long-term reduction in monitoring program efficiency and effectiveness and ultimately may jeopardize support for the program (Lindenmayer and Likens 2009). All of this is in addition to the often substantial cumulative financial costs of repeatedly spending "trivial" amounts to monitor "X".

A sequence for planning

Statistical thinking requires starting with clear specification of the program's objectives, then working through the necessary analyses, measurement details, and survey design (Fig. 2.1). This sequence may appear backwards, but the primary monitoring goals and objectives largely determine the required analyses. Taken together, the objectives and analyses broadly determine *what to measure.* These decisions then combine with ecological understanding and logistical constraints to influence *how to measure* as well as *where and when to measure.* Each of these decisions has the potential to introduce error into the resulting information products and thus should be made collaboratively by field scientists and an applied statistician (or two!).

Planning often requires iterating among these decisions, refining objectives and methodology decisions until arriving at a monitoring program plan that is *effective,*

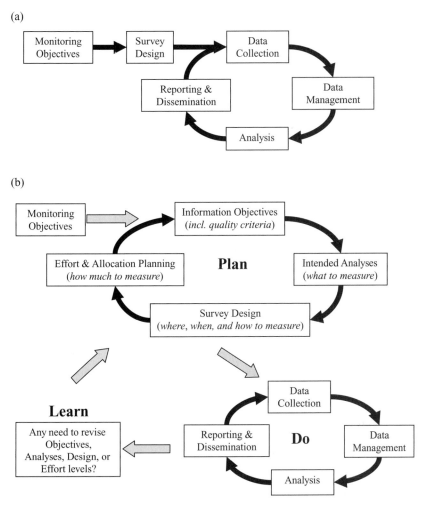

Figure 2.1 The monitoring process: as (a) the commonly perceived implementation sequence, and (b) the recommended planning and implementation sequence to ensure effectiveness, efficiency, and feasibility.

e.g. produces the required information with the required accuracy and precision (Olsen *et al.* 1999, Lyons *et al.* 2008); *efficient*, e.g. does so without excessive use of staff, equipment, or funds; and *feasible* with available staff (including technical capacity), equipment, and funds.

Unfortunately, thorough planning may not occur if data collection is viewed as the most time-sensitive component, fostering an attitude of urgency "to get out into the field and figure out the analysis later" (Fig. 2.1a). This attitude guarantees failure or at least inefficiency. Thorough planning is sometimes viewed as requiring an overly large initial investment of staff time and money just to start a long-term monitoring program. Experience suggests it is a very small investment, with great return, relative to the cumulative effort expended in the average lifetime of a long-term monitoring program (Lindenmayer and Likens 2009, 2010b).

This chapter reviews the core decisions required when planning a long-term monitoring program and their associated statistical concerns. I organize my discussion around three general issues: linkages among components (Fig. 2.1b), common errors, and guidance to relevant resources. In the process, I review many statistical concepts and terms necessary for understanding the rest of this book. An additional objective is to encourage field biologists and other natural resource scientists to seek out experts in statistical thinking early on in the development process. My broader goal is to improve monitoring by raising awareness of the need for sustained dialog among managers, field scientists, and applied statisticians when developing a monitoring program.

Statistical considerations when specifying monitoring goals and objectives

A monitoring program will fail without explicit, well-defined objectives (Elzinga *et al*. 2001: 247–270, Reid 2001, Yoccoz *et al*. 2001). Objectives provide the criteria for determining the appropriate analyses, which then largely dictate the data needed and the appropriate survey and monitoring design options (Olsen *et al*. 1999). Applied statisticians are trained to see these linkages, and therefore are essential participants in the process of formulating objectives.

Monitoring objectives need to be *s.m.a.r.t.* (Williams *et al*. 2007): *specific* and unambiguous, *measurable* with appropriate field data, *achievable*, *results*-oriented, and applicable over the relevant *time*-frame.

Program managers often do not recognize an ill-defined objective until the analysis stage, when it becomes obvious that the resulting information does not actually answer the intended question. Consider the following real example. Seabird abundance was monitored for decades in a set of bays on an island. Analysis was postponed until technical assistance became readily available. Eventually, after much effort collecting, managing, and analyzing the data, trends in annual density were estimated for each of 15 species in each bay. "Suddenly" it became apparent that the program could not directly answer the motivating question – are wintering densities of species "X" increasing or decreasing around the island? The survey was designed to assess trends only at the spatial scale of individual bays. In decades of monitoring, this spatial scale of interest was never sufficiently clarified between the managers and the field biologists.

Objectives must specifically identify the scope of inference – the totality of resources, conditions, locations, or environments to which the conclusions are to be applied and thus for which the results must provide scientifically defensible inference (Olsen *et al*. 1999). Monitoring "changes in species 'X'" is not a well-defined, *s.m.a.r.t.* objective. The following almost is: monitor brown bear density on the Togiak National Wildlife Refuge with a survey effort level and monitoring frequency expected to be sufficient for detecting a population decline in density which, if sustained over the next 30 years, would meet the IUCN "vulnerable" criterion (IUCN Standards and Petitions Subcommittee 2010) with a Type I error rate of 0.10 and Type II error rate of 0.20 (Reynolds *et al.* 2011).

As in this example, objectives need to specify (Elzinga *et al.* 2001: 248):

- the attribute(s) of the resource to be monitored (density, phenology);
- the spatial domain and time-frame of the monitoring;
- the type of change one is interested in detecting;
- the magnitude of change to be detected;
- the precision required in estimating the change; and
- the error rates acceptable when assessing the change (Box 2.3).

Unless the desired standards for information quality are specified, you won't be able to assess whether the monitoring program is doing a good job. Defining these standards is difficult because it requires identifying how the information will be used. The brown bear monitoring objective remains vague in this regard. When monitoring is to inform a specific management decision, the objective should also specify the potential management actions that will be taken as a result of the change assessment (Lyons *et al.* 2008). This specification of the decision-making context ultimately determines most of the other aspects of the objective, e.g. the acceptable error rates, required precision, and magnitude of change of interest.

Even when monitoring does not focus on informing specific management decisions (e.g. surveillance monitoring; see Chapter 3), the program still will be improved by thoughtful consideration of how the information will be used. Such consideration helps clarify both the data's potential uses and limitations, invariably instilling more rigor even if just by helping remove false expectations. For example, detecting declines in a bear population by itself will not provide insight into the cause of the decline (Nichols and Williams 2006). Ultimately, any assessment of change will raise questions regarding causes of change. Clarifying how the data will be used forces this discussion to occur before data collection begins and can help identify likely drivers of change that potentially should also be measured.

The objectives must be specific enough to allow broad identification of the required analyses. Is the objective to estimate a summary of current resource status over the study region (e.g. mean density), assess whether status exceeds some pre-defined threshold, calibrate a model for mapping or predicting resource values across the region, or collect data for comparing two competing models giving distinct management guidance? These are all *status* objectives, but they require very different analyses and potentially very different survey designs (de Gruijter *et al.* 2006: 15–25 is a good, albeit technical, discussion). As discussed below, similar remarks apply to objectives focused on assessing "resource change".

It can be challenging to develop well-defined objectives for a monitoring program with extremely broad scope, or that must be responsive to unforeseen issues (Overton and Stehman 1996, Lindenmayer and Likens 2009). It is useful to distinguish a monitoring program's primary objectives, which usually remain constant across the program's life, from its secondary objectives, which may have intermediate or even short-term time-frames and change throughout the program's life (Lindenmayer and Likens 2009).

Box 2.3 Why bias and uncertainty matter[3]

"It is far easier to put out a figure than to accompany it with a wise and reasoned account of its liability to systematic and fluctuating errors. Yet if the figure is . . . to serve as the basis of an important decision, the accompanying account may be more important than the figure itself." J.W. Tukey (1949: 9)

As described below, sample estimates *without* associated assessments of bias and uncertainty are usually uninterpretable. Such data can be worse than no data (Lurman *et al*. 2010). Due to the psychology of decision making, exposure to any quantitative information, no matter how limited or caveat-laden, can bias later decisions even if someone is told to ignore the earlier information (Tversky and Kahneman 1982).

How can bias lead to arbitrary and capricious decisions?

Evidence from a biased sampling distribution does not estimate the true attribute value. Rather, the value it estimates differs from the true value by some amount, the bias. If the bias is known or estimated its effect can be removed and sound decision-making can occur. If only a limit on the bias' magnitude is known, or only its direction (positive or negative) but not its magnitude, then limited interpretation and decision making may still be possible. However, if the bias is unknown or un-bounded, the evidence (e.g. an estimate of a trend in the population) will not be interpretable, and any decision making based on it will be arbitrary and capricious.

  Bias can also lead to confounding. Confounding occurs when change is estimated (across time, space, or some other controlled factor such as management treatment) from values subject to bias, the biases are not corrected, and the biases differ (at the different times, locations, treatments, etc.). The estimated change then reflects both the underlying true change and the change in biases. Without knowledge of the change in the biases, the estimated change cannot be interpreted as an estimate of the true change. Ignoring the confounding leads to an arbitrary and capricious interpretation of the evidence. Eliminating the confounding requires acknowledging, estimating, and adjusting for the biases.

Type I and Type II errors

The existence of sampling uncertainty often entails the use of statistical hypothesis testing when assessing whether or not the evidence suggest the true value differs from some reference value. For example, in an analysis of population change, we may be assessing whether the true population change differs from the reference value of zero. (While there are other statistical approaches to this problem and important limitations to the hypothesis-testing framework, it provides the simplest framework for presenting the following fundamental issues.)

---

[3] Adopted from Lurman *et al*. (2010).

Statistical hypothesis testing introduces two types of decision-making errors that must be considered in planning and conducting scientific data collection: Type I errors occur when the evidence suggest an underlying difference when in fact there is none; Type II errors occur when the evidence fails to suggest an underlying difference when in fact there is one.

Assuming unbiased estimators, because of sampling uncertainty some samples will result in evidence relatively far from the sampling distribution's center at the true value. When the true value and reference value are the same, then these "extreme" samples will also be far from the reference value. Observing such a sample suggests a difference between the true value and reference value. In classical statistical hypothesis-testing, the decision-maker explicitly sets the acceptable probability of these Type I errors, usually denoted by α. A Type I error probability of $\alpha = 0.10$ means the decision-maker is willing to mistakenly conclude there is a difference (when in fact there is not) 10% of the time. Deciding on the appropriate α value is a policy choice (Shrader-Frechette and McCoy 1992, Field *et al.* 2004, Taylor *et al.* 2007). The smaller the acceptable Type I error probability, the larger the difference required between the observed evidence and reference value in order to assert a difference between the true value and the reference value.

Now consider the case when the true value and reference value differ by a specified amount. Because of sampling uncertainty, some samples will result in evidence relatively far from the center of the sampling distribution and closer to the reference value. Being closer to the reference value, observing one of these samples might suggest no difference between the true value and the reference value. The probability of such Type II errors is often denoted by β; a test's statistical power is $1 - \beta$, the probability of detecting the difference. Power is discussed further in the text.

Consider the Forest Inventory and Analysis (FIA) Program of the US Forest Service (http://www.fia.fs.fed.us/), a 75-year-old nationwide program whose overarching goal is to monitor and report on the status, condition, and trends of all forests in the USA regardless of ownership (Gillespie 1999, Smith 2002). The FIA Program has both very specific primary objectives for some classes of information and, so as to accommodate the changing needs of the broad user community, more flexible secondary objectives. Having tiered objectives has allowed the FIA to meet its core mission while continually adapting to meet evolving information demands – a sign of a successful adaptive monitoring program (Lindenmayer and Likens 2009). The primary objectives include core variables that must be collected by all regional units using the same protocols. The secondary objectives are either "core optional" – variables that a region may choose to collect, but only using specific protocols – or "regional optional" – variables that the region chooses to collect however it sees fit. The secondary objectives are identified in a series of annual meetings with stakeholders at local, state, regional, and national scales. Further, the program has instituted a formal change proposal process to guarantee adopted objectives and modifications are *s.m.a.r.t.* (http://socrates.lv-hrc.nevada.edu/

fia/jct/proposed_changes.htm). I will revisit the FIA example repeatedly to illustrate various concepts.

Developing monitoring objectives requires sustained dialog between managers, field scientists, and statisticians. This dialog usually benefits from using conceptual ecosystem models (e.g. Fancy *et al*. 2009, Lindenmayer and Likens 2009) and decision analysis tools, such as structured decision making and rapid prototyping (Lyons *et al*. 2008). Conceptual ecosystem models are particularly useful when the broad monitoring objective is to track critical aspects of system conditions, e.g. the National Park Service's Vital Signs program (Fancy *et al*. 2009; Chapter 22). Incorporating statistical thinking is perhaps most effective at this early stage so as to help guarantee that the identified objectives really are *s.m.a.r.t*. These dialogs will be improved if participants recognize that field scientists and managers tend to focus on *nouns* (the topic(s) of interest or concern) while statisticians tend to focus on *verbs* (the methods required to answer the question at hand; Easterling 2010). Further guidance on developing sound monitoring objectives is provided by Elzinga *et al*. (2001: 247–270), Maddox *et al*. (2001), Noon (2003), Williams *et al*. (2007), and, from a broader perspective, Ford (2000: 73–102).

Statistical considerations when specifying intended analyses

It is crucial that the analyses required to meet at least the primary objectives be identified, to an adequate degree, before designing the measurement and survey details. Different analyses can require different data (*what and how measured*) collected in different ways (*where measured*). Consider, for example, wildfire burn severity, a measure of fire intensity and fuel reduction (Murphy *et al*. 2008). This is important both as a measure of a fire's effects and for predicting remaining fuel loading and thus near-term susceptibility to another fire. Two approaches to estimating mean burn severity are to sample sites and then either estimate the mean directly or use the sampled sites with remotely sensed data to fit a model and predict burn severity at any location in the fire from the associated remote sensing data. Model fitting requires collecting both the ground-based measures of burn severity and the remote sensing data. However, it is also best approached using a very different method of selecting the field sampling locations than would be used simply to estimate mean burn severity (Murphy *et al*. 2008): stratifying on the remote sensing values to optimize model fitting versus, perhaps, a spatially balanced sample of ground locations. The intended analysis must be determined before you can identify the most appropriate survey design.

With a well developed *s.m.a.r.t*. objective, broad details of the analysis should follow directly. For example, it will be obvious whether the required analysis involves estimating current status (e.g. mean burn severity across the region), temporal change, or some other type of analysis. Current status might be estimated using classic survey design methods while estimating change or making predictions inherently require fitting some form of model except for the simplest situations (e.g. de Gruijter *et al*. 2006: 228). These are different inferential approaches with different implications for data collection (detailed later in the chapter).

Returning to our extended example, the FIA Program's objectives include estimating both status and trends for a variety of attributes at each of a series of nested spatial scales (National Forest, state, region, national) as well as investigating the ecological relationships among attributes and modeling forest dynamics. The need for status estimates at multiple spatial scales was the dominant consideration for the survey design arising from the intended analyses, eventually leading to adoption of a common spatial sampling design across the largest spatial unit of interest – the USA. The design traded efficiency in satisfying any one objective for the ability to adequately satisfy multiple objectives. Trend estimation and supplemental analyses either accommodate this data-collection design or require supplemental data collection. As discussed below, the trend objectives led to selection of permanent sampling plots to improve the precision of trend estimates. Deciding what attributes to measure at each site was based on extensive consideration of the primary and secondary objectives at each spatial scale, the logistic constraints of data collection, and expected budgetary constraints.

Although even with this book there are differing opinions about the value of objectives focused on detecting trends (e.g. Chapters 3, 18), monitoring programs often have an objective of assessing change. Therefore, the rest of this section focuses on the case of fitting a model to estimate change and the analysis details that should be considered prior to data collection. Note that the general steps equally apply to any model-fitting analysis.

Modeling change – step 1: specifying plausible form(s) of change

Both the type of change to be detected (its form and time scale) and its potential causes influence the type of model chosen for analysis (e.g. Philippi *et al.* 1998). This, in turn, can influence the survey design choice, hence the data-collection costs, feasible sample sizes, and expected precision and ability to detect changes. Different analyses and potentially different designs will be required depending on whether the objective is to detect change between two time points, exceedance of a pre-determined threshold, a trend, or changes in the relationship between the resource attribute and some unspecified covariate, such as maximum temperature.

For example, an objective to assess the difference in status between two points in time must specify whether the information required is an estimate of change over the whole region, i.e. *net change*, or of average change per unit, i.e. *gross change* (Duncan and Kalton 1987, Fuller 1999, McDonald *et al.* 2009). These entail different analyses and survey designs (for example, estimating gross change requires visiting the same sample units each time period). If the objective involves estimating change across more than two observation periods, is the expected change a shift in level (step change), a linear trend, or a specific form of nonlinear change? These require fitting different types of models (e.g. Chapters 11, 12) and each model may best be fit using data from different survey designs; see the statistical literature on *optimal designs for model fitting* for linear regression (e.g. Draper and Smith 1981), nonlinear regression (Seber and Wild 1989), etc.

Modeling change – step 2: specifying the (plausible) drivers or predictors of change

The simplest model of resource change uses time as the only predictor. This is the least efficient and least informative model of resource change (Nichols and Williams 2006). It is inefficient at detecting change because the extreme variability of many natural systems, in space and time, implies that the model may not be able to detect a change until a very long time passes and very large changes occur. As it provides no insight as to the likely causes underlying a detected change, it fails to advance system understanding and improve management efficiency (Nichols and Williams 2006; Chapters 4, 18).

The ability to detect changes may be improved by incorporating more system knowledge into the monitoring analyses by either including predictors or using a more structured model than standard multivariate regression; for example, structural equation modeling (Chapter 15) or mechanistic models (e.g. Raftery *et al.* 1995). Such models generally require greater levels of statistical expertise to select, develop, fit, and assess and will likely increase the demands on survey design and measurement as they likely require observation of more variables.

Incorporating more system knowledge into the model can also increase insight as to the potential underlying causes of change, thus advancing research into possible causes and, in turn, improving the prediction of future resource states. Increasing both statistical power and insight increases the quality of information produced by the monitoring program, speeding up both learning about the system and improvement of system management. For example, the FIA Program's objectives include exploration of relationships among observed ecological attributes and modeling forest dynamics through time, both of which seek to improve understanding of forest systems. In turn, greater understanding will allow design of improved forest monitoring methods.

Conceptual ecosystem models are effective tools for identifying relevant drivers and predictors (see, for example, Maddox *et al.* 2001, Noon 2003, Lindenmayer and Likens 2009). The decision then is whether to try to account for these drivers through the survey design, e.g. by stratification, or to include them as covariates in the data collection and analyses (Olsen *et al.* 1999). This is an important decision that requires consideration of the trade-offs between potential precision improvements from the more complicated design against such a design's more complicated analysis and possible resulting restriction on unplanned analyses, i.e. lack of flexibility (Overton and Stehman 1996).

A common alternative to synthesizing prior knowledge and identifying important drivers or covariates is to "measure everything and hope the statistical machinery will miraculously identify the relevant predictors and model". This is a poor strategy. Even if one of the variables happens to be an important predictor, forethought is usually required to identify the most appropriate measurement method (e.g. Nusser *et al.* 1998), as discussed below. Further, this strategy increases the risk of producing spurious results (Type II errors; see Box 2.3) that have little ecological basis for interpretation (Anderson 2008). Success is much more likely if you devote sufficient time BEFORE data collection to identifying plausible model forms and predictors based on current system understanding (Burnham and Anderson 2002: 15–19, Nichols and Williams 2006, Anderson 2008).

The "measure and hope" strategy also increases costs for data collection, data management, and analysis, thus reduces *how many* samples you can gather. Smaller sample sizes, in turn, reduce the precision of both estimates and predictions and likely reduce management effectiveness and efficiency. There is always a cost for measuring more variables "just in case, since we are already here", so it is important to consider the trade-offs in measuring additional factors (without a clearly defined objective) versus engaging in thorough planning to identify these secondary objectives and analyses. Statistical thinking can help clarify these trade-offs by identifying potential uses, or perhaps more importantly, restrictions on usage, of specific additional attributes.

This raises a critical general issue regarding costs: the long-run cost-effectiveness of thorough planning. Synthesizing prior knowledge to identify drivers and potential model forms has associated costs – for planning, data collection, data management, and analysis. However, those costs must be weighed against the potentially much greater savings resulting from the gains in system insight, predictive ability, and management effectiveness and efficiency (Rhodes *et al.* 2006, Lyons *et al.* 2008).

Modeling change – step 3: specifying the intended model fitting and assessment process

When planning, it is important to specify at least the intended primary analyses to the level of detailing the estimation process (Fuller 1999; Hale 1999, 2000; Abhat 2009). In doing so, consider not only the implied financial costs, but also the technical expertise, equipment, and software for each stage of the survey, dissemination of the resulting information summaries, and documenting and archiving the data for future uses. Specify the level of statistical expertise required for the primary analyses, whether the primary analyses can be completed with existing methods and tools, and how the major assumptions of the primary analyses – methodological and biological – will be assessed. Generally, the more complicated the model, the more complicated the model fitting and assessment, and thus the more time expected to be required to complete the analysis to generate the information from the data. This impacts the total cost of information and the most efficient allocation of resources among planning, collection, and analysis.

Statistical considerations when specifying measurement details

Having identified the factors to monitor, one still must identify the measurable attribute(s) that will be used to represent each factor, i.e. the variables. These decisions include further refining what to measure as well as how intensively to measure it, the latter referring to both the measurement process's recording/response scale (nominal, ordinal, quantitative) and the precision and accuracy of the measurement method (Ford 2000: 131–136, Elzinga *et al.* 2001: 265–270). These are important decisions with subtle and potentially severe implications for quality and cost, as well as attainment of monitoring objectives. Because these decisions have implications both for the types of analyses to which the data can be subjected and the constraints on feasible sampling designs, it is another topic that benefits from incorporating statistical thinking before any data collection occurs. Again, these decisions are best made through collaboration between managers, field scientists, and applied statisticians.

An attribute or measurement choice must effectively represent the intended system feature, which requires accuracy, precision, and *selectivity* (Olsen *et al.* 1999, Ford 2000: 133–136). A selective attribute and measurement process results in measurements that "depend only on the intended ecosystem property" (Olsen *et al.* 1999: 30) and thus are clearly interpretable. Insufficient selectivity is usually identified by ambiguous interpretation of the results, i.e. the results could have arisen for many different reasons, a problem that should have been identified in planning. Attributes must also have reasonable costs, technical requirements, and inter-observer error rates (Elzinga *et al.* 2001: 250–252). Collectively, this is a lot of features to have and often one's initial choice of attributes may not meet all these standards.

Consider the broad management objective of comparing importance to shorebirds of different river deltas in a major staging area. How should "importance" be measured? It might be reasonable to assess importance via "use", but how should use be measured? Ground-based counts of the number of shorebirds along a transect is a feasible but not selective measurement because other factors affect this measurement besides the focal process of habitat selection: variation in available habitat due to tidal or wind influences, shorebirds' ready response to disturbance, their general rate of movement when feeding, variation in shorebird abundance across the season, etc. In this case, the best measure may be to take regular, repeated counts at different times of the day across different days of the season in order to estimate a total shorebird-days or shorebird-hours. This would be a substantial increase in effort and logistics compared to one or two transects a season. Note how thinking about the measurement gets into details of how to get the numbers that lead to the seasonal summary as well as into details of how the summary will be used. These details are part of the *response design* discussed later in the chapter. Developing a selective attribute can require much thought and experimentation (Ford 2000: 131–168, Elzinga *et al.* 2001).

Failing to select appropriate measurements is a common problem when the intended analyses have not been adequately specified prior to data collection. Even a selective measurement process may entail severe analysis or information quality costs. For example, one might decide to use a broadly defined ranking system to measure overall "level of disturbance" on a vegetation plot rather than take quantitative measurements of numerous specific attributes; the measurement both reduces field effort and avoids potentially complicated multivariate analyses. However, if the objective requires estimating trend in disturbance rating – here an ordinal response measure – the resulting analysis will be complex. Identifying unexpected analysis implications of potential measurement choices is another important reason to seek statistical advice long before entering the field.

If feasible, it can be very beneficial to adopt a formal attribute selection process that incorporates technical reviews from field scientists (who understand what is feasible in the field), statisticians, and data-management experts. The FIA uses such a process for reviewing proposed attributes, measurement methods, and other aspects of the data collection (http://socrates.lv-hrc.nevada.edu/fia/jct/proposed_change.htm). Their review groups consist of technical experts from within and outside FIA. The process ensures a thorough technical review from a whole problem perspective and promotes interaction between the technical experts in different regions. In turn, this encourages compatibility over large spatial scales and improves coordination on national program issues and projects.

Implications of measurement details for survey design

Besides clearly specifying the attributes to measure on each variable (e.g. size, fecundity, presence/absence, % cover), the primary measurement details must be specified sufficiently to allow identification of:

- the appropriate sample units (e.g. scale, a part of the *response design* as discussed below);
- considerations in selecting sample units (the *sampling design*);
- how to measure each attribute on the selected sample units (the *response design*);
- when to revisit or re-measure sample units (the *revisitation design*). These decisions are discussed further below.

Statistical considerations when specifying the survey design

Poor survey design is a predominant cause of monitoring program failures (Reid 2001, Legg and Nagy 2006). Why? Although apparently not widely recognized outside statistical circles, valid inference usually requires that analyses account for the way the data were collected, i.e. the survey design (Little 2004). The exception occurs when the data collection meets the technical requirements of being *ignorable*. [See Little (2004) or Valliant *et al.* (2000: 36).] For example, taking a stratified random sample, but calculating estimates using the formulas for a simple random sample, will give biased and uninterpretable estimates (except for some specific situations). In light of earlier comments, this implies that in selecting the survey design you must consider the intended analysis and in conducting the analysis you must account for the survey design – analysis and design are intimately linked. A flawed survey design can produce data from which no amount of complex analysis will be able to extract interpretable information (Lurman *et al.* 2010). Alternatively, a well-thought-out design can provide much richer information while perhaps even decreasing field costs per unit of precision. This makes it imperative to obtain statistical advice when developing these components of the monitoring program. You can only collect these data once, so best to do it right (ideally) or at least not wrong!

The temporal aspect of long-term monitoring adds a dimension of further design decisions (reviewed below) and raises the costs of flawed decisions – due to organizational inertia and delayed recognition of flaws. Developing an effective and efficient survey design for natural resource monitoring is even more challenging because resources tend to vary in space and time (sometimes relatively rapidly), can be mobile, are often hard to detect, or do not occur in distinct naturally defined units. All of these features present challenges that can undermine standard survey designs if care is not taken.

It is helpful to distinguish three elements of the survey design – the process used to select units to sample (the *sampling design*), the decision of what to measure on each selected unit and how to measure it (the *response design*; Stevens and Urquhart 2000), and, in monitoring, the decisions regarding the schedule for measuring units (the *revisitation design*). Note that some references use the term "sampling design" as "survey design" is used here: to refer to the collective set of decisions. These three elements entail numerous very important concepts and considerations. The following sections aim to explain the importance of these concepts. Although the concepts are neither difficult nor complex, they can be subtle in certain applications. Their implications can remain hidden until one attempts to address non-textbook problems, such as in most natural resource monitoring! Failure to consider each of these issues can leave one with uninterpretable observations.

Statistical considerations when specifying the sampling design

Why is sampling design important?

This question is tied directly to the fundamental purpose of monitoring. You monitor because you want to answer a question about an attribute of the world (Olsen *et al.* 1999, Nichols and Williams 2006, Lyons *et al.* 2008): are management actions maintaining or improving the state of attribute "X"? Has "X" exceeded acceptable levels? Even surveillance monitoring implies an underlying question regarding change. Answering the question invariably requires assessing the difference between the attribute's value and a reference value (Abelson 1995). For example, does the change in population abundance differ from 0?

Usually the attribute's value must be estimated from a sample (i.e. a subset selected for measurement) of the population of interest since measuring the complete population is too expensive or logistically impossible or unnecessary for obtaining information of sufficient quality. Sample estimates, however, are subject to bias and uncertainty (Box 2.2). If the decision-maker does not account for these features in the assessment then the resulting decision will be unjustified and possibly incorrect (Box 2.3). Valid inference from sample-based evidence thus requires knowing or estimating its bias and uncertainty and accounting for them in the assessment and decision making. Estimating the bias and uncertainty of sample-based evidence *requires* statistical data collection and inference methods, so their use is a necessary condition for valid inference (Deming 1950: 9–14, Peterson *et al.* 1999).

Sampling design considerations

The sampling design requires explicitly defining a number of features, many of which require consideration of the program objectives. Suppose the moose composition (bull : cow sex ratio) on the refuge you manage is a consideration in your decision whether to allow a subsistence harvest next season. The *Resource Population of Interest* is all the moose within the boundaries of the refuge over the time period of the survey. More generally, it is the group of elements (items, organisms, sites, etc.) about which one would like to learn something. Assume there is a 2-week window in the fall to do the survey, after most leaves have fallen and while the moose are still aggregated for the rut, and it will take the full 2 weeks to complete. It is important to clearly specify and document the spatial and temporal dimensions of this population of interest (Refuge Y, these 2 weeks per year) to make sure this matches the information objectives and to maintain consistency across years.

Note the assumptions already needed to force the real world into this conceptual structure: it takes 2 weeks to conduct the survey, but the sex ratio could change during that period from fatalities, emigration, or immigration. The plausibility of assuming a constant sex ratio depends on the refuge shape and area, its habitat quality relative to the surroundings, the presence of moose near the boundary, differential predation rates, etc. The plausibility of this assumption should be documented in the survey protocol so other users of the data can judge for themselves.

As with many natural resources, we cannot directly select from and measure the elements (e.g. moose) of the resource population of interest (Fuller 1999, Stevens and Urquhart 2000). We do not know how many moose there are, so we have no way of making a list or map of them from which to select a sample. Even if we could make such a list, we would not know where to find each selected sample moose in order to record its sex. Instead, we are forced to define some intermediary entities (*Sample Units*) for which we can identify the full *Statistical Population of Interest*, select a sample, and locate and measure the selected sample. For the moose survey, we define a grid of 5 km × 5 km cells over the conveniently rectilinear 100 km × 500 km refuge. While we want to learn about the moose population, we have to do so by making inference on the statistical population (sometimes termed *Target Universe*[4]) of 2000 5 km × 5 km sample units covering the refuge. Classical survey design assumes a finite set of discrete sample units; modern methods allow for sampling more general objects (points in space, etc.). We'll use aerial surveys from a small plane, so our *Sample Frame*, the set of sample units from which we can actually sample, is all 2000 grid cells. The surveying will occur throughout the 2-week period identified. Thus the target universe and sample frame are identical for this survey.

Why is it important to do this pedantic documentation of the population of interest, universe, and frame?

Statistical inference flows from the sample to the sample frame. In natural resource monitoring there are often differences between the sample frame and the target universe, termed a *frame error* (Lessler and Kalsbeek 1992, Lesser and Kalsbeek 1999). Frame errors can cause bias. All of the units in the target universe may not be available for selection (*undercoverage*) or the sample frame may include units not actually in the target universe (*overcoverage*). For example, logistic or safety concerns limit the portion

---

[4] Confusingly, some authors use "target universe" as used here and others use it as "resource population of interest" as used here.

of a cliff-nesting seabird colony that is available to be sampled to just those sections that curve inland and allow safe viewing from the opposing cliff face. Yet this restriction may introduce bias if those cliff sections differ from the rest of the colony with respect to the measured attribute. For example, reduced exposure to storm waves could plausibly increase average productivity rates of those nests relative to others on more exposed cliffs. Alternatively, perhaps the reduced exposure to the seas increases opportunities for terrestrial predators? A supplemental study may be required to assess these issues.

If frame errors result in a bias, and that bias changes across monitoring occasions – perhaps because the sample frame was not explicitly documented and is unknowingly modified, trend estimates can become confounded with temporal changes in the frame errors (Box 2.3). Frame errors in the temporal dimension can be a subtle source of confounding. For example, female migratory waterfowl can arrive on their molting grounds with non- and failed-breeders arriving earlier than successful breeders. If the timing of banding surveys changes over the years with respect to the timing of staging such that more recent annual surveys occur earlier in the staging and thus before many successful breeders have arrived on the grounds, then estimates of changes across the years in survival of successful breeders will be confounded with changes across the years in their underlying availability for recapture (see also Chapter 11). Frame errors are best prevented by thorough documentation of the sample unit or support, population of interest, target universe, and sample frame, including both spatial and temporal boundaries of each.

Sample selection

The validity of any inference from the observed sample to the unobserved units in the frame depends on the relationship between these two sets. That relationship is determined by the sample selection method. If inherent biases in the sample selection are not properly accounted for in the analyses, the results will be misleading and management actions based on them may not be defensible. For example, Greenpeace successfully sued the US National Marine Fisheries Service over critical habitat designation for Steller sea lions in part because of the biased data used to develop the initial designation (discussed in Lurman *et al.* 2010: 256–257). Given the central importance of the sample selection process to estimation and analysis, the approach should be decided through sustained dialog between field scientists and an expert in statistical sampling in that field of application so that the chosen approach best accounts for knowledge of the system under study, field logistics, and other constraints.

Three general approaches to sample selection are discussed here. Extrapolating from a *judgment sample* relies solely on the assumption that the sample is "representative" with respect to the attributes of interest, meaning it provides unbiased estimates. Note the plural – multiple assumptions of representativeness are required if multiple attributes are measured on each sample unit. History has repeatedly shown this to be an unreliable sampling method and that "the sampler's . . . selection is in fact subject to all sorts of biases, psychological and physical" (Yates 1935: 202, quoted in Kruskal and Mosteller 1980: 190). Nor do judgment samples provide any logical basis for estimating the sampling uncertainty associated with their estimates, leaving the estimates uninterpretable.

Haphazard and convenience sampling, discussed in Chapter 5, have similar negative characteristics.

Probabilistic selection methods select the sample following some probabilistic method or design that introduces randomness into the selection process. The randomness provides statistical protection against unrecognized biases from the selection. If you were to repeat the sample selection a large number of times and actually generate the sampling distribution of the estimates, the average estimate would be the true population value (i.e. this is a *design-unbiased* approach). A wide variety of sample selection methods have been developed to utilize different types of information (e.g. on characteristics of the statistical population, on the logistics of sampling) to meet different types of sampling goals (e.g. reducing estimation uncertainty, minimizing sampling cost, providing estimates for specific subpopulations as well as the overall statistical population). Chapters 5–6 focus on designs commonly recommended for long-term monitoring; the volume by de Gruijter *et al.* (2006: 73–80) provides a key for identifying appropriate probabilistic sampling designs. New methods continue to be developed for specific sampling objectives, e.g. adaptive sampling (Chapter 17).

Model-based selection methods assume a mathematical model of the statistical population of sample unit attribute values, e.g. a model of distribution of the resource values and, perhaps, their relationships across space or time such as a geo-statistical model (e.g. Ver Hoef *et al.* 2006, Diggle and Ribeiro 2007). One can then select the sample units to observe by determining which ones will provide the most precise estimates of the unknown model parameters. For example, in the burn-severity example mentioned earlier, if you assume the ground-based burn severity measure is linearly related to the remote sensing index, then selecting half the sample from sites with the smallest remote sensing index values and half from sites with the largest index values would give the most precise model parameter estimates (Draper and Smith 1981). Developing a mathematical model can require a sufficiently studied and understood system.

Inference from the selected sample to unobserved units

Design-based inference uses the probabilities of selection for each of the sample units selected to calculate an estimate for the full statistical population, along with the estimate's uncertainty (see any standard survey sampling text such as Cochran 1977, Scheaffer *et al.* 1986, Lohr 1999, or Thompson 2002). Once you know the correct formula for calculating an estimate and its uncertainty from the chosen design, the formulas apply to every attribute measured (assuming all the attributes are measured on the same data scale: nominal, ordinal, count, or continuous). This inference approach obviously can only be applied to samples that were probabilistically selected. Among its benefits is the absence of any assumptions regarding the distribution of attribute values among the statistical population. The analyses are usually simpler than most model-based methods.

Model-based inference methods use the observed sample to estimate the values of an assumed model's unknown parameters and extrapolate by assuming the model is an adequate description of the population (de Gruijter *et al.* 2006: 15–19). The methods rely on two assumptions: that the model accurately describes the distribution of the attribute and that the sample selection process is ignorable, that is, independent of the attribute being measured (Valliant *et al.* 2000). These approaches "maximize the leverage of data," allowing strong inferences from relatively limited sample information (Stevens *et al.* 2007), but generally at the expense of more complicated analysis and planning and much stronger requirements regarding system knowledge. There are a number of methods for estimating uncertainty for model-based inferences, some analytical, many computationally intensive, but the specific formulas and technical details must be worked out anew for each model.

Model-based inference still requires all the other design decisions; it simply provides an alternative when probabilistic sample selection is not used and relies on model fitting rather than being driven by selection probabilities. In particular, it remains essential to clearly define the target universe and sample frame in order to avoid the potential for confounding from changing frame errors.

Even in model-based inference, it is commonly recommended to use some form of probabilistic selection to avoid selection bias, although the analysis ignores the selection probabilities (Särndal *et al.* 1992: 534). Note that in such cases the model fitting should account for any special structure in the sampling, such as stratification or clustering, unless it is formally ignorable. This need to account for such sampling structure in the model, and hence the analysis, is one reason to thoroughly plan both the desired analyses and survey design. Since estimating change generally requires a model, the most efficient estimates of change when using a fairly structured survey design may entail fairly complicated models, hence complicated analyses.

Considerations in choosing between design-based and model-based inference

Design-based inference methods are appropriate for situations when the thing we want to learn about can be viewed as fixed during the survey period and the uncertainty arises from our inability to measure every unit of interest (sampling variation). This is known as *static indeterminism* (Neyman 1960). It assumes that the resource value associated with each sample unit is constant during the sampling period and is measured perfectly (Särndal *et al.* 1992: 537). Either assumption may be untenable. If ignored, measurement errors can bias both the survey estimates and their estimates of uncertainty (Särndal *et al.* 1992: 605–614). Measurement-error processes can either be modeled and incorporated into the design-based inference, such as in distance sampling (Buckland *et al.* 2001) or you can use a full model-based inference approach, such as wildlife abundance methods which jointly model the population distribution and the observation process (Borchers *et al.* 2002).

Many resources of interest are dynamic processes – air pollution levels, demographic processes such as survival, etc. Because of their inherent dynamics, they present at least two sources of uncertainty – the uncertainty from sampling *and* the uncertainty from having to estimate the underlying dynamic relationships. This *dynamic indeterminism* (Neyman 1960) is the realm of stochastic models, hence model-based inference.

Design-based inference methods estimate global summaries over the full study region, such as moose composition over the entire refuge. Estimates for subsets of the region can be calculated but do not use measured values of the attribute of interest from the rest of the study region. Model-based inference can estimate global summaries and also allows one to borrow strength from across the study region to improve inference about specific subregions, termed *small area estimation*, by fitting the model to all the observations but then making predictions to just the specific subregion of interest (Rao 2003). As mentioned earlier, prediction is an inherently model-based task, as is estimation of change or trends in most situations.

Sampling for multiple attributes

When designing a sampling design focused on a single attribute, usually the more information about that attribute that can be incorporated into the design, the more precise the estimates. However, when multiple attributes are to be measured on each sample unit, tailoring the sampling design to improve estimation for one attribute usually entails worsening estimation of another attribute. This is why many monitoring programs focused on multiple attributes tend to use relatively simple sample selection designs, such as some form of systematic sampling in space (Bellhouse 1988). Chapter 5 discusses this issue further.

The FIA Program is a good example of an effort which seeks information about many attributes at numerous spatial scales. The target universe is the total land area of the USA, including water bodies. For the purposes of reporting estimates, this total land area is divided into subpopulations typically defined by county boundaries, groups of counties, or large public ownerships such as National Forests (Reams *et al.* 2005). The FIA sample is based on a global hexagonal sampling frame (Overton *et al.* 1990, White *et al.* 1992), where each hexagon is approximately 6000 acres. The program uses a systematic design of randomly selecting plots ("Phase 2 plots") within each hexagon. This guarantees even spatial coverage at each of the spatial scales of interest and allows for a wide range of potential analyses. Such designs allow for relatively simple design-based analyses for estimating status of each measured attribute, while also providing flexibility for more complicated planned and unplanned model-based analyses.

Implications of the sample selection method

The choice of a sample selection method is directly linked to other critical statistical considerations and decisions:

- The way you select the sample can determine the basis for inference. Inference from a judgment sample rests solely on your assertion; inference from a model-based selection rests on the model for its justification; design-based inference rests on the method of probabilistic selection.
- Different selection methods require different analyses to assess uncertainty. Judgment samples provide no assessment of uncertainty.
- Selection methods differ in their accommodation of different sampling objectives, such as spatial balance (Chapter 6), logistical constraints, and knowledge of attribute characteristics.
- Under either the design-based or model-based inference, different methods of selecting the sample will differ in the precision of the resulting estimates, i.e. in their *efficiency*.

Time spent planning the sample selection method and exploring feasible alternative designs may produce large savings from the cost of data collection and/or produce higher quality and more useful information.

- As discussed earlier in the chapter, the analysis must consider the sample selection, so you have to think through the desired analysis, the inference method, and the survey design before embarking on data collection.

Statistical considerations when specifying the revisit design

Compared to short-term surveys, monitoring introduces its own set of design decisions regarding the relationship between sample selections at different survey events, termed the *revisitation design*. These are in addition to decisions about the sample and response designs. A *panel* refers to sample units measured on the same occasion. The additional decisions in monitoring are (i) which sample units belong to which panels, and (ii) which panels are visited each time period (McDonald 2003). These decisions determine the trade-off between precision of trend estimates, which is maximized by revisiting the same panel every time (Duncan and Kalton 1987, Overton and Stehman 1996, McDonald 2003), and the accuracy of status estimates, which is maximized by selecting a new sample each time. This latter approach is the easiest to analyze but least efficient for detecting temporal change (Duncan and Kalton 1987, Fuller 1999; Chapter 7).

Visiting the same panel each time is subject to problems if new sample units are added to the sample frame as time passes, because these units were unavailable when originally selecting the panel. Conversely, units may drop from the frame as time passes, potentially causing the panel to shrink. For example, monitoring colony attendance of crevice nesting seabirds often focuses on permanent plots within the colony (e.g. Renner *et al.* 2011). Colonies are originally established on talus scree slopes; eventually vegetation colonizes and thrives given the nutrients in the seabird guano, reducing suitability for auklet nesting. Eventually, such permanent plots likely will cease to host seabirds, regardless of local population trends, as the colony is forced to move elsewhere. Frequent visits to the same panel may also lead to observer impacts (Chapter 7), causing sample units to be unrepresentative of the target population.

Alternative designs exist that seek to balance these trade-offs: rotating panel designs, serially alternating panels, augmented panel designs, etc. (Duncan and Kalton 1987, Urquhart *et al.* 1998, Urquhart and Kincaid 1999, McDonald 2003; see also Chapters 7, 10). The complexity of these other designs usually requires active participation of a statistician for the analysis (Schreuder *et al.* 1993: 178–182, Urquhart *et al.* 1998). Efficiently estimating temporal change involves fitting models that explicitly account for the dependence among observations from revisited units. Selection of a revisitation design should consider the relative magnitude of the different factors influencing the variance of the chosen measurement of change (Chapters 7, 8).

As with the choice of how best to select a sample, the choice of a monitoring program's temporal component involves trade-offs among information objectives and logistical constraints, including cost. For the temporal component, a major trade-off with regard to information objectives is one of precision in estimating current status versus change, such as trend (Chapter 7). These considerations can become very complex for large-scale programs with a wide variety of attributes and spatial scales of interest, such as the FIA program. That program chose to use permanent plots for more precise estimates of trend and change, among other considerations. The plots are distributed among 5 panels in the eastern half of the continent and 10 panels in the west. Each panel is a spatially balanced subset of all the plots (Reams *et al.* 2005). Panels are sampled each year through a non-overlapping design. This sampling scheme trades some precision in estimating status of particular targeted populations, because it covers all ownerships and landcover classes, for greater flexibility in addressing other issues, including unanticipated questions (Bechtold and Patterson 2005, Scott *et al.* 2005, Patterson and Reams 2005).

Statistical considerations when specifying the response design

The *response design* describes both the specific sample unit (or *support*) used for the study – its shape, orientation, and size – and how the attributes of interest are to be measured on the selected sample units. This can be very important in natural resource surveys, because often there is no inherent sample unit and thus one has to be defined and imposed. These seemingly simple decisions can have a large impact on accuracy and precision of measurements, adequacy of analysis assumptions, and even the types of feasible analyses (Larsen *et al.* 2001). For example, numerous field studies in vegetation ecology have shown the impact of sample unit size and geometry on estimates (e.g. Bellehumeur and Legendre 1998, Keeley and Fotheringham 2005).

Consider the earlier example focusing on estimation of moose sex ratio through surveys of 5 km × 5 km sample units. A narrower strip might have been easier to fly and guarantee coverage by the observers. However, too narrow a strip would introduce concerns over moose moving into or out of the sample unit during the survey (recall the assumption that the value of each sample unit remains constant during the survey's duration). Narrow strips would decrease the ability of the observers to track the sample unit's boundary and only count moose that occur within the selected unit. There could be frame errors if the sample units do not fully cover the resource population of interest with no intrusions from non-study areas (Stevens and Urquhart 2000). Regardless of the sample-unit geometry in this example, the study would need to address potential bias due to imperfect detection and imperfect identification of males from the air.

Developing the response design requires statistical thinking and is, again, best developed through dialogue between the field scientists and an applied statistician. Data-collection methods should be tested via pilot studies prior to final protocol development (Olsen *et al.* 1999). Barring that, consideration should be given to incorporating measurement protocol experiments as a research component of any monitoring program that employs new methods in order to improve long-term efficiency and information quality (as frequently done, for example, in the US National Park Service long-term monitoring program; see Chapter 22). Similar side studies should be incorporated to assess key assumptions. Modifications of methods must consider effects on long-term comparability of data.

Box 2.4 Common challenges: lots of errors

If one repeatedly conducted the survey, drawing all possible samples and calculating their resulting estimates, one could assess how much each estimate differs from the true attribute value. The *total survey error* summarizes an average of these differences. Specifically, total survey error is the square root of the mean squared difference (Kish 1965: 510). Total survey error has two components: (i) bias, the systematic or directional errors that arise from factors such as imperfect detection, frame errors, etc., and (ii) variable errors, the random or not-consistently directional errors that are caused by differences in observer skill, variation in the process under study, etc.

The mathematics work out such that *Total Error* $= \sqrt{VE^2 + Bias^2}$ where VE is the variable errors (Kish 1965). As stated above, Total Error is equal to the square root of the Mean Squared Error (MSE), a summary of overall error used in other chapters. The MSE equals the statistical variance plus Bias^2.

If the only variable errors were from sampling instead of a census, then VE^2 would just be the sampling variance. However, there are a lot of other potential error sources since every decision and action in the course of monitoring can contribute to error in the final estimates (Table 2.1). The magnitude of contribution from most of these other error sources can only be estimated from explicitly designed studies. Unlike sampling error, most of the other variable errors will not be reduced by increasing sampling effort. In addition, biases are not reduced by increasing the sampling effort. Thus total error will still be high with a very precise but biased survey. Chapter 3 provides additional discussion of trade-offs in minimizing bias vs. variable errors. Total survey error is controlled by thorough planning, appropriate effort allocation, and instituting QA/QC procedures to maintain implementation consistency and improvement.

The response design also must be considered when decisions are made about allocation of available sample effort. The response design, sample selection, and sample size are major factors of total survey error over which we have control (Box 2.4). For a given total data-collection budget and sampling design, there is a trade-off between number of sample units measured per survey (sample size), the relationship of the sample units (the survey design), and the effort spent measuring each selected unit (response design). A response design requiring less effort at each selected sample unit allows for sampling a larger number of units. In some cases, measurement error will be sufficiently large relative to spatial variation that overall uncertainty would be reduced by employing a more intensive response design at fewer sites (Chapter 18). In other cases the opposite will hold (Chapter 8). Similarly, one must consider the relative costs of measuring a sample unit versus traveling to each sample unit when determining the most efficient allocation for a fixed data-collection budget.

Developing the FIA's response design required much time, thought, and testing, unsurprisingly given the numerous attributes under consideration, the logistical constraints on field crew size and time available at each site, and safety and budgetary constraints. The effort that went into developing this design, and its complexity, were major motivators for the program's technical review process for any proposed modifications to the current design. The current response design systematically distributes 4 subplots within an area slightly larger than 1 acre at each Phase 2 sample-unit plot. A variety of attributes are collected, including those related to the site, trees, understory vegetation, habitat, disturbance agents, etc. (http://socrates.lv-hrc.nevada.edu/fia/dab/databandindex.html#4._Current_National_Core_Field). In addition, 1/27 of the Phase 2 plots (1 per ~160 000 acres) are sampled for detailed measurement of forest health indicators (Woodall *et al*. 2010).

Effort planning and allocation for monitoring

Because of all the decisions that go into developing a monitoring program, it is important to determine whether the effort levels required to attain adequate quality information are feasible. This is done by conducting planning studies, including power analyses or simulation studies (e.g. Renner *et al*. 2011). Planning calculations allow one to explore how the myriad decisions that form the monitoring program coalesce to determine the program's expected information quality and costs. One can use the calculations to investigate trade-offs in effort levels, survey design components, analysis plans, or other features. Such planning also can be used to improve monitoring *efficiency* by investigating trade-offs and identifying an effort allocation among measurements, sites, and years that meets the information quality requirements for the least cost. Such planning also helps garner institutional support since it provides a gauge of potential success and can be used to assess the potential effects of declining support. Perhaps most importantly, planning studies inform managers and the organization of the effort levels, and hence costs, required to meet the information-quality objectives. Although developing and conducting such planning calculations requires statistical expertise, the cost is generally negligible relative to the overall costs of implementing and maintaining a monitoring program for even just a brief period (e.g. Reynolds *et al*. 2011).

Failure to conduct such calculations and just "do what you can with available resources" often wastes resources on efforts with very little chance of detecting differences of interest when they exist (Taylor and Gerrodette 1993, Legg and Nagy 2006, Taylor *et al*. 2007). This can provide a false sense that "everything is fine and under control" when really management just doesn't have information of adequate quality to detect system changes (Fairweather 1991). For example, a review of marine mammal surveys concluded that even precipitous declines (50% decrease over 15 years) in most whale, dolphin, porpoise, and ice-hauling pinniped populations would go undetected given recent levels of survey effort (Taylor *et al*. 2007). Inconclusive evidence from a study with low power does not provide an information basis for making a judgment (Shrader-Frechette 2007: 5), and thus may need to be ignored in decision making to avoid mistaking "absence of evidence" as "evidence of absence".

Conducting a planning study

Statistical power depends on six factors: the actual difference between the true and reference values (termed the *minimum detectable effect*), the Type I error probability chosen by the decision-maker, the underlying population variation, the survey design, the sampling effort, and the analysis method. Everything else being held constant, a larger sample size results in less sampling uncertainty and thus higher power, and a smaller minimum detectable effect requires a larger sample size or acceptance of a larger Type I error rate. Of these factors, the decision maker controls the Type I error probability, the survey design, the analysis, the sample effort, and the desired minimum detectable effect. Statistical planning uses estimates of population variation components to investigate how power changes with changes in one or more of the factors under the decision maker's control (Taylor and Gerrodette 1993, Field *et al.* 2005; Chapters 7–10).

The first step in effort planning is to specify the information quality required by the objective (Maddox *et al.* 2001, Field *et al.* 2004, Lyons *et al.* 2008), i.e. the decision-maker's risk tolerances (acceptable Type I and II error rates). This can be difficult to determine, requiring much dialog between the manager or decision-maker(s), the field scientists, and applied statisticians to identify and translate the information needs into the relevant accuracy and precision requirements. Note that this dialogue automatically forces clarification of the program objective. Structured Decision Making is a very effective method of identifying this aspect of the objective (Gregory and Keeney 2002). Universal standards cannot be established for setting power and Type I error levels appropriate for all decision-making contexts. The levels must reflect both the knowledge base of the relevant body of science and the risks associated with the alternative actions stemming from the potential decisions (Shrader-Frechette and McCoy 1992, Field *et al.* 2004). In the absence of other clear standards, formal or informal regulatory standards can be useful. For example, monitoring of biological populations could use the IUCN standards for classifying a population as vulnerable (IUCN Standards and Petitions Subcommittee 2010); see, for example, Renner *et al.* (2011).

With these requirements in mind, one can conduct investigations to identify (i) the most efficient survey design for the given setting; (ii) the necessary effort levels to attain the quality requirements using that design; and (iii) the efficient allocation of effort across measurements, sites, and years (Larsen *et al.* 1995, 2001, Field *et al.* 2005; Chapters 7–10). Ideally, one incorporates cost constraints into these calculations (e.g. Cochran 1977, Field *et al.* 2005; Chapter 8). The calculations require specifying all the previously discussed components of the monitoring program, e.g. the planned analyses, the attributes being measured and the survey design, as well as both the major sources of variation in the attribute of interest (the components of variance), e.g. spatial, temporal, etc., and their magnitudes (Larsen *et al.* 2001; Chapters 7–9). Estimates of the components of variance must come from prior knowledge, pilot studies, or similar systems. For example, the Manager's Monitoring Manual website provides a compilation of year-to-year coefficients of variation from published avian and amphibian studies, searchable to taxonomic level and target species as well as attribute, study duration, and analysis (http://www.pwrc.usgs.gov/monmanual/management.htm, accessed 8 June 2011).

The generally simpler analyses associated with status monitoring entail relatively simpler effort calculations. Sample-size formulas are widely available for standard design-based surveys (e.g. Cochran 1977, Scheaffer *et al.* 1986). Alternatively, simulation is often intuitive and can be tailored to the specific application, e.g. the WiSP package for R (Zucchini *et al.* 2007). Model-assisted design-based (e.g. distance sampling) and model-based surveys require specialized calculations (e.g. Mackenzie and Royle 2005), although software supporting such calculations is increasingly available for specific methods (e.g. Bailey *et al.* 2007, Thomas *et al.* 2009).

Effort calculations for change or trend monitoring are more complex because they must explicitly account for the temporal revisitation design as well as an assumed model of how the resource of interest is actually changing (for example, see Renner *et al.* 2011). Gerrodette (1993) provides software for doing these calculations when the analysis is simple linear regression against time, the revisitation design selects a new sample each time, and a consistent survey design and monitoring frequency are used. Other chapters discuss approaches (Chapters 7, 8) or software (Chapter 9) relevant to some study-design situations. More complicated designs and/or analyses require simulation-based studies (e.g. Rhodes *et al.* 2006, ter Braak *et al.* 2008, Renner *et al.* 2011; Chapters 10, 17).

Effort calculations often require values for parameters about which there is very little information at hand. Such situations reinforce the importance of adopting an attitude of adaptive monitoring (Lindenmayer and Likens 2009). For example, I recommend updating the planning calculations after a few survey periods to see if effort levels are adequate, need to be increased to be effective, or can be reduced to improve program efficiency. Such learning is essential in long-term monitoring.

Learning to learn: the importance of interim analyses and effort re-assessments

"A process should be run so as to generate product *plus information on how to improve the product*" (Box 1957, emphasis in original).

An effective monitoring program should produce both the information required to meet the objectives and also information that improves system understanding and can be used to improve the monitoring process itself. For example, monitoring within an adaptive management framework explicitly seeks to improve system understanding by updating and revising the suite of competing system models (e.g. Lyons *et al.* 2008; Chapter 4).

A monitoring program should explicitly strive to improve its efficiency. As data accrue, the relative error contributions of different component decisions should be re-assessed and used to guide improvement of the survey design, response design, and effort allocation (Deming 1950: 24–26, de Gruijter *et al.* 2006: 38–40). Analysis methods should be revisited in light of both improved system knowledge and methodological advances. These can be major technical undertakings and will only occur if program resources are explicitly set aside to do this. For example, the FIA program continually investigates analytical methods that can expand the information produced, reduce uncertainty, and better utilize remotely sensed data to reduce field costs (e.g.

Nelson *et al.* 2007, Opsomer *et al.* 2007, USDA Forest Service 2007, Goward *et al.* 2008). In the US government's 2009 fiscal year, 7% of the FIA program's total of 381 federal employee person-years were devoted to such techniques research, with an additional 28% devoted to analysis and information management (Gretchen Moisen, pers. comm., August 2010).

Employing such a "learning to learn" perspective requires planning, as it may impact data management and collection, as well as commitment of technical capacity and resources. Is it worth it? Consider these examples. The survey design and sample effort for an annual survey of crevice-nesting seabirds was re-assessed using the most recent decade of data. The assessment identified a more efficient survey design and suggested that the current sample size could be reduced and still meet the information quality objectives (Renner *et al.* 2011). In another setting, data from a program for monitoring sockeye salmon escapement on the Newhalen River in Alaska were used to assess alternative variance estimators appropriate for the program's "systematic sample in time" survey design. The study identified an estimator appropriate for the type of temporal dependence observed in salmon escapement data. Use of this estimator would reduce estimates of uncertainty by two orders of magnitude (Reynolds *et al.* 2007).

Programmatic investigations to improve program effectiveness and efficiency should include all sources of errors affecting the overall quality of information, not just those associated with sampling or the measurement process (Box 2.4). Those arising from sources other than sampling may be more easily and effectively controlled (e.g. through improvements in data recording and other aspects of data collection and management). The validity and repeatability of all analytical steps need to be considered. Unfortunately, it is not unusual to encounter "Type III" errors – employing the wrong analysis – especially with complicated sampling designs (e.g. Overton and Stehman 1996).

This argues for incorporating a quality assurance (QA) program throughout the monitoring process (Ferretti 2009). Currently full-fledged QA programs are found mainly in large-scale environmental monitoring such as the US Environmental Protection Agency's Quality System (USEPA 2009a), the US Forest Service's FIA program, and the US National Park Service's Vital Signs program (Chapter 22). Monitoring programs need organizational processes for accommodating changes in measurement methods, budgets, and questions (Overton and Stehman 1996, de Gruijter *et al.* 2006: 40–41), such as the technical review processes of the FIA program mentioned earlier. In the long run, re-assessing a monitoring program's survey design and analysis decisions as information accrues should result in improved management efficacy and efficiency, earlier problem detection and resolution, improved production efficiency, development of "outside" clientele and program champions, greater institutional support, and improved institutional awareness of the true costs of effective monitoring.

Summary

Monitoring is an information-producing process whose every component raises statistical concerns (Table 2.1). There is a pervasive need for statistical thinking throughout the

Table 2.1 Common errors in natural resource surveys, with some potential remedies (developed from Deming 1950: 26–50, Kish 1965: 509–547, and Lesser and Kalsbeek 1999).

Source	Comments/examples	Potential remedy
Vague objectives	Includes failure to identify specific information required and information quality required	Peer review, thorough planning
Frame errors	Target universe and sample frame differ	Explicit documentation and assessment prior to sample selection; redefine target universe; improve sample frame; supplemental study to assess possible bias
Measurement method	Bias – e.g. from inability to detect all objects actually present	Revise measurement method to eliminate bias or account for it via model (e.g. distance sampling)
	Variation – e.g. the sample unit's value changes during the measurement process, as with the number of shorebirds within 50 m of a 100-m transect line	Improve measurement process; multiple measurements per site per survey; composite or cluster sampling
Observer errors	Flawed protocols – vague directions and coding guidance, over-reliance on observer memory or subjective decisions, etc.	Thoroughly tested and documented protocols; peer review; scheduled assessment and revision of protocols; observer training and assessment; QA/QC
	Bias or variation – e.g. random (non-directional) errors in estimation by observers	Improve training; refine measurement process and protocol; field guides (e.g. ocular estimates)
Non-response or availability bias	For example, inaccessible sample units; objects hidden from view of observers because of intervening vegetation or because the objects are underwater	Revise survey timing and measurement method to reduce availability bias; revise protocol to reduce survey burden (e.g. in surveys of human subjects) and increase participation (e.g. by land-owners controlling access); gather auxiliary information to estimate expected response ("imputing"); additional sampling effort to identify common features leading to problem (e.g. availability bias only occurs in riparian habitat of > 70% cover); modify sample selection probabilities to reduce effect of bias; modify sample frame to eliminate the problem "strata"; conduct supplemental studies to estimate bias
Data recording, transcription, management	For example, coding errors in the field or in later stages of transcription; lost records; loss of data recorder power during survey; etc.	Automate aspects of data collection; improve QA/QC procedures, including "in field" error checking; digital data collection to reduce transcription stages and automate aspects of QA/QC; thorough accountability for data management, including metadata and archiving; dedicated data management staff; regular assessment of data collection and management procedures

(cont.)

Table 2.1 (*cont.*)

Source	Comments/examples	Potential remedy
Analysis	For example, conducting the wrong analysis for the design under which the data were collected; analysis coding errors; human errors in conducting, interpreting and presenting results	Peer review, including statistical review; thorough analysis protocols; software testing, debugging and documentation; adopting "reproducible research" methods and software (http://reproducibleresearch.net/)
Confounding due to changing biases	For example, assessing population change between two surveys but failing to account for changes in detection rates	Thoroughly document protocols and adhere to them; document and assess impact of any changes in protocols, including changes to training, equipment, and measurement methods; consistently employ thorough QA/QC procedures; conduct timely analyses in order to detect problems quickly; employ an analysis that accounts for changes in relevant covariates (which requires identifying them ahead of time and measuring them!)
Sampling variation (sampling errors)	Uncertainty in the final estimate arising from measuring only a subset of the sample frame	Increase sample size; use a more efficient sample selection design; use an analysis that accounts for variation among sample unit values attributable to variation in specific covariates
Selection bias	Systematic differences between the sampled and unsampled portions of the sample frame; biased result arising from not accounting for the sample selection probabilities in the analysis	Use probabilistic selection; properly account for selection probabilities in the analysis
Process variation	Changes, usually temporal, in the sample frame's total value arising due to changes in the underlying physical, biological, or ecological process. For example, year–to-year variation in annual growth rate. If these are not explicitly accounted for by the analysis they will tend to increase the uncertainty of the estimated change	Reduce by using a less variable attribute (e.g. estimate demographic rates or habitat rather than abundance); incorporate associated covariates in the analysis or employ a more mechanistic model rather than a simple "change across time" model

planning of a monitoring program because the final product – information – is the result of a series of variation-contributing processes (Snee 1990) and its quality is measured in inherently statistical terms of bias, uncertainty, and efficiency. Moreover, statistical concerns underlie any discussion of minimizing cost through efficient control of bias and uncertainty. All too often, this need for statistical thinking remains unrecognized until after monitoring has begun, data are in hand, and the program coordinator seeks advice on appropriate statistical analyses. Unfortunately, calling in the statistician *after* the study may be no more than asking for an autopsy to identify the cause of death (Fisher 1938).

Program managers and field scientists need to adopt a "whole-problem" perspective, recognize the inherent linkages and statistical considerations, and include an applied statistician (or many) on the development team from the start. The resulting monitoring program will be more effective, more efficient, and more successful. The last century has been termed the Statistical Century (Efron 2003) for the rapid rise of statistical thinking and its broad adoption in field after field: agriculture, demographics, education, industrial processes, medical research, astronomy, laboratory sciences, and so on. Environmental monitoring is moving toward membership in that list, but won't fully join until there is much broader recognition among decision-makers and field scientists of the field's myriad statistical considerations.

Acknowledgments

This chapter reflects thoughts from collaborations and many long discussions with Heather Renner, Emily Silverman, Ken Newman, Alice Shelly, Bill Thompson, Melinda Knutson, Eric Taylor, and Hal Laskowski, among others, though they are not responsible for the presentation. Special thanks to two anonymous reviewers for their thoughts and suggestions, Gretchen Moisen for her FIA insights and encouragement, and Robert Gitzen for his comprehensive review. My deepest gratitude to the field biologists of the National Wildlife Refuges in Alaska for their dedication to conservation, their willingness to indulge in some statistical thinking, and their patience when all they wanted was a quick answer to "a simple question". The findings and conclusions in this article are those of the author and do not necessarily represent the views of the US Fish and Wildlife Service.

3 Monitoring that matters

Douglas H. Johnson

Introduction

Monitoring is a critically important activity for assessing the status of a system, such as the health of an individual, the balance in one's checking account, profits and losses of a business, the economic activity of a nation, or the size of an animal population. Monitoring is especially vital for evaluating changes in the system associated with specific known impacts occurring to the system. It is also valuable for detecting unanticipated changes in the system and identifying plausible causes of such changes, all in time to take corrective action.

Before proceeding, we should define "monitoring." One definition of "monitor" (Microsoft Corporation 2009) is "to check something at regular intervals in order to find out how it is progressing or developing." The key point here is "at regular intervals," suggesting a continuing process. Some definitions do not indicate the repetitive nature of monitoring and are basically synonymous with "observing." Most monitoring, in the strict sense of the word, is intended to persist for long periods of time, perhaps indefinitely or permanently. Similarly, Thompson *et al.* (1998: 3) referred to the "repeated assessment of status" of something, but noted that the term "monitor" is sometimes used for analogous activities such as collecting baseline information or evaluating projects for either implementation or effectiveness. For their purposes, they restricted the term to involve repeated measurements collected at a specified frequency of time units. Let us adopt that definition, recognizing that repeated measurements imply collecting comparable information on each occasion.

In this chapter, I give a personal perspective on some issues associated with monitoring (Box 3.1), including issues addressed throughout this volume. I have had the good fortune to have worked with many biologists and statisticians involved with monitoring activities, mostly of animal populations. Also, I have personally engaged in some monitoring activities as well as analyzed data generated by monitoring programs. Some of what follows are suggestions; these generally are straightforward. Others are just issues that should be considered; I have no ready solutions other than the trite answer, "It depends." I hope that framing some of these issues in this chapter will lead to further thought and discussion about them, and ultimately to satisfactory resolutions.

Design and Analysis of Long-term Ecological Monitoring Studies, ed. R.A. Gitzen, J.J. Millspaugh, A.B. Cooper, and D.S. Licht. Published by Cambridge University Press. © Cambridge University Press 2012.
This chapter is a work of the United States Government and is not protected by copyright in the United States.

Box 3.1 Take-home messages for program managers

I define monitoring to be repeated measurements made consistently on multiple occasions over time. Despite recent criticisms of monitoring in the absence of hypotheses to be tested, I argue that such "surveillance" monitoring has value for detecting unanticipated changes in time to take remedial action. In my experience, some features critical to the success of any long-term monitoring program, as well as challenges frequently faced by these programs, include the following:

- The objectives of any monitoring program should be clearly stated. From many possible competing objectives, decide on the primary objectives and downplay or dismiss others. Specifying *how* resulting data will be used is critical in determining the objectives.
- One common purpose of monitoring is to estimate a trend, but "trend" is not a well-defined term. Different, but equally reasonable, definitions can lead to divergent conclusions. Even more importantly, estimates of trend, however defined, are highly dependent on the time period over which trend is estimated; what seems to be a consistent trend during one time interval may appear to be normal fluctuations over a longer time scale.
- A good monitoring program needs to be sustainable, which involves considering current and future budgetary limitations when designing the program, maintaining participant involvement, ensuring continued access to monitoring sites (which can be problematic for sites on private lands), and reporting regularly to participants and managers.
- With continued advancements in available quantitative methodology and measurement techniques/technology, a difficult question often arises: should an ongoing, long-term monitoring program be abandoned for a newer, better-designed one? Such decisions should consider the costs : benefits of alternative choices (see also Chapter 23). It can be argued that continuity can be maintained while adopting a new program if both the old and new programs are conducted for several occasions and results are used to calibrate the new program. If a strong correlation between the monitoring programs is found, are there advantages to switching? If the two programs do not provide comparable results, then the choice is between the continuity provided by retaining the old program and whatever advantages the new program confers. Foremost in deciding whether or not to switch to a new monitoring program are the objectives of monitoring and evidence that the current program is not satisfying them as well as a new program would.

Although my opinion is that the term "monitoring" should not be applied to shorter-term targeted data collection (e.g. to assess the effects of a management action), I also note that the adaptive monitoring paradigm of Lindenmayer and Likens (2009) could accommodate both surveillance monitoring and hypothesis-driven components. An "adaptive monitoring program" would be designed to address well-defined and tractable questions that are specified at the start of the program. As these questions are resolved and others arise, the monitoring design may be modified, but the integrity of long-term "core" measures is maintained, a feature that is central to adaptive monitoring.

Monitoring with a purpose

First things first: the objectives of a monitoring program should be both clear and clearly stated. As also emphasized in other chapters (e.g. Chapters 2, 18), this is a critical issue, one that always involves consideration of the following questions:

- Why are you doing it? Many objectives for any monitoring program can be proposed, but deciding on the primary objectives, and delegating others to secondary status or ignoring them, is important. A monitoring design will involve trade-offs among competing objectives, so agreeing to primary objectives will make subsequent decisions easier.
- What features will you monitor? The number of individuals in a specified population? Species richness? Age composition? Breeding status? Body condition?
- What area? North America? A state or province? A national wildlife refuge? Aquatic habitats within a refuge? Areas within a national park that are readily accessible by humans?
- What resource? If birds are the focus, is it just breeding birds? Diurnally active birds? Adult birds? Birds that emit sounds during some survey period?
- What metric or attribute of the resource? Most commonly, monitoring focuses on population size or some related metric such as presence/absence or frequency of occurrence. For this reason, and due to my own experience, I will focus on population size, but the issues discussed can be translated to other ecological metrics.

Note that addressing these questions requires a clear understanding of the objectives of the monitoring program. Despite the need to identify primary and possibly secondary objectives, it may be useful to collect ancillary information for which there is no immediate application, but potential uses can be envisaged and such information is easy to gather without compromising the stated objectives.

Implicit in stating the objectives of monitoring is consideration of the use of resulting data. If one is monitoring a population of some species, for example, is it anticipated that a certain result will trigger some management action to remedy the situation? If so, what result, and what actions are feasible? Sometimes these questions cannot be addressed at the initiation of a monitoring program; typical results and "normal" fluctuation may be unknown.

Monitoring and hypothesis testing

Nichols and colleagues (Nichols 1991, 2000; Yoccoz *et al.* 2001; Nichols and Williams 2006) have argued that monitoring should have a very specific purpose; that is, it should be designed either to test a specific hypothesis or to inform a specific management program, analogous to Platt's call for strong inference in scientific investigation (Platt 1964). They criticized surveillance monitoring, which they defined as monitoring not guided by a priori hypotheses and their corresponding models (Nichols and Williams 2006).

Nichols and Williams (2006) said that a popular approach to use for surveillance monitoring in conservation practice requires a two-phase process before useful action can be taken, potentially reducing the efficiency of monitoring and its effectiveness in helping managers act before it is too late to reverse an undesirable change (see also Chapters 4, 18). First, a population decline must be detected from the results of the monitoring program. Second, either some action is taken to remedy the situation, or more focused research is called for to identify the cause of the population decline (Nichols and Williams 2006). As an alternative strategy, they proposed that decisions about actions to take be viewed as a problem in structured decision making. That process requires hypotheses and associated models of how a system (e.g. a population of animals) will respond to management actions. These models will predict the behavior of the system. Monitoring then can provide data against which predictions are compared. Models whose predictions are borne out are viewed as more credible than those that fail to predict accurately. By iterating this process, one gains confidence in the understanding of the system and can make better decisions about managing it. Such an approach of monitoring in an adaptive resource management context is the focus of Chapter 4 in this volume.

These recommendations cannot be faulted whenever some treatment or impact is anticipated. To understand how a system operates, or to assess the effects of management intervention, the system definitely should be observed both before and after any interventions are made. Those observations generally need to be taken only during a defined time period following the intervention, not indefinitely. As Marsh and Trenham (2008) noted, programs intended to learn about a system usually have shorter lifetimes than "surveillance" programs (see also Chapter 23).

Because learning about a system is typically an iterative process, alternating between hypothesis generation and evaluation, one might argue that continuous observation of the system is required, ergo, monitoring. However, the learning process might well generate predictions at different stages that can be examined only by observing different aspects of the system. For example, attempting to understand the dynamics of a population might initially involve estimating population size over a period of time to determine how it varies naturally or changes in response to some management action. Later in the learning process, predictions about the proportion of young in the population might be needed to evaluate competing hypotheses. This likely could require a change in the observational process, so it no longer fits the customary definition of long-term monitoring. Even later in the learning process, information about the sex composition or genetic traits might be needed, thus necessitating further changes to the observational process. In contrast, monitoring as we use the term implies consistency of methods over time.

My definition of monitoring above involves long-term, consistent observation of a system. By this definition, pre- and post-intervention observations do not constitute monitoring. Moreover, I suggest that the term "monitoring" not be applied to observational processes designed specifically to evaluate competing hypotheses or models. To reduce confusion, I believe the term "monitoring" is best restricted to what Nichols and Williams (2006) called surveillance monitoring, but I realize that mine is a lonely viewpoint.

Nichols and Williams (2006) criticized surveillance monitoring for providing "low information" (*sensu* Platt 1964) observations, resulting in inefficient and ineffective learning. Nevertheless, especially in multispecies monitoring programs, such omnibus monitoring does seem appropriate. Consider the North American Breeding Bird Survey (Robbins *et al*. 1986), which monitors populations of hundreds of species, some fairly well, some poorly. It seems inconceivable that hypotheses and models describing the population dynamics of all these species could be developed and evaluated through an adaptive learning process. Instead, it seems necessary to follow the admittedly inefficient path noted by Nichols and Williams (2006): first to use monitoring results to detect, retrospectively, changes in populations of some species, then either to take remedial action or to conduct more focused research to identify causes of the changes.

Fortunately, surveillance monitoring programs can in fact form the basis of an evaluation of management effects (e.g. Johnson 2000). For example, suppose that a park-wide breeding bird-monitoring program has been established in a national park, and that a specific management activity, say a prescribed burn, is planned for a part of the park. It may be unlikely that the extant monitoring program will adequately estimate the effects of the burn. One could develop another study design that focuses directly on lands affected by the burn. Instead, perhaps the monitoring program already in place can be augmented to address the question of bird responses to the burn. It is likely that adding more monitoring sites in areas likely to be affected, and similar sites that would not be affected ("control" or comparison sites) would suffice (thus constituting a Before–After, Control–Impact or BACI design). These sites would be used for as many years as necessary to estimate the effect of the burn, and then terminated or, if sufficient resources were available to continue monitoring them, they could be designated as separate strata. Another set of supplemental sites might be established to assess a management action elsewhere in the park. The key is that a continuous monitoring program is in place to detect any changes in bird populations caused by *unplanned* activities (e.g. hurricane, wildfire, oil spill) while also establishing a framework for evaluating the effects of planned activities, when needed and where needed.

Further, surveillance monitoring intended for a specific purpose often has even more valuable applications to other issues. The various monitoring programs established to track waterfowl populations in North America (e.g. Nichols 1991) were created basically to assess population changes and provide information for harvest management. From the data that accumulated during those programs, we have also learned much about the migrational behavior of waterfowl (e.g. Johnson and Grier 1988), density dependence (Viljugrein *et al.* 2005), utility of waterfowl as umbrella species (Skinner and Clark 2008), and other topics.

Similarly, the North American Breeding Bird Survey (BBS) was developed to assess changes in population sizes of a broad array of avian species – i.e. essentially as a surveillance monitoring program. In addition to addressing that objective, data from the BBS have provided more information about the ranges of species; facilitated the development of spatially explicit models of bird occurrences; identified species, species groups, and ecosystems most deserving of conservation attention; and much more (USGS 2007). A 2002 report (Peterjohn and Pardieck 2002) identified 270 scientific papers and theses that relied on data produced by the BBS.

Another example of unplanned uses of monitoring programs involves surveys of greater sage-grouse (*Centrocercus urophasianus*). Western states individually initiated surveys of sage-grouse during the early to mid 1900s primarily to provide information for harvest management. Methods differed among the states, and changed over time, in efforts to make them more consistent among states (Connelly and Schroeder 2007, Johnson and Rowland 2007). Now data from those surveys have been used to address a variety of important questions largely unrelated to harvest (e.g. Knick and Connelly 2011).

A non-avian example was provided by van Mantgem *et al.* (2009). They synthesized information about tree mortality from 76 forest plots in the western United States to conclude that mortality rates have been increasing almost uniformly across that area during the past few decades. The authors suggested that regional warming was a likely causal factor; mean annual temperatures in the western United States have risen by about 0.4°C per decade in the past 40 years and snowpack in the study region has diminished (van Mantgem *et al.* 2009). Those forest plots originally had been established for a diversity of purposes; few if any of them were intended to address climate change questions.

Aldo Leopold collected dates of phenological events occurring near his "shack" in central Wisconsin, during 1935–1945 (Leopold and Jones 1947). Such data are useful, he said, to permit an interpolation of contemporaneous events for any phenological event or calendar date, and to permit one to adjust for early or late seasons by translating calendar dates into "phenological dates" (Leopold and Jones 1947). It is very doubtful that he conducted that monitoring with hypotheses in mind about the effects of a changing climate on plant and animal activities. Yet, those data, when compared with similar records taken during 1976–1998, have proven highly valuable for demonstrating that some natural events are cued by temperature whereas others are driven by day length (Bradley *et al.* 1999).

Many agencies have stated objectives that involve monitoring ("monitoring" in the sense of long-term consistent observation). For example, the US National Park Service has an Inventory and Monitoring Program charged with monitoring park ecosystems to better understand their dynamic nature and condition and to provide reference points for comparisons with other, altered environments (Chapter 22). The US Environmental Protection Agency has an objective to estimate the current status, trends, and changes in

Table 3.1 Hypothetical example of counts
made of some organism on five occasions
and the proportional rate of change from one
occasion to the next.

Occasion	Count	Rate of change
1	1000	NA
2	900	0.900
3	800	0.889
4	700	0.875
5	1000	1.429

selected indicators of condition of the nation's ecological resources on a regional basis with known confidence (USEPA 1993).

Consider the medical world. Physicians routinely monitor patients' blood pressure and body temperature. Generally they do so not to detect a specific disease, which would be analogous to testing a specific hypothesis, but instead they use abnormal values as cues to investigate reasons for those departures. Clearly, this is an example of surveillance monitoring, and an effective one.

I do not want to leave the impression that monitoring for the sake of monitoring is a worthwhile endeavor. Certainly when developing a monitoring program one should consider possible events (treatments, impacts, etc.) that would have an effect on whatever is being monitored. A monitoring plan for a fire-susceptible forest, for example, should be able to assess the consequences of a fire, whether prescribed or not. Monitoring sea life near a route plied by numerous oil vessels should be able to determine the consequences of an oil spill, even though none is planned. Still, "surveillance monitoring" is essential for detecting unexpected changes in time to take remedial action.

What is a trend?

One common purpose of monitoring is to estimate the trend in whatever is being monitored (Chapters 7, 11). "Trend" is generally defined as a general tendency, direction, or movement. A more statistically oriented definition was provided by Easton and McColl (1997): "Trend is a long term movement in a time series. It is the underlying direction (an upward or downward tendency) and rate of change in a time series, when allowance has been made for the other components." Kendall and Buckland (1971: 155) termed a trend a "long term movement in an ordered series . . ."

With those vague definitions in mind, consider the following contrived example (suggested by J. D. Nichols, USGS Patuxent Wildlife Research Center, Laurel, MD USA, personal communication 12 December 2009). Assume that counts are made of some organism consistently on five successive occasions. Counts on Occasions 1–5 were 1000, 900, 800, 700, and 1000, respectively (Table 3.1). The rate of change is defined as the count on one occasion divided by the count on the previous occasion. Corresponding

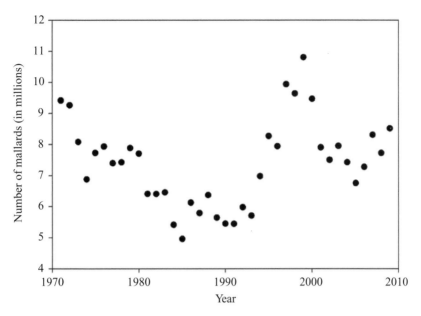

Figure 3.1 Estimated number of mallards, in millions, in the North American May Waterfowl Survey area, 1971–2009 (USFWS and USGS 2010).

rates of change between each pair of successive years were 0.900, 0.889, 0.875, and 1.429, respectively.

What is the trend of the counts? Because the final count was the same as the initial count, the data could be said to show a flat trend. That conclusion is echoed by the geometric mean of the annual rates of change, 1.0. The arithmetic mean of the rates, in contrast, is 1.023, indicative of an increasing trend. Performing a regression of Count on Occasion, however, generates a regression coefficient of –20.0 (SE = 46), suggesting a decreasing trend. Even in this simple data set, which has no uncertainty about the accuracy of the count values, we can reach divergent conclusions about its "trend," depending on how it is defined.

Interpretation of real data may be even messier. In the mid 1980s, there was concern about the status of the North American mallard (*Anas platyrhynchos*), a common and widespread duck that is prized by waterfowl hunters. The mallard population is annually monitored throughout its primary North American breeding range (Martin *et al.* 1979). The extensive survey has used basically the same design and procedures since at least 1971, and estimated populations had declined steadily from 9.4 million mallards in 1971 to 5.0 million in 1985 (Fig. 3.1). The trend (here, the correlation between estimated population size and year) during the 1971–1985 period was steep and strongly negative ($r = -0.88$, $P < 0.0001$). Despite the obvious nature of this decline, skeptics questioned the reality of the decline per se, or whether or not it reflected anything more than a cyclic decline in wetlands during that period. The issue was controversial enough to warrant detailed analyses (Johnson and Shaffer 1987). Johnson and Shaffer (1987) concluded from three different analyses that the decline appeared real and could not be explained solely by changes in wetland numbers. The authors cautioned, however, that conclusions about whether or not the decline was "real" were highly dependent on the choice of endpoints, which was rather arbitrary.

In the subsequent 23 years, mallard numbers indeed rebounded and even exceeded the highest early counts. The trend during 1985–2009 was markedly positive ($r = +0.59$, $P = 0.002$). For the entire 39-year period there has been no overall trend; the correlation coefficient between mallard count and year was $+0.15$ ($P = 0.35$). The recovery of the population was not a mystery; wet conditions returned to the prairies while the Conservation Reserve Program had restored huge areas of good nesting cover for mallards (Reynolds *et al.* 2001).

So, in addition to conflicting conclusions that different definitions of trend might reach, estimates of trend are highly dependent on the time period over which trend is estimated. Such strong dependence on the choice of time period has never been more contentious than in the recent debates about global climate change. Individuals most concerned about the change note that recent average global temperatures are the highest recorded, whereas skeptics argue that recent temperatures are not so extreme when viewed from a longer perspective. In their definition, Kendall and Buckland (1971): 155) anticipated exactly such disagreements when they observed that "a movement which is a trend for one purpose may not be so for another; e.g. a systematic movement in climatic conditions over a century would be regarded as a trend for most purposes, but might be part of an oscillatory movement taking place over geological periods of time."

More fundamentally, metrics other than trend may be more appropriate cues to stimulate management. A period of consistently low values may be a useful trigger. For example, management of the North American midcontinental population of sandhill cranes (*Grus canadensis*) currently is intended to maintain a three-year running average of estimates of the March population between 349 000 and 472 000 cranes (that is, within 15% of the 1982–2005 average; Kruse *et al.* 2010). Alternatively, a constriction (or expansion, depending on the species) in the range of a population may suggest the need for management intervention.

Sustainability

A good monitoring program needs to be sustainable; a program that ends prematurely constitutes a failure and likely a substantial waste of effort. Sustainability starts at the design phase. I have been involved in too many planning sessions for monitoring in which the planners were told not to worry about costs; managers would either come up with the funds or modify the plans to accommodate the available funds. One problem with designing a Rolls Royce when only a used Yugo can be afforded is that, should a decision-maker not be a strong believer in the need for monitoring, he or she may simply say they cannot afford the Rolls Royce, and so no program is put in place. Another concern is that, if a plan is modified, the modification may not be done by the most informed individuals, which typically would be the original planning group. Also, it may be that the agency can afford a Rolls Royce now, but in a few years budgets will tighten or priorities change, and payments for the Rolls Royce cease. The monitoring program then falters or fails. Conversely, the funding available might not support even a minimal monitoring effort, in which case it may be better not to initiate it. It is far better to plan with realistic expectations of both costs and benefits (see also Chapter 22).

Interest and acceptance by participants

Sustainability of participant involvement is also critical. Sometimes monitoring programs are received with great enthusiasm, possibly because the resources being monitored are of great current interest. If interest wanes over time, participants – especially if they are unpaid volunteers – likely will drop out and jeopardize the continuity and value of the program.

Timely reports from monitoring programs are essential in maintaining support for the efforts. Feedback about results of the program and evidence of their utility help maintain the interest of participants as well as agency managers. Reports to participants can be informal; for example, participants in the North American BBS receive an annual report that includes, among other things, anecdotes submitted by participants describing strange or amusing incidents that occurred while doing their surveys. Reports to managers and agencies will put the information in the hands of those who can best make use of it, thereby justifying continuation of the program.

Besides being essential for programmatic sustainability, buy-in by participants in a monitoring program is critical for other reasons. This is true for paid participants and especially so for volunteers. They need to understand the objectives of the program and the value of its results. They also need to appreciate and follow the standards and protocols of the program. The better that they understand the objectives and value of the program, the less likely they will be to lose interest and perform poorly. Also, well-motivated participants will more closely follow the proper procedures.

In a similar vein, participants should be allowed or encouraged to record information that they believe is important, even if such data will not routinely be used in the program. In a bird survey, for example, asking participants to note interesting birds that they notice, but are beyond 100 m, even though records only within 100 m will be used, will reduce their possibly unconscious desire to include a "cool" bird 120 m away by convincing themselves that it was just inside the 100-m range.

It can be a significant challenge to balance what is needed to ensure sound and efficient statistical inference with the need to keep participants motivated and interested (Box 3.2). Asking observers to do "too much" conceivably could cause burnout or reduce coverage that could be achieved with available effort. Many voices (e.g. Burnham 1981, Johnson 1995, Thompson *et al.* 1998, Nichols *et al.* 2000, Anderson 2001, Williams *et al.* 2002, Yoccoz *et al.* 2001, Farnsworth *et al.* 2002, MacKenzie and Kendall 2002, Rosenstock *et al.* 2002, Ellingson and Lukacs 2003, White 2005, Buckland 2006; see also Chapters 4, 18, and 19) have called for accounting of imperfect detectability in surveys of wildlife, fish, and other organisms. They suggested repeat visits to sites, use of two or more participants at a site, or measurements of additional and often difficult-to-determine

Box 3.2 Common challenges: Sound design vs. the sound of frogs

In monitoring programs that depend on volunteer participation, it may be a challenge to balance the need to maintain volunteer interests and motivation with the need to implement a statistically sound monitoring protocol (see also Chapter 21). One monitoring program that met with some reluctance from participants was the North American Amphibian Monitoring Program. Designers wanted survey routes and stops along routes selected as randomly as possible without regards to whether calling amphibians were abundant there. Already-existing programs, such as Wisconsin's Frog and Toad Survey, had many years of data collected under a protocol that allowed volunteers to choose stops along routes where amphibians were known to be present. It was recognized that volunteers would have little interest in driving back-country roads at midnight several times a year while recording few if any frogs or toads. No easy advice can be given to reconcile the disparate interests of planners of monitoring programs and volunteers who carry them out. One potential remedy would be conducting a hybrid program, with both random sites and sites with a history of occupation. A comparison of results from the two sets of sites should be of interest to both groups, as well as in its own right.

variables such as distance to each animal. All such recommendations increase the effort needed to conduct a survey and reduce the number of sites that can be monitored. The overall value of attempting to account for imperfect detectability has been questioned (Hutto and Young 2002, 2003; Johnson 2008). Once again, the costs and benefits relative to the monitoring objectives need to be evaluated (see also Chapter 11). In particular, are indices to the population satisfactory or are actual estimates needed?

Sustained access to monitoring sites

The sustainability of a program and its ability to support long-term inference about the target population of interest can be affected by changes in access to monitoring sites over time. For example, the North American BBS is based on roadside routes (Robbins *et al.* 1986). If traffic along a road becomes too heavy, the safety of the observer as well as their ability to hear birds may render that route unsuitable. When such changes occur, routes are moved to a nearby area where traffic is – at that time – less problematic. Of course, as urbanization expands and associated traffic increases, routes are moved to less-developed areas, with the net result of the BBS becoming less representative of the survey area (O'Connor *et al.* 2000). At a minimum, such a shift of monitoring locations should be recognized as causing a discontinuity in the monitoring program.

Sometimes access to a monitoring site can be restricted due to changes in habitats. For example, in a program to monitor wetland birds in areas where wetlands are highly dynamic, a wetland-edge location from which all birds on a wetland can be seen in one year might be totally inundated and virtually inaccessible the next (Johnson *et al.* 2009).

If some monitoring sites are on privately owned land, permission to access those sites may be granted initially but denied in the future. Owners may change attitudes or simply grow weary of having monitoring staff on their property. Lands change hands. At a minimum, denial of access reduces sample sizes and increases uncertainty. Importantly, however, consequences to results could involve large biases if denial of access was associated with changes occurring on the property. For example, imagine that part of a forest is owned by an individual who allows monitoring sites for birds to be located on his property. Suppose that birds have been monitored at these and other sites in the forest for a while, and then the landowner sells to a developer, who no longer allows monitoring. Then monitoring sites on that property are lost, and any changes in bird populations associated with the development are not recorded. The program would fail to detect a major change that occurred.

Some other sampling design challenges

Stratification by habitat

For many ecological populations, it is tempting to stratify the area to be monitored by land-use category or habitat type. That is, the area may be divided into non-overlapping categories or subareas and each area sampled separately, eliminating among-stratum variance from resulting estimates. For example, many animals have dramatically different densities in different habitats, and habitats often can be delineated and recorded in geographical information systems. Hence the conditions for stratification to be most valuable are met; specifically, counts of the animals or other ecological measurements are likely to be more similar within a stratum than they are between strata (Cochran 1977).

Problems can arise, however, if boundaries of those habitat-based strata change. Trees may invade a grassland, for example, blurring the distinction between woodland and grassland strata. If the original strata were permanently marked, and sample sites remain positioned as they were initially, the only negative consequence of the habitat change would be reduced precision of estimators. For trees invading grassland, for example, differences in animal populations between the woodland and grassland strata will be reduced, as will the benefits of the stratification.

The problem often is merely confusion that arises because strata are named after major habitat types. Strata are areas from which random samples are selected and for which estimates are made. When trees invade grassland, a plot may be woodland by habitat but still be in the grassland stratum.

Attempting to "chase" the changing habitats by moving stratum boundaries is appealing but can result in more serious problems. The originally selected monitoring sites may no longer be representative of the modified strata, and if they had been randomly chosen, they no longer reflect a random selection from the modified strata. Without reselecting monitoring sites from the revised strata, it is not clear how valid inferences can be drawn.

Selecting a new sample of sites will result in a loss of continuity and possibly a major change in the population estimate associated simply with sampling variation.

When major changes in land use or habitat are unlikely, drawbacks to using these attributes to form strata are likely to be minimal, whereas potential gains are substantial. Potential drawbacks can be reduced, however, if similarly useful strata can be drawn by using features less likely to change, such as soil type, geologic landform, or watershed.

Expanded target population

Here, by "target population," I mean the statistical population of units from which a sample is selected for measurement (see also Chapters 2, 5). The situation I refer to arises when the statistical population no longer comprises all units on which those measurements are relevant. I illustrate the issue by an example in which units are sites on which certain animals are counted, but the issue is more general.

Some monitoring surveys were designed initially with a target population encompassing the entire area used by a target organism. For example, virtually all sandhill cranes in the North American midcontinental population migrate in spring through the State of Nebraska and stop for a period along the Platte and North Platte rivers (Benning and Johnson 1987). To monitor the size of that population, biologists identified a spring staging area that contained nearly all sandhill cranes in late March of each year. That survey area was divided into strata, based upon the extent of crane use in previous years. Then a systematic sample of transects was drawn from each stratum. Biologists annually counted cranes within those transects during aerial surveys. However, as the crane population grew, and land-use changes reduced the amount of preferred food (mostly waste corn) available in the survey area, biologists noted increasing numbers of cranes outside the boundaries of the surveyed area. The survey could no longer be assumed to cover a representative portion of the area used by cranes, and results from the survey are considered minimal estimates of the population size.

In the sandhill crane example, biologists were able to recognize when non-surveyed areas became occupied by cranes. A different situation arises when target organisms are discovered at a non-surveyed site, and it is not known if the site is newly occupied, or had been long occupied but was not known to biologists. Piping plovers (*Charadrius melodus*) in the northern Great Plains population nest on barren sand and gravel islands and shorelines of large rivers, lakes, and alkaline wetlands (Elliott-Smith and Haig 2004). These birds are surveyed at known breeding sites throughout their range every five years (Elliott-Smith *et al.* 2009). Frequently, however, breeding plovers are found at sites not previously known to host the birds. It cannot always be ascertained whether the birds were using that site for the first time, or they had occupied it previously, but had gone unnoticed.

Three scenarios come to mind (Table 3.2): (i) populations could be expanding from the surveyed areas into larger areas; (ii) unknown to biologists, the population has existed in a larger area "always" (i.e. since the beginning of the survey); or (iii) the population is shifting, with reduced numbers in the surveyed area and larger numbers

Table 3.2 Scenarios in which target organism occur elsewhere than the surveyed area, consequences to estimators of population size, and appropriate resolutions of the problems, depending on when investigators become aware of the situation.

Situation	Consequences	Awareness	Resolution
Population expanding into larger area	Estimator increasingly biased low	Always	Separate estimate for expansion area
		Recently	Redesign, acknowledge discontinuity and bias in early estimates
		Never	None; value of survey is reduced
Population "always" (relative to when monitoring commenced) existed in larger area	Estimator "always" biased low	Always	Design is suboptimal; impute missing values (Chapter 13) or obtain separate estimate for expanded area
		Recently	Redesign, impute missing values or acknowledge discontinuity and bias in early estimates
		Never	None; value of survey is reduced
Population shifting from surveyed area into larger area	Estimator increasingly biased low and some survey effort wasted	Always	Redesign, acknowledge discontinuity
		Recently	Redesign, acknowledge discontinuity and bias in early estimates
		Never	Estimator becomes more biased with time

outside of it. Such a shift could be caused by deteriorating habitat conditions at the known sites (e.g. Schmelzer and Harris 2009). In any of these situations, the estimator of the population size is biased low; if the population is expanding or shifting, the bias increases over time. Available solutions to these problems depend on when the survey designers become aware of the situation: early (from the beginning of the survey), only during the recent few occasions, or never. Potential resolutions include obtaining a separate estimate for the expansion area or redesigning the survey to accommodate known changes. In the latter case, there will be one or more discontinuities in the series of estimates, which should be acknowledged, as should any bias in estimates obtained before the redesign (Table 3.2).

The fallacy of unbiased estimators

In many monitoring situations, there exist large areas where the target organism is absent or uncommon. With spring-migrant midcontinental sandhill cranes in North America, for example, some birds may have not yet arrived at, and others may have already departed from, the Platte River survey area at the time of the survey. Such cranes are within the migration corridor – and are part of the resource population of interest – but are not available to be counted when data are collected from the survey area. The current protocol for dealing with such cranes is to have other biologists check out likely sites for cranes on the date of the survey and to encourage anyone who sees cranes elsewhere to report their numbers. Numbers of cranes observed through these ad hoc methods are simply added to the number estimated from the formal survey. Assuming that these supplemental surveys do not regularly cover all suitable sites and therefore on average miss some birds each year, the estimator of the total number of cranes using the migration corridor each year will be negatively biased. An alternative would be to sample probabilistically a number of sites outside the current survey area, seeking an unbiased estimator. This alternative approach likely would result in numerous surveyed sites where no cranes occur. A drawback of this approach is illustrated in the next example.

Consider a large park that comprises mostly coniferous forest and rocky mountaintops. Suppose there is interest in monitoring the population size of a species of bird that is fairly common in the forest but rare in the rocky mountaintops. The temptation, especially if most mountaintops are very difficult to access, is to develop a plan that surveys sites within the forest but ignores mountaintops. Doing so, of course, would result in a biased estimate of the park population.

The bias could be eliminated by stratifying the park into two non-overlapping subsets, a forest stratum and a mountaintop stratum, drawing random samples of sites within each stratum, and monitoring the bird of interest at all selected sites. Many more sites, relative to total area, would be drawn from the forest stratum than from the mountaintop stratum. Nonetheless, considerable effort would be expended to access the mountaintop sites to count the birds there. Nearly all such visits would result in counts of zero birds, but occasionally a single bird might be encountered. It is reasonable to ask if the effort to sample from the mountaintop stratum in order to eliminate the bias is worthwhile.

Understanding what bias really means may be helpful in addressing that question. As reviewed in Chapter 2, an unbiased estimator is one for which the average over *all possible samples* is the true value. However, in reality, only one of all the possible samples is taken, and that single sample may be very non-representative of the entire group. Let's get specific with our example. Suppose the mountaintop stratum is 50 times as large as the forest stratum and would comprise 1000 possible sample sites, versus 20 for the forest stratum. Assume the average number of target birds in the forest stratum is 40 per site, whereas only 10 sites out of 1000 in the mountaintop stratum contain birds, and only one each (i.e. average abundance in the mountaintop stratum is 0.01 per site). Hence the total number of birds in the park is 810, the number we wish to estimate.

Let's take random samples of six sites in the forest stratum and two sites in the mountaintop stratum, recognizing the much higher density of birds in the former stratum. Most likely, neither of the two sample units in the mountaintop stratum will contain any birds [probability of no birds in first mountaintop unit × probability of no birds in second mountaintop unit $= (1 - 10/1000)^2 = 0.9801$]. In that case the estimate of total population size in the park will be about $40 \times 20 + 0 \times 1000 = 800$ (the mean abundance per sampled forest site × total forest sites in the park + mean abundance per sampled mountaintop site × total mountaintop sites in the park; for this example we ignore variation among samples from the forest stratum). If a single mountaintop sample site happens to contain a bird (probability $= 0.0198$), the estimate will be about $40 \times 20 + (1/2) \times 1000 = 1300$. In the very unlikely event (probability $= 0.0001$) that both mountaintop sample sites contain a bird, the estimated total would be 1800. The bottom line is that, despite all the effort needed to sample from the stratum in which the species is rare, you are most likely to get a number slightly below the actual total, but you have a small chance of obtaining estimates that are wildly different from the correct value.

The unbiasedness of the estimator is relevant only conceptually, as if you were to repeat the entire procedure many times. However, the survey is done only once, so most likely a slightly low estimate will be obtained, but occasionally a misleadingly high estimate could result. The fact that the conceptual overall average of all possible samples is accurate is of little consolation. More relevant than bias as a metric for an estimator's quality is a combination of bias and variance, such as Mean Squared Error (bias^2 + variance; see Chapter 2). Statisticians routinely accept a small amount of bias (as with ratio estimators) to reduce the total error. Drawing a small sample from a large stratum is unlikely to reduce the total error, as noted above. It may be better to reduce the target area by excluding strata that are difficult to survey but known to contain few of the surveyed organisms. With this approach, if the currently very low proportion of the total population occurring in these strata could increase over time, the situation would raise issues similar to that of an expanding target population, discussed above.

Replacing an existing survey

As science has progressed, and as greater demands have been placed on existing ecological monitoring programs, more attention has been given to the design of surveys and the objectivity and reliability of their results. For birds, the development of conservation plans for various taxa across large spatial expanses has led to calls for greater coordination and consistency across monitoring programs (US NABCI 2007). In addition, concerns about roadside bias and imperfect detectability have spawned interest in modifying or replacing ongoing surveys. A challenging question then arises: should an existing, long-term monitoring program be abandoned for a newer, better-designed one? Or should an existing survey be markedly modified to improve it even though future results may be incompatible with earlier results? What about the investment already made in the existing program? Or, would the revised program perhaps include a subset of data that would be directly comparable to the existing survey?

For example, many national parks, forests, and wildlife refuges individually have developed monitoring surveys for birds and some other taxa. A recent emphasis has been on adopting methods that are comparable across parks, or forests, or refuges. The US Forest Service for many years employed various monitoring programs for birds in many national forests and grasslands in the West. The agency is now contemplating a different program that would be consistent across all of its lands (Robert C. Skorkowsky,

US Forest Service, 9 January 2009 memo). The new program is also expected to reduce any biases from roadside surveys as well as to incorporate privately owned land. However, as mentioned earlier, incorporating private land in a monitoring program poses risks to the integrity of the program.

Grassland birds were identified as a group with species that appeared to be suffering population declines, based first on anecdotal reports then on BBS results, but the BBS does not have intensive coverage in much of the Great Plains, where most of the grassland remains. Biologists considered novel methods to monitor those populations. However, Dale *et al.* (2005) found that simply adding more BBS-like routes in grassland areas provided a viable alternative that was consistent with already ongoing monitoring.

It can be argued that continuity can be maintained while adopting a new program if both the old and new programs are conducted for several occasions and results are used to calibrate the new program. Several occasions, however, will be needed to estimate the relation between results from the two programs. Let's say we have two measurement techniques, and during a series of calibration years we collect measurements with each technique, and then calculate the correlation coefficient between the resulting pairs of data. We do this after two years, and get a correlation coefficient of 1.0 (or −1.0), seemingly indicating that we have perfect comparability between the two techniques. Unfortunately this has nothing to do with their consistency. For example, assume the first technique yielded measurements of 3 and 9, respectively, for years 1 and 2. Regardless of whether the second technique yielded measurements of (3.1, 8.9) for the two years or measurements of (3.1, 2000), the correlation between these hypothetical measurements from the two techniques is 1.0. If we simulate this "calibration" by repeatedly generating random numbers to use as the two years of data with each technique, each resulting correlation coefficient will always be +1 or −1. With three years of such random "data", on average the absolute value of the correlation coefficient will be 0.64. The expected average for completely random pairs of observations remains fairly large even as the number of occasions increases, indicating that conducting the old and the new programs for several occasions may not be sufficient to determine accurately the true relation between the two.

Possibly the results from different spatial units could be used for calibration. For example, if there are, say, 10 strata, then a single monitoring occasion will yield 10 pairs of numbers (old program, new program) that can be correlated. Of course, all 10 pairs had been obtained under the same conditions (e.g. weather, number of animals, their distribution, possibly observer, etc.), so that the generality of the association between the two programs may be very questionable. In any event, it likely is important to compare the two programs under a variety of conditions, which will require a large number of occasions to become confident about any relationship between results from the programs. Especially troublesome would be a nonlinear relationship. The costs of conducting both programs on multiple occasions should be considered while making a decision about adopting a new one. Or possibly all that is needed is an acknowledgment that a discontinuity exists between time periods associated with the two monitoring programs.

If a strong correlation between the monitoring programs is found, one might ask if indeed there are advantages to switching. Such a decision should involve other considerations, such as costs and features that differ between the programs (see also Chapter 23). If the two programs do not provide comparable results, then the decision boils down to a choice between the continuity provided by retaining the old program and whatever advantages the new program confers.

Foremost in deciding whether or not to switch to a new monitoring program are the objectives of monitoring and evidence that the current program is not satisfying them as well as a new program would. As obvious as this statement about objectives appears, it is not often fully taken into account. If, for example, the overriding objective of a monitoring program is to track populations of a certain species or group of species at a site, continuity will be more important than consistency with programs used at other sites. Conversely, if the goal is to compare populations at different sites, consistency of methods among those sites will be an overriding concern.

Adaptive monitoring

Lindenmayer and Likens (2009) suggested an approach they termed adaptive monitoring. Their paradigm is intended for long-term monitoring *and* research and would seem to accommodate both surveillance and hypothesis-driven monitoring. They recommended that a monitoring program be designed to address well-defined and tractable questions that are specified before monitoring begins. They also advocated careful statistical design, a monitoring design based on some model of how the system to be monitored works, and a plan driven by management interests. They suggested that, after such a monitoring program has functioned for a period of time, initial questions might have been resolved, but new ones arisen. Also, improved technology or analytic capability may make improvements possible; if so, the program can be adapted accordingly. Key to their paradigm, however, is maintaining the integrity of long-term "core" measures.

Future research and development

Several issues worthy of further investigation are apparent. Some are very general, such as how trend should be defined. However, most issues are program-specific and depend upon the objectives and methods of a particular monitoring program. Among these are:

- How can surveillance-monitoring programs be augmented to address specific planned events such as a management action in a prescribed area, or particular questions not already adequately resolved? Can results of the surveillance monitoring be used to inform both the design of the augmented effort and interpretation of results from that effort? Can results of the augmented effort be useful to the long-term monitoring program?
- Which metrics emanating from a monitoring program are most appropriate for triggering an action? "Trend" is often adopted as an informative metric, but, as shown

above, it is not well-defined and is susceptible to how one specifies the interval over which trend is estimated. Metrics such as the level of a population index may be more reasonable, but variability and uncertainty associated with estimation should be acknowledged.

- What issues should be considered when contemplating the substitution of an ongoing monitoring program with a new one? How do benefits of the new program offset the loss of continuity associated with termination of the ongoing one? These issues arise generally, but I suspect a resolution of them will be specific to individual programs. It would be useful to evaluate a large number of monitoring programs in relation to how these issues were addressed, and compare the successes of the programs.

Summary

Here I consider monitoring to involve repeated measurements made comparably at a specified frequency of time, in contrast to assessments made for a shorter duration intended to evaluate the effects of specific events. Such surveillance monitoring is useful for determining the overall status of the system being monitored as well as its temporal dynamics, estimating – although usually imprecisely – the effects on the system of an unanticipated event, and suggesting aspects of the system worthy of more detailed study. A monitoring program should be designed with sustainability in mind, with regards to funding and other necessary resources, participation by observers, access to monitoring sites, and data management. Stratification of the area being monitored typically produces substantial gains in the precision of estimators. If stratification is based on features that can change over time, such as vegetation type, consequences for overall estimates will be minor, but stratum-specific estimates may be markedly affected. If the area monitored is intended to comprise the entirety of the area used by a target organism at the time of the survey, the existence of organisms outside that area can be problematic. The consequences, and potential remedies, depend on when the organisms first occupied the outlying areas and when they were first noticed (Table 3.2). If target organisms occur but are uncommon in large areas where monitoring at random or representative sites is difficult, it may be pointless or even misleading to attempt for an unbiased estimate by including "just a few" opportunistic or cherry-picked sites in such areas. Drawing a small sample from these large portions may eliminate bias, but it is unlikely to reduce the total error. It may be better to reduce the target area by excluding the problem areas. As survey and analytic techniques continue to advance, it is tempting to abandon ongoing monitoring programs in favor of "new and improved" methods. Before doing so, costs and benefits of both the existing and proposed programs should be considered carefully. In particular, do the perceived deficiencies in the existing program lead to erroneous conclusions and decisions? Are the benefits of the new program adequate to compensate for the loss in continuity caused by a change? Will the new and improved program need to be replaced in 10 or 20 years, when newer and even-more-improved methods become available?

Acknowledgments

P. H. Geissler, R. A. Gitzen, J. E. Gross, R. L. Hutto, J. D. Nichols, G. A. Sargeant, J. R. Sauer, W. L. Thompson, and an anonymous referee offered valuable comments on an earlier draft of this chapter, but should not be held responsible for views expressed here. The use of trade names does not imply endorsement by the US government.

4 Maximizing the utility of monitoring to the adaptive management of natural resources

William L. Kendall and Clinton T. Moore

Introduction

Data collection is an important step in any investigation about the structure or processes related to a natural system. In a purely scientific investigation (experiments, quasi-experiments, observational studies), data collection is part of the scientific method, preceded by the identification of hypotheses and the design of any manipulations of the system to test those hypotheses. Data collection and the manipulations that precede it are ideally designed to maximize the information that is derived from the study. That is, such investigations should be designed for maximum power to evaluate the relative validity of the hypotheses posed.

When data collection is intended to inform the management of ecological systems, we call it *monitoring*. Note that our definition of monitoring encompasses a broader range of data-collection efforts than some alternative definitions – e.g. Chapter 3. The purpose of monitoring as we use the term can vary, from surveillance or "thumb on the pulse" monitoring (see Nichols and Williams 2006), intended to detect changes in a system due to any non-specified source (e.g. the North American Breeding Bird Survey), to very specific and targeted monitoring of the results of specific management actions (e.g. banding and aerial survey efforts related to North American waterfowl harvest management). Although a role of surveillance monitoring is to detect unanticipated changes in a system, the same result is possible from a collection of targeted monitoring programs distributed across the same spatial range (Box 4.1). In the face of limited budgets and many specific management questions, tying monitoring as closely as possible to management needs is warranted (Nichols and Williams 2006). Adaptive resource management (ARM; Walters 1986, Williams 1997, Kendall 2001, Moore and Conroy 2006, McCarthy and Possingham 2007, Conroy *et al.* 2008a) provides a context and specific purpose for monitoring: to evaluate decisions with respect to achievement of specific management objectives; and to evaluate the relative validity of predictive system models. This latter purpose is analogous to the role of data collection within the scientific method, in a research context.

Design and Analysis of Long-term Ecological Monitoring Studies, ed. R.A. Gitzen, J.J. Millspaugh, A.B. Cooper, and D.S. Licht. Published by Cambridge University Press. © Cambridge University Press 2012.

Box 4.1 Take-home messages for program managers

Managers and administrators face multiple trade-offs in making monitoring decisions. For those decisions related to informing or evaluating management actions, these trade-offs are easier to evaluate when the decision process itself is clearly laid out, including objectives, alternative actions, and prediction of consequences. Adaptive resource management (ARM) provides a framework for transparent and disciplined consideration of the decision process, in the face of various sources of uncertainty. Using this framework, managers can evaluate the direct costs of intense monitoring (e.g. equipment, travel, and personnel time), against the indirect opportunity costs of inadequate monitoring (e.g. biased perception of the status of the system, partial or complete failure to achieve objectives, misinformation about factors that affect the system being managed).

Administrators also face the question of how much to invest in management and monitoring to achieve specific resource objectives, versus investing in surveillance monitoring for assessment of general trends and identification of future threats (see also Chapters 22, 23). We suggest investigating the possibility of adapting a collection of management-focused monitoring programs, spread across the landscape, to also provide more general surveillance for other species or for other aspects of the landscape. We believe in most cases it would be more effective to add a mission that is more general to one that is designed to answer a particular question (e.g. assign pilots flying waterfowl surveys in support of harvest management to also count other water birds) than to use surveys designed for general surveillance to answer a specific management question, perhaps on a different spatial scale.

As discussed extensively in Chapter 2, there are many types of decisions to be made in setting up any monitoring program. These include which attributes of the system to monitor, how to measure those attributes, whether additional information is needed to adjust raw measures (e.g. for detection probability), and how intensively to monitor. Each of these decisions results in benefits (e.g. precise information for decisions) and costs (e.g. monetary, opportunity costs related to competing needs, possibly bias in measures). Some costs are transparent (e.g. monetary cost of conducting surveys) and others can be latent (cost in terms of management objectives due to poor monitoring). Therefore, trade-offs naturally exist in making decisions about monitoring. In this chapter we expound on this notion, and how ARM can provide a structure for evaluating these trade-offs.

We will begin with a brief outline of ARM as a structure for making informed management decisions over time, and then discuss the role of monitoring within that structure. In the next section we will show that the ARM framework dictates the purpose, design, frequency, and scale of monitoring. Following that, we will discuss how monitoring design affects management performance: directly, via its effectiveness in providing information needed for state-dependent decisions, and less directly, through its effect on the rate of learning about system function. We then will illustrate these points using the adaptive harvest management of waterfowl in the United States as an example. Finally, we will provide a summary of these points, and suggest some future directions.

The theme of this volume is long-term monitoring. The long term is also appropriate for our discussion of the role of monitoring in a decision context. Many decision problems involve long-term management objectives, for a system to be managed perhaps in perpetuity, with decision points periodically during that time frame.

Monitoring in the context of ARM

What is adaptive resource management?

The concept of adaptive management of natural resources was comprehensively treated in the seminal book by Walters (1986). The ARM concept is based on the principles of decision analysis, or management science, which go back even farther, at least to the Second World War (Bellman 1957; see Williams *et al.* 2002). Popular lore about what defines ARM varies, from "willingness to change what you are doing" to "experimental management" to "actively adaptive optimization".

We follow the characterization by Williams and Johnson (1995) of ARM as a form of making recurrent management decisions "in the face of uncertainty, with an emphasis on its reduction". So a premise of ARM is that uncertainty exists. We can categorize this uncertainty into four types: environmental variation, partial controllability, partial observability, and structural uncertainty. We will discuss each of these in more detail as the chapter progresses. For now we assert that the first three types should be acknowledged in any approach to decision making. Acknowledging and striving to reduce structural uncertainty, or uncertainty about key features or mechanisms of system dynamics, is the defining feature of ARM, relative to other approaches to decision making where structural uncertainty either is not a concern (e.g. a decision to play the lottery) or cannot be resolved through subsequent decision opportunities (e.g. a decision to permanently breach a specific impoundment). We can add an addendum to the premise listed above: reduction of structural uncertainty over time could produce better future decisions – i.e. decisions with greater expected return on the desired objective of management. A final premise of ARM is that decisions must be made in the face of uncertainty. To forego or delay a decision is a decision to continue the status quo.

The ARM framework begins with the elements of an informed decision process:

(i) *management objectives*;
(ii) a set of *alternative management actions*;
(iii) *prediction(s)* about the consequences of each candidate management action, both with respect to management objectives and system dynamics; and
(iv) a *monitoring program* to assess the state of the system for making a decision, and to evaluate the chosen action and prediction(s). That is, how good was the decision with respect to management objectives and how accurate were the predictions of system response?

Each component is discussed below. Based on the first three components, an optimal decision is sought, which is then evaluated by the fourth component, monitoring. This process calls for transparency, clarity, and specificity. The more specifically each of the first three components is developed and articulated, the easier it will be to see how monitoring can inform the decision process. To the extent that the first three elements are hidden or obscured, the potential effectiveness of a monitoring program will be reduced.

Management objectives

The success of wildlife management is assessed against objectives set by managers or a collection of stakeholders. Because of this, objectives must be articulated precisely, or there is no basis for evaluating management. For example, assume the management objectives of a coastal refuge are stated as follows: "We want to promote the reproductive output of both loggerhead sea turtles (*Caretta caretta*) and piping plovers (*Charardius melodus*)." Suppose further that one candidate management action is expected to produce an average 20% hatching rate in sea turtle nests and 50% nest success in piping plovers, whereas an alternative action would produce an average 50% hatching rate in sea turtle nests and 20% nest success in piping plovers. Given how the objectives were phrased, it is not clear which action is preferred. However, if the objectives were expressed as "maximize piping plover nest success, subject to the constraint of at least 20% hatching rate in sea turtles", then the first action would clearly be the preferred alternative. In this latter expression, the objectives were expressed more clearly, and could be expressed mathematically, as an *objective function* with constraints. For our purposes, this objective function identifies attributes to be measured in order to evaluate management performance (i.e. hatching rate of sea turtles nests and nesting success of piping plovers), and therefore gives specific direction to a monitoring program. These same metrics will need to be included in predictive models, in order to evaluate management alternatives.

The development of objectives can take numerous paths. Trade-offs can be established by focusing through maximization or minimization on one objective primarily, with competing objectives handled through constraints, as in the example above. Alternatively, trade-offs could be handled by reducing competing objectives to a common currency and assigning relative weights to each objective. For example, we could standardize nest success of plovers and hatching rate of sea turtles to a score from 0 to 1 and maximize $0.7 \times$ (piping plover score) $+ 0.3 \times$ (sea turtle score). Regardless of the approach, if objectives are sufficiently explicit then the attributes that monitoring programs should measure will be clear.

Alternative management actions

Resource management is about choosing among alternative actions to pursue a stated objective. It is up to stakeholders and managers to identify these alternatives, based on the objectives that have already been identified. The set of reasonable candidates could vary over time, and be dependent on the system state (current conditions). For example, the set of candidate actions available to a manager when a species is abundant is probably different from the set of actions when the species is on the brink of extirpation.

If candidate actions are applied indirectly to the system, or there is a chance a chosen action will not be implemented as planned, their immediate impacts on the system are subject to *partial controllability*. This term refers to the variability of immediate response of the system to the implemented action. For an example of indirect action, perhaps an impounded wetland is to be drawn down to expose 10% of the area in mudflat, through the removal of boards from a water control structure (Lyons *et al.* 2008). The management objective might be to maximize food availability for migrating shorebirds, but the indirect treatment might not have its intended effect in meeting this objective. For example, board removal might fail to create the desired habitat conditions. An example in which the implementation of a decision is distinct from a chosen decision is where full development of a proposed wind farm is approved by a government agency but the project is abandoned shortly into development. In each example, monitoring the immediate impact of a decision, including the degree to which it has been implemented, will be desirable.

Predictive model(s)

At each decision point for each management action being considered, a manager must predict the consequences of the decision, with respect to both management objectives and system dynamics. This prediction comes from a *model* of the system, or the average across a set of models, which in turn are based on hypotheses about key factors that drive system dynamics. A predictive model might be as simple as the articulated intuition of a manager or expert on the system (e.g. "if we choose option 1 then consequence A will happen, whereas if we choose option 2 then consequence B will happen"). The model may be something more complex, hopefully supported by multiple rich data sets. Often the predictive model is something in between. Regardless, the necessity of prediction implies that modeling is not optional in making informed decisions.

As stated above, a defining feature of ARM is its explicit consideration of structural uncertainty; i.e. competing hypotheses about how the system works and its expected response to a management action. These competing hypotheses are common in ecology, and can arise from different formal theories in ecology, from uncertainty that arises from analyses of sparse data, or simply from differences in opinion between managers or system experts about a specific aspect of a system.

These competing hypotheses are articulated through specific predictive models. This model set can be composed of a discrete set of models (e.g. two expert predictions about the dynamics of the population), uncertainty about the value of one or more parameters for a given functional form (e.g. slope parameter for a linear-logistic model of density-dependent recruitment), or both.

The conduct of science is devoted to resolving structural uncertainty about a system, even if just for academic purposes. In a management context, resolving this uncertainty becomes important only if that resolution leads to better management decisions in the future – i.e. if that resolution improves the attainment of management objectives. However, it is important that predictions from at least one model in the candidate set, or the average prediction across models, be reasonably accurate across time and conditions. Otherwise, new predictive models should be sought, perhaps in a cycle of "double-loop learning", as described below.

Once a set of competing models is developed, each model will have some influence over the decision to be selected. That influence is expressed through a prior distribution of belief about the models in the set. For a model set consisting of a discrete set of models, this prior can be expressed as relative weights on each model, which sum to 1.0. Where the model set is continuous and defined based on one or more parameters, the prior is defined by a continuous distribution on each uncertain parameter.

An analysis of the model set is carried out to indicate a "best" decision given the current observed state of the system and given current understanding about the system, as reflected by the models and their respective priors. Often, this analysis is a formal dynamic optimization, but satisfactory decisions may be available through heuristic or simulation-based approaches (Williams 1989).

Finally, as we describe in more detail below, system learning occurs through a recurrent process of comparing model predictions to their monitored outcomes. As this activity carries forward through time, learning also occurs with regard to the decision process itself. At periodic intervals, all elements of the decision framework – from the management objectives to the monitoring design and protocols – may be evaluated and revised to address shortcomings identified in the decision structure. Thus, one purpose of this process of "double-loop learning" could be to develop new predictive models if all the members of the current model set have been consistently poor predictors of monitored outcomes.

To summarize, ARM is about putting structure on a decision process that is iterative (i.e. multiple decision points over time or space), and dynamic (the optimal decision at a given point in time or space could be dependent on actual current conditions, not the conditions that were expected to occur). As the dynamics of the system play out over multiple time periods and decisions, there is an opportunity to reduce uncertainty about system dynamics, and thereby make better subsequent decisions. Monitoring plays key roles in this process.

Roles of monitoring in ARM

Given the elements and process of ARM, four roles arise for monitoring in this process: (i) to assess system state before a state-dependent decision is made; (ii) after a decision is implemented, to measure variables related to objectives; (iii) to measure variables related to learning; and (iv) to gather data used to develop additional predictive models in the future. The first three roles are central to the process, and the fourth arises out of practicality. We now expound on these roles.

Assess system state

When the consequences of each candidate management action are predicted using the identified system model(s), the optimal decision with respect to the specified objective function is often dependent on the current status or state of the system. For example, if objectives relate to providing an ideal composition of seral stages of forest across a landscape to benefit a bird species, the optimal decision at time t could depend on the current composition of seral stages (Moore and Conroy 2006). Under one set of conditions, no action might be needed to achieve the objective, according to the models, whereas under other conditions burning or thinning might be required. In this case a monitoring program needs to include assessment of the composition of seral stages at each decision point.

Measure variables related to objectives

Once a decision has been implemented, the first question is how effective the decision was. Therefore, variables related to the objectives need to be monitored. Measuring system state could again be useful here if achieving a desired system state was part of the objective (e.g. the target composition of seral stages discussed above, or a specified minimum abundance or age composition of an endangered species). In addition to system state, there might be a return of interest. If an objective was to maximize harvest, or minimize incidental take, or maximize park visitor satisfaction, then the outcome of the decision (realized harvest, or take, or satisfaction) should be monitored. By collecting these data, progress toward the set of management objectives is measured.

Measure variables related to learning

The next question, after a decision has been implemented, is how well the set of system models predicted the outcome. Therefore, the variables predicted by those models (e.g. abundance, species presence/absence, habitat composition) should be measured, to compare results with prediction. Bayes' Theorem is then used to update relative faith in each of the candidate system models, using the monitoring data collected to update the prior relative belief in each model to a posterior belief. The details of Bayes' Theorem are discussed below. Over time, if a robust predictive model is in the set, the weight on that model should move toward 1.0. This use of monitoring should be the most familiar to scientists that are not focused on management problems, because it exactly parallels the scientific method: the system is observed, hypotheses are generated, a study is designed, data are collected and analyzed, and inference is made. The analogy is even more obvious under a multiple hypothesis approach to science (Chamberlin 1890, Platt 1964).

Develop new system models

A final use of monitoring data in ARM is to build predictive models of system dynamics for future use. This is not a core role of monitoring in principle, but an artifact of practicality and limited budgets. An accurate description of system dynamics is best done through experimentation, where the system is manipulated specifically to test for mechanisms and how they affect dynamics. In fact, the strategy of conducting short-term management experiments followed by a focus on management objectives is one option for conducting adaptive resource management. However, in many cases monitoring data sets are collected during periods where the system is not being intentionally manipulated to increase scientific knowledge. Therefore, the models built from such data might not be as robust as desired. Nevertheless, we return to our point that decisions must be made in spite of this drawback, based on predictions. Practically speaking, monitoring data is a logical place to start for building those predictions. The "double-loop learning" process, introduced earlier, is a natural setting for periodically exploiting the wealth of accumulated monitoring data in order to assess and improve the model set.

Attributes of the monitoring program dictated by the ARM framework

In considering and developing a monitoring program, many decisions must be made. These include what attributes of the system to measure, how often to measure a given attribute, and the spatial scale and grain at which attributes should be measured. Once these questions are answered, other decisions include the sampling design to be used, the effort to expend, and in some cases whether raw measures (e.g. abundance, presence/absence, species richness, survival) should be corrected for biases due to imperfect detection or non-representative sampling. We will address the first set of issues in this section, and the second set of issues in the next section. The answers to these questions are driven by the decision structure and the role monitoring plays in that structure, which we have outlined up to this point.

The attributes to be measured through monitoring are determined by the set of system states, the set of objectives, and the set of models. In dynamic decision making the optimal decision is often state-dependent, in which case the current state of the system needs to be assessed. The state could include variables that management is intended to manipulate (e.g. population size, species diversity, seral stage of forest). It could also include variables that are not under management control, but whose status could affect the management action to be taken. For example, if management actions include drawing water down in an impounded wetland to create good feeding conditions for shorebirds, the proper duration for removing boards from the water control structure could depend on weather conditions such as temperature and wind, which affect evaporation rates. In this case measuring these weather variables would be important to the decision.

Measuring attributes of the system state would also be important where system state is a component of management objectives (e.g. maximize persistence of a species; maintain acceptable range of population size; maximize species richness or diversity; minimize proportion of sites with a disease pathogen or an invasive species). In this way the consequences of a management decision can be evaluated with respect to how close the system state is to its target. Similarly, as stated earlier, if objectives involve a return for each decision over time (e.g. maximize harvest, minimize take, maximize park visitor satisfaction, minimize cost, maximize number of young produced), this attribute also needs to be measured.

Selection of attributes also must consider the role of monitoring data for learning about the comparative predictive ability of each candidate model, or whether any of the candidate models have much predictive power. Monitoring data are compared with the predictions of each model, and through Bayes' Theorem relative faith in each model is updated over time. This use indicates which attributes to measure. Monitoring system state is again important here, because system models predict system state. However, there could be additional measures that are important for evaluating the predictive models. This is especially true if system models are composed of submodels. For example, a population model might consist of submodels describing death and recruitment and be divided into seasonal components. If structural uncertainty is driven or largely represented by uncertainty about a submodel (e.g. is post-harvest survival density-dependent or not), monitoring variables related to that submodel (e.g. post-harvest survival and density) would increase the speed of learning. In addition, if partial controllability is a big source of variability (e.g. pulling boards for x hours results in highly variable quantity of mudflat), then the proximate impact of the decision (e.g. the quantity of mudflat) should be monitored in addition to the ultimate consequence of the decision (e.g. use by feeding shorebirds).

The frequency of monitoring effort is also dictated by the decision context. At a minimum – and unless monitoring effort is formally considered as a decision variable (Hauser *et al.* 2006) – monitoring frequency and timing should match the decision frequency and timing as closely as possible. Because of the role monitoring plays in making state-dependent decisions, monitoring should provide the latest information on system state in time for the decision to be made, if possible. If decisions are made annually, such as with hunting, then annual monitoring is needed. In some cases decision points are less frequent. For example, if success of translocation of seabird chicks in year t to a new colony is based on the proportion that return and successfully breed within 10 years, then it is logical to monitor the number of breeding pairs at year $t + 10$, and not before.

The appropriate scale of monitoring also comes from the decision structure, specifically the spatial scale referred to by the objectives, at which decisions are being implemented, and at which the system models predict. If decisions are being deliberated and implemented on a scale limited to two wetlands within a wildlife refuge, then monitoring is focused on the response to management in those two wetlands. If instead decisions are being developed and implemented based on a larger spatial scale (e.g. based on objectives for all wildlife refuges in the northeastern USA), then the monitoring program should be designed to produce inference at that scale. Of course, whether the scale of management and monitoring is large or small, issues of sampling design such as sample size, random locations, stratification, etc., are germane.

The questions that must always be addressed when designing a monitoring program (what to monitor, how often, at what scale, where) are difficult to answer in detail or with confidence when the objective of monitoring is to address a broad or vaguely defined issue. For example, selection of appropriate temporal, spatial, and biological resolutions may not be straightforward questions when the purpose of monitoring is to provide a sentinel to detect impacts of a general class of environmental threats or to detect general trends. These questions are much easier to answer when the decision problem and the structure of that decision are well-defined. The objectives and general design features of the monitoring program arise naturally out of an ARM context. This context also permits an opportunity to evaluate the cost of poor monitoring (imprecise or biased estimation) in terms of lost return on management objectives, and even an opportunity to make monitoring decisions adaptively over time. We discuss this in more detail in the next section.

Monitoring design affects management performance

The purpose of ARM is to provide guidance for decision making when structural uncertainty about the system hinders the selection of an action. Decision-making designs that take into account structural uncertainty make predictions of decision consequences under competing scenarios of how the system works. Therefore, if structural uncertainty is substantial, designs that account for uncertainty should provide greater management returns over the long run than designs that do not. That is, the manner in which uncertainty is addressed in a decision-making design influences the design's expected *management performance*.

Monitoring design has a profound effect on management performance, and it works through two points in the decision-making process. First, at the time of decision making, decisions based on accurately assessed conditions and on informative ancillary data are likely to yield greater management return than those based on poorer information. Below we will address the three other forms of uncertainty that commonly confront natural resource decision making (partial observability, environmental stochasticity, and partial controllability); in all cases, monitoring can be fashioned to diminish their performance-reducing effects. Second, following the making of a decision, learning occurs and structural uncertainty is reduced on the basis of how well each model's prediction matches the measured resource outcome. The rate of learning, and consequently the long-term management performance, is directly affected by the quality of measurement of this outcome, which is, of course, determined by monitoring design.

Structural uncertainty is the specific form of uncertainty that is addressed through ARM. However, the other three forms confront nearly all decision making, adaptive or not, in natural resource management. Each of them reduces management performance because they either reduce our ability to assess the true, current condition of the system, or they potentially steer a path of management off course through uncontrollable and largely unpredictable events.

Monitoring design in relation to partial observability

Partial observability is the inability to accurately or precisely observe those parts of the system that are critical to making a decision. For example, the decision whether to apply herbicide in a native meadow may be determined in part on seed bank density of an invasive species (e.g. Reinhardt Adams and Galatowitsch 2008), and any assessment of this variable will be measured with some degree of sampling error. In another example, the decision about where to implement hydrological manipulations in a group of wetlands may be based on abundance of a key salamander species in each wetland (e.g. Martin *et al.* 2009b). In this example, not only is the variable of interest measured with sampling error, but it is also measured with negative bias if detectability of the salamander is not taken into account. Bias would also result where a complete census is achieved within sampling units over space, without accounting for proportion of the area sampled, or

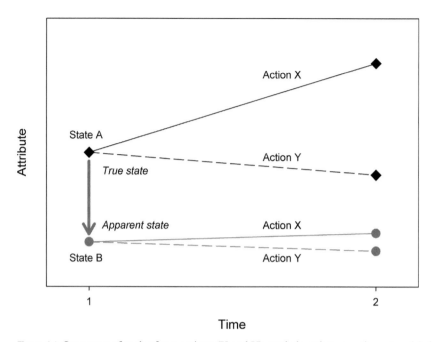

Figure 4.1 Outcomes of each of two actions (X and Y) carried out between times 1 and 2 depend on state of the resource at time 1. *Partial observability* is the circumstance by which the true nature of the resource state is misinterpreted as another, apparent state. In this case, knowledge that state A is the true state of the resource implies that the two actions produce divergent outcomes at time 2 (diamonds). Suppose instead that the true but unknown state of the resource is state A, but measurement error produces an assessed value at state B. This misinterpretation of state suggests that the two actions are expected to produce similar results, potentially leading to an inappropriate decision.

where sampling occurred only within those spatial units identified by a GIS (Geographic Information System) layer or similar model as "suitable habitat".

In these examples, the decision maker is to some degree blinded about the true state of the system, and the decision is thus based on an apparent rather than a real state (Fig. 4.1). For example, in the wetland example, a severe undercount of the population of salamanders could lead the decision maker to an action that would have been different had the true abundance been known. Where decisions are made on the basis of hard-to-count animal or plant species, the temptation is often strong to rely on an indirect index of the population rather than on a direct estimate of abundance, because the cost of obtaining an index is usually less than that required to estimate abundance. However, reliance on an unvalidated index rather than a true assessment of abundance can lead to suboptimal decision making, and the cost savings in foregoing abundance estimation should be balanced against the opportunity cost of misdirected management (Moore and Kendall 2004). Even if an estimator of state conditions is unbiased, poor precision in the estimator can also lead to suboptimal decision making. For example, a sample of seed bank density may produce an estimate of overall seed density that is unbiased but has high variance. In this case, it may be quite common to draw a sample that seriously misrepresents the true state of the seed bank and therefore produces a poor decision. Thus, partial observability leads to loss in management performance, the size of which depends on sensitivity of the system to state conditions and the accuracy and precision of the state measurement.

Reducing partial observability and diminishing its effect on management performance is accomplished through sound monitoring design. As we stated at the outset of this chapter and in the previous section, the decision context defines the monitoring design. That is, the suite of state variables on which the decision is based determines what system attributes are to be monitored and at what scales. For example, if horseshoe crab (*Limulus polyphemus*) harvesting is to be regulated annually based on status of a bird that subsists on crab eggs during its migratory stopover, then the aim of monitoring should be to provide an estimate of bird abundance in the region of interest on an annual basis (McGowan *et al.* 2011). Within those general requirements, there exists any number of ways to conduct monitoring, and many potential levels of desired precision. There are no set guidelines for conducting ARM. In fact, even sparse information provides a basis for decision making, but as we have suggested above, management performance suffers as quality of information declines. Therefore, the principles of randomization, replication, and where relevant estimation of the detection process should be employed within the monitoring design to increase information quality to the degree practicable. One way to determine a reasonable intensity for monitoring is to conduct a computer simulation of the decision process under alternative monitoring designs. By doing so, it is possible to quantify the relationship between monitoring effort and management performance (see Moore and Kendall 2004, Moore and Conroy 2006).

Monitoring design in relation to environmental stochasticity

Two other forms of uncertainty, environmental stochasticity and partial controllability, are different processes that have the same implication for decision making: an inability to predict the exact outcome of a proposed action beyond some mean value. Environmental stochasticity refers to all the random demographic and environmental effects that intervene between successive decision points, making exact prediction of an outcome impossible (Fig. 4.2). For example, consider the management of a wildflower for which an annual decision is made to either burn, fertilize, or rest the area on the basis of plant abundance. Even if the plant is easy to detect and enumerate, predicting the exact outcome of a management action is unrealistic because all of the processes that affect plant survival and productivity (e.g. rainfall, drought, herbivory, demographic stochasticity, pollinator dynamics) are themselves not predictable. At best, we can predict a mean response to the management action, averaged over all of the random outcomes, but any realized response could depart substantially from the predicted mean response. In this case, management performance has been reduced because the resource has been driven to an unintended state.

Generally speaking, environmental stochasticity cannot be reduced. However, there are strategies to anticipate the effects of one or more stochastic factors. These strategies are often informed by some element of monitoring. One approach may be to stratify the

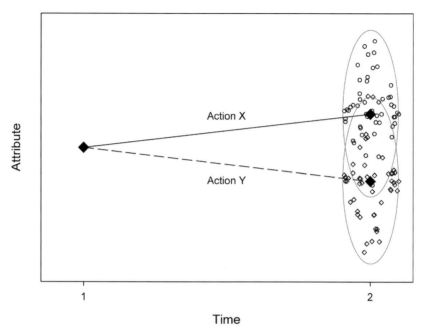

Figure 4.2 Outcomes of each of two actions (X and Y) carried out between times 1 and 2 yield a mean result at time 2 (filled diamonds), but any single realization of an action produces an unpredictable response around the mean. *Environmental stochasticity* implies that reliable prediction of exact outcome of an action is not possible, and that a management decision can result in a resource outcome driven away from its intended direction.

managed resource in some way so that management performance is "quarantined" from between-stratum variability. For example, if wildflower productivity responds to rainfall, but the nature of the response differs between drier and moister sites, then it might make sense to stratify the area into two or more moisture regimes and develop a separate decision model for each stratum. By doing so, between-stratum variation in response to rainfall is removed from the decision context, and variation due to rainfall has been made smaller in at least one of the strata. Because different prediction models are used in the strata, potentially different decisions could occur among strata at the same decision opportunity. Developing the stratification scheme would not itself require recurrent monitoring, but it may require a one-time assessment of certain system attributes (e.g. spatial mapping of an area's moisture regime).

An alternative strategy that does require recurrent monitoring is the capture of an ancillary state variable that provides information about a stochastic parameter. For example, in the wildflower management problem, suppose that plant herbivory by small mammals is a highly variable driver of plant productivity, but it is known that the variability is due in part to availability of an alternate food source. Assessment of the alternate food source could therefore serve as a "leading indicator" of the likely impact of herbivory on the wildflower. The decision model would thus incorporate status of

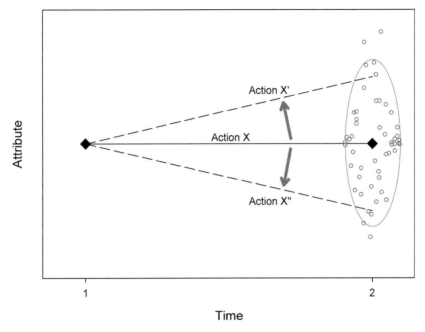

Figure 4.3 Expected outcome of a management action (X) carried out between times 1 and 2 is known, but ability to precisely control the execution of the action results in uncertainty about resource response at the time the action is planned. *Partial controllability* refers to the variability in the immediate, proximal outcome of the action and how it contributes to the variability of the resource outcome. Here, action X is only partially controllable, with realizations X′ and X″ directing the resource outcome (small circles) away from the intended direction (filled diamond) at time 2.

the alternate food source as a predictor of herbivory, and the decision would require knowing both current wildflower abundance and current levels of the alternate food source. To drive this decision model, a monitoring program that captures both attributes must therefore be in place.

Even if neither strategy can be employed, targeted monitoring could be used to collect data on the stochastic processes of interest. These data could be used in a post hoc, double-loop learning effort to refine prediction models and to gain greater understanding about the process.

Monitoring design in relation to partial controllability

As described earlier, partial controllability is a factor when the decision maker is unable to carry out an action exactly as intended (Fig. 4.3) – for example, realizing a harvest rate of 15% when 10% was intended. Like environmental stochasticity, partial controllability results in the resource being driven to a state where it was not intended to go, with a consequent reduction of management performance. Unlike environmental stochasticity, the effect of partial controllability can often be reduced through different forms or intensities of management control. For example, exerting greater control over harvest rate might require implementing check-station access and increasing enforcement patrols. However, there may be situations where it cannot be reduced. In such cases, the problem should be viewed in the same light as that of environmental stochasticity, with possible use of the stratification or ancillary state variable strategies described above and monitoring support as required.

A related but often overlooked component for ARM-focused monitoring design is a scheme to document the realized management action. For example, a habitat management program might record attributes about a prescribed fire (area burned, intensity of burn, etc.), a harvest management program might record attributes about the hunting season (number of hunters, total kill, etc.), and a species reintroduction program might record attributes about the release (number of animals released, age/sex composition of release cohort, etc.). A record of the realized action permits model updating to occur accurately when actions are only partially controllable. The long-term database that results from these records allows periodic refinement and improvement of the decision models.

Monitoring design in relation to structural uncertainty

Partial observability, environmental stochasticity, and partial controllability all hinder management performance, but monitoring design can play a large role in limiting their effects. Now we turn our attention back to ARM specifically, in which the decision is faced always in the context of multiple, competing predictive models that represent structural uncertainty. How are monitoring design and the reduction of structural uncertainty related?

Like the other forms of uncertainty, structural uncertainty, if unresolved, reduces management performance. Our decisions are always more informed if we could operate under a single model that we somehow *knew* to be appropriate for our system, rather than having to consider the possibility that any of several models *could be* appropriate (Fig. 4.4). Using the process of decision making to move incrementally away from this latter situation to the former is one of the ultimate aims of adaptive resource management.

Reduction of structural uncertainty comes about by comparing predictions of decision outcomes (generated by the models) against realized outcomes (measured by the monitoring program). In other words, we compare what we predicted would happen against what actually occurred. Models that make better predictions inherit a greater share of credibility, or belief, than those that do not, and their increased influence is recognized in later rounds of decision making. Models that lose credibility have less influence on future decision making. If one model consistently predicts outcomes better than all others, it eventually gains enough credibility to effectively drive decision making on its own, thus maximizing management performance. Of course, we would like this to happen sooner than later. That is, we would like to maximize the *rate of learning*.

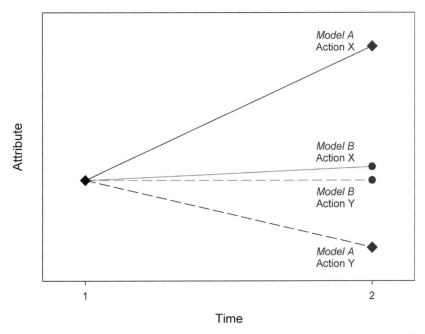

Figure 4.4 Outcomes of each of two actions (X and Y) carried out between times 1 and 2 depend on choice of a system model (*A* or *B*). Decision outcomes are greatly divergent under Model *A* relative to Model *B*, suggesting possibly different best decisions under each model. *Structural uncertainty* is uncertainty about relative confidence in each model as an appropriate representation of the system. Increased confidence about one model relative to the other suggests reduced uncertainty about choice of appropriate action.

There are many factors that affect rate of learning, including degree of structural difference between models, range and number of decision alternatives, and ability to predict stochastic environmental conditions. In addition, monitoring design has a fundamental effect on the rate of learning because the content and quality of information provided by monitoring is incorporated directly into the mechanism used to reallocate credibility among models, Bayes' Theorem:

$$p_{t+1,j} = \frac{p_{t,j} L_j(x_t)}{\sum_i p_{t,i} L_i(x_t)}. \tag{4.1}$$

For decision model j, $p_{t,j}$ and $p_{t+1,j}$ are its current and new levels of credibility, respectively; the summation in the denominator is over all candidate models in the model set. If structural uncertainty is represented by a probability distribution over a continuous parameter rather than by discrete models, then model credibility measures $p_{t,j}$ are replaced by probability densities and summations by integrands (Link and Barker 2010). The term $L_j(x_t)$ is the model's *likelihood* computed for the current data x_t provided by the monitoring program. The likelihood is a measure of probability of the data having arisen from model j, and the function can take on familiar distributional forms. For example, for continuously scaled predicted quantities such as seed density, where it is possible to obtain field estimates of the mean and variance of this attribute, the normal or gamma distributions might be appropriate likelihood functions. Alternatively, if models generate predictions of counts (animals, occupied plots, etc.), then the Poisson or negative binomial distributions may be appropriate.

The model credibility measures $p_{t,j}$ are also represented as probabilities, and because they are required to add to 1 across models, they reflect relative degrees of credibility among models. Therefore, Bayes' Theorem merely reallocates the total share of credibility among models in proportion to model likelihoods. To the extent that a model's prediction matches the observed value *and* the observed value is measured precisely, the likelihood increases. If the observation does not support the prediction, or if the observation is measured with low precision (despite possibly high agreement between the observation and prediction), the likelihood decreases (Box 4.2).

The role of the likelihood in this reallocation process makes it clear that monitoring design greatly influences the rate of learning. For example, probabilistic selection of sampling units guards against systematic sources of bias in the mean or variance estimate (Chapters 2, 5). Bias in the point estimate is particularly troublesome because it can consistently support an inappropriate model and thus sabotage learning. This concern is highly relevant for monitoring programs which count or measure the occurrence of organisms that are not perfectly detectable. To be most effective in ARM, such programs need to account for potential biases due to incomplete detectability: they should provide direct estimates of detection rate, or they should otherwise provide evidence that the rate of detection is relatively constant across all population sizes and other relevant decision attributes. More broadly, regardless of the specific natural resource being managed and monitored, design and implementation of the monitoring program must consider numerous potential sources of bias that can misdirect learning (Chapters 2, 3, 5, 11).

Even if a model's predictions are consistent with field data, the model may be slow to accrue credibility if the data yield estimates with poor precision. In extreme cases, poor precision has an effect similar to bias, causing credibility to accumulate on inappropriate models (Moore and Conroy 2006). Increasing replication is the most direct way of increasing precision, but strategies such as stratification and ratio estimation (Cochran 1977) may be effective when replication cannot be further augmented.

So far we have described a conventional ARM process where a resource is monitored, a decision is made about the resource based on its observed status, and monitoring is repeated to assess and update the prediction models. Under this arrangement, management and learning are pursued simultaneously over the recurrent steps of the decision process, and varying the monitoring design is the primary approach to increasing the rate of learning. We can consider two alternatives to this basic arrangement in which

Box 4.2 Monitoring and learning in ARM

The following example illustrates the role of monitoring in the updating of knowledge during a cycle of decision making within the ARM framework. Suppose a lake is managed for production of a fish subject to harvest, and managers contemplate a single type of action each year believed to increase the abundance of the juvenile class. However, there is uncertainty about the effect of the management action due to uncertainty about the degree of negative feedback related to population density of the adult classes. Therefore, given input about current adult population density and perhaps other state variables measured as part of an annual population monitoring program, there are at least two models that produce competing predictions about the response of juvenile density to the action. Model A, which includes an adult density-dependent effect, predicts a lower observed response than does Model B, which lacks the density-dependent effect. Assume the likelihood under each model is the normal probability density function:

$$L_j(x) = \frac{1}{\sigma\sqrt{2\pi}} e^{-\frac{(x-\mu_j)^2}{2\sigma^2}},$$

where μ_j represents the mean predicted observed juvenile density under each model ($j = A$ or $j = B$), and σ^2 is the variance common to both models. For this example, assume that Models A and B predict 4.0 vs. 5.0 juveniles per unit volume, respectively, as the response to the current management treatment. If uncertainty is initially so great that either model is equally plausible, then a reasonable initial weight to assign to each model is $p_{0,A} = p_{0,B} = 0.5$. Given these weights, predictions of observed juvenile density by each model (μ_A and μ_B), an observation of the outcome (x) and a variance that come from the monitoring program, and the form of the likelihood function, then updating weights on the models is straightforward through application of Bayes' Theorem:

$$p_{1,j} = \frac{p_{0,j} L_j(x_0)}{p_{0,A} L_A(x_0) + p_{0,B} L_B(x_0)}.$$

The calculations are illustrated below for two cases of monitoring precision:

	Model	
	A	B
Initial model credibility ($p_{0,j}$)	0.5	0.5
Model predictions (juveniles/unit volume) (μ_j)	4.0	5.0
Observation ($x = 4.9$ juveniles/unit volume)		
Case 1: High precision monitoring ($\sigma = 1.0$)		
Likelihood [$L_j(x)$]	0.266	0.397
Posterior credibility ($p_{1,j}$)	0.401	0.599
Case 2: Low precision monitoring ($\sigma = 3.0$)		
Likelihood [$L_j(x)$]	0.127	0.133
Posterior credibility ($p_{1,j}$)	0.489	0.511

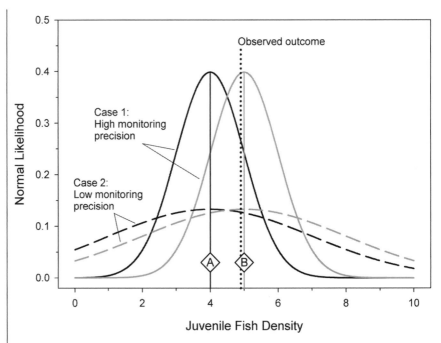

Normal likelihood for Models A (black, predicted density $= 4.0$) and B (gray, predicted density $= 5.0$), where the variance parameter in the normal likelihood corresponds to monitoring precision. The difference between models in the likelihood of an observed observation of 4.9 is much greater in Case 1 than Case 2.

In this example, the observation seems to lend support to model B because the observation is closer to μ_B than μ_A. However, we must take into account the size of the difference with respect to monitoring precision (see figure above). When precision is taken into account, the information provided by the observed value is quite different in the two cases. With low precision monitoring in this case, the posterior credibilities of the two models have not diverged much; i.e. little learning has occurred. Clearly, to the extent that monitoring precision can be increased through better or more intensive sampling design, rate of learning about managed systems increases correspondingly.

learning is pursued somewhat apart from the pursuit of management goals, and the gained knowledge is then folded back in to more efficiently pursue management objectives.

The first approach involves formal experimentation in which some crucial experiment is designed to test among the competing hypotheses. Evidence gathered from the experiment could then be used to assign credibility values to the competing models, and adaptive resource management as described above could proceed forward from that point. In essence, this approach is a way to jump-start the adaptive process by using experimentation to quickly arrive at more informative credibility values. Because all emphasis is placed on learning during this phase, management returns are "put on hold" while the experiment runs, but this may be a tolerable short-term sacrifice in some settings. If management cannot be put on hold, then a workable alternative might be to carry out experimentation on some logical portion of the managed resource; for example, a set of plots chosen from an entire area of management interest, or a sample of stream segments in a watershed. Conventional adaptive resource management could occur on the non-experimental portion of the resource. Model credibility values could be updated on the basis of information coming from both the experimental and non-experimental units.

The second approach involves treating an element of monitoring design, such as level of replication, as a formal decision variable considered jointly with the decision variable pertaining to the managed resource (e.g. Hauser *et al*. 2006). For example, in a harvest problem, an annual decision might address not only level of harvest but also intensity of monitoring effort. This approach suggests the tracking of a state variable that measures the current potential to gain management performance with added monitoring effort; in other words, the current *value of information*. Value of information is an assessment of what one sacrifices (in units of the resource) to manage under continuing structural uncertainty (Lindley 1985). High value of information indicates great structural uncertainty and a large benefit to be gained by resolving it. Thus, an indication that value of information is currently high should lead the decision maker to increase monitoring effort. The payoff of doing so should be a reduction in structural uncertainty and a gain in management performance. Consequently, as current value of information decreases, monitoring effort should decrease. Conceptually, this approach is similar to the cost : benefit framework (based partly on a monetary valuation of information) outlined in Chapter 23 for making optimal choices among potential uses of monitoring dollars.

Example: adaptive harvest management

Adaptive resource management is being explored or adopted for an increasing number of natural resource management problems. One of the first applications in wildlife management, where the decision implemented is derived directly from the decision model, is the adaptive harvest management of mallard ducks (*Anas platyrhynchos*). Since 1995 the US Fish and Wildlife Service (USFWS) has set annual hunting regulations (season length and bag limit) for midcontinent mallards based on a fully developed ARM decision model (USFWS 2010a). We outline the elements of this decision model (Johnson *et al*. 1997, Kendall 2001, Johnson 2011), and how monitoring programs are integrated into the process.

Objective function

The primary objective of mallard harvest management is to maximize cumulative harvest over an infinite time horizon. This long time horizon makes conservation of the species an implicit objective. However, an explicit population objective to maintain at least 8.5 million mallards annually, based on the goals of the North American Waterfowl Management Plan, is imposed through a constraint on harvest. If the predicted population size under a candidate management action is less than 8.5 million, the harvest that would result is penalized. This penalty is proportional to the difference between predicted population size and 8.5 million. Given these objectives, the outcome of a harvest decision is evaluated based on resulting harvest and population size. Therefore, it is important to monitor these two variables each year.

The USFWS invests considerable effort and funds in harvest surveys (Raftovich *et al.* 2010) and population surveys (USFWS 2010b). A comprehensive sampling frame of hunters is developed when hunting licenses are issued. Hunters are later surveyed about the species and number of birds shot, and the total harvest is then estimated. The population survey, a collaborative effort between the USFWS and Canadian Wildlife Service (CWS), as well as biologists from individual states and provinces, is one of the most extensive in the world, covering 3.6 million km^2 of breeding habitat throughout the north-central USA, central Canada, and Alaska (Smith 1995). A large sample of transects were allocated at the inception of this survey, based on a stratified systematic sample with a random start. These transects are flown, and all ducks and ponds are counted. On a subset of segments within these transects ground crews conduct a complete count, in a double-sampling scheme, to correct for visibility bias. Population estimates are highly precise (Coefficients of Variation $< 5\%$; USFWS 2010b). This survey serves as an archetype for a monitoring program: it is precise, adjusted for bias, and designed to directly inform a management decision.

Management actions

Candidate management actions include four regulations packages, each of which specifies a particular combination of season length and bag limit. Partial controllability is accounted for by acknowledging variability in realized harvest rate across years under a given regulatory package. This information is monitored through an extensive pre-season banding program for waterfowl throughout the range of the midcontinent mallard, conducted each summer by USFWS, CWS, state, and provincial personnel. Thousands of birds are banded each year, with bands that include a toll-free phone number for reporting to the USGS Bird Banding Laboratory. Harvest and survival probabilities are estimated using band recovery models (see Brownie *et al.* 1985).

Population models

Harvest management of mallards is currently based on four population models, with predicted breeding population size in year $t + 1$ as a function of harvest in year t and two state variables observed in year t: breeding population size and number of ponds in the Prairie Pothole region of the USA and Canada (USFWS 2010a). The annual prediction is derived from submodels based on parts of the annual cycle (Johnson *et al.* 1997, USFWS 2010a), which includes sex-specific summer survival rate, recruitment of young into the fall population, sex- and age-specific harvest rates, and winter survival rate. The four population models derive from combinations of two models for survival of the hunting season, and two recruitment models. The effect of hunting on annual mortality is assumed to be either completely additive (total annual survival is reduced proportionally for each bird killed) or completely compensatory up to the point of natural mortality (harvest kill is not counted against annual mortality because some other mechanism such as increased food availability per surviving mallard compensates for it). Recruitment is assumed to be weakly or strongly density-dependent, where spring population size serves as the measure of density.

This harvest management problem is an excellent example of the fourth role of monitoring in adaptive management: to build system models. Each component of the population models being used was developed from theory parameterized by monitoring data (Johnson *et al.* 1997). Survival models were built upon the extensive banding programs mentioned earlier. Within the population model, recruitment occurs in the fall rather than spring, because recruitment data come from a "parts collection" survey conducted as part of the harvest survey discussed above. In this survey, a subset of hunters identified from the harvest survey sampling frame are provided with envelopes with which to mail wings from the birds they harvest. Age is assigned to each wing by experts at "wing bees" in each of the four administrative flyways. Raw age ratio data are corrected for differential vulnerability by age/sex class, using band recovery data.

Inputs for the four population models are breeding population size, number of ponds, and harvest decision (regulatory package of season length and bag limit) for the current year. The first two variables are the state variables that need to be monitored, as discussed above. Output variables from the models are predicted breeding population size and number of ponds for the following year. Predicted pond numbers do not vary by model, so for the purposes of updating the weights on each model only breeding population size is used. However, because band recovery data and parts collection data are collected every year, parametric uncertainty within the individual survival and recruitment models could be reduced over time. Operationally, uncertainty about the distribution of harvest rates under each candidate regulations package is refined each year with new band recovery data (USFWS 2010a).

The use of monitoring data to build system models is logical in this case. These monitoring programs (population survey, harvest survey and parts collection, banding program) had been in place for decades before the initiation of adaptive harvest management in 1995, thus providing data over a range of conditions for building models. Throughout most of their history these surveys were designed to support the same focused decision process that adaptive harvest management is addressing. Therefore they include the right type of data needed to answer the management question.

Finally, because they were designed to support management at a mid-continental scale, they are collected at that scale, and therefore are useful to predict abundance at that scale. There was and is a wealth of waterfowl research projects being conducted on this population, including population models (e.g. Johnson *et al.* 1987). However, these efforts were at spatial scales smaller than the scale of harvest management. Extrapolating findings from research or even management at smaller scales to make predictions at larger scales often requires strong assumptions (Walters 1986). In addition, the data needed to drive models at smaller scales are often not available at the larger scale. For example, if predator density is found to have a strong effect on recruitment from a research project at a smaller spatial scale, this is not useful to modeling at the scale of interest unless predator density information is available or could be inferred at that spatial scale. We do not mean to negate the usefulness of smaller-scale studies to inform management. Such research studies are very useful for identifying hypotheses to be considered for modeling and management at a scale such as the midcontinent. If an important causal factor is identified from research at the smaller spatial scale, the ability to measure that factor at the appropriate scale could be investigated (e.g. via remote sensing), and if successful a model including that mechanism could be added to the model set for ARM.

The ideal approach for building robust models would be experimental manipulation at the scale of management (Walters and Holling 1990). This approach was implemented for midcontinent mallards in the 1980s, where regulations were stabilized into a set number of restrictive and liberal years, respectively, regardless of the trajectory in population size (Brace *et al.* 1987). Nevertheless, even after that experiment there was still substantial uncertainty about the effect of harvest on survival. Balancing the value of the information gained from experimentation, versus the short-term cost with respect to management objectives, is one of the steps in the ARM process. This balance can be formally incorporated into the decision algorithm using an actively adaptive, or dual control, approach to optimization (Williams 1996).

In summary, monitoring data have played a direct and integral role in the adaptive harvest management of mallards, and were designed to do so. These roles include providing data for construction of population models, assessment of system state (population size and habitat variable) for making decisions, evaluation of outcome of decisions with respect to objectives (harvest and population size), and model evaluation (population size, survival rates, harvest rates, recruitment rates). In addition, monitoring data continue to be used to evaluate the predictive ability of the model set as a group, and therefore can identify when better models should be sought. Johnson (2011) provides an excellent synopsis of the progress of this adaptive harvest management process over the last 15 years. Moreover, the large-scale, long-term waterfowl monitoring component provides a good example of a program that is focused on management questions, but could also be used as surveillance for other environmental or anthropogenic effects (e.g. climate change or spread of avian influenza).

Future research and development

Monitoring in an ARM framework provides clear re-occurring benefits for informing management and conservation. However, managers and agencies frequently may also want or have a mandate to address questions about long-term changes beyond those tied closely to a well-defined management decision-making framework (Chapters 3, 7, 22). Because adaptive management involves iterative decision making, in many cases essentially in perpetuity, the time scale for monitoring in this context is long-term. In many cases, surveillance monitoring is unlikely to be sufficient to directly inform management decisions, either due to inadequate locations of sampling points or collection of the wrong metrics. However, might the converse be effective?

> **Box 4.3** Common challenges: having it all
>
> Adaptive management provides a context and structure for planning and designing monitoring programs to inform management decisions. In this context both the costs and benefits of investment in monitoring effort are directly measurable. Nevertheless, agencies are also often interested in general surveillance monitoring, to detect general trends or in hope of detecting the effect of an unanticipated threat. In many cases, decision processes apply to the long term, and therefore the time horizons for both decision-focused monitoring and surveillance monitoring are comparable. This creates a potential trade-off that must be resolved at the agency level. Given this situation, an agency might consider somehow combining approaches.
>
> In the ARM context, there is a clearly articulated use of monitoring to support specific decisions. For cases where active management of systems is occurring across the range of interest for surveillance, we would favor attempting to create a network for surveillance from individual management-oriented monitoring programs. This would entail adding responsibilities (e.g. counting additional taxonomic groups) to the monitoring mission identified by the adaptive management process.
>
> Conversely, if an agency wanted to have a single region-wide monitoring program, then part of the design optimization process, in addition to minimizing overall variance, would be to incorporate the collection of monitoring objectives that arise from the various adaptive management programs in the region. The balancing of these ARM-focused monitoring objectives, as well as a general surveillance objective, would itself be a formidable decision problem. Nevertheless, it would represent a rigorous approach to balance the competing monitoring needs.

Perhaps a collection of management-focused monitoring programs (e.g. on refuges which are monitoring decision outcomes), distributed over a large range when considered together, could provide surveillance for unanticipated factors driving the larger system. We believe this is worth considering (Box 4.3); if feasible, such an approach would give some benefits of surveillance monitoring without compromising the well-defined role of monitoring in the ARM framework.

Summary

Our objective was not to present or review particular techniques or approaches to design or analysis of monitoring programs. Instead, our intent was to discuss monitoring within the larger framework of structuring and making informed resource management decisions, which monitoring is intended to support. We outlined adaptive resource management as a cohesive framework for structuring management problems and producing informed decisions in the face of various sources of uncertainty.

Using this framework, we showed that when the decision context, management objectives, candidate actions, and predictive system models are clearly defined, the core roles and contributions of monitoring programs to that decision process become clearly defined. Specifically, monitoring programs should measure system state to inform the decision, and measure the results of the decision to evaluate management performance and reduce scientific uncertainties about the system. In addition, monitoring data can be used in an inductive way to build future predictive models for the system. Furthermore, given these clear roles of monitoring, one can evaluate candidate monitoring decisions (sampling design, sampling intensity, decisions about analysis such as relying on indices or correcting for detection probability) not only in terms of the monetary costs of more monitoring effort, but the opportunity costs (in terms of other management objectives) of reduced monitoring effort. Therefore, this framework provides a basis for evaluating the relative merits of various sampling and estimation techniques discussed in the rest of this volume.

Acknowledgments

We thank Jim Nichols, Julien Martin, an anonymous reviewer, and the editors for helpful comments on previous drafts of the manuscript. The manuscript also benefited from numerous discussions about the role of monitoring with various colleagues through the years.

Section II

Survey design

5 Spatial sampling designs for long-term ecological monitoring

Trent McDonald

Introduction

The spatial survey component (i.e. sampling design, Chapter 2) of a long-term ecological monitoring study dictates *where* sample locations are placed in a study area. The primary purpose of the spatial design is to select a sample of locations in such a way that valid scientific inferences can be made to all geographic regions of interest. The spatial design does not, in general, specify the size or shape of sample units, what or how target variables are measured, nor how sample units are revisited through time. Those specifications are left to other parts of the larger survey design. Because the spatial design dictates where samples are located, the terminology surrounding spatial designs is slightly different than that in classical finite population sampling, and reflects the fact that two-dimensional geographic locations are of primary interest. Table 5.1 translates the general definitions of a few key terms in classical sampling theory to more specific terms applicable to spatial surveys.

The purpose of this chapter is to discuss key characteristics of good spatial designs and introduce readers to some common spatial design examples. A major alternative to the designs discussed here is the generalized random tessellation stratified (GRTS) design presented in Chapter 6. In fact, a GRTS design should probably be favored over some of the designs in this chapter for implementation in real-world large-scale and long-term ecological monitoring programs due to the GRTS's added flexibility and spatial coverage assurances. Despite this, the designs of this chapter are important because they will continue to be used in studies that have focused purposes and where relatively simple designs are required. The designs of this chapter can also be combined with other designs, including GRTS, and implemented in two or more levels of nested sampling.

This chapter starts by discussing the differences between scientific and non-scientific survey designs. All designs discussed subsequently are scientific designs. Next, the differences between research and monitoring studies are discussed. Designs generally appropriate for each type are then described in separate sections. Final comments and summary are presented at the end (see also Box 5.1).

Design and Analysis of Long-term Ecological Monitoring Studies, ed. R.A. Gitzen, J.J. Millspaugh, A.B. Cooper, and D.S. Licht. Published by Cambridge University Press. © Cambridge University Press 2012.

Table 5.1 Traditional finite population sampling terms and their equivalents in the context of spatial environmental surveys.

Traditional finite population terms		Spatial sampling terms	
Term	Definition	Term	Definition
sample unit	The smallest entity on which data will be collected	*sample site*	A special case of *sample unit* that is tied to a geographic location. Can be a point or a polygon (e.g., pixel)
target population	A collection of *sample units* about which inference is sought	*study area*	A collection of *sample sites* about which inference is sought, usually contiguous
sample	A subset of *sample units* in the *target population*	*spatial sample*	A subset of *sample sites* in the *study area*

Scientific vs. non-scientific designs

As background for subsequent discussions, this section defines what are and what are not *scientific* designs. It is my view that only scientific designs should be implemented in monitoring studies because they provide data that impartially represent the study area. Objectivity and impartiality of sample site placement is paramount in monitoring studies because consumers of the survey's information will rely on estimates to accurately portray the study area. Consumers of estimates expect them to unbiasedly estimate known parameters, and to apply to all parts of the study area. The only way to achieve unbiased estimates and proper inferential scope is to implement a scientific design. It is not possible to guarantee these characteristics when data have been collected under a non-scientific design, no matter the analysis (e.g. see Chapters 2, 11).

Scientific designs

The primary difference between scientific and non-scientific designs is that the probability of including any particular site in the sample can be quantified in the former, but not the latter. The probability of including any particular site in the sample is called the *first-order inclusion probability*, and when first-order inclusion probabilities are knowable, the design is said to be a *probability sample*. Scientific designs, as defined here, are simply those that draw some sort of probability sample (see also Chapters 2 and 6).

In order to draw a probability sample, the algorithm selecting the sample must be repeatable and involve some sort of stochastic component (e.g. random draws, random starting point, etc.). In most monitoring situations, it is best to use probability samples that result in equal first-order inclusion probabilities, but this is not absolutely necessary. Equal first-order inclusion probabilities will help insure that adequate data exist to estimate a multitude of current and future parameters. Being able to adequately estimate

Box 5.1 Take-home messages for program managers

Most natural resource monitoring programs seek to estimate characteristics and dynamics of a study area based on data collected at a small subset of sites within the area. The spatial sample design specifies *where* sample sites are placed, not *when* or *how* they are sampled. In selecting a sample design, the goal is to choose and implement a design that allows inference to all regions of the study area, sampled and un-sampled. *Scientific* environmental sampling designs – those that allow objective extrapolation from sample sites to the entire study area – must collect some sort of *probability* sample. Probability samples are defined as those where the probability of including any particular unit of the population in the sample is knowable and is > 0. Probability samples are selected via a repeatable process that includes some type of objective stochastic element (i.e. a random draw of some sort). In contrast, *non-scientific* designs, such as *judgment sampling*, *haphazard sampling*, and *convenience sampling*, do not draw probability samples. Inclusion probabilities are unknown under non-scientific designs, and it is therefore impossible to make direct statistical inferences to the entire study area using the collected data. Conclusions about the broader study areas therefore, are based on assumptions or speculation, and such designs can be easily discredited if there is disagreement about the conclusions.

The objectives of a study should guide the choice of a sample design. Many commonly used sampling schemes are most appropriate for *research* studies that are designed to answer a specific question in a short period of time. In contrast, *monitoring* studies usually target a wide range of variables, and are usually long-lived (10–30 years). The spatial design of monitoring studies must provide adequate data to address a wide range of anticipated and unanticipated estimation tasks over a much longer time frame, while ensuring good spatial coverage of the area (i.e. spatial balance).

To meet these requirements, most ecological monitoring programs will consider the following designs. *Systematic* sampling (e.g. sampling a grid of sites across the study area) is simple and generally appropriate for long-term large-scale monitoring studies of two-dimensional resources that collect a wide range of variables. The *General Random Sampling* (GRS) design draws equal or unequal, ordered or unordered, random and systematic samples in one dimension and is appropriate for monitoring resources such as river segments. However, one difficulty with systematic and general random samples is how to replace unsuitable sites or add sites while maintaining spatial balance. These difficulties are generally alleviated with the designs discussed in Chapter 6. The designs of this chapter can also be combined with other spatial designs to maximize flexibility for meeting a program's objectives. For example, stratification – dividing the study area into two or more non-overlapping groups and sampling each subarea independently – may be appropriate when separate estimates for each stratum (e.g. management unit) are desired for the life of the monitoring program.

a large number of parameters, some of which cannot be foreseen at design time, increases the project's general utility and will help ensure the project remains relevant to data consumers. If first-order inclusion probabilities are unequal, they must be strictly greater than zero. In some studies, it may be preferable to use unequal inclusion probabilities if, for instance, the population is stratified, or to increase the odds of obtaining sites in certain areas. However, care must be taken; subsequent design-based analyses must incorporate this aspect of the design (Chapters 6, 11, 14), and it may not be clear how to properly account for unequal inclusion probabilities in some model-based analyses (e.g. Chapter 7).

Non-scientific designs

The types and characteristics of non-scientific designs are varied and great; however, non- scientific designs are generally defined as those that do not draw probability sam- ples. That is, the design is not repeatable or does not involve some sort of stochastic component. Without a probability sample, there is no statistical basis by which infer- ences can extend beyond the sampled sites. Without a scientific design, the applicability of estimates to areas beyond those actually sampled is based on professional opinion regarding the "representativeness" of sample sites. This might be acceptable except that the definition of "representativeness" is nebulous (see Chapter 6) and arguments over the accuracy of estimates can rage for some time. If some controversy is associated with the monitored resource, it is easy to cast doubt on results derived from non-scientific designs, usually by accusing the researchers involved of some sort of bias (i.e. conduct- ing the study such that a predetermined outcome is obtained). Without a repeatable and quantified sampling design, it is impossible to refute or prove accusations of bias, and debates about the study's results often cannot be resolved. In practice, the only way to settle these types of debates is to conduct a second study that implements a scientific design.

The perils of non-scientific designs are well known (Edwards 1998, Olsen *et al.* 1999, McDonald 2003). Yet, non-scientific designs remain seductive because they are easier, quicker, and less expensive in the short term to implement than scientific designs. In particular, three types of non-scientific designs are popular: *judgment*, *haphazard*, and *convenience* samples.

Judgment samples

Judgment samples are those in which researchers, usually familiar with the resource, make a judgment or decision regarding the best places to locate sample sites (see also Chapter 2). Researchers may place sample sites where "the most change is anticipated", or "in critical habitat", or "where the impact will be", or "where interesting things will happen". In fact, these researchers are making judgments about what will happen, or are making tacit assumptions about relationships among parameters. These judgments and assumptions are wrong or only partly correct, and the study can be severely biased or fail to uncover otherwise important patterns. If a judgment sample is implemented, assumptions should be acknowledged. In limited cases when a research study is tightly scoped, judgment samples may be warranted. However, even in these cases, there is no basis for extrapolating results beyond the sampled sites.

Haphazard samples

Haphazard samples are those that are collected without a defined or quantifiable sampling plan. Under haphazard sampling, the placement of sample sites is not usually planned out in advance. Many times, haphazard samples are placed and taken where and when field crews have extra time after performing other duties. In these cases, field crews may be instructed to place sample sites wherever they happen to be. When these types of spatial samples are used in monitoring studies, meaningful general-purpose analyses are very difficult to conduct. Haphazard and unplanned sample sites may provide interesting observations, but should not be considered part of scientific monitoring plans.

Convenience samples

Convenience samples are those that include sites because they are easy and inexpensive to assess. Forethought may be placed into the design, and the design may call for a probability sample, but the probability sample only selects sites from regions of the study area that are convenient to visit, despite the fact that the entire study area is of interest. Because budgets are always limited, convenience samples are a tempting and popular way of reducing costs. Convenient sample sites are often located close to research facilities, roads, or other access ways, such as trails.

The problem with convenience samples is that they do not represent the entire target population. Assuming a probability sample has been taken, this is not a problem unless inference to the entire study area is desired. For example, drawing a probability sample near roads is perfectly acceptable as long as it is clear that estimates apply to regions of the study area that are near roads ("near" must be defined). Identifying regions near roads and drawing a probability sample results in a scientific sample of the near-road population; if treated as such, it is not a convenience sample. It is an inferential mistake, however, to represent estimates constructed using a near-road probability sample as applying to a larger study area.

Monitoring vs. research

In addition to understanding scientific and non-scientific designs, it is important to make some distinctions between *monitoring* and *research* studies. This distinction is important because the spatial sampling designs of both types generally meet different goals. Designs are later classified as either research or monitoring, and this section defines both types, starting with *research*.

The primary defining characteristic of research studies is that they are focused on a specific question. Research studies generally focus on one or two target variables (albeit potentially ones with numerous attributes of interest, as in community-level studies) measured over a relatively brief period of time (1–5 years). Research studies usually address an immediate problem or crisis, or inform an impending management decision. When that problem, crisis, or decision is resolved, motivation to continue the study drops. For this reason, it is usually paramount to use a spatial design that provides the highest possible statistical precision to estimate a target parameter(s) in the shortest amount of time. Many typical research studies are convenient for graduate students because study objectives are usually narrowly focused enough that the study can be performed during a graduate student's normal tenure.

In contrast, a key characteristic of most monitoring studies is that they are not designed just to answer a specific, narrow question. As in Chapter 3, the objectives of what this chapter refers to as "monitoring" studies usually boil down to *watching* an environmental resource. Watching the resource does not necessarily mean focusing estimation on many different parameters. Long-term studies that focus on a few key parameters (e.g. within an Adaptive Resource Management Framework, Chapter 4) may be called "monitoring" because those parameters usually function as indicators of the system's overall health. Compared to research studies which address very concrete questions, the objectives of monitoring studies unfortunately often can seem vague (e.g. "... to estimate current status and detect trends of aquatic resources ..."). Monitoring studies tend to be long in duration, typically 10–30 years, because annual variation in the system of interest can be large and trend detection therefore requires an extended period (see Chapter 7). Monitoring studies also often are large-scale in their desired inference scope, where "large" means the study incurs significant costs traveling between sample sites (e.g. see examples in Chapters 6, 14, and 20). Monitoring studies are often funded by agencies with management authority of the resource, and maintaining funding tends to be difficult because these studies generally do not address a management crisis.

As far as spatial sampling design is concerned, monitoring studies generally cannot focus on a single variable or objective. For a monitoring study to survive 1–3 decades, its spatial design needs to be robust enough to address multiple issues, easy to implement and maintain, and able to provide data on unforeseen issues when they arise. Real-world monitoring studies almost always have multiple objectives, and it is therefore very difficult to optimize the placement of locations for estimation of any one parameter. Unfortunately, much of the statistical literature on spatial design is relevant only when interest lies in a single parameter. Experimental design books such as Steel and Torrie (1980) and Quinn and Keough (2002) discuss techniques such as blocking and nesting of plots as a way to maximize the study's ability to estimate a treatment's effect on a single parameter. Stratification, discussed in texts such as Cochran (1977), Scheaffer *et al.* (1979), and Lohr (2010), is usually thought of as a technique designed to improve estimation of a single parameter. Maximum entropy (Shewry and Wynn 1987, Sebastiani and Wynn 2000), spatial prediction (Müller 2007), and Bayesian (Chaloner and Verdinelli 1995) sampling methods are designed to locate sites such that a criterion (such as information gain) is maximized. The criterion used by these latter designs is a function of a single variable and thus these procedures optimize a design for one parameter at a time. This is undesirable for most monitoring studies, with the exception of monitoring focused on specific system parameters, because a design that is optimal or near-optimal for one particular parameter

Table 5.2 Common spatial designs suitable for *Research* and *Monitoring* studies, along with references and a brief description/comment. See corresponding section of the text and references for a more thorough description.

Study type	Sampling design	References	Description/comment
Research	Simple random	Cochran (1977, Ch. 2) Scheaffer *et al.* (1979, Ch. 4) Särndal *et al.* (1992, Ch. 3) Lohr (2010, Ch. 2)	Sites selected completely at random. Covered by many sampling texts and articles
	Two-stage	Cochran (1977, Ch. 11) Scheaffer *et al.* (1979, Ch. 9) Särndal *et al.* (1992, Ch. 4) Lohr (2010, Ch. 6)	Large or *primary* sites selected first. Small or *secondary* sites within the selected primary sites are sampled second. In general, design and sample size can vary by stage
	Stratified	Cochran (1977, Ch. 5; 5A) Scheaffer *et al.* (1979, Ch. 5) Särndal *et al.* (1992, Ch. 3) Lohr (2010, Ch. 3)	A special case of *two-stage* sampling where strata define primary sites, and all primary sites are sampled (census at stage one). A special case is *one-per-strata* sampling (Breidt 1995). In general, design and sample size can vary by strata
	Cluster	Cochran (1977, Ch. 9; 9A) Scheaffer *et al.* (1979, Ch. 9) Särndal *et al.* (1992, Ch. 4) Lohr (2010, Ch. 5)	A special case of *two-stage* sampling that occurs when all secondary sites within a primary site are sampled (census at stage two)
Monitoring	Systematic	Cochran (1977, Ch. 8) Scheaffer *et al.* (1979, Ch. 8) Särndal *et al.* (1992, Ch. 3) Lohr (2010, Sec. 2.7)	Also, *grid* sampling. Suitable for 2D resources. Geometry of grid is usually square (square cells) or triangular (hexagonal cells)
	General Random Sample (GRS)	This chapter	Suitable for 1D resources. Produces weighted or unweighted, ordered or unordered, simple random and systematic samples of fixed size

is suboptimal for other variables unless spatial correlation between the variables involved is strong.

The main part of the remainder of this chapter consists of two sections and relevant references in Table 5.2. One section (*Research Designs*) discusses four common designs that are more likely to be applied to research studies unless integrated into a larger design suitable for ecological monitoring. However, the designs in this section can also be applied on their own to monitoring studies under certain circumstances (e.g. few monitored variables). These designs will be familiar to many readers, and consist of *simple random sampling*, *two-stage sampling*, *stratified sampling*, and *cluster sampling*. The second section (*Monitoring Designs*) describes two designs that are relatively more likely to satisfy the goals of a monitoring study. These designs are the *systematic*, which is applicable to two-dimensional resources, and the *General Random Sample* (GRS), which is applicable to one-dimensional resources (e.g. picking a sample segment along a stream). An important characteristic of both these designs is that they can ensure a high degree of spatial coverage.

Research designs

This section contains brief descriptions of four common spatial sampling designs that will probably be appropriate for research studies. In general, these designs are appropriate when one or a very few variables are targeted for inference and the study does not last longer than 5 years or so. As noted above, a long-term study's interest may lie in a single variable, and in this case the designs of this section could be applied in a monitoring setting. Also, certain of the sampling plans, whether designated here as research or monitoring designs (i.e. simple random, systematic, and GRS) can be applied at different stages of a larger design. For example, it is possible to draw a systematic sample of large collections of sampling units at one level, followed by a simple random sample of units from within the large collections. In a monitoring context, such complex designs are frequently used in large-scale studies focused on status and change at numerous spatial scales (e.g. national to local; Nusser *et al.* 1998).

With the exception of simple random sampling, it is important to keep in mind that a significant amount of a priori information about the target variables and the study area is needed. For instance, to implement a stratified design, regions of the study area must be classified into categories. This categorization requires knowing or estimating an auxiliary variable upon which strata are defined. If a maximum entropy design is to be implemented, some information about the magnitude and structure of spatial covariance must be known. If a cluster design is to be implemented, the size and configuration of clusters must be known throughout the study area.

Simple random sampling

Simple random sampling is well-known, being mentioned by nearly every text and article on spatial sampling theory. Simple random samples, sometimes called just *random samples*, are simple mathematically and are often a basis by which other designs are compared (e.g. Chapter 17), but they are rarely actually used in large studies.

Simple random samples of a geographic study area are drawn by first bounding the study area with a rectangular box to delineate the maximal extent of the horizontal (x) and vertical (y) coordinates. Random sample site coordinates are then generated as random coordinate pairs one-at-a-time within the bounding box. Random coordinate pairs can be generated by choosing a random deviate from a uniform distribution over the range of the x coordinates, and an (independent) random deviate from a uniform distribution for the y coordinates. When generated, the random point (x, y) will fall somewhere within the study area's bounding box, but not necessarily inside the study area. If the random point (x, y) falls outside the study area, it is discarded and another point generated. The process of generating points inside the study area is repeated until a desired sample size is achieved.

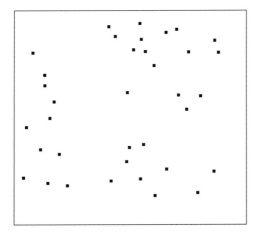

Figure 5.1 Example *simple random sample* of size 36 from a square study area.

A realized example of a simple random sample from a rectangular study area appears in Fig. 5.1.

In many cases, simple random sampling is an excellent approach for providing an objective assessment of the sampled study area. If a particular sample site is inaccessible or unsuitable, additional sites are easily generated and the sample will retain its basic statistical properties. Simple random sampling is the only design listed in this section that does not utilize a priori information on target variables, and therefore can be easily and rapidly implemented in most studies. Simple random sampling, however, suffers from the fact that it does not guarantee uniform coverage inside the study area. It is possible for significant "clumps", and corresponding "holes", to develop in the spatial coverage of simple random samples. For example, in the realized sample of Fig. 5.1 there is a conspicuous vertical strip approximately 1/3 of the way from left to right in the horizontal direction that is devoid of sample sites. Assuming some degree of spatial correlation exists in the study area for the variables of interest, these clumps and holes leave some regions of the study area over represented while other regions are underrepresented. If no spatial correlation exists, all areas are equally represented from a statistical point of view regardless of sample site locations. The probability of realizing significant clumps and holes declines quickly as sample size increases; nevertheless, simple random samples are generally unpopular for real studies because researchers desire strong assurances of good spatial coverage.

Two-stage designs

Two-stage designs contain two nested levels of sampling. Two-stage designs are implemented when sites can be naturally grouped into large collections, and it is relatively easy to select the large collections. These large collections of sample sites are called *primary sample units* and are defined to exist at level one of the study. When geographic areas

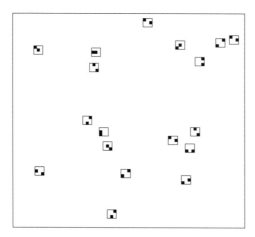

Figure 5.2 Example *two-stage sample* of 18 primary units and 2 secondary units from within each, for a total of 36 sample sites. Each primary unit was defined to be a 3 × 3 block of 9 secondary units. Primary and secondary units were chosen by simple random sampling.

are sampled, the primary units are usually large polygons such as watersheds, sections, counties, states, etc. In general, primary units could be any structure containing multiple sites or sample units. Under this design, primary units are selected using one of the basic spatial designs like simple random sampling, systematic sampling, GRS, or even GRTS sampling.

Once primary sample units are defined and selected via some design, the actual locations for sample sites are selected from within each selected primary unit. In this setting, the actual sample sites are called *secondary sample units* and are said to exist at level 2 of the study. Secondary units are selected from each primary unit using one of the basic spatial samples like simple random, systematic, GRS, or GRTS. It is not necessary for the sampling design or sample size to be consistent across primary units, but usually they are. An example two-stage sample appears in Fig. 5.2. It is possible to implement a multi-stage design with three or more levels when sample units naturally nest within one another. For example, when sampling the entire USA, it may make sense to select states, then townships within states, then sections within townships, then sample site within townships. Chapter 8 discusses variance component estimation for multi-stage sampling. Many texts on finite population sampling theory (such as Cochran 1977, Scheaffer *et al.* 1979, Särndal *et al.* 1992, Müller 2007, and Lohr 2010) discuss 2-stage designs, but usually assume a simple random sample at both stages.

Two-stage designs with less than a census of primary units at level 1 are relatively rare in ecological studies. More commonly, every primary unit at stage 1 is selected (i.e. a census is taken at stage 1) and all sampling is done at stage 2. In this case, the design is said to be *stratified* (next section). However, if there are many large primary units, and the levels are practical and logical, selecting a spatially balanced sample of primary units, followed by a spatially balanced sample of secondary units, may be a good design for either research or monitoring purposes.

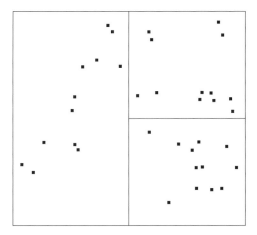

Figure 5.3 Example *stratified sample* of 36 sites from 3 strata. In each stratum, 12 sites were selected using the simple random design.

Stratified sampling

Stratified designs are special cases of two-stage designs that occur when all primary units are selected and a sample of secondary units is drawn from each (Fig. 5.3). When all primary units are sampled, the primary units are renamed *strata*. Collectively, the strata partition the study area into mutually exclusive sets of secondary units (sites). Like two-stage sampling, sample size and even design can vary between strata.

In the example shown in Fig. 5.3, a simple random sample of size 16 was drawn from each of the three strata. Because the strata in this example differ in size but the same number of samples were selected in each stratum, the density of sites in each stratum varies. When this happens, the sample is said to be *non-proportionally allocated*. If the density of sites is equal in all strata, or is proportional to stratun size, the sample is said to have *proportional allocation*. In Fig. 5.3 the left-hand stratum is twice as big as the other two. Proportional allocation would have been achieved if sample sizes were $36 \div 2 = 18$ in the large stratum and $36 \div 4 = 9$ in each of the smaller two strata.

Proportional allocation can be important if *post*-stratification of the sample is anticipated. In stratified sampling, the strata membership of each potential sample unit is determined when the sample frame is constructed and is a permanent part of the sample design. With post-stratification, the population is (re-)stratified for the purposes of estimation. Post-stratification can be a useful and flexible approach as long as appropriate analyses are implemented (e.g. Särndal *et al*. 1992, Chapter 7). If the sample design used stratification with proportional allocation, post-stratification is easy because all sites are included with equal probability. If allocation is not proportional, post-stratification introduces unequal probabilities into the analysis, which is valid statistically but complicates the analysis.

Stratified samples are popular designs for ecological studies. In ecological studies, strata are usually geographic regions of interest or are based on the perceived level of a target variable. For example, a study of fish populations in rivers might stratify based on elevation. Another fish study might attempt to classify stream segments into "low productivity", "medium productivity", and "high productivity" strata and draw samples from each.

Theoretically, stratification can be used to reduce the variation or uncertainty of final parameter estimates relative to not stratifying. Technically, this is correct. If average variation of a variable within strata is substantially less than variation of the variable's average among strata, population level parameter estimates will be more precise under stratified sampling than not. However, if variation within and between strata does not follow this pattern, for example when strata are chosen poorly, it is possible for stratification to inflate uncertainty in the final parameter estimates relative to not stratifying. It is this potential for harm and the inability to correctly stratify on all variables at once that makes variance reduction through stratification nearly impossible in real studies. Therefore, stratification as a means to reduce variation is not usually an achievable or worthwhile goal for monitoring studies (Box 5.2).

Stratification, however, is very useful when a guaranteed number of sample sites are required in each stratum. A guaranteed number of sites in certain areas is desirable when separate estimates are required in each stratum. For example, estimates may be required for each state in a study spanning multiple states. It makes sense in this case to stratify on state boundaries to guarantee adequate sample size from each state.

When constructing strata, it is best to base strata boundaries on factors that either do not change or change very slowly (see also Chapters 3, 22). Examples of good strata boundaries are geographic (mountain ranges, elevation, etc.) or political boundaries (states, counties, etc.). Examples of poor strata boundaries are those based on habitat classifications or distance from roads because these criteria produce boundaries that have the potential to change rapidly. The definition of "good" habitat can change when new studies are published; new roads can be built, and old roads can be closed. It is desirable to define non-changing strata because boundaries and site inclusion probabilities are forever fixed once samples are drawn. If a particular site is found to be misclassified into the wrong stratun, the site cannot be moved to the correct stratun (Chapter 3). In this case, it is possible to post-stratify the sample to correct the misclassifications, but this introduces unequal inclusion probabilities and complicates the analysis unless proportional allocation was used.

Cluster sampling

Cluster samples are special cases of two-stage samples that occur when all (i.e. a census of) secondary units are selected from primary units. Like two-stage sampling, cluster sampling draws a sample of primary units at stage 1. At stage 2, however, a cluster design selects all secondary units within each primary unit (e.g. Fig. 5.4).

In ecological studies, cluster sampling can be useful when all sample units of a collection (cluster) are relatively easy to collect. For example, in some situations it may be relatively easy to capture an entire group of gregarious animals. In these cases, it is reasonable to define the group as a cluster, and individuals within the group as the secondary units. Cluster sampling is not popular for sampling geographic locations,

Box 5.2 Common challenges: gray areas in sampling

A common problem in monitoring studies is stratification. When strata are defined by variables like habitat or perceived density of organisms, a stratification scheme designed to reduce variance can actually inflate variance. Changing the stratification scheme after the initial sample has been taken greatly complicates analyses (e.g. long-term trend detection). This author has rarely seen stratification achieve its variance reduction goals, although others have achieved this goal in their particular situations (e.g. Chapter 3). The experiences of this author indicate that too little is usually known about the response of interest to construct proper strata. When this is the case, the primary utility of stratification is to provide separate estimates of parameters in each stratum.

A gray area surrounding the design of spatial samples is whether it is useful to intentionally implement a design that yields unequal probabilities of inclusion. Unequal probabilities of inclusion are a stochastic form of stratification, and most of the same hazards are present. While it is technically possible to reduce the variance of estimates through judicious use of variable inclusion probabilities, this author suspects that this goal will rarely be achieved because too little is known about the relationship between inclusion probabilities and the variable of interest. Moreover, up-weighting or down-weighting inclusion of a certain site may be appropriate for one variable, but may not be appropriate for others. Like stratification, the best use of differential weighting is to assure a certain number of samples with certain characteristics (e.g. from a small portion of the study area of special importance that would be underrepresented with equal probability sampling). Otherwise, the safest and most robust designs for monitoring are those that give all sites equal probability of inclusion in the sample.

Lack of access to sample sites is a common problem in monitoring studies. When access is denied, the denial should be noted and another site selected for inclusion. If possible, the additional sites should be included in a way that maintains spatial balance of the sample. At analysis time, researchers are faced with two paths forward. First, researchers can exclude the denied areas from the study area, assuming the locations of denied areas are known, and make inference only to the accessible portions of the study. The second path is to seek a variable that is known on the denied areas (e.g. a variable contained in a public map) and construct a correlative model between it and the variable of interest using data from the accessible areas. If such a model is constructed, values for the variable of interest in the denied area can be predicted using the correlative model. Neither one of these paths is extremely satisfying, and researchers should do their best to obtain access or to exclude known difficult-to-access areas a priori.

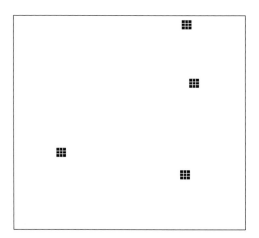

Figure 5.4 Example *cluster sample* of size 4. Each primary unit was defined to be a 3 × 3 block of 9 secondary units. A total of 9 × 4 = 36 sites were chosen.

except for the special case of *systematic* sampling (next section), unless the number of clusters sampled is high.

Monitoring designs

While there are many designs generally applicable to research studies, and even more methods for improving or optimizing the spatial design for a single parameter (see Müller 2007), there are relatively few spatial designs that perform well in general purpose monitoring studies. The primary characteristic of the few designs that perform well in monitoring studies is that they ensure broad spatial coverage of sites (Kenkel *et al.* 1989, Nicholls 1989, Schreuder *et al.* 1993, Munholland and Borkowski 1996, Stevens and Olsen 2004; Chapter 6). Ensuring broad spatial coverage works well in monitoring studies because spatial variation is often one of two large sources of variation relevant to nearly all environmental monitoring efforts (the other large source is temporal variation).

Of the designs that ensure good spatial coverage, three have the highest likelihood of satisfying all objectives of a real-world monitoring study: *systematic*, *GRS*, and *GRTS*. Both systematic and GRS designs are covered here. The third design, GRTS, is covered at length in Chapter 6. Other designs can work in monitoring studies, but they are usually more difficult to apply in a way that satisfies all objectives.

Systematic sampling in two dimensions

Systematic samples, also called *grid* samples, are special cases of cluster sampling wherein each "cluster" is spread out over the entire study area (i.e. secondary units are not contiguous) and only one "cluster" is sampled. Systematic samples derive their

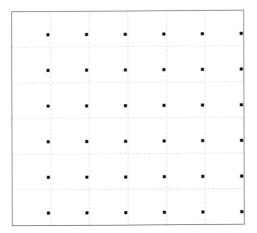

Figure 5.5 Example of a 6 × 6 *systematic sample* with random start. Dashed lines indicate *blocks* of secondary units (sites) from which a single unit is selected. The set of sites in the same relative position in each block constitutes a *cluster*.

name from the fact that sample sites are systematically placed. That is, after an initial random start, systematic samples choose secondary units in the same relative location within groups or *blocks* of units. The size and shape of each block is called the design's *step size* or *grid spacing* and defines the number of samples drawn (i.e. sample size). In general, systematic samples do not yield a fixed number of sites per draw. That is, the "clusters" do not all contain the same number of sites unless the study area's size is evenly divisible by the intended sample size (in all directions). While varying sample sizes can pose problems logistically, it is known that varying cluster sizes do not pose any statistical problems (Särndal *et al.* 1992). This section discusses two-dimensional systematic samples. One-dimensional systematic samples are possible and should be implemented as GRS, discussed in the following section.

A systematic sample of a rectangular geographic region is shown in Fig. 5.5. The dashed lines show boundaries of blocks of secondary units from which a single unit is selected. The width and height of these blocks is the horizontal and vertical step size. The size of blocks is determined by the desired sample size and dimensions of the study area. Figures 5.5 and 5.6 each show samples of 36 sites from a study area divided into 36 blocks, but the block configurations are different in these two examples. Once block sizes are determined, a random location within the first block is generated. The site at that location is then selected and the remainder of the sample is filled with sites in the same relative position inside other blocks.

If the study area is an irregular polygon, a systematic sample can be drawn by laying the randomly placed and usually randomly oriented grid of sample sites over an outline of the study area and choosing grid points that fall inside. This process of intersecting a randomized grid and a non-rectangular study area produces unequal sample sizes because different numbers of sites fall in the study area depending on the randomization. Other

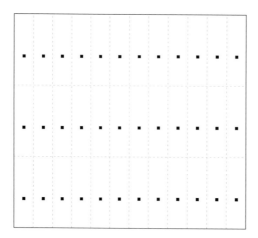

Figure 5.6 Example of a 3 × 12 *systematic sample* with random start. Dashed lines indicate *blocks* of secondary units (sites) from which a single unit is selected. The set of sites in the same relative position in each block constitutes a *cluster*.

than logistical difficulties, varying numbers of sample sites do not pose any problems for analysis. Under systematic sampling, inclusion of units is not independent. Given the identity of one unit in the sample, the identities of all other units in (or not in) the sample is known.

Systematic samples are excellent designs for monitoring studies. They are useful for monitoring because the realized sample has extremely good spatial balance. There are no clumps of sample sites or holes. However, systematic sampling suffers because realized sample size is usually variable and it can be difficult to replace sample sites that are inaccessible or inappropriate. If a site is inaccessible or inappropriate, and researchers wish to replace it with another, it is difficult to decide where the additional, or "off grid", site(s) should be placed. Replacement is difficult because any placement of less than a full grid reduces spatial balance of the sample. Statistically, replacement of systematically chosen sites does not cause any more inferential problems than replacement of sites under other designs.

Spatial balance can also suffer if budget and time run out before the systematic sample is completed. If budget or time constraints prevent data from being collected from all sample sites selected by a systematic design, or some other designs, it is possible for large sections of the study area to go unsampled unless researchers purposefully order site visits to preserve spatial balance for the portion of the sample visited up to that point. Guarding against spatial imbalance under foreshortened field seasons means researchers cannot visit sites in a systematic order, and this reduces logistical efficiency. With GRTS designs, sites can be added or subtracted without compromising the spatial balance of the sample (Chapter 6), assuming sites are visited in the sequence produced by the reverse hierarchical ordering algorithm, but this also means that sites cannot be visited in an arbitrary or most logistically efficient order.

General random samples of one dimension

General random samples encompass multiple approaches to sampling from one-dimensional sample frames. The algorithm presented in this section can be used to draw (weighted or unweighted, ordered or unordered) simple random samples or systematic samples over one-dimensional resources. Samples drawn by the algorithm have fixed size. While GRS will, in general, be difficult to implement on two-dimensional resources, one-dimensional resources encompass a large number of natural resource types that include entities such as streams, beaches, lake edges, forest edges, unordered sample frames, etc. Both finite and infinite one-dimensional resources (such as points on a line) can be sampled using the GRS algorithm, assuming the infinite resources can be discretized into a finite list. To use this algorithm on a two-dimensional resource, locations must be mapped from 2D to 1D in a meaningful way, which is exactly the main contribution of the GRTS algorithm presented in Chapter 6.

Assume that N sampling units exist in a population, and that a fixed size sample of n units is desired. Assume that it is desirable for first-order inclusion probability for sample unit i to be proportional to the values x_i. The vector of x_i values, i.e. $\mathbf{x}$, could be constant – i.e. all values identical – or could contain values such as size of the unit, distance from a particular location, or anticipated level of a target variable. When $\mathbf{x}$ is constant, the design is said to be equi-probable. When $\mathbf{x}$ is not constant, the sample is said to be drawn with unequal probabilities. As stated earlier, equi-probable designs are generally more appropriate for monitoring because equi-probable designs provide flexibility and adequate data for unenvisioned estimation tasks, but either could be used.

GRS algorithm
Drawing a GRS involves the following four primary steps:

(i) *Scale vector* $\mathbf{x}$. Scale values in $\mathbf{x}$ to sum to n and use the result as first-order inclusion probabilities. That is, set the first-order inclusion probability vector $\boldsymbol{\pi}$ to

$$\boldsymbol{\pi} = n\frac{\mathbf{x}}{J'\mathbf{x}} \tag{5.1}$$

where J is a vector of 1's. In non-vector form, the ith element of $\boldsymbol{\pi}$ is

$$\pi_i = n\frac{x_i}{\displaystyle\sum_{j=1}^{N} x_j}.$$

(ii) *Check for inclusion probabilities > 1 and re-scale if necessary*. If all $\pi_i \le 1$, this step can be skipped. Otherwise, set the indicator vector $\mathbf{e}$ to the positions of those $\pi_i > 1$. For example, if the 3rd and 12th elements of $\boldsymbol{\pi}$ are greater than 1, set the vector $\mathbf{e} = [3, 12]$. Let $\boldsymbol{\pi}_{\mathbf{e}}$ be the set of elements of $\boldsymbol{\pi}$ that are >1 and $\boldsymbol{\pi}_{\mathbf{e}'}$ be the set of elements of $\boldsymbol{\pi}$ that are ≤ 1. Let $\mathbf{x_e}$ be the elements of $\mathbf{x}$ that produced $\pi_i > 1$, and $\mathbf{x_{e'}}$ be the elements of $\mathbf{x}$ that produced $\pi_i \le 1$. Let $\|\mathbf{e}\|$ be the number of elements in $\mathbf{e}$. Modify $\boldsymbol{\pi}$ by re-setting $\pi_{\mathbf{e}} = 1$, and re-scaling the remaining elements:

$$\boldsymbol{\pi}_{\mathbf{e}'} = (n - \|\mathbf{e}\|)\frac{\mathbf{x_{e'}}}{J'\mathbf{x_{e'}}}. \tag{5.2}$$

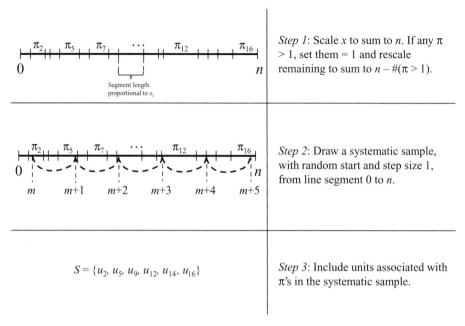

Figure 5.7 Heuristic pictorial representation of drawing a general random sample (GRS) of size $n = 6$ from a population of size $N = 16$.

Repeat this step until all elements of π are ≤ 1. At the end of this step $\sum \pi_i = n$ and $0 < \pi_i \leq 1$ for all i.

(iii) *Choose a random start.* Draw a random number between 0 and 1. Let this random number be m. That is, set the variable m to a single random deviate from the Uniform(0, 1) distribution.

(iv) *Draw a systematic sample from π using random start m.* Select, as the jth element of the sample, the ith population unit if,

$$\sum_{k=0}^{i-1} \pi_k \leq (m + j - 1) < \sum_{k=1}^{i} \pi_k \tag{5.3}$$

where $\pi_0 = 0$. The systematic sample has step size 1.

A pictorial presentation of the GRS algorithm is given in Fig. 5.7. Program R code (R Development Core Team 2011) for drawing a GRS appears in Appendix 5.1 and as an online supplement.

Note that the GRS algorithm given above is silent regarding the ordering of units in the population. Because they are parallel vectors, the ordering of units in the population effects the ordering of elements in π. The ordering of π affects the partial sums in the last step of the algorithm, thus giving the GRS different properties. In fact, the GRTS sampling process orders π using a two-dimensional hierarchical randomization such that units close in space tend to be close in π. If all elements in $\mathbf{x}$ are equal, and the order of units in the population is *randomized* prior to selecting the GRS, the resulting sample is a *simple random* sample. If all elements in **x** are equal, but the order of units in the population is fixed (not randomized), the resulting sample is a fixed-size *systematic* sample. This type of 1D systematic sample is appropriate when it is desirable to order units according to some auxiliary variable, such as distance from a geographic location, elevation, easting, or northing. For example, it might be desirable to order stream segments by river mile (distance from the mouth) to assure that sample sites are located in all parts of the river. It might be desirable to order a sample frame of quadrats by elevation to ensure that samples are collected across a range of elevations. If elements in **x** are not equal, and the order of units is randomized, the resulting selection is a simple random sample with probability of inclusion proportional to **x**. If elements in **x** are not equal, but the order of units is fixed, the resulting sample is a systematic sample with probability of inclusion proportional to **x**.

Inclusion probabilities for GRS

At analysis time, it is important to consider the properties of the sampling design under which data were collected (see also Chapters 6, 14). Principally, this involves computing or estimating the design's first-order and second-order inclusion probabilities (third- and higher-order inclusion probabilities are usually ignored). First-order and second-order inclusion probabilities are important because they are integral to Horvitz–Thompson (Särndal *et al.* 1992) and other estimation techniques. For example, to properly estimate the total or mean of a parameter, the reciprocals of first-order inclusion probabilities are used to weight observations (Chapter 14). Units with small probability of inclusion receive more weight than those with larger inclusion probabilities in the Horvitz–Thompson estimator. Second-order inclusion probabilities influence the variance of the Horvitz–Thompson estimator, and are integral to variance estimation (see Särndal *et al.* 1992). To estimate variance, the reciprocals of second-order weights are used to weight functions of sample unit pairs. Other forms of estimation, such as regression and trend estimation, also require knowledge of first- and second-order inclusion probabilities.

Mathematically, first-order inclusion probabilities are defined as

$$\pi_i = Pr(u_i \in S) \tag{5.4}$$

where u_i is the ith unit in the population and S is the realized sample. Second-order inclusion probabilities are the probabilities of including *pairs* of units in the realized sample. The probability of including both the ith and jth units in the sample is defined as

$$\pi_{ij} = Pr(u_i \in S \cap u_j \in S). \tag{5.5}$$

First-order inclusion probabilities can vary among units in the population, but must be > 0 for all units. If $\pi_i = 0$, unit i is not in the population by definition. Second-order inclusion probabilities can also vary among pairs of units, and can equal 0 for some pairs. For example, in systematic samples, the probability of including a unit and its neighbor is 0 unless step size is very small.

The first-order and second-order inclusion probabilities for most spatial designs are either known or easy to compute. For example, the first- and second-order inclusion probabilities of a simple random sample are,

$$\pi_i = \frac{n}{N},$$
$$\pi_{ij} = \frac{n(n-1)}{N(N-1)}. \tag{5.6}$$

With cluster sampling, the second-order inclusion probability for two units in the same cluster is the probability of selecting their cluster at level 1 of the design.

For the GRS algorithm, first-order inclusion probabilities are easy to obtain. First-order GRS inclusion probabilities are the scaled interval lengths used in the final systematic sampling step. That is, the first-order GRS probability of unit i being selected is its associated π_i. The second-order inclusion probabilities for a GRS are more difficult to obtain. In theory, second-order GRS inclusion probabilities are computable. In practice, however, the computations required are too time-consuming to be useful. Consequently, second-order inclusion probabilities are usually approximated, either by formula or by simulation. Stevens (1997) gives formulas for special cases of GRTS samples that can also be used to approximate second-order inclusion probabilities for GRS samples. Second-order inclusion probabilities in other cases will usually be approximated by simulation. The simulation approach is conceptually simple, but requires an exceptionally large amount of computer resources for big populations. A "brute force" simulation to approximate second-order inclusion probabilities replicates the GRS algorithm and tallies the number of times each pair of units occurs in the sample. R code to implement the "brute force" simulation appears in Appendix 5.1 and as an online supplement. This R code produces a $N \times N$ matrix containing all possible second-order inclusion probabilities, and will work for relatively small populations. Large populations may require a large amount of memory, specialized programming, simplifying assumptions, or use of parallel processing to speed the calculations. In many applications, it is only necessary to compute inclusion probabilities for units in the sample. In these cases, it is possible to reduce memory requirements by storing the $n \times n$ matrix of realized second-order inclusion probabilities instead of the full $N \times N$ matrix of theoretical probabilities.

Future research and development

A primary goal of the spatial design is to maintain good spatial balance of sample sites. When sample units are inaccessible or inappropriate and must be replaced, it can be challenging to maintain spatial balance. GRTS designs allow units to be added in a way that maintains spatial balance, but other designs do not. Because systematic sampling is popular and will continue to be used in the foreseeable future, recommendations for maintaining the spatial balance of systematic samples when sites are replaced would of practical value in many studies.

In many studies, the amount of time field crews spend collecting data at a particular sample site is either highly variable or unknown prior to initiation of the study. Field sampling budgets are usually fixed and a highly variable field sampling effort means that the number of sites that can be measured is variable. In these cases, maintenance of spatial balance and field efficiency is challenging under any design. Maintenance of spatial balance in foreshortened field seasons may dictate that field crews follow some type of randomized visit plan, but such a visit plan can mean increased travel costs as crews visit sites in the "random" order. From a field perspective, it would be more efficient if crews could transit to the next nearest neighboring site once a particular site is finished. A sample design that maintains spatial balance when field seasons are foreshortened yet reduces travel costs relative to the designs available today would be most welcome.

With increases in field data collection technology (automated collection, remote collection, video, audio, etc.) more types of environments will be sampled in the future. In particular, the desire and ability to sample environments that are three-dimensional (or more) will increase. Robust, spatially balanced, and easy to implement designs that select sites in three or more dimensions are currently needed. More generally, sampling designs that sample both two- or three-dimensional space and time simultaneously in a balanced manner potentially could be valuable. The panel rotation concepts discussed in Chapter 7 are adequate and appropriate for current long-term monitoring studies, but an explicit sampling design for time would make rotation of effort more statistically rigorous.

Summary

This chapter discusses selection of spatial samples, and some popular spatial sampling designs. Concepts included the definition of *scientific* and *non-scientific* designs. Scientific designs are those that draw probability samples, while non-scientific designs are those that do not. A case was made that environmental studies, and monitoring studies in particular, should always use scientific designs. Another concept discussed was the distinction between *research* and *monitoring* studies. This distinction was made because the spatial design needs of both studies are generally different. The spatial design of research studies must usually provide the highest possible statistical precision for a particular parameter or hypothesis. The spatial design of monitoring studies must provide adequate data to address a wide range of anticipated and unanticipated estimation tasks over a long time frame. Four spatial designs generally appropriate for research studies were described. These designs included *simple random*, *two-stage*, *stratified*, and *cluster sampling*. Two spatial designs generally appropriate for monitoring studies were described. These designs were *systematic sampling* for two-dimensional resources and *general random sampling* for one-dimensional resources. A third highly useful monitoring design is discussed in the next chapter.

Appendix 5.1. R code for GRS sampling

This appendix contains two R functions, one to draw a GRS, the other to approximate second-order inclusion probabilities for a GRS. Inputs, outputs, and example calls are contained in the header comments of both functions.

General Random Sample function

```
F.grs <- function(id, n, x=NULL, sort.by=NULL, seed=NULL){
#
#    F.grs - R function for drawing a General Random Sample.
#
#    Inputs:
#    id= either a scalar or vector. If id is a scalar, it is
#        assumed to be population size (N), and a equal
#        probability sample of size n is drawn from a population
#        of size N. The ID's in this case are 1:N. If id is a
#        vector, the values in ID are used as ID values for
#        units in the population. Population size is the length
#        of id in this case.
#    n = scalar, the desired sample size.
#    x = a vector of weights to which inclusion probabilities are
#        proportional. Length of x must be same as length of id. Or,
#        x=NULL if id is a scalar.
#    sort.by = a vector parallel to id and x used to sort units
#        in the population. If sort.by is null, no sorting is
#        performed.
#    seed = an integer scalar to use for the initial random seed
#        If missing, set.seed is not called.
#
#    Output:
#    A data frame with dimension nx2 containing the general random
#    sample. The data frame contains the variables $id = ID's of
#    units in the sample and $pi = inclusion probability for units
#    in the sample.
#
#    Examples:
#    s <- F.grs(100,20,sort.by=runif(100)) # draw a simple random
#                                          # sample of 20 from 100
#    s <- F.grs(100,20) # draw simple systematic sample of 20 from 100
#    s <- F.grs(1:100,20,1:100)  # draw systematic sample of units
#                                # with ID's 1 to 100 and probabilities
#                                # proportional to ID values
```

```
#    s <- F.grs(1:100,20,1:100,runif(100)) # draw randomized sample
#                                          # of with probability
#                                          # proportional to ID values

    if( length(id)<= 1){
        id <- 1:id
        x <- rep(1,length(id))
    }
    if( length(x)<= 0 & is.null(x)){
        x <- rep(1, length(id))
    }
    if( !is.null(sort.by) ){
        ind <- order(sort.by)
        id <- id[ind]
        x <- x[ind]
    }
    if( !is.null(seed) ){
        set.seed( seed )
    }

    p <- n * x / sum(x)
    while( any( p > 1) ){
        gt.1 <- p > 1
        p[gt.1] <- 1
        p[!gt.1] <- (n - sum(gt.1)) * x[!gt.1] / sum(x[!gt.1])
    }

    x <- cumsum(p)
    m <- seq(runif(1), n, by=1)
    ind <- findInterval(m,x)+1
    s <- id[ ind ]
    p.s <- p[ ind ]

    ans <- data.frame( id = s, inclusion.prob = p.s )
    ans
}
```

Function to generate 2nd-order inclusion probabilities

```
F.2nd.order <- function(id, n, x=NULL, randomize=FALSE, seed=NULL, reps=1000){
#
#    F.2nd.order - R function to approximate 2nd order inclusion
#    probabilities of a General Random Sample.
```

```
#
#    Inputs:
#    id= either a scalar or vector. If id is a scalar, it is
#         assumed to be population size (N), and a equal
#         probability sample of size n is drawn from a population
#         of size N. The ID's in this case are 1:N. If id is a
#         vector, the values in ID are used as ID values for
#         units in the population. Population size is the length
#         of id in this case.
#    n = scalar, the desired sample size.
#    x = a vector of weights to which inclusion probabilities are proportional.
#         Length of x must be same as length of id. Or,
#         x=NULL if id is a scalar.
#    randomize = a boolean value specifying whether or not
#         to randomize the population before drawing the
#         sample. TRUE = randomize, FALSE = do not randomize.
#    seed = an integer scalar to use for the initial random seed
#    reps = scalar specifying the number of iterations to
#         do in the simulation.
#
#    Output:
#    A symmetric matrix of size NxN containing 2nd order inclusion
#    probabilities. Cell (i,j) is the approximate probability
#    that unit i and unit j both appear in the sample.
#
#    Note: Inputs id, n, and x are exactly the same as the
#    inputs to F.grs. To implement a design ordered by another
#    variable, sort the vectors id and x before calling
#    this routine.
#
#    Examples:
#    pi2 <- F.2nd.order(100,20,randomize=T)# 2nd order inclusion
#                                          # probs for simple
#                                          # random sample.
#    pi2 <- F.2nd.order(100,20)  # 2nd order probs for systematic
#    pi2 <- F.2nd.order(1:100,20,1:100) # 2nd order probs for
#                                       # unequal prob systematic
#    pi2 <- F.2nd.order(1:100,20,1:100,T)  # 2nd order probs
#                                          # for randomized unequal
#                                          # prob design.
    if( length(id)<= 1){
        id <- 1:id
        x <- rep(1,length(id))
    }
```

```
N <- length(id)
pi2 <- matrix(0,nrow=N,ncol=N)
id.ord <- sort(id)

for( i in 1:reps ){
    if( randomize ){
        s <- F.grs(id, n, x, sort.by=runif(N))
    } else {
        s <- F.grs(id, n, x)
    }
    ind <- id.ord %in% s$id
    ind <- matrix(ind, ncol=1) %*% matrix(ind, nrow=1)
    pi2 <- pi2 + ind
}

pi2 <- pi2 / reps
pi2
}
```

6 Spatially balanced survey designs for natural resources

Anthony R. Olsen, Thomas M. Kincaid, and Quinn Payton

Introduction

A common objective for a monitoring program is to characterize an environmental resource based on inference from the sites selected to be monitored to the entire target population. The scale of monitoring ranges from local studies, to regional monitoring, to nationwide monitoring programs. Rarely can these monitoring efforts monitor at all locations, or sites, within the study region. Consequently, a major consideration is how to select representative sites from which it is possible to make inferences to the entire study region.

In addition to the different spatial scales of interest across monitoring efforts, these studies may focus on different environmental resources. A state or province may be interested in all small lakes (e.g. < 10 ha) with the objective of classifying the lakes as meeting designated uses (i.e. having values of designated water quality attributes that do not exceed a specified threshold), partially meeting designated uses, or not meeting designated uses. In this case, the elements of the environmental resource are individual lakes and site selection is based on selecting a subset of lakes from the target population. Alternatively, a state may be interested in all perennial streams and rivers within the state to determine the total stream length that meets a nutrient criterion. In this case, the elements of the environmental resource are all possible locations on the stream and river network within the state and site selection is based on selecting sites on the stream network to be monitored. The stream network is the target population and is viewed as a continuous linear network. Finally, a state may be interested in monitoring a single large estuary within the state (e.g. Puget Sound in Washington State, USA) to determine the proportion of the estuarine area that has sediment contamination exceeding a criteria. In this case, the environmental resource, i.e. the target population, is an area and sites are selected from all possible locations within the estuarine surface area. Similar examples can be given for terrestrial environmental resources.

Some examples of US national-scale environmental resource monitoring programs are:

Design and Analysis of Long-term Ecological Monitoring Studies, ed. R.A. Gitzen, J.J. Millspaugh, A.B. Cooper, and D.S. Licht. Published by Cambridge University Press. © Cambridge University Press 2012.

- National Aquatic Resource Surveys (NARS) conducted by the US Environmental Protection Agency and states to monitor the condition of lakes, reservoirs, streams, rivers, wetlands, estuaries, and near shore coastal waters (http://water.epa.gov/type/watersheds/monitoring/nationalsurveys.cfm).
- Forest Inventory and Analysis (FIA) program of the US Forest Service focusing on the Nation's forests (http://fia.fs.fed.us/) (discussed in Chapter 2).
- National Wetland Status and Trends program of the US Fish and Wildlife Service monitors wetland acreage changes in the conterminous United States (http://www.fws.gov/wetlands/StatusAndTrends/index.html).
- Vital Signs Monitoring program of the National Park Service to track the overall condition of natural resources in parks (http://science.nature.nps.gov/im/) (Chapter 22).

In all of these cases and in all other environmental resource monitoring programs, an important feature of their respective environmental resources is that they are distributed over geographic space. When probability sampling designs explicitly use spatial location in the selection of the sites, we define the resulting designs as "spatially explicit", or simply "spatial", survey designs. These designs must be able to address the large variation in scale and the three fundamental types of target populations – those that consist of points, lines, or polygons.

This chapter first discusses (i) the concept of representative samples in this spatial context, building on the broad overview of spatial sampling provided in Chapter 5. Our chapter then (ii) describes a specific type of spatially balanced survey design, generalized random tessellation stratified (GRTS) survey design (Stevens and Olsen 2004); (iii) introduces a definition for spatial balance and proposes spatial balance metrics; and (iv) compares the spatial balance properties of simple random sampling to a GRTS survey design, using these spatial balance metrics.

Representative sample and representative sampling process

The scientific literature and specifically the environmental monitoring literature include numerous uses of the expression "representative sample", or expressions similar to "the sample is representative of the population". In many cases the meaning of these expressions is not explained and it is left up to the reader to interpret what the authors meant. Collectively, scientists have a general sense of what is meant when they read "representative sample", although on closer examination the term is found to have multiple interpretations. In a series of three landmark papers in 1979, Kruskal and Mosteller (1979a, 1979b, and 1979c) reviewed the non-scientific writing and the scientific (excluding statistics) and statistical literature to classify and illustrate the various meanings of "representative sample" and "representative sampling". They classified the meanings for "representative sampling" in the statistical literature as follows:

- General acclaim for data.
- Absence of selective forces.

- **Miniature of the population.**
- Typical or ideal case(s).
- **Coverage of the population.**
- Vague term, to be made precise.
- **Representative sampling as a specific sampling method.**
- Representative sampling as permitting good estimation.
- Representative sampling as good enough for a particular purpose.

The first six meanings also occur in non-scientific and scientific (excluding statistics) literature. We focus on the meanings (in bold above) that are particularly relevant to monitoring design.

What is a "miniature of the population?" Conceptually, it is a small portion of the population that arises from a perfect mixing of the population. An example is a sample of blood taken during a physical examination to ascertain particular chemical concentrations in a person's blood. If the population had two categorical variables, such as sex, two age groups, and two income groups, then a miniature of the population would have the same percent of individuals in the $2 \times 2 \times 2 = 8$ categories as the entire population.

A representative sample as a miniature of the population is appealing in monitoring because the characteristics of the sample then would apply to the entire resource being monitored. Constructing a sample that is a miniature of an environmental resource is extremely difficult unless the number of factors involved is very small. For example, in monitoring lakes, factors that may be considered could be five lake area categories, two origin categories (natural versus human-made), and three elevation categories. In this example, all possible combinations of these factors generate 30 categories, assuming all combinations are present in the target population. A miniature of the population would require at least a sample size of 30 in order to just have a single site in each combination. A single site in each combination is unlikely to be a miniature of the population since the number of lakes in each combination is unlikely to be the same. Hence, the number of sampled sites would need to be proportional to the number of lakes in each combination which may require a very large sample size if the number of lakes in the categories are very different. Even if the sample size is sufficient to select the number of lakes required for achieving proportional allocation of lakes in the sample, a procedure for selecting lakes within each category is required.

One approach would be to use professional judgment to select representative sites within each combination. Generally samples selected in this way fall short of being miniatures due to bias in professional judgment occurring (known or unknown), i.e. selection bias (Chapters 2, 5, 11). In the context of representative sampling as a miniature of the population Kruskal and Mosteller (1979c: 251) quote the following definition of representative sample by Stephan and McCarthy (1958: 31–32): "A *representative sample* is a sample which for a specified set of variables, resembles the population. . . . [in that] certain specified analyses. . . . (computation of means, standard deviations, etc. . . .) will yield results. . . . within acceptable limits set about the corresponding population values, except that. . . . [rarely] the results will fall outside the limits . . . the mere statement or claim that a sample is representative of a population tells us nothing." A representative sample that is a miniature of the population is useful when inferences to the entire population are the focus. Thus the issue is: how do we know that the sites are a miniature of the population?

Coverage of the population is the concept that a representative sample is best achieved by getting the sample from as many parts, or partitions, of the population as possible. In that sense it is similar to the concept of a miniature of the population. The difference is that coverage of the population does not imply that the sample has the same relative frequencies in the sample as the population as does the "miniature of the population" concept of representative sampling. This type of representative sampling may be useful in the context of building models where the partitions ensure that the range of each variable being considered in the model is in the sample. When the monitoring objective is inference to the entire population, a representative sample that only provides coverage of the population is not sufficient.

A third meaning is representative sampling as a specific sampling method. Note that this meaning focuses on a process to generate a sample – not to a specific sample. Fundamental to this meaning is that each element of the population has a chance of appearing in any realization of the representative sampling process. The simplest representative sampling process is a simple random sample where each element of the population has an equal chance of being selected. For example, in a study of all lakes (N of them) in the State of Oregon, USA, each lake would have a chance of being selected as one of the lakes in a sample of size n. In this case each lake has a probability of n/N of being selected in the sample. This is a simple example of what is called a probability survey design (see Chapters 2, 5). Kruskal and Mosteller (1979c: 257) recognize that while simple random samples are intuitively representative in a long-run average sense, specific realizations can be "wildly unrepresentative". Note that the likelihood of "wild realizations" decreases as the sample size becomes larger. For example, a single realization of a simple random sample of lakes in Oregon may lead to all lakes being located in western Oregon and include only lakes < 10 ha. It is unlikely that such a sample would be accepted as a representative sample of lakes in Oregon. That is, a representative sampling process does not necessarily produce a representative sample.

The appeal of a representative sampling process is that statistical inferences are then available that allow quantification of the uncertainty associated with any estimates (Chapter 2). Snedecor and Cochran (1967) define probability sampling (i.e. a representative sampling process) as having the properties that (i) every unit of the population has a known probability (> 0) of being included in the sample; (ii) the sample is selected by some method of random selection, or systematic selection or systematic selection with a random start, consistent with probabilities; and (iii) the probabilities of selection (weighting factors) are used in making inferences from the sample to the population in question. Note that representative sampling processes may purposely select samples that are not miniatures of the population (e.g. by using stratification) and then account for this when statistical estimation is completed. In this case of stratification, the representative sampling process applied in each stratum is intended to result in a representative sample that is a miniature of the population within the stratum.

Obtaining a representative sample in environmental monitoring

In monitoring designs, what can be done when implementing a representative sampling process to improve the chance that each realization of the process results in a representative sample, i.e. is a miniature of the population? One approach is to stratify by characteristics of the population, then sample each stratum proportional to its occurrence in the population and within each stratum select sites with equal probability. For example, lakes may be stratified by four classes based on lake surface area, four categories of ecoregions, three elevation categories, and as natural versus anthropogenic lakes. The stratification guarantees that the sample will include samples from each of the strata. The proportional sampling ensures that each stratum is sampled in proportion to its occurrence in the population. The equal probability of selection reduces the chance of selection bias within a stratum. However, stratification requires information upon which the strata can be defined for the entire population. Such information may not be readily available, or the number of strata desired may make it impossible to proportionally allocate the sample to the strata given the sample size available. For example, natural versus anthropogenic lake information is typically not available. For the example, a total of 96 strata are required which will make it impossible to proportionally allocate a sample size of 100 to the strata.

Another approach is to construct strata that partition the population spatially so that the sample is a miniature of the population spatially. Constructing explicit spatial strata and sampling them proportional to the number of lakes in each spatial stratum also leads to the same problem when the number of spatial strata is large. An alternative way to use spatial strata underlies the spatial survey design approach described in this chapter.

Our view is that environmental monitoring designs should incorporate the concept of a representative sample being a miniature of the population and that the representative sample should be the result of a representative sampling process (Box 6.1). In addition, since environmental resources are intrinsically spatial in nature, monitoring designs should explicitly use spatial location in the selection of the sites. Conceptually, we postulate that spatial distribution is a useful surrogate to using combinations of characteristics of the population to get a representative sample that is a miniature of the population. That is, *representative samples with the same spatial distribution as the population are more likely to be miniatures of the population than samples that do not have the same spatial distribution as the population.* Space is not always sufficient on its own. For example, it is possible that a lake sample that has the same spatial distribution as the population could consist of only small lakes. Our contention is that spatial survey designs are more likely to minimize these occurrences than not. Spatial survey designs incorporate these principles.

To illustrate, assume we have a sample of lakes in Oregon that has the same spatial distribution as all lakes in Oregon. Intuitively, the perception is that if the sample has the same spatial distribution as the population of lakes then the sample is more likely a miniature of the population. Our perception that the sample is a miniature of the lake population is likely increased if we know that the sample is the result of a representative sampling process. The representative sampling process reduces the

---

Box 6.1 Take-home messages for program managers

The "GRTS" spatially explicit survey designs for natural resources were created to address several issues that are commonly faced when a monitoring program based on probability surveys is designed. First, natural resources occur as discrete objects (e.g. whole lakes), linear networks (e.g. streams), or collections of areas (e.g. areas within a collection of habitat patches), represented in Geographic Information Systems (GIS) as collections of points, linear networks, or polygons. The GRTS designs have options for sampling any of these three types. Second, the GRTS spatial survey design process is a representative sampling process that is constrained so that each potential sample generated is a representative sample, a miniature of the natural resource in space. That is, each sample has the same spatial distribution as the natural resource population. While this has significant statistical advantages compared to alternatives such as simple random or systematic sampling, it also has the advantage that when the sample is displayed graphically viewers are likely to accept the sample as being representative – and therefore more likely to accept the value of the estimates of status or trends resulting from the sample. The availability of a software program to create GRTS spatial survey designs makes these designs easily accessible to those designing monitoring programs. Hundreds of such designs have been created at local to national scales and for aquatic and terrestrial natural resources (e.g. Chapters 10, 16). Our recommendation is that GRTS designs should be the default choice whenever probability sampling is used in monitoring programs.

Implementing a GRTS spatial survey design, regardless of the design's complexity, requires that statistical analyses using data from the design incorporate the properties of the design, e.g. stratification or unequal probability of selection. That is, the statistical analysis must match the spatial survey design. This may be a change in culture for an organization using a spatially explicit survey design for the first time. Selecting the design depends on the monitoring program having a set of clearly defined, quantitative objectives (Chapters 2, 3, 18). Developing these objectives is typically the most difficult task in designing the monitoring program. It requires that managers and those developing the monitoring program discuss and agree on exactly what the program will produce.

---

likelihood of selection bias and increases our perception that characteristics of the lakes that are important to the study will be proportionally represented in the sample of lakes.

The concept of incorporating spatial regularity in sampling environmental populations is well established. Systematic sampling using a regular grid (or variations thereof) has been studied and used extensively (Bickford et al. 1963, Olea 1984, Gilbert 1987; Chapter 5). These are common in studies of areal resources such as forests or contaminated areas. Stevens (1997) developed generalizations of grid-based systematic designs, i.e. random tessellation stratified (RTS) designs, and applied them to areal resources. Subsequently, Stevens and Olsen (2004) generalized the RTS designs, making them applicable to point and line environmental resources as well as areal resources. These generalized random tessellation stratified (GRTS) designs are spatial survey designs that explicitly incorporate the concept of spatial balance as a means of increasing the likelihood that each realization of the representative sampling process results in a miniature of the population. A GRTS spatial survey design is a representative sampling process that results in representative samples that are more likely to be miniatures of the population than other common spatial survey designs.

Generalized random tessellation stratified (GRTS) survey designs

Stevens and Olsen (2004) present the theory behind GRTS designs and give an example of its application. A software implementation of the GRTS algorithm [function *grts*()] is available in the "spsurvey" package (Kincaid and Olsen 2012) developed for the R statistical software (R Development Core Team 2011). The "spsurvey" package is available on the R website (http://www.r-project.org/) and the US Environmental Protection Agency's Aquatic Resource Monitoring (ARM) website (http://www.epa.gov/nheerl/arm/). The latter site also includes information on designing monitoring programs for aquatic resources. Our objective in this section is to describe the GRTS algorithm in accessible terms to users. Use of the "spsurvey" *grts*() function is demonstrated in Appendix 6.1. Readers should also consider additional implementation issues discussed in Box 6.2.

Basic GRTS algorithm: GRTS equal probability spatial survey design

To describe the GRTS spatially balanced algorithm, we use a simple illustrative example of selecting a sample of size 8 from a population of 1957 lakes in Oregon. In this example, the primary monitoring objective is to estimate the current status of the lakes in terms of a benthic macroinvertebrate index of biotic integrity (IBI; Stoddard *et al.* 2008) and report on the number or proportion of lakes having specific ranges of the IBI score. Conceptually, each lake could have a benthic macroinvertebrate sample collected for the lake using an appropriate field design, taxonomic identification could be completed, and a value for the index computed for the lake. In this case, lakes are considered a point environmental resource because the objective is to report on the number of lakes in terms of their biological integrity where biological integrity is based on a single lake-wide measure of integrity. Assume the target population is all lakes that are greater than 1 ha, more than 1 m deep, and with at least 10% of their surface as open water. The GRTS algorithm requires a list of all lakes in the target population along with their location as *x-,y*-coordinates, usually in an equal area projection (e.g. Albers) rather than geographic coordinates. We use the lakes within Oregon which are greater than 1 ha in the National Hydrography Dataset (http://nationalmap.gov/) as that list, i.e. the sample frame from which the lakes will be selected.

The core concept of the GRTS algorithm is the creation of spatial strata by constructing a grid that satisfies specific requirements. The following describes what the algorithm and software that implements it does. A map showing the sample frame with a square grid overlaid is shown in Fig. 6.1a. The grid is constructed such that (i) the number

---

Box 6.2 Common challenges: implementing GRTS

The main focus of this chapter is providing the conceptual basis of spatial survey designs and their implementation using the GRTS process applied to natural resources modeled as points, linear networks, or polygons. The implementation using the "spsurvey" package in R requires the preparation of sample frame as a GIS layer which is imported to R as an ESRI shapefile (Appendix 6.1). This requires users to have access to ArcGIS, which is less of an issue than a few years ago. However, the creation of a single GIS layer that has the attributes required for the spatial survey design and identifies the objects in the layer that should be included in the sample frame typically requires more effort than users expect.

   The first major decision is whether to model the natural resource as a point, linear network, or polygons based on the monitoring objectives. For example, the user may have a polygon lake GIS layer but the monitoring objectives may require that lakes be modeled as points. In this case the polygon layer must be converted to a point layer. The coordinate system used by the GIS layer also matters. In most cases, a GIS layer in geographic coordinates (latitude, longitude) is not appropriate for use in spatial survey design. The reason is that the distance represented by a degree of longitude is not the same distance represented by a degree of latitude. Typically, an area preserving projection should be used, e.g. an Albers or UTM projection.

   A monitoring program based on a complex spatial survey design also requires using an appropriate statistical analysis of the monitoring data. Not doing so can result in incorrect estimates. Even when the spatial survey design is a non-stratified, equal probability design, the statistical analysis can be improved by using more complex variance estimation procedures. Since spatially balanced designs reduce the probability of unusual samples, a variance estimator can be defined that is unbiased and that produces smaller estimates than the usual simple random sample variance estimator. Stevens and Olsen (2003) defined this local neighborhood variance estimator and showed that it performs better than alternatives. Unfortunately, the estimator cannot be computed without the use of the R software (or other software that computes generalized inverses for matrices). For spatial survey design data, the "spsurvey" package includes functions that will calculate means, totals, percentiles, and cumulative distribution functions as well as their local neighborhood variances.

---

of rows and columns must be a power of 2, i.e. 2^m where $m > 0$ and in this case is 2, and (ii) it has an extra row and column of empty grid cells at the top and right. The choice of m is discussed later, although it depends on the sample size desired and the spatial distribution of lakes. The user does not need to specify m as the GRTS algorithm computes it. The extra row and column are used to ensure that any two lakes may occur in different cells. This is accomplished by randomly shifting the sample frame by selecting a single random x and random y from uniform distributions over their ranges defined by the boundaries of the lower left cell and moving all lakes (i.e. the sample frame) by adding the random shifts to their coordinates. Figure 6.1b shows the resulting shift. The

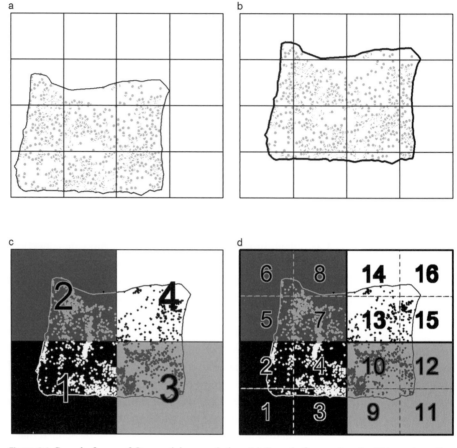

Figure 6.1 Sample frame of Oregon lake population. (a) Sample frame placed within systematic grid with extra row and column of empty grid cells with the grid size determined by the sample size and distribution of the elements to be sampled; (b) sample frame randomly shifted; (c) first-level hierarchical ordering of grid cells ($m = 1$); (d) second-level hierarchical ordering of grid cells ($m = 2$).

reason for requiring that any two lakes may occur in different cells is that this allows the possibility that both lakes could be in a sample that is selected. Technically, this ensures that their joint inclusion probability is greater than zero, a requirement for the Horvitz–Thompson variance estimator (Horvitz and Thompson 1952).

Next, the cells are numbered based on m hierarchical levels associated with the 2^m by 2^m grid of cells. At hierarchical level $m = 1$, the cells are numbered 1 to 4 and for illustration purposes colored from dark to light gray in Fig. 6.1c. At hierarchical level $m = 2$, each of the cells from the previous level ($m = 1$ in this case) are subdivided into four sub-cells. For simplicity the cells are numbered from 1 to 16 (Fig. 6.1d). Technically, a base 4 hierarchical numbering scheme is used to identify the cells and track their hierarchical location. Note that if another hierarchical level were required ($m = 3$), the number of cells would be $2^3 \times 2^3 = 64$. The hierarchical level is chosen so

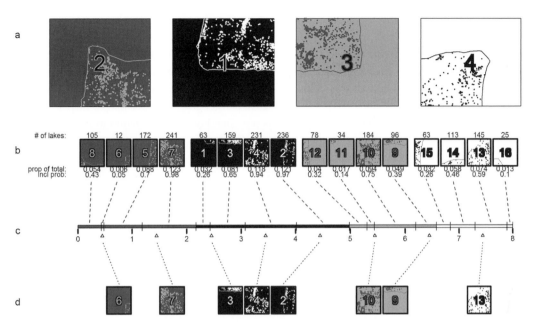

Figure 6.2 GRTS cell selection process. (a) Random ordering of first hierarchical level, $m = 1$.
(b) Random ordering of second hierarchical level, $m = 2$, within first hierarchical level cells. (c,
top) Creation of line with segments proportional to inclusion probabilities, p_i, and a total length
of 8, which equals the desired sample size. (c, bottom) Selecting cells using a systematic sample
with a random start of 0.45, and the triangles indicating the systematic set of points of 0.45, 1.45,
2.45, . . . , 7.45. (d) The 8 cells selected based on their line segment containing one of the
systematic set of points.

that when the sample is selected it will include at most one lake from a cell. Thus the
hierarchical level m depends on the sample size required and the spatial distribution of
the lakes. The number of lakes that occur in a cell depends on the spatial distribution
of lakes which impacts the probability of one or more lakes being selected in a cell. In
our example with a sample size of $n = 8$ and the lake distribution given by the sample
frame, the number of hierarchical levels required is $m = 2$, whereas a sample size of
$n = 50$ using the same sample frame requires $m = 4$.

The next step is to complete a hierarchical randomization process. The process begins
at hierarchical level $m = 1$ by randomly ordering the cell numbers and placing them in
that order on a "line" (Fig. 6.2a). In this case the cell order is 2, 1, 3, and then 4. Then
the next hierarchical level ($m = 2$) is randomly ordered within the previous hierarchical
ordering. That is, the four sub-cells in each of the cells at hierarchical level $m = 1$ are
randomly ordered independent of the other cells at that level (Fig. 6.2b). In this case,
level $m = 1$ cell 2 has randomly ordered sub-cells 8, 6, 5, and 7; cell 1 has randomly
ordered sub-cells 1, 3, 4, and 2; cell 3 has randomly ordered sub-cells 12, 11, 10, and 9;
and cell 4 has randomly ordered sub-cells 15, 14, 13, and 16. These randomly ordered
cells are placed on a "line" (Fig. 6.2c). Note that this hierarchical ordering process places
the cells on the line so that portions of the state that were together at the first level remain

close together at the next level, as illustrated by the gray-scale color coding. If a third hierarchical level were required, the same process would be repeated for each of the 16 sub-cells in the second row and similarly for any additional hierarchical level. The basic idea of this hierarchical randomization process is to map two-dimensional space onto a line of cells where cells that are next to each other on the line tend to be next to each other in the original two-dimensional geographic space and at the same time introduce as much random ordering as possible. For an introduction to this concept of mapping a sample frame to a "line", see Chapter 5, which illustrates this process for one-dimensional sampling.

For this example, assume for simplicity of illustration that we desire a sample size of eight and that the spatial survey design specifies that the lakes should have an equal probability of being selected. This would be a simple random sample design if we were not interested in a spatially balanced GRTS design. Instead, we chose an equal probability GRTS design. How do we select eight lakes from the 16 cells and make sure that they are spatially distributed across Oregon in the same way that the lakes in the sample frame are distributed across Oregon? First, we determine the number of lakes in each cell and their proportion of the total number of lakes in Oregon. Then we multiply these proportions by the desired sample size of 8 to obtain the inclusion probability of each cell. That is, the inclusion probability $\pi_i = n\frac{e_i}{E}$ for cell i where e_i is number of lakes (extent) in cell i, E is the total number of lakes in the sample frame (i.e. total extent), and n is the sample size. Note that the inclusion probability for the cell is the sum of the inclusion probabilities for all lakes, i.e. sample units, in the cell. In Fig. 6.2b all inclusion probabilities ("inc prob") are <1. Each cell is mapped to a line segment of a length that is proportional to the inclusion probability of the cell. In this case, the cells are used to define segments of a line of length 8 (the desired sample size). Those cells with a small inclusion probability (fewer lakes) have shorter line segments and those with a large inclusion probability (more lakes) have longer line segments (Fig. 6.2c).

The next step is to select eight cells from the line using a systematic sample process with a random start. Since the line has length equal to the sample size, we chose a random number between 0 and 1, say 0.45 to obtain a systematic set of points on the line at 0.45, 1.45, 2.45, ..., 7.45 (indicated by triangles in Fig. 6.2c). Cells associated with the line segments with points from the systematic sample are selected (cells 6, 7, 3, 4, 2, 10, 9, and 13), Fig. 6.2d. Each of these cells will contribute a lake to the sample. The line segment associated with a selected cell is composed of smaller segments representing each lake. In this case because lakes are being selected with equal probability these segments are the same length (Fig. 6.3a). Individual lake line segments are randomly placed on the cell segment and the lake is selected where the systematic point occurs on the segment (at 0.45 or fifth lake in this example, Fig. 6.3b). This is completed for all cells to obtain the sample of eight lakes (Fig. 6.3c). Table 6.1 summarizes the data and computations that are used to select the eight cells and then to select the specific lake within each cell (assuming the lakes are randomly ordered with the cell). The combination of the hierarchical randomization process and the use of a systematic sample of the line based on it ensures that the sample will be spatially distributed across Oregon with

Table 6.1 GRTS selection of cells and lakes within cells for Oregon lake example.

Cell	# Lakes in cell	Proportion of lakes in cell	Cell inclusion probability	Cell line segment start	Cell line segment end	Systematic sample point	# Lake segments in cell to point	Lake selected within cell
8	105	0.054	0.429	0.000	0.429			
6	12	0.006	0.049	0.429	0.478	0.45	5.08	6
5	172	0.088	0.703	0.478	1.181			
7	241	0.123	0.985	1.181	2.167	1.45	65.71	66
1	63	0.032	0.258	2.167	2.424			
3	159	0.081	0.650	2.424	3.074	2.45	6.33	7
4	231	0.118	0.944	3.074	4.018	3.45	91.96	92
2	236	0.121	0.965	4.018	4.983	4.45	105.58	106
12	78	0.040	0.319	4.983	5.302			
11	34	0.017	0.139	5.302	5.441			
10	184	0.094	0.752	5.441	6.193	5.45	2.21	3
9	96	0.049	0.392	6.193	6.586	6.45	62.83	63
15	63	0.032	0.258	6.586	6.843			
14	113	0.058	0.462	6.843	7.305			
13	145	0.074	0.593	7.305	7.898	7.45	35.46	36
16	25	0.013	0.102	7.898	8.000			
Total	1957	1.000		8.000				

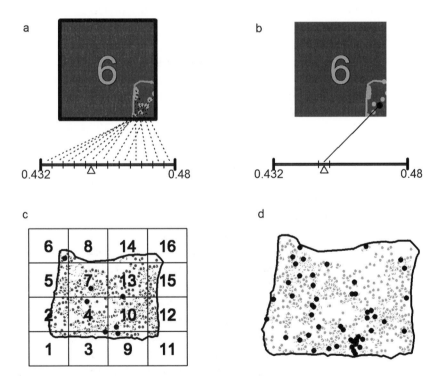

Figure 6.3 Illustration of the random selection of a lake within a selected cell. (a) Random ordering of lakes on line with equal length assigned to each lake. (b) Selection of lake when random location of point occurs within the lake's line segment. (c) GRTS sample of size eight. (d) GRTS sample of size 50 where number of hierarchical levels determined so that total inclusion probability of each grid cell was <1.

a spatial distribution that is similar to the spatial distribution of lakes. This is better illustrated in Fig. 6.3d with a sample size of 50. Note that even though two lakes are close together, it is possible for both of them be in a sample due to the random shift of the sample frame (see lakes selected in cells 3 and 9). Also the selection process within a cell is the same as selecting a simple random sample of size one from the lakes in the cell.

How is the hierarchical level, i.e. "m", determined? The value for m is selected so that all $2^m \times 2^m$ grid cells in the hierarchy have an inclusion probability π less than 1. The inclusion probability depends on the total sample size, the number of lakes within the cell, and the total number of lakes in the population. Because the grid is randomly located, the inclusion probability can only be calculated after the randomization, as the number of lakes in a cell is unknown until then. If we assume the lakes are uniformly distributed in space and the desired sample size is n, then an initial value for m is the next larger integer of $\log_4(n)$. If n equals 100, then m would be equal to 4. When the lakes are clustered, then m will mostly likely need to be greater to ensure the inclusion probability is less than 1. The determination of m is automatically calculated in the "spsurvey" *grts()* function so the user does not need to specify m.

Basic GRTS algorithm for linear networks and polygons

We used lakes as a collection of points in space to illustrate the GRTS algorithm. The same basic algorithm is used for a linear network, e.g. streams, roads, or trails, and for polygons, e.g. forests, estuaries, or lakes sampled as polygons. The process is the same through the steps illustrated in Fig. 6.1 and Fig. 6.2a,b except that instead of points in cells we now have linear segments or polygons in cells. In the step illustrated by Fig. 6.2c, the algorithm differs in the determination of the line segment length associated with each cell and then is the same for the step illustrated by Fig. 6.2d. The subsequent step of selecting a sample point on the linear network within a cell or selecting a sample point within a polygon in a cell differs from that illustrated for lakes. Note that linear networks, such as stream networks, are digitally represented by straight line segments of different lengths. While the sample unit is a point for discrete natural resources, i.e. a lake in the example, the sample unit for linear networks or polygons is any location on the linear network or any point within the polygons. In both cases the number of sample units is infinite as opposed to finite for a point natural resource.

What differs in the step illustrated by Fig. 6.2c, i.e. in constructing the line? In the lake example, we determined the number of lakes in each cell. We call this the extent of the sample frame within the cell. For a linear network the extent is the total length of the segments that make up the linear network within the cell. If a segment extends outside the cell it is clipped at the cell boundary and only that portion of the segment length is used. The rest of the segment is included in the extent computations of the neighboring cell. The inclusion probability is $\pi_i = n\frac{e_i}{E}$ for cell i, where e_i is extent (linear network length) in cell i, E is the total extent (total linear network length in the sample frame), and n is the sample size. For an areal sample frame (polygons), the extent is the total area of the polygons that are within the cell. If a polygon extends outside the cell it is clipped at the cell boundary and only that portion of the polygon area within the cell is used. Extent is in units of numbers for a point sample frame, length for linear network sample frame, and area for a polygon sample frame. The line segments associated with each cell for a linear network or areal sample frame have length equal to the extent within each cell. Otherwise the construction of the line is the same as for points as is the selection of the systematic sample of points on the line.

How is a random point selected on the linear network within a cell? These segments within a cell are randomly ordered and then placed on the line in that order with length equal to their inclusion probability (which is proportional to their length). Using the example based on lakes, the segment on the linear network selected from the first cell would be the line segment that included 0.45. For illustration assume it is the first line segment and that it has an inclusion probability of 0.03. The cell (6) begins its line segments at 0.429 and ends at 0.478 and the first linear-network segment begins at 0.429 and ends at 0.46. The selected point would be $100 \times (0.45 - 0.429)/(0.46 - 0.429) = 66.7\%$ up the line segment from the beginning of the line segment. Given we know the percent we can determine its location on the linear-network segment and then its x-,y-coordinates. Note that sample points may be selected anywhere on the linear network.

How is a random point selected in the polygons within a cell? Areal sample frames are collections of polygons of different areas. The polygons within a cell are randomly ordered and then placed on the line in that order with length equal to their inclusion probability (which is proportional to their area). Using the example based on lakes, the polygon selected would be the polygon that included 0.45, assume it is the first polygon and that it has an inclusion probability of 0.03. The cell (6) begins its line at 0.429 and ends at 0.478 and the line segment associated with the first polygon begins at 0.43 and ends at 0.46, consequently the first polygon would be selected. Then within that polygon a random point would be selected. The latter is accomplished by selecting random x- and y-coordinates while ensuring the coordinates occurred within the polygon. While it may seem simpler to just select random x- and y-coordinates within the entire selected cell and ensuring the random point is in a polygon, this has drawbacks. First, the cell may include only one polygon that is very small compared to the size of the cell. Picking a random x, y-coordinate pair based on the cell will have a low probability of having the point fall within the small polygon requiring many random draws before the point falls within the cell. Second, more complex designs may require some polygons to have a probability of being selected that is not proportional to their area (unequal probability designs, discussed in next section).

GRTS applied to stratified and unequal probability spatial survey designs

Environmental monitoring programs typically have monitoring objectives or other requirements that require the use of complex survey designs. Examples of complex designs applied to aquatic resources are the US EPA National Aquatic Resource

Surveys (see http://water.epa.gov/type/watersheds/monitoring/nationalsurveys.cfm) for lakes (points), streams and rivers (linear network), coastal waters (polygons), and wetlands (polygons). Four basic design options (reviewed in Chapter 5) are simple random sample, stratified random sample, unequal probability sample, and stratified unequal probability sample designs. Each of these is also an option with GRTS designs. We have already described the GRTS equal probability spatial survey design. It is equivalent to a simple random sample except that it ensures each sample realization has a spatial distribution that is similar to the spatial distribution of the environmental resource. Next we discuss stratification and then unequal probability sampling. The fourth basic design, a stratified unequal probability spatial survey design, simply combines stratification with an unequal probability design.

A stratified GRTS spatial survey design is a simple extension of the basic GRTS algorithm. Stratification divides the sample frame into independent sample frames that collectively equal the entire sample frame. Since each stratum has its sample selected independently of the other strata, the GRTS algorithm is simply applied to each stratum. For example, see Appendix 6.1 for *grts*() commands used if the population of Oregon lakes was stratified into "mountain" and "xeric" region groups.

Unequal probability survey designs are another basic survey design and can be used as an alternative to stratified survey designs when the objective is to reduce variance of estimates (Lohr 1999). An unequal probability GRTS spatial survey design involves a simple modification to the basic GRTS algorithm to incorporate the unequal probability information. The basic GRTS algorithm requires that inclusion probabilities be determined for each cell. For the lake example these are proportional to the number of lakes in a cell since each lake was to be selected with equal probability. Suppose that instead of placing lakes in mountain vs. xeric areas in two different strata, we wanted to select "mountain" and "xeric" lakes with unequal probability, say a xeric lake with twice the probability of a mountain lake. Then when calculating the inclusion probability we would assign a line-segment length value of 1 to all mountain lakes and a line-segment length value of 2 to all xeric lakes. Then we would sum those values for all lakes in a cell to obtain the cell extent e_i and the total extent E. Note that now E is no longer equal to the number of lakes, but is equal to the number of mountain lakes plus twice the number of xeric lakes. Otherwise the selection process remains the same. For linear networks, the process is similar except that a segment length is multiplied by the values to get the inclusion probabilities. For polygons, the polygon areas are multiplied by the values to get the inclusion probabilities. Typically a user will know the number of samples desired for mountain and xeric lakes. For example, the user may want the expected sample size to be 50 for mountain lakes and 50 for xeric lakes for a total sample size of 100. (The "spsurvey" *grts*() algorithm uses the sample-size information for the two categories of lakes to determine the appropriate values to use when calculating the inclusion probabilities; see Appendix 6.1.) Note that in contrast to a stratified sample the unequal probability sample does not guarantee exactly 50 lakes in each category – only that on average over repeated sample draws 50 lakes will be in each category with the total always being 100 lakes. This is an inherit feature of unequal probability sampling.

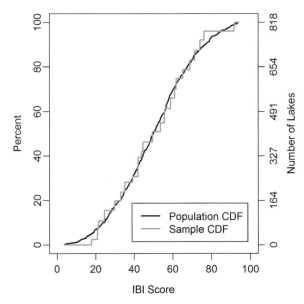

Figure 6.4 Illustrative cumulative distributions (CDFs) for a simulated index of biotic integrity (IBI) score variable for a target population of lakes and a miniature of the population (sample) using a sample of size 25. See Chapter 14 for additional discussion of CDFs.

Spatial balance and spatial balance metrics

Earlier we stated that monitoring designs should be based on a representative sampling process that results in a representative sample that is a miniature of the target population and that one way to achieve this was to use a process that resulted in the sample having the same spatial distribution as the population. We then described GRTS spatial survey designs and claimed that they satisfy these requirements by selecting spatially balanced samples. What do we mean by "a sample having the same spatial distribution as the population"? In this section, we first define "spatial balance" to give specific meaning to this phrase and then we define specific metrics to determine if a specific realization (a sample) is spatially balanced.

First, we give a quantitative definition measuring what it means for a representative sample to be a miniature of the population. Assume a single variable, x, is of interest. The population is completely characterized by its cumulative distribution function $F(x)$. A sample from the population can be used to construct a sample distribution function $\widehat{F}_n(x)$ (Fig. 6.4; Chapter 14). One possible metric to measure representativeness can be based on the Kolmogorov–Smirnov statistic $D_n = \max |\widehat{F}_n(x) - F(x)|$. This metric measures the maximum deviation of the sample distribution function from the population distribution function. When D_n is small the sample empirical distribution function is "close" to the population cumulative distribution function. Consequently, inferences based on the sample empirical distribution function will be the same as for the population. Most monitoring programs measure many indicators so that this condition would need to

Figure 6.5 Dirichlet tessellation for a sample of size 25 from Oregon lake population.

be satisfied for all indicators and preferably their multivariate empirical distribution function. While this metric defines what we mean by a representative sample being a miniature of the population, it is not useful in practice since the population distribution function is unknown.

Given the difficulty in determining whether a sample is a miniature of the population, instead we focus on metrics that measure the spatial balance of an equal probability, non-stratified sample of size n. Our spatial balance metrics are based on the Dirichlet tessellation (also called Voronoi or Thiessen polygons) associated with the sample of size n. The Dirichlet tessellation is the set of polygons over the study region created such that all locations within any given polygon are closer to one of the sample points than to any other sample point (Fig. 6.5). These polygons are then used to calculate the relative extent (i.e. the number, length, or area) or proportion p_i of the target population that is within each polygon associated with site i. If the sample is spatially balanced we would expect that the proportion of the target population in each polygon would be $1/n$. Figure 6.5 illustrates the Dirichlet tessellation for a sample of size 25 for the Oregon lake population.

Based on the proportions p_i from the Dirichlet tessellation of the sample, we define three alternative spatial balance metrics of a selected sample:

(i) The J_p spatial balance metric is Pielou's evenness index, which is based on the Shannon–Wiener index. It is defined as $J_p = -\sum_i^n p_i \ln p_i / \ln n = H/H_{\max}$, where H is the Shannon–Wiener diversity index and $H_{\max}$ is its maximum value (Legendre and Legendre 1998). This is one of several diversity indices used to measure diversity in categorical data. The maximum for H, $H_{\max}$, occurs when all p_i are equal resulting in the maximum value being equal to $\ln n$. We define spatial balance using Pielou's evenness index where perfect spatial balance has $J_p = 1$, i.e. perfect evenness of the target population associated with all sample sites.

(ii) The χ_p^2 spatial balance metric is equal to $\chi_p^2 = \sum_{i=1}^{n} (p_i - 1/n)^2/(1/n)$. This metric is simply the common chi-square statistic $\sum (O - E)^2/E$. Perfect spatial balance occurs when $\chi_p^2 = 0$.

(iii) The S_p spatial balance metric is defined as $S_p = \sqrt{\sum_{i=1}^{n} (p_i - \bar{p})^2/(n-1)}$, the standard deviation of proportions. In this case, perfect spatial balance occurs when $S_p = 0$.

We conducted a simulation study on the spatial-balance properties for a GRTS equal probability survey design vs. those of a simple random sample (IRS) survey design that does not use location as part of the sample selection process. Five alternative target populations were used: (a) lakes in Oregon as a point population using lake centroid as the point; (b) hydrologic units for Oregon (see http://www.ncgc.nrcs.usda.gov/products/datasets/watershed/ for definition of hydrologic units) as a point population using unit centroid as the point; (c) perennial streams in Oregon as a linear network population; (d) hydrologic units in Oregon as polygon population; and (e) lakes in Oregon as polygon population. For each target population 1000 samples of size 5, 10, 25, 50, and 100 were selected using GRTS and IRS designs, i.e. $5 \times 5 \times 2 \times 1000 = 50\,000$ samples. For each sample we computed the Dirichlet tessellation and calculated the value of the three spatial balance metrics. In this chapter, we only report the results for the Oregon lake population treated as points to compare the spatial balance properties of GRTS and IRS designs. The results for the other populations were similar.

The spatial balance properties for the GRTS and IRS simulated samples are presented in Fig. 6.6. The GRTS evenness metric median is greater than the IRS median for all sample sizes as are the 75th percentile and the maximum (Fig. 6.6a, left). For sample sizes of 25, 50, and 100 the evenness metric values for GRTS samples are almost entirely greater than 75% of the values for IRS samples. In all cases, larger values of the evenness metric reflect better spatial balance. For sample sizes of 5 and 10, the GRTS evenness distribution is better in terms of spatial balance than for IRS. For both GRTS and IRS samples the variability of the evenness metric increases as the sample sizes become smaller, although the increase is greater for IRS than for GRTS. Overall, the simulation shows that any single realization from a GRTS spatial survey design is more likely to result in a sample that is spatially balanced than any single realization from an IRS spatial survey design. The long tail for IRS compared to that for GRTS emphasizes the relative poor performance of IRS in achieving spatial balance compared to GRTS. The chi-square and standard deviation metric simulation results are consistent with those from the evenness metric, although small values now indicate better spatial balance (Fig. 6.6a, center and right). The relationships among the three metrics for a sample size of 25 are shown for both GRTS (Fig. 6.6b) and IRS (Fig. 6.6c). The three metrics are highly correlated and have a nearly linear relationship, but we present all three metrics for consideration by readers who may want to implement a similar examination of comparative spatial balance. Variability increases among the metrics as spatial balance decreases, although this is less so for the chi-square and standard deviation metrics.

Figure 6.7 shows the most and least spatially balanced sample realizations from the 1000 simulations for GRTS and IRS, based on the evenness metric. It is visually difficult

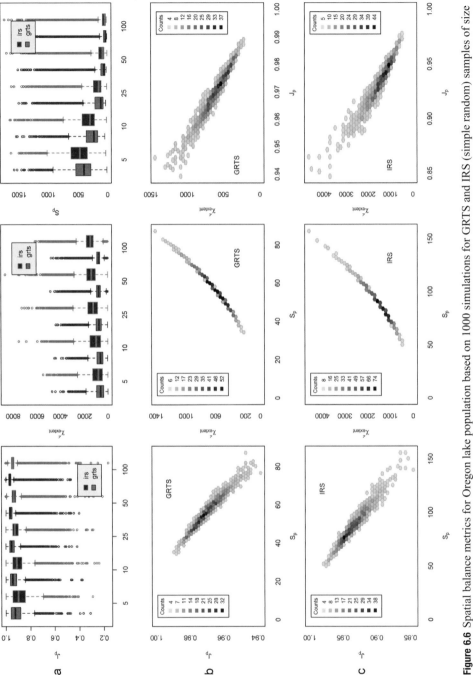

Figure 6.6 Spatial balance metrics for Oregon lake population based on 1000 simulations for GRTS and IRS (simple random) samples of size $n = 5, 10, 25, 50,$ and 100. (a) Results for GRTS vs. IRS for (from left to right) evenness metric, chi-square metric, and standard deviation metric. (b) Pairwise relationships among metrics for GRTS, $n = 25$. (c) Pairwise relationships among metrics for IRS, $n = 25$.

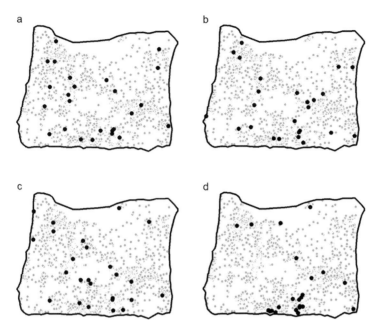

Figure 6.7 For sample size of 25, realization of a sample from 1000 simulated samples that is (a) the most spatially balanced GRTS sample ($J_P = 0.99$); (b) the most spatially balanced IRS (simple random) sample ($J_P = 0.98$); (c) the least spatially balanced GRTS sample ($J_P = 0.93$); (d) the least spatially balanced IRS sample ($J_P = 0.85$).

to distinguish the most (a) and least (c) spatially balanced samples for GRTS. We would expect this to be the case since the evenness metric values are very similar, $J_p = 0.99$ and $J_p = 0.93$, respectively. The most spatially balanced IRS sample is similar visually to the GRTS most and least spatially balanced samples; its evenness metric value is 0.98. The IRS least spatially balanced sample ($J_p = 0.85$) is noticeably different with the sample not reflecting the spatial distribution of the Oregon lakes.

Overall, Figs. 6.6 and 6.7 illustrate the difference in spatial balance between GRTS and IRS. Both can produce a sample that has good spatial balance. The spatially balanced samples that are possible using the GRTS representative sampling process are a subset of all the samples that are possible using the IRS representative sampling process. Consequently, the IRS samples with low spatial balance, as indicated by the J_p evenness metric, do not occur for GRTS.

Additional issues

Spatial balance properties when entire sample is not used

A spatial survey design implemented as part of an environmental monitoring program typically encounters anticipated as well as unanticipated problems. For example, the sample locations from the design may be (i) determined to not be a member of the target population; (ii) found to be inaccessible as a result of a landowner denying access; (iii) found to be physically inaccessible due to safety concerns or excessive cost to access; or (iv) not sampled due to insufficient time to complete the study within the allowed time or as a result of a reduction in funds. Each of these problems leads to a smaller sample size than what was originally planned. The smaller sample size resulting from the first three problems can be addressed by constructing the design so that additional sample locations are available if needed. As an example, if the study planned to collect data from 50 sample locations and only 40 of the 50 selected sample locations could be sampled, then if additional sample locations were available they could be used to obtain data from 10 replacement sample locations. The objective is simply to have the study result in data for the planned sample size of n, i.e. 50 sample locations in the example. Meeting the desired sample size does not address other issues arising from the "non-response" of locations not sampled (such as bias; see discussion in Chapters 3, 5). The question is whether it is possible to augment the original "base" survey design with a supplemental design of "over sample" sample locations.

One approach is to implement a spatial survey design where the sample size is the total of the desired sample size n (base sample) plus an additional sample size o (over sample) that would only be used if necessary. This can be done for an IRS spatial survey design, simply by selecting a sample of size $n + o$, using the first n sample locations first, and then using the o locations in the order they appear in the randomized list from the design. For a GRTS spatial survey design, the sample locations selected will be ordered according to the spatial hierarchical order used in the GRTS algorithm. Heuristically, if $n = 50$ and $o = 25$, then the first 50 sample locations will be located in only 75% of the study region. Stevens and Olsen (2004) define a reverse hierarchical ordering process that addresses this problem. It is not part of the basic GRTS spatial survey design process and is only relevant when the entire sample size is not used (e.g. Olsen and Peck 2008, Olsen *et al.* 2009). Note that the "spsurvey" *grts*() function demonstrated in Appendix 6.1 automatically does the reverse hierarchical ordering of the selected sample.

We investigated the spatial balance properties of IRS and GRTS spatial survey designs when only a portion of the entire sample ($n + o$ from above) was used. This addresses the question of whether the reverse hierarchical ordering algorithm results in a spatially balanced sample when the entire sample is not used. For each simulated sample, the sample points were used in the order they appeared in the sample. For GRTS simulated samples this was after the reverse hierarchical ordering was applied. Sample points were added 1 point at a time up to the maximum sample sizes of 50 or 100. The results for Oregon lakes are given in Fig. 6.8. For any specific subsample size, the summary is similar to the boxplot summaries in Fig. 6.6a, except that here the minimum, 25th percentile, median, 50th percentile, and maximum values are plotted. When the subsample size is close to the planned sample size, the spatial balance properties are similar to the full sample size of 50 (Fig. 6.8a) and of 100 (Fig. 6.8b). As the subsample size becomes smaller, the spatial balance properties of both IRS and GRTS spatial survey designs deteriorate, although GRTS continues to be better than IRS until very small sample sizes where it is similar to IRS.

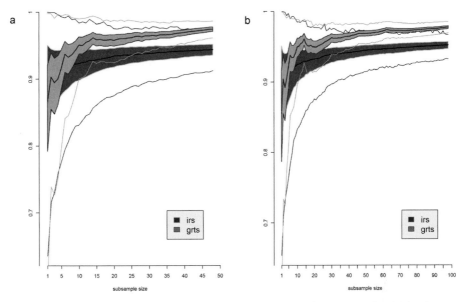

Figure 6.8 Spatial balance properties as a function of sample size for GRTS and IRS (simple random) samples based on 1000 simulations of the Oregon lake population. The y-axis is the evenness spatial balance metric (J_P). Sample points were added one point at a time up to the maximum sample sizes of (a) 50 or (b) 100. GRTS samples were arranged in reverse hierarchical order (see text). The shaded area includes the middle 50% of the simulated metric values (25th to 75th percentiles), dark line is the median (50th percentile), and the outer lines the minimum and maximum metric values.

GRTS applied to temporal designs

Chapter 7 discusses temporal (revisit) designs that are useful for trend detection. The most direct use of a spatial survey design for detecting trends is simply to select a spatially balanced sample, collect measurements on indicators of interest in the first time period, and then revisit the same sites in subsequent time periods. This is commonly called an annual revisit (or always revisit) panel design. (A panel is a collection of sites that have the same revisit pattern over time.) Another trend design is to never revisit the same sites in subsequent time periods. For example, if the monitoring program is planned for 20 years, then each year a new panel of sites is visited for a total of 20 panels. Other panel designs are discussed in Chapter 7.

Regardless of the panel design, a spatially balanced sample for trend detection can be selected as follows. Assume each panel will have n_i sites for $i = 1$ to P panels. The total number of unique sites is $n = \sum_{i=1}^{P} n_i$. A spatially balanced sample of n sites is selected, and the reverse hierarchical ordering algorithm is used. Then, the first n_1 are assigned to panel 1, the next n_2 are assigned to panel 2, and so on. Each panel of sites is a spatially balanced sample. Note that each panel is a spatially balanced sample for the population so that status estimates are possible for each time period as well as trend and change estimates across and between time periods.

Future research and development

Our description of spatial balance focused on spatial balance with respect to the target population. If the population is clustered, then the spatially balanced sample will be clustered. In some situations population sample units that are close together in space may be expected to be similar in their characteristics, i.e. they are correlated. It may be desirable to have a different type of spatial balance in this case – spatial balance with respect to geography. In the Oregon lake example, this means that the sample would be more uniformly spread across the state. Geographic spatial balance can be defined, although it is more difficult to implement. The "spsurvey" *grts*() algorithm can be applied but the user must create the inclusion probabilities based on a two-dimensional spatial density estimation of the population. Further research is required before this will be available.

Summary

In this chapter we have focused on spatial survey designs, and specifically spatially balanced survey designs and GRTS, and their use in monitoring programs for natural resources. Natural resources occur in two-dimensional space and can be modeled as points, linear networks, and polygons. Building on the concept of a representative sample as a miniature of the population, we described how spatially balanced survey designs can achieve that representativeness. We also distinguished a representative sampling process from a representative sample, noting that a representative sampling process does not necessarily result in a representative sample for every sample realization. We then defined spatial balance metrics and showed how a GRTS spatial survey design results in better spatial balance than a simple random sample (IRS). As GRTS spatial survey designs can also include stratification and unequal probability of selection, our conclusion is that GRTS spatial survey designs should be used for all natural resource monitoring programs where a probability survey design is the appropriate design to meet the monitoring objectives.

Acknowledgments

Marc Weber provided us with the GIS shapefiles used in our simulations. We appreciate reviews provided by Phil Larsen and two anonymous referees. The information in this document has been funded by the US Environmental Protection Agency (USEPA). This manuscript has been subjected to review by the National Health and Environmental Effects Research Laboratory's Western Ecology Division and approved for publication. Approval does not signify that the contents reflect the views of the Agency, nor does mention of trade names or commercial products constitute endorsement or recommendation for use.

Appendix 6.1. Implementing GRTS spatial survey designs in R

The R package "spsurvey" implements the GRTS spatial survey design algorithm and includes vignettes illustrating how the function *grts()* within the package is used to select a spatially balanced sample given a design specified by the user. The "spsurvey" package is available from the R software website (http://www.r-project.org/) or the aquatic resource monitoring website (http://www.epa.gov/nheerl/arm/). The function *grts()* selects samples from sample frames based on points, linear networks, or polygons to obtain a GRTS spatial survey design. In addition to equal probability GRTS spatial survey designs, the function *grts()* includes options for stratified GRTS spatial survey designs, unequal probability GRTS spatial survey designs based on user specified categories, or unequal probability GRTS spatial survey designs where the unequal probabilities are proportional to a user-specified continuous variable.

The following is an example of using the R package "spsurvey" to select the sample of 50 lakes. Three basic steps are required: (i) reading in the sample frame, (ii) specifying the design, and (iii) selecting the sample:

```
att <- read.dbf("Oregon_Lakes")
dsgn <- list(None=list(panel=c(Base=50), seltype= "Equal"))
lakes <- grts(design=dsgn,
DesignID='ORLakes',
type.frame='finite',
src.frame='shapefile',
in.shape='Oregon_Lakes',
att.frame=att,
shapefile=TRUE,
out.shape="OR_Lake_Design_Sites",
prj="Oregon_Lakes")
```

The first line reads in the attribute file that is part of the ESRI shapefile of Oregon lakes. It contains attributes associated with each lake, such as name, area, etc. The second line specifies the GRTS spatial survey design requirements. In this case "None" means the design is not stratified, "Base = 50" means that the sample size is 50 and all are assigned to a panel named "Base", and "seltype = 'Equal'" means that the lakes should be selected with equal probability. The *grts()* function provides the information the algorithm requires to select the lakes using the design "dsgn". "DesignID" specifies a prefix that will be added to the sample number assigned to the lakes and used as a "siteID". The term "type.frame" indicates that the sample frame is a collection of points (as opposed to a linear network or set of polygons); "src.frame" specifies that the source of the sample frame is an ESRI shapefile and "in.shape" gives the name of the ESRI shapefile. The term "att.frame" specifies that the attribute information is contained in "att", an R data frame; "shapefile = TRUE" indicates that the selected sample should be written out as an ESRI point shapefile and "out.shape" gives the name to use for that shapefile. The term "prj = 'Oregon_Lakes'" tells *grts* that the map projection of the output shapefile is the same as the input ESRI shapefile. The R object "lakes" contains the same information as the output shapefile and can be used within R.

If the lakes were stratified into mountain and xeric strata with sample size of 50 in each, this would be specified in a design requirement statement as follows:

```
dsgn <- list(mountain=list(panel=c(Base=50), seltype= "Equal"),
xeric=list(panel=c(Base=50), seltype= "Equal"))
```

The only change to the *grts*() function would be to add a line with "stratum ='stratum'" where 'stratum' would need to be the name of one of the columns in the attribute portion of the shapefile and each lake would have a stratum attribute of either "mountain" or "xeric". Note that exactly 50 lakes would be selected in each of the two strata.

For an unequal probability sample of lakes where the expected sample size is 50 for mountain lakes and 50 for xeric lakes, the design statement used in spsurvey would be:

```
dsgn <- list(None=list(panel=c(Base=100), seltype= "Unequal",
caty.n=c(mountain=50, xeric=50)))
```

where "seltype" now specifies that unequal probability of selection is to be used to select the sample of size 100 and that the inclusion probabilities should be constructed within *grts*() so that the expected sample size is 50 for mountain lakes and 50 for xeric lakes. The *grts*() algorithm uses the sample sizes and the sample frame information for the two categories of lakes to determine the appropriate values to use when calculating the inclusion probabilities. The *grts*() statement requires the addition of "mdcaty = 'laketype'" where 'laketype' is an attribute in the shapefile with each lake having an attribute of either "mountain" or "xeric".

7 The role of monitoring design in detecting trend in long-term ecological monitoring studies

N. Scott Urquhart

Introduction

Context

All monitoring is done in some context. The approach and topics considered in this chapter do not apply to all situations, so for clarity I define a context for this chapter. Monitoring programs will almost always be interested in many responses.[1] The example presented in this chapter has about 30 responses, a fairly typical number; some of the discussion will touch on the problems arising from various responses having different sorts of variation and design criteria. However, most of my discussion will focus on analyses of one response at a time, symbolized by Y with suitable subscripts to further identify it.

Monitoring, as used here, refers to studies intended to describe the geographic extent of a resource, general response size, and possible trends therein. Such studies are not sufficiently detailed in either time or space to support ecological modeling of underlying dynamic processes. Resources are *always* limited. When legal or regulatory requirements specify that inferences must apply to large areas, and budgets are limited, ecological modeling is neither feasible nor appropriate. This chapter is directed toward detection of trend in such situations, but this focus is not intended to imply that monitoring is done exclusively for the purpose of trend detection. In my experience, monitoring has always had many objectives. When trend is detected, a logical next step is to seek to characterize the dynamic process which causes it, assuming time and money are available to do that (see also Chapter 3).

This chapter's context assumes interest in a well-defined population over which the response is defined and varies. Such populations could be all lakes in the Northeastern USA, all streams in the USA, all of the grasslands in the Pawnee National Grasslands, all of the wetlands in a mountainous national park, or all of the archeological sites along the Colorado River in Grand Canyon National Park. So what are distinctive characteristics of this list? They all are located in space. It may be possible to sample

---

[1] What are called responses here are called *indicators* by the Environmental Monitoring and Assessment Program of the US Environmental Protection Agency (USEPA), *vital signs* in the US National Park Service's monitoring program (Chapter 22), *variables* in some contexts, etc.

Design and Analysis of Long-term Ecological Monitoring Studies, ed. R.A. Gitzen, J.J. Millspaugh, A.B. Cooper, and D.S. Licht. Published by Cambridge University Press. © Cambridge University Press 2012.

them from a list, but they may have to be sampled in one- or two-dimensional space (see Chapters 5, 6). Plots (sites), or more generally evaluation units, can be defined in them; with effort, it is possible to return to the same site at a future time. The responses relate to sedentary features of the site, like plant growth, soil erosion, physical habitat, or to organisms which move small distances, like small mammals or macroinvertebrates, or possibly some kinds of fish. This context may *not* be suitable for more mobile animals, from butterflies to birds to moose, because they ordinarily have little site identity (unless interest focuses on a specific point in their lifecycle). This context assumes that a response can be identified with a site, and that site can be revisited over time. Most ecologically interesting responses are influenced by a variety of local factors like elevation, slope, aspect, soil type, geologic factors which greatly influence soil and water chemistry, etc. Many such factors are characteristics of a site; by returning to the same site over time, those influencing factors remain nearly constant. This means that a return to a site is much like remaining in the same block of a randomized block experimental design. It makes no difference what makes the plots in a block similar, so long as they respond much more similarly than plots in another block. Likewise, what constitutes a site's identity is of no concern for the detection of trend. (Of course, it may be of great interest for other purposes.)

This context will speak of time in terms of years, because many interesting biological responses have annual cycles, and can be re-evaluated over many years. Many responses need to be evaluated during an appropriate time or index window each year during which the response changes very little as a function of time. Determining this time window may be difficult, and may differ between desired responses. Although time will be viewed as years, actual time periods could be longer or shorter than years; the essential thing is to return to the site at the same point in the temporal cycle of the responses and the site. Although most of the examples will have annual increments, most of the results will apply if field work were done every other year, or even less frequently. We will find that elapsed years are a major factor in the power to detect trend. Missing an occasional scheduled revisit, or even a whole year, proves to be only a minor statistical irritation, not a fatal mistake in a study.

Monitoring surveys usually have multiple objectives. Two frequently specified objectives relate to the estimation of status and trend. In a general way, status characterizes the size of the response of interest at a particular point in time. Trend, the major focus of this chapter, concerns a general increase or decrease of the response of interest over years. A trend can be quite nonlinear (e.g. Chapter 11), but a response which generally goes up, or generally goes down, across years can be detected by searching for linear trend. The detection of linear trend does *not* imply that the actual trend is linear. One of the advantages of framing trend detection with a general linear model, as here, is that it allows a more complex trend model when appropriate. On the other hand, if the year factor has a major effect on the response(s) of interest, we will find that trend is hard to detect. The presence of a substantial year effect on top of any systematic trend, namely a substantial year-to-year change in the true response size, effectively shifts an aspect of sample size from the number of sites to the number of years elapsed.

Do not confuse trend with pattern. To contrast pattern to trend, consider these examples of very real and pronounced patterns: photosynthesis ordinarily has a substantial diurnal

---

Box 7.1 Take-home messages for program managers

Several general issues should be considered by ecological monitoring programs seeking to assess the presence and magnitude of trends. Sites for conducting monitoring should be selected using established statistical sampling methods. This chapter shows how trend can be detected using data from sites selected in this way. Moderate amounts of trend can be detected if studies are continued for a substantial time using the same data-collection protocols. Frequent visits to the same site are *not* essential, and may even induce effects due to observer impacts (e.g. trampling) at the sites. Trend detection takes longer if the amount of trend is small, and if there is substantial year-to-year variation not related to the trend. Trend takes longer to detect with few sites, such as the case with monitoring of small National Park properties. However, trend is easier to detect if sites are eventually revisited, say at intervals as long as 10 years, provided the same data acquisition and responses are continued in use. Eventual revisits allow the statistical analysis to remove the effect of site differences which usually occur, regardless of the responses evaluated or frequency of revisits. Moreover, some irregularities, such as some missed visits to sites or some years with no monitoring, do not pose serious problems for the approach advanced here.

---

pattern, up during daylight, but down at night; length of day at a specific latitude has a pronounced annual pattern; the temperature of a body of water often has a pronounced annual pattern. Carbon dioxide measured in the atmosphere on Mauna Loa has a distinct annual pattern as well as a long-term trend which is nearly linear (see http://www.esrl. noaa.gov/gmd/ccgg/trends/). The approach advanced here is inappropriate for studying patterns which ordinarily are short-term relative to the trends.

The general perspective here is that the monitoring survey should have a spatial selection of sites according to an established probability survey process (Box 7.1; Chapters 5, 6). Statistical analyses for status should be done using the (sampling) weights resulting from that sampling design (Chapters 6, 14). However, survey methodology for the detection of trend in long-term monitoring studies has not progressed far enough to accommodate moderately complex, but appropriate, (mixed) methods for estimating trend. A defensible alternative is to apply a model-assisted approach based on a general linear model with appropriate random components to address the question of trend detection. This chapter develops and illustrates two such models. Both accommodate trend across years, have random contributions from the randomly selected sites, and have random contributions from years; the magnitude of these random contributions is characterized by variance components. The two models differ in that one estimates fixed population-wide linear trend, while the other models a different trend at each site. As sites are randomly selected, the variation in those trends in the latter case also is characterized by a variance component.

One of the practical problems in applying these concepts is to decide on the reasonable magnitude of the components of variance. These may be estimated from similar surveys, but similar surveys rarely exist. Use of "available data" is often fraught with definitional ambiguity. For example, the methodology for obtaining responses may have changed over time relative to exact collection point (e.g. deepest part of a lake, vs. outflow), sample handling procedures, and response evaluation protocols, to name a few I have experienced. The example presented here has been gathered over 11 years in the State of Oregon, USA, using consistent definitions. The responses characterize physical characteristics of salmon spawning streams. References point to other illustrations concerning both biological and chemical responses of lakes. Chapter 8, as well as Chapters 2, 9, and 10, further discuss potential sources of information and approaches for estimating variance components.

Many of the concepts discussed herein evolved from my long-time interaction with the US EPA's Environmental Monitoring and Assessment Program (EMAP); the acknowledgment section identifies people and projects which have contributed to my understanding of the situation discussed here.

Previous work

A substantial amount of work in statistical aspects of trend detection was summarized in the second issue of the *Journal of Agricultural, Biological and Environmental Statistics* in 1999 (volume 4). The papers entitled 'Environmental surveys over time' (Fuller 1999) and 'Design of supplemented panel surveys with application to the national resources inventory' (Breidt and Fuller 1999) are especially significant. These papers complement the work my colleagues and I published in 10 papers listed in the literature citations (Urquhart *et al.* 1993, 1998; VanLeeuwen and Urquhart 1994; Larsen *et al.* 1995, 2001, 2004; VanLeeuwen *et al.* 1996; Urquhart and Kincaid 1999; Courbois and Urquhart 2004; Kincaid *et al.* 2004). The latter have focused primarily on aquatic resources while the former focused on land use. Although the statistical approaches are rather different, all arrived at similar conclusions. Some aspects of the different modeling approaches are discussed subsequently.

Panel plans

Panel plans specify the temporal designs for site visits and revisits. The sampled population units, i.e. sites, are partitioned into one or more panels. Each of the sites in the same panel will have the same temporal pattern of revisits. We consider five specific panel plans in this chapter: always revisit (or pure panel; Fuller 1999), never revisit, rotating panel, augmented serially alternating, and partially augmented serially alternating, referred as Design 1 through Design 5, respectively. Their temporal patterns are displayed in Table 7.1. Design 1 specifies that there is simply one panel of sites visited during each time period; Design 2 specifies as many panels as time periods. The rotating panel (Design 3) specifies visits to all sites in a panel for several consecutive times; after that the panel receives no more visits. The US Department of Agriculture's National Agricultural Statistics Service (NASS) uses a rotating panel design with visits for 5 consecutive years (Cotter and Nealon 1989). A serially alternating plan specifies that sites in a panel are visited every rth year, where r is an integer. A complete serially alternating

Table 7.1 Schematic specification for five planned revisit panel plans.

Panel	Size	1	2	3	4	5	6	7	8	9	10	11	12	
														Time periods (= years)
DESIGN 1 = ALWAYS REVISIT = SAME SITES														
1	60	X	X	X	X	X	X	X	X	X	X	X	X	⋯
DESIGN 2 = NEVER REVISIT = NEW SITES														
1	60	X												
2	60		X											
3	60			X										
4	60				X									
5	60					X								
6	60						X							
7	60							X						
8	60								X					
9	60									X				
10	60										X			
11	60											X		
12	60												X	
⋮	⋮													⋱
DESIGN 3 = ROTATING PANEL														
−3	12	X												
−2	12	X	X											
−1	12	X	X	X										
0	12	X	X	X	X									
1	12	X	X	X	X	X								
2	12		X	X	X	X	X							
3	12			X	X	X	X	X						
4	12				X	X	X	X	X					
⋮	⋮					⋱	⋱	⋱	⋱	⋱				
DESIGN 4 = AUGMENTED SERIALLY ALTERNATING														
1	50	X				X				X				⋯
2	50		X				X				X			⋯
3	50			X				X				X		⋯
4	50				X				X				X	⋯
COMMON	10	X	X	X	X	X	X	X	X	X	X	X	X	⋯
DESIGN 5 = PARTIALLY AUGMENTED SERIALLY ALTERNATING														
1	50	X				X				X				⋯
2	50		X				X				X			⋯
3	50			X				X				X		⋯
4	50				X				X				X	⋯
5	5	X	X											⋯
6	5		X	X										⋯
7	5			X	X									⋯
8	5				X	X								⋯
9	5					X	X							⋯
10	5						X	X						⋯
11	5							X	X					⋯
12	5								X	X				⋯
13	5									X	X			⋯
14	5										X	X		⋯
15	5											X	X	⋯

design will contain r panels. The augmented serially alternating plan displayed in Table 7.1 (Design 4) utilizes a serially alternating plan with $r = 4$ plus an additional panel that is visited every year. The partially augmented serially alternating plan displayed in Table 7.1 (Design 5) utilizes a serially alternating plan with $r = 4$ plus additional panels that are similar to the original panels but include at least one pair of visits in consecutive time periods. Chapter 10 provides additional examples of panel plans.

Note that serially alternating plans without augmentation are "unconnected" in the experimental design sense (Searle 1987). A lack of connectedness can be illustrated by this simple case: Suppose all of the data for sites 1, 2, and 3 were collected in years 1, 2, and 3, but those for sites 4, 5, 6, and 7 were collected in years 5 and 6. The two sets of data can be analyzed separately, but there is no connection between the two sets which allows the comparison of the first three sites with the last four; likewise for the first three years and the last two. This situation can cause serious but often unrecognized computational problems. In the augmented and partially augmented serially alternating plans, the additional panels result in these plans being connected in the experimental design sense, thereby supporting estimation of components of variance using linear model methodology.

To compare sample allocation to panels, the size of each panel for Designs 1–5 is provided as a column in Table 7.1 for a total sample size per year, n, equal to 60 site visits, except during startup years. Exact results depend on the stated sample sizes, but the same pattern of results occurs for smaller or larger numbers of annual site visits (e.g. to panels having more or fewer sites than specified in Table 7.1).

Statistical models

A statistician's natural inclination is to start with the general case, then investigate special cases. However, a general formal statement of the models and underlying sampling framework may be an obstacle for many readers of this chapter. The next subsection illustrates the essentials of the approach without reference to the general case. Many formalities are omitted; refer to the subsequent subsection for more complete explanations.

A special case

Suppose a monitoring program had the good fortune to visit every one of a set of sites (subscripted by $i = 1, \ldots, s$) in each of a set of years (subscripted by $j = 1, \ldots, t$). Let any of the responses evaluated during the visit be denoted by Y_{ij}. A statistical model to describe such data could be denoted as follows:

$$Y_{ij} = \mu + S_i + \beta j + T_j^* + E_{ij}. \tag{7.1}$$

This signifies a response reflecting an overall mean response, a site effect, a linear time trend over years, a further time effect not captured by the linear trend (sometimes called a nonlinearity or lack-of-fit term), and a residual. The random components are assumed to be uncorrelated other than from the possible consequences of finite population sampling.

No assumption of normality is necessary for the following. Characterize the random variation in the random components by components of variance: $S_i \sim (0, \sigma^2_{SITE})$, $T_j^* \sim (0, \sigma^2_{YEAR})$, $E_{ij} \sim (0, \sigma^2_{RESIDUAL})$. If you estimate the linear trend (slope) at each site, and average those estimates, then:

$$var(\hat{\beta}) = \frac{\sigma^2_{YEAR} + \frac{\sigma^2_{RESIDUAL}}{s}}{\sum (j - \bar{j})^2}. \tag{7.2}$$

This formula illustrates a very important fact: interannual variation across years not characterized by linearity can substantially decrease the precision of the estimate of trend; the number of sites visited (s) has no effect on this component.

Does revisiting the same sites have much effect? Yes. So suppose instead that new sites were visited each year. Equation (7.2) changes to:

$$var(\hat{\beta}) = \frac{\frac{\sigma^2_{SITE}}{s} + \sigma^2_{YEAR} + \frac{\sigma^2_{RESIDUAL}}{s}}{\sum (j - \bar{j})^2}. \tag{7.3}$$

Note that the site component of variance appears in Equation (7.3), but not in Equation (7.2); this is because sites are revisited in the first case, but not in the second. A comparison of these equations shows the major statistical consequence of eventually revisiting sites: the removal of the site component of variance from the variance of the estimated slope.

As the sites represent a random sample of possible sites, the trend could plausibly change from site to site. Thus now consider the more complex case of a different slope at each site, i.e. the slope is random: $B_i \sim (\beta, \sigma_B^2)$, so we are now dealing with this model:

$$Y_{ij} = \mu + S_i + B_i j + T_j^* + E_{ij}. \tag{7.4}$$

In this case, Equation (7.2) changes to:

$$var(\hat{\beta}) = \frac{\sigma^2_{YEAR} + \frac{\sigma^2_{RESIDUAL}}{s}}{\sum (j - \bar{j})^2} + \frac{\sigma_B^2}{s}. \tag{7.5}$$

Note that the variance in the slopes could substantially increase the variance of the estimated (average) slope. Further note that the first denominator would increase with increasing years, thereby reducing the effect of the year and residual variance components, but the impact of variance in slopes is not similarly reduced. Only more sites will decrease its effect. You might ask, "Why bother with this more complicated model? It only makes variance look worse." If there really is variation in slopes across sites, and that is ignored, much of the variance between slopes would end up hidden in the residual variance.

The general case

This subsection continues, but expands, the notation of the previous section. Consider a finite population of sites $\{S_1, S_2, \ldots, S_N\}$ and a set of response values for the population of sites at time t $\{Y_1(t), Y_2(t), \ldots Y_N(t)\}$, a finite population of time series. Although time is continuous, suppose that only one sample can be observed at any site in any year and sampling can occur only during an index window of, say, 10% of the year. Let $\bar{Y}_i(\bullet)$ denote the average value of the response at site i over the time domain of interest, let $\bar{Y}_\bullet(t)$ denote the average response value for time t, and let $\bar{Y}_\bullet(\bullet)$ be the overall average response value. Then the components of variance, following the perspectives of Cornfield and Tukey (1956), are given by:

$$\sigma^2_{SITE} = var\{\bar{Y}_i(\bullet)\} = var\{\bar{Y}_i(\bullet) - \bar{Y}_\bullet(\bullet)\}$$
$$\sigma^2_{YEAR} = var\{\bar{Y}_\bullet(t)\} = var\{\bar{Y}_\bullet(t) - \bar{Y}_\bullet(\bullet)\} \qquad (7.6)$$
$$\sigma^2_{RESIDUAL} = var\{Y_i(t) - \bar{Y}_i(\bullet) - \bar{Y}_\bullet(t) + \bar{Y}_\bullet(\bullet)\}.$$

Note that these variance components are population values, not sample values. One of the reviewers of the chapter commented that "This perspective on mixed effects, as expressed in the equations given, is relatively old-fashioned, not what you usually see these days in discussions of random effects." True, but this perspective is appropriate when the models must cover the case of a finite population, as of lakes, and it supports the inference directly to the sampled population.

A random sample will select sites; this could be represented by random subscripts. We choose to index the sample values consecutively, so the response values become random variables. Consequently, the components in Equation (7.6) represent variation over random samples from finite populations of deviations. Let $\{Y_{ij}\} = \{Y_i(t)\}$, where i indexes sites and j time period, e.g. years. Then a statistical model is given by:

$$Y_{ij} = \bar{Y}_{..} + (\bar{Y}_{i.} - \bar{Y}_{..}) + (\bar{Y}_{.j} - \bar{Y}_{..}) + (Y_{ij} - \bar{Y}_{i.} - \bar{Y}_{.j} + \bar{Y}_{..})$$
$$= \mu + S_i + T_j + E_{ij}$$
$$= \mu + S_i + \beta j + (T_j - \beta j) + E_{ij} \qquad (7.7)$$
$$= \mu + S_i + \beta j + T_j^* + E_{ij}$$

where $S_i \sim (0, \sigma^2_{SITE})$, $T_j^* \sim (0, \sigma^2_{YEAR})$, $E_{ij} \sim (0, \sigma^2_{RESIDUAL})$, and the random variables are otherwise approximately uncorrelated; normality is neither assumed nor essential. The last line of Equation (7.7) emphasizes we will examine linear trend across time. Of course, sampling a finite population induces a slight negative correlation among the response values from the selected sites, but for the population and sample sizes of the magnitude foreseen here, these correlations are negligible. This becomes problematic only when the number of annual site visits gets below 20 and serious when that number falls below 10.

Because I am discussing the general case in this section, I will note that Urquhart *et al.* (1993) outlined the general variance–covariance structure encompassing Equations (7.6) and (7.7) completely, but the statement of the linear model without the explicit inclusion of a linear term has confused some readers. There were statistical reasons for this, but subsequent papers more explicitly included the trend term, as in Equation (7.7). On a related note, the general variance–covariance structure outlined by Urquhart *et al.* (1993) allows for correlation among year effects to be > 0 (i.e. autocorrelated year effects) and for the correlation between responses within sites in consecutive years to be less than

1.0. Such correlations logically exist, but practically I suspect they cannot be estimated with data sets of the size produced by monitoring programs. If the site correlation in consecutive years is quite high, using 1 as a default value is defensible. When I have investigated the impact of site correlations being somewhat less than unity, power has decreased a bit, but the important patterns have not changed. Likewise, low correlations between year effects have led to the same patterns of power as a default value of 0.

Let p be the index panel and n_{pj} indicate the number of sites in panel p for year j; $n_{pj} > 0$ for panels visited in year j, but $n_{pj} = 0$ otherwise. For each p, n_{pj} is either a constant or zero; that constant can change with p. Design 1 has only a single panel ($p = 1$) so $n_{1j} = s$ and $s = 60$ for the example. The average of Equation (7.7) across all sites in this one panel for each year will contain the mean, an average site effect, the year effect for that year, a contribution from the trend, and an average residual effect. Thus the slope contrast evaluated across these averages will have

$$var(\hat{\beta}_{DESIGN\ 1}) = \frac{\sigma^2_{YEAR} + \frac{\sigma^2_{RESIDUAL}}{s}}{\sum (j - \bar{j})^2}. \tag{7.8}$$

In Design 2, a different set of sites will appear in each panel-year average, so

$$var(\hat{\beta}_{DESIGN\ 2}) = \frac{\frac{\sigma^2_{SITE}}{s} + \sigma^2_{YEAR} + \frac{\sigma^2_{RESIDUAL}}{s}}{\sum (j - \bar{j})^2}. \tag{7.9}$$

Note that the site component of variance appears in Equation (7.9), but not in (7.8); this is because sites are revisited in Design 1, but not in Design 2.

More generally for panels visited in year j, the vector of cell means has a covariance matrix Σ:

$$cov(\bar{Y}_{pj}) = \Sigma(\sigma^2_{SITE}, \sigma^2_{YEAR}, \sigma^2_{RESIDUAL}, n_{pj}). \tag{7.10}$$

Let X denote a regressor matrix with a column of ones and a column containing the number of the time period (j) for panels with $n_{pj} > 0$. Then an estimator of the regression coefficients and its variance–covariance matrix is given by:

$$\hat{\beta} = (X'\Sigma^{-1}X)^{-1}X'\Sigma^{-1}\bar{Y}$$
$$cov(\hat{\beta}) = (X'\Sigma^{-1}X)^{-1}. \tag{7.11}$$

An estimate of trend is given by the second element in $\hat{\beta}$ and its variance by the second diagonal element in $cov(\hat{\beta})$. See Urquhart *et al.* (1993) for details about the above equations.

Incorporating variation in trend across sites

The model stated in Equation (7.7) provides the basis for examining one trend across an entire region, namely the same trend for each site. As the sites examined represent a random selection of the available sites, the trend could vary from site to site. The following model incorporates the possibility of a regional trend (β) as well as site-level deviations therefrom:

$$
\begin{aligned}
Y_{ij} &= \bar{Y}_{..} + (\bar{Y}_{i.} - \bar{Y}_{..}) + (\bar{Y}_{.j} - \bar{Y}_{..}) + (Y_{ij} - \bar{Y}_{i.} - \bar{Y}_{.j} + \bar{Y}_{..}) \\
&= \mu + S_i + T_j + E_{ij} \\
&= \mu + S_i + B_i j + T_j + (E_{ij} - B_i j) \\
&= \mu + S_i + \beta j + ((B_i - \beta) j) + (T_j - \beta j) + (E_{ij} - (B_i - \beta) j) \\
&= \mu + S_i + \beta j + B_i^* j + T_j^* + E_{ij}^*.
\end{aligned}
\tag{7.12}
$$

The new random component introduces another variance component, $B_i \sim (\beta, \sigma_B^2)$, or equivalently $B_i^* \sim (0, \sigma_B^2)$, with others as before, and this component of variance has to be incorporated into Equation (7.10). The last line of Equation (7.12) displays each of the random contributions with an expected value of 0. Note that the effect of different trends across sites, if present and modeled using an analytic model incorporating different trends, would reduce the residual variance, a feature illustrated in the example. Dawn Vanleeuwen investigated statistical features of this model as part of her doctoral research (VanLeeuwen 1993, VanLeeuwen and Urquhart 1994, VanLeeuwen et al. 1996), but the aquatic data then available were so poor that we decided not to emphasize analyses of real data at that time. The data now available for consideration in this chapter support estimation of the additional component of variance very well.

The introduction of this additional component of variance, of course, changes the variance of the estimated slope for Design 1, given in Equation (7.8) to:

$$
var(\hat{\beta}_{DESIGN\ 1}) = \frac{\sigma_{YEAR}^2 + \frac{\sigma_{RESIDUAL}^2}{s}}{\sum (j - \bar{j})^2} + \frac{\sigma_B^2}{s}.
\tag{7.13}
$$

In Design 2, a different set of sites will appear in each panel-year average; their variance will increase with years because the random slopes are multiplied by the year number. Consequently the estimate of trend will be a weighted average of the year averages with less weight on the later years than simple regression would give. The actual function, $var(\hat{\beta}_{DESIGN\ 2}) = f(\sigma_{SITE}^2, \sigma_{YEAR}^2, \sigma_B^2, \sigma_{RESIDUAL}^2, s, years)$, is similar to a combination of Equations (7.13) and (7.9), and the variance of a slope from weighted regression.

Subsequent analyses will present estimates of components of variance using Equations (7.7) and (7.12). Two other relevant estimates of residual variance will be presented for comparative purposes. The estimate of variance assuming simple random sampling (SRS) is the familiar simple estimate of variance which assumes no structure in the sampling; if a model is of any relevance, it should yield a lower estimate than this. The local estimate of variance proposed by Stevens and Olsen (2003) utilizes local spatial correlation, which frequently exists in many ecological/environmental investigations. It is based on accumulating variance information from a window which moves across the landscape, and can incorporate variable probability (density) sampling, if present (see also Chapters 6, 14). As of this writing, the local estimates of variance have not incorporated information from repeated visits to the same sites across time. Both of these additional variance estimates presented in the tables were computed entirely ignoring years.

Power

To detect trend we test the null hypothesis, H_0, that no trend is present in the response values; the alternative hypothesis, H_A, is that a trend equal to β^0 is present. The ability of a panel plan to detect trend can be expressed as power (see also Chapter 8). Because Equations (7.1) and (7.4) can be divided by $\sigma_{RESIDUAL}$, power will be evaluated in terms of these ratios:

$$\frac{\sigma^2_{SITES}}{\sigma^2_{RESIDUAL}}, \quad \frac{\sigma^2_{YEARS}}{\sigma^2_{RESIDUAL}}, \quad \text{and} \quad \lambda = \frac{\beta^0}{\sigma_{RESIDUAL}} \tag{7.14}$$

where the denominator in the ratios depends on the sampling design and the entire evaluation process.

A detailed examination of the multiplier matrix, $(X'\Sigma^{-1}X)^{-1}X'\Sigma^{-1}$ [Equation (7.11)], reveals that, in the absence of a year effect, $\hat{\beta}$ is a weighted average of estimates of $\hat{\beta}$ coming from each panel. By applying a somewhat generalized version of the central limit theorem, the estimate of the trend has an approximately normal distribution, regardless of the distribution of the individual response values. Thus the second element of $\hat{\beta}$ is approximately distributed as $N(\beta^0_1, \sigma^2_{\hat{\beta}})$ under H_A, so power of a two-sided test (using $\alpha = 0.10$) can be evaluated at the point $\beta_1 = \beta^0_1$ in H_A as

$$Power\left(\beta^0_1\right) = \Phi\left(-1.645 - \frac{\beta^0_1}{\sigma_{\hat{\beta}_1}}\right) + 1 - \Phi\left(1.645 - \frac{\beta^0_1}{\sigma_{\hat{\beta}_1}}\right) \tag{7.15}$$

where $\Phi(\bullet)$ denotes the standard normal (cumulative) distribution function.

Examples

Data source and characterizations

Two previous publications (Larsen et al. 1995, Urquhart et al. 1998) presented estimates of components of variance for indicators of chemical and biological conditions in lakes, mainly in the Northeast USA. The data set used here also reflects aquatic systems, but mainly of physical properties of streams in Oregon. It was collected by the Oregon Department of Fish and Wildlife (ODF&W) under the auspices of the Oregon Plan for Salmon and Watersheds (http://oregonstate.edu/dept/ODFW/freshwater/inventory). Further detail about the data set is available at http://oregonstate.edu/dept/ODFW/freshwater/inventory/orplan/orplnhab.htm.

As part of broader-scale studies, surveyors collected information on channel size, flow, substrate composition, large wood, habitat complexity, and riparian characteristics. The study area was divided into five regions called Gene Conservation Groups (GCG). About 50 sites were visited in each of the five coastal monitoring areas (North Coast, Mid-Coast, Mid-South, Umpqua, and South Coast) each year. Approximately 25% of the sites were visited annually, 25% visited every three years, 25% every nine years, and 25% visited one time only to balance the project's ability to measure trends and describe conditions (status) across each geographic area. The response names and a brief description of each are listed in Appendix 7.1. The original data set contained 3302 records. Of these, 782 were not surveyed, which for the present purposes we will regard as missing, and 366 were in regions with too few observations to establish trends. The remaining 2154 observations were nearly equally distributed across the five GCGs. The data and programs used in the analyses reported here are available as an online supplement for this chapter and also at http://www.stat.colostate.edu/starmap/data.

The target populations of streams for the study were based upon a hydrography data layer developed by the US Geological Survey at the 1 : 100 000 scale. Streams upstream of large dams that blocked anadromous fish passage were removed from the selection (sample) frame. A random tessellation stratified (RTS) design (Stevens 1997; see also Chapter 6) was used to select potential sample site locations within the population of stream segments. Stevens and Olsen (1999) described the RTS survey design as applied to the integrated monitoring of habitat, adult spawners, and juvenile salmonids for the ODF&W. The advantage of the RTS selection protocol was the selection of sites spread randomly across the landscape, better representing habitat conditions within a GCG for Coho salmon, and reducing overall sample variance compared to SRS. In all GCGs surveyed, sample sites were selected using a continuous sampling model (Stevens and Olsen 1999) according to stream length; sampling rates were varied by GCG to provide an approximately equal number of sample sites (50) in each year.

Note that the analysis of this set of data is intended to illustrate how such analyses can be done, and provide ranges of values for the subsequent power evaluations. It is *not* intended to provide subject matter inferences for the study from which it came.

Estimates of components of variance

Estimates of components of variance are displayed in Table 7.2 for the model with one slope, and in Table 7.3 for the model having as many slopes as sites. The original data were scaled so the estimate of the residual variance component was either a decimal value, or had one nonzero digit in the units position (Appendix 7.1). The other components of variance are presented as ratios with the residual component as a denominator, consistent with the form displayed in Equation (7.14). Based on the range of estimates for each variance component, representative values (specified in the following discussion) were selected for use in example power analyses.

Although most of this discussion will focus on components of variance, one column of information was included to show that data of the sort illustrated can actually identify trend when it is present. Eleven of the 33 responses demonstrated trend, having significance between $P < 0.05$ and $P < 0.005$ (all of the P-values are approximate). The scaled estimates of the site component of variance varied from slightly negative values (two values) to 15+. All of the positive estimates were significant at $P < 0.005$. (Note that the negative estimates of components of variance are not to be a source of concern; they are to be expected when their associated parameter is very small or even zero – see Chapter 9 for further discussion.)

Table 7.2 Scaled estimates of components of variance, scaled by residual variance (displayed in last column), based on a statistical model with one slope. Data are from a study conducted by the Oregon Department of Fish and Wildlife (State of Oregon, USA) as described in the text. Responses definitions are explained in Appendix 7.1.

Variable	n	Trend[a]	Site/Res	Year/Res	Local/Res	SRS/Res	Res
PRICHNLL	2154	***	5.788	0.029	2.792	6.871	0.994
SECCHNLL	2081		2.532	0.016	1.773	3.544	0.321
PRICHNAREA	2154		7.756	−0.002	4.024	8.768	1.875
SECCHNAREA	2081		2.207	0.018	1.417	3.205	7.141
PCTSCCHNLA	2081		1.130	0.008	1.171	2.134	0.158
GRADIENT	2154	*	15.279	0.024	6.581	16.345	1.999
VWIRCH	2154	**	0.975	0.005	1.010	1.994	0.445
WIDTH	2154	*	6.640	−0.005	3.450	7.647	1.443
ACW	2154		6.550	0.012	3.070	7.568	6.063
ACH	2154	*	1.439	0.015	1.383	2.498	3.044
NOPOOLS	2154		1.595	0.044	1.282	2.630	0.413
PCTPOOLS	2154		3.355	0.036	1.649	4.400	1.547
PCTSCPOOL	2154		2.846	0.048	1.701	3.906	1.232
PCTSWPOOL	2154	*	1.674	−0.004	1.146	2.672	1.060
SCRPOOLD	2061		2.696	0.046	1.672	3.768	3.056
RIFFLEDEP	1882		1.767	0.141	1.605	2.899	0.301
LRGBLDR	2133	*	−0.333	−0.003	0.414	0.670	0.682
PCTSNDOR	2154		5.223	0.008	1.892	6.225	0.924
PCTGRAVEL	2154		1.551	0.021	1.214	2.569	0.812
PCTBEDROCK	2154		7.124	0.005	3.189	8.098	0.220
POOL1P_KM	2154	***	2.633	−0.000	1.853	3.650	2.448
CWPOOL	2102		1.051	0.005	1.323	2.065	7.497
PCTSHADE	2154		1.925	0.059	1.429	2.954	0.933
PCTEROSION	2117		0.369	0.043	0.786	1.404	1.610
PCTUNDERC	2112		0.615	0.026	0.867	1.630	0.312
LWDPIECE1	2146		1.954	0.036	1.324	2.982	0.462
LWDVOL1	2146	***	1.101	0.003	1.080	2.141	3.128
KEYLWD1	2146	***	1.184	−0.000	1.208	2.230	0.772
RESIDPD	2075		2.540	0.038	na[b]	3.592	2.111
LRGBLDR1	2133	***	−0.210	−0.001	0.477	0.803	0.680
CON_20PLUS	2154		1.053	0.036	0.910	2.078	0.245
CON_36PLUS	2154		0.656	0.029	0.901	1.671	4.778
BVR_DAM	2154		0.833	0.006	0.971	1.844	1.618

[a] Significance of linear trend: * means significant at $P < 0.05$; ** < 0.01; *** < 0.005.
[b] Computational error code of an unidentified nature.

Estimates of the year component of variance varied from slightly negative (three values) to 0.141. Sixteen of the 33 estimates of the year component of variance were significant at $P < 0.005$ including all of the estimates greater than 0.020; a few slightly smaller values were significant due to differing sample sizes and revisit patterns. A total of 19 estimates were significant at $P < 0.05$. A substantial reason for presenting these estimates is to determine plausible values for illustrations of power to follow. The scaled

Table 7.3 Scaled estimates of components of variance, scaled by residual variance (displayed in last column), based on a statistical model with as many slopes as sites. Data are from a study conducted by the Oregon Department of Fish and Wildlife (State of Oregon, USA) as described in the text. Responses definitions are explained in Appendix 7.1. The numbers of observations used in estimation are the same as shown in Table 7.2.

Variable	Trend[a]	Site/Res	Year/Res	SiteSlope/Res	Local/Res	SRS/Res	Res
PRICHNLL	***	7.676	0.029	0.022	3.587	8.828	0.774
SECCHNLL		2.050	0.009	−0.001	1.754	3.507	0.325
PRICHNAREA		10.912	−0.008	0.014	4.736	10.320	1.593
SECCHNAREA		2.281	0.018	−0.015	1.136	2.568	8.911
PCTSCCHNLA	***	0.697	−0.006	−0.012	0.985	1.796	0.187
GRADIENT	*	23.143	0.030	0.035	9.581	23.795	1.373
VWIRCH	***	−0.277	−0.005	−0.010	0.879	1.736	0.511
WIDTH	***	8.005	−0.009	−0.002	3.381	7.495	1.472
ACW		11.679	0.015	0.021	3.917	9.657	4.752
ACH	***	1.724	0.011	0.009	1.540	2.781	2.734
NOPOOLS		1.112	0.049	−0.001	1.260	2.584	0.420
PCTPOOLS		2.444	0.060	0.010	1.852	4.942	1.377
PCTSCPOOL		1.946	0.070	0.006	1.828	4.198	1.147
PCTSWPOOL		3.495	−0.004	0.057	1.983	4.622	0.613
SCRPOOLD		2.944	0.055	−0.006	1.554	3.503	3.287
RIFFLEDEP		1.729	0.161	0.004	1.676	3.028	0.288
LRGBLDR		−1.004	0.012	−0.018	0.317	0.513	0.891
PCTSNDOR		5.820	0.017	0.005	2.012	6.622	0.868
PCTGRAVEL		2.318	0.029	0.005	1.297	2.746	0.760
PCTBEDROCK		10.940	0.012	0.019	3.967	10.073	0.177
POOL1P_KM	***	3.744	−0.005	0.015	2.210	4.352	2.053
CWPOOL		3.056	0.040	0.060	2.326	3.631	4.264
PCTSHADE		2.313	0.093	0.004	1.511	3.124	0.882
PCTEROSION		0.651	0.039	−0.001	0.776	1.385	1.632
PCTUNDERC		0.454	0.030	−0.000	0.865	1.626	0.313
LWDPIECE1	***	2.467	0.043	−0.002	1.291	2.908	0.474
LWDVOL1	***	2.289	0.007	0.004	1.136	2.252	2.973
KEYLWD1		1.364	−0.000	−0.006	1.113	2.053	0.839
RESIDPD		3.709	0.065	0.010	na[b]	4.036	1.878
LRGBLDR1		−0.949	0.013	−0.018	0.368	0.620	0.882
CON_20PLUS		1.599	0.018	0.001	0.927	2.117	0.241
CON_36PLUS		2.369	0.010	0.011	1.032	1.914	4.173
BVR_DAM		0.303	0.000	0.006	1.052	1.998	1.493

[a] Significance of linear trend: * means significant at $P < 0.05$; ** < 0.01; *** < 0.005.
[b] Computational error code of an unidentified nature.

site variance ranged from 0 to more than 10, so we will use values of 0, 1, and 5 in the illustrations. Similarly, the scaled year component ranged from 0 to 0.059, so we will use values of 0.000, 0.010, 0.025, and 0.050. The estimates of the residual variance component, defined previously as the variance of the E_{ij}, ranged from 0.3 to 7.5. The other components of variance presented in Tables 7.2 and 7.3 were scaled by this residual variance; this was motivated by the scaling presented in Equation (7.14). If we had added a scaling of σ^2 to Equation (7.14), the residual variance would be scaled to 1, the value we will use in power evaluations. Scaled trend varied from 0 to over 0.05, so we will investigate values of 0.01, 0.02, and 0.03.

The SRS estimate of variance is much larger than the residual estimate unless the site component of variance is negative (seen in two cases), the expected outcome. In many cases it is 2–8 times larger than the residual estimate. This indicates that model Equation (7.7) clearly accounts for many features of the situation which are ignored by SRS. In many cases the local estimate of variance is substantially larger than the residual. If there is local spatial variation, then there should be fairly substantial differences in responses across sites, an observed outcome. This is especially noticeable by examining those responses for which local estimate divided by residual estimate > 2, but recall that year effects were ignored in the local estimates presented here.

The results with a potentially different slope for each site (Table 7.3) are very similar to those in Table 7.2, except now there is a variance component for slopes across sites. Fourteen of these estimates were significant at $P < 0.05$, and 10 at $P < 0.005$. The values range from 0.000 to 0.057, so the power illustrations will be based on values of 0.000, 0.010, 0.025, and 0.050.

Power curves

The previous section developed parameter values for the power curves to be displayed and discussed here. Figure 7.1 shows 28 of more than 300 power curves which were evaluated for this discussion. Those displayed are typical of the entire body of curves; patterns are completely consistent across all of the computed curves. As each part of Fig. 7.1 is discussed, the parameter values underlying it will be specified. All of the patterns of visits/revisits are based on 60 visits each year. The comparative patterns remain unchanged with other sample sizes. Of course, power goes up faster with larger sample sizes and more slowly with lower sample sizes. Fig. 7.1a–d (top half of the figure) are based on the statistical model having one (fixed) slope; Figures 7.1e–h are based on the statistical model allowing random slopes, one for each site.

- Figure 7.1a ($\sigma^2_{SITE} = 1, \sigma^2_{YEAR} = 0.01, \sigma^2 = 1, \lambda = 0.02$) shows the effect of the temporal (revisit) design. Design 3 clearly loses because its revisits occur over only five years. Designs 1, 4, and 5 perform very similarly in all contexts, and consistently have the greatest power. They have planned revisits over the entire period of evaluation. As Designs 4 and 5 are more complicated than Design 1, you might ask, why use any other than it? Two factors need to be recognized. Designs 4 and 5 plan for visits to about four times as many sites as Design 1. This has no effect on trend detection, but it greatly broadens the base of inference for other purposes. This chapter considers only trend, but if we were to consider measures of status (size of the response) we would find Designs 4 and 5 provide far superior estimates of status compared to Design 1.

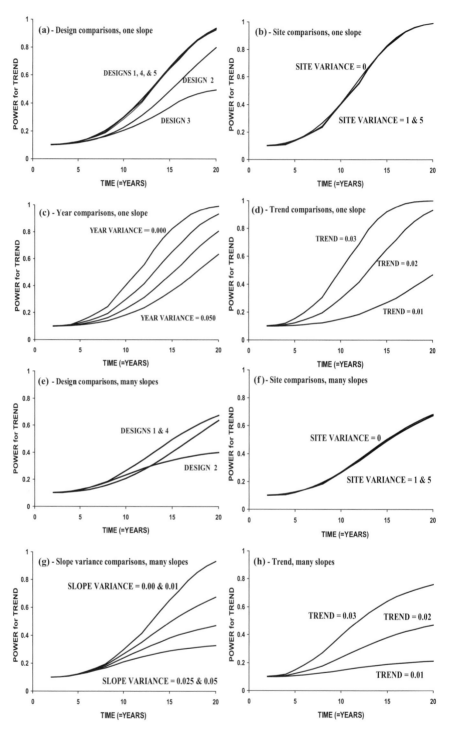

Figure 7.1 Power curves for 28 conditions discussed in the text.

Secondly, revisits can impact the research sites simply by people repeatedly accessing the site; visits could become the effect you measure! Designs 4 and 5 suffer from this problem far less than does Design 1.

- Figure 7.1b ($\sigma^2_{YEAR} = 0.01, \sigma^2 = 1, \lambda = 0.02$, Design 4) shows that the site variance (σ^2_{SITE}) has virtually no effect on power for trend detection. Revisits make no difference if there is no site effect; the upper curve is quite smooth. There are very small corners in the power curves for nonzero site variances at 4, 8, and 12 years. Power for trend is slightly lower with a nonzero site effect just before these times, just before the next cycle of revisits begins.
- Figure 7.1c ($\sigma^2_{SITE} = 1, \sigma^2 = 1, \lambda = 0.02$, Design 4) shows the substantial effect of even small amounts of year variation (σ^2_{YEAR}). Effectively, as the year variance increases, an aspect of sample size migrates from the number of sites to the number of years.
- Figure 7.1d ($\sigma^2_{SITE} = 1, \sigma^2_{YEAR} = 0.01, \sigma^2 = 1$, Design 4) shows the effect of the amount of actual trend (λ), of the standardized amount of 0.01, 0.02, and 0.03 per year. These curves show exactly what should be expected: the greater the amount of trend actually is, the greater the chance of detecting it.
- Figure 7.1e ($\sigma^2_{SITE} = 1, \sigma^2_{YEAR} = 0.01, \sigma^2_B = 0.01, \sigma^2 = 1, \lambda = 0.02$) compares Designs 1, 2, and 4. Design 5's power curve is indistinguishable from Design 4's, and Design 3 is such a loser that it will not be considered further. The power curves in Fig. 7.1e are substantially lower than the corresponding ones in Fig. 7.1a; this results because the component of variance for random slopes substantially decreases power; the value illustrated is modest. Power falls further for larger values, as illustrated in Fig. 7.1g. Design 1 has higher power than Design 4 for an obscure reason: Design 1 bases its inference on 60 site-specific trend lines while Design 4 bases its inference on 210 lines, 200 estimated by very few points. Design 2 loses even worse because it includes no revisits, making estimation of trends very uncertain.
- Figure 7.1f ($\sigma^2_{YEAR} = 0.01, \sigma^2_B = 0.01, \sigma^2 = 1, \lambda = 0.02$, Design 4) again shows that the site variance (σ^2_{SITE}) has virtually no effect on power for trend detection, very similar to Fig. 7.1b. Again, power is decreased by the presence of random slopes.
- Figure 7.1g ($\sigma^2_{SITE} = 1, \sigma^2_{YEAR} = 0.01, \sigma^2 = 1, \lambda = 0.02$, Design 4) shows the substantial effect of increasing amounts of variation in the slopes (σ^2_B). Figure 7.1g is has no analog in the upper set of four graphs, but clearly illustrates how increasing amounts of variation in the multiple slopes has a major effect on power.
- Figure 7.1h ($\sigma^2_{SITE} = 1, \sigma^2_{YEAR} = 0.01, \sigma^2_B = 0.01, \sigma^2 = 1$, Design 4) shows the effect of the amount of actual trend (λ) of the standardized amounts of 0.01, 0.02, and 0.03 per year. Figure 7.1h illustrates again how the presence of different (random) slopes across sites reduces power from when it is present.

Computational considerations

The estimates of the components of variance using the one slope model were obtained using the following sort of SAS (2008) code:

Box 7.2 Common challenges: issues of normality

Many statistical treatments assume that responses are normally distributed. Unfortunately, many statisticians assume that this assumption is no problem in applications; many ecologists assume normality is essential for any defensible statistical analysis. Unfortunately, both perspectives frequently make substantial errors in this regard. Most statistical inference using linear models can be defended with only second moment assumptions – assumptions about the model, variances, and covariances. Use of tables derived using normality assumptions can be shown to be reasonable approximations under these much weaker conditions, facts known for more than 50 years. Because "nice" statistical theory can be developed starting from normality assumptions, these are the developments graduate students in statistics are taught. These students come away with the impression that normality, or similar distributional assumptions, is necessary for all applications. The estimation of components of variance using software such as SAS reflects the same perspectives; careful choice of options are essential if you don't want to assume, perhaps incorrectly, normality. For example, this was the reason for the option `method=type3` on the SAS procedure statement described in the *Computational Considerations* section.

```
PROC mixed DATA=ODFW2008.DATA method=type3 noclprint;
      class ID_NUM YEAR;
      model VariableName = TIME/cl covb s;
      random ID_NUM YEAR;
```

For the multiple slopes models, the term "ID_NUM*TIME" was added to the random statement. "ID_NUM" is the site identifier, "YEAR" is Julian year, and "TIME" is YEAR-1997 so the start year is scaled to be 1. This code was embedded in a macro which looped across the 33 response variables, and the five regions and all data together. "DATA" statements organized the estimates of components of variance so they could be easily exported for inclusion in tables. The local estimates of variance were obtained using the R (R Development Core Team 2011) *total.est()* function in the R package "spsurvey" (Kincaid 2008; see also Chapters 6, 14). The original SAS and R code is accessible as an online supplement and at the web site cited in the data section.

Note that the options used in the SAS procedure, specifically as indicated by the term "method = type3", led to estimation using second moment assumptions, much weaker assumptions than have to be made to use procedures utilizing maximum likelihood or Restricted Maximum Likelihood (REML) estimation (see also Chapters 8 and 9 for additional examples and discussion). Similar procedures are now available in R. In particular, the approach here has carefully avoided any distributional assumptions, normality or otherwise, about the responses (Box 7.2).

For those readers familiar with various "types" of sums of squares often defined for classical linear models (e.g. Searle 1987), note that Type 3 sums of squares (SS) were used here rather than Type 2. Type 2 SS are an artifact left over from the days of hand computation. Type 2 SS adjust main effects for interactions, yet interactions are designed to account for what main effects fail to explain. Type 3 SS adjust main effects for other main effects, but not interactions, and interactions are adjusted for other interactions having the same number of factors. Furthermore, for this application, the sums of squares are used only to generate coefficients on the expected mean squares, a use little influenced by the type of sum of squares. Readers wishing to explore the effect of such options and rerun the analysis using different options can access the data and SAS code, as described above.

The power curves displayed herein were computed using an old matrix-oriented language named APL. It is very simple to compute Equations (7.10) and (7.11) in this language, but it is no longer very widely available. Tom Kincaid, USEPA (pers. communication), has developed and used R code for power of the sort described here; there are plans to eventually incorporate that code into the "spsurvey" R library (Kincaid 2008). Frank Deviney of the University of Virginia developed SAS code some years ago for evaluating power in the one slope case. When it is generalized, it will be made available on the data web site mentioned earlier. I know of no currently existing code for evaluating variances needed for the multiple-slope case other than that used here, although it may be incorporated into "spsurvey" in the foreseeable future.

Discussion

Long-term ecological monitoring studies must be accomplished in the face of substantial variation, but that variation has a well-defined structure which can be used to advantage in designing studies. One defensible approach is to characterize that variation through components of variance in a mixed linear model which incorporates the effects of sites, years, residual, revisit panels (the temporal plan), trend, and possibly site differences in trend. This discussion is based on my examination of more than 1000 cases of values of these effects, more than 300 done just for this chapter. Ability to detect trend can be characterized by statistical power, and demonstrated in graphs over years. This analysis demonstrates four things.

First, elapsed time is a friend. Regardless of other effects, power to detect trend increases with years. The rate of increase and its eventual value depend on other parameters and the temporal plan. Non-zero year effects, even very small ones, reduce power compared to the case of zero year effects. In the absence of year effects, the number of sites is a major factor in the variance of the estimate of trend: witness its effect in Equation (7.2). However, the year variance is not divided by the number of sites in this variance expression, only by the usual regression divisor, which is driven by the number of years spanned by the data. Consequently, as the year effect increases, an aspect of sample size moves from the number of sites toward the number of years. If there is a well-known (linear) model for part of the year effect, an advantage of the linear model approach is that it can be incorporated into the model, thereby reducing or even eliminating the year component of variance. An example could arise in working with things like some salmon runs which have three-year cycles. The magnitude of the run 3 years previous could be used as a "covariate" on this year's run size, thereby reducing a year effect, perhaps substantially.

Second, the temporal or revisit plan has a major effect on power. Eventual revisits to a site support the statistical estimation of the effects of the individual sites, thereby removing their effect from power. If sites are not revisited, as with Design 2, the site component of variance remains in the variance of the estimated slope, as illustrated in the comparison of Equations (7.2) and (7.3). Design 1 is what many ecologists assume is necessary to evaluate trend. It is the best design for trend detection by a very small amount, but it has two other severe limitations. Although this chapter is intended to address trend, most long-term studies also have substantial interest in estimating the size of the response, often called status. Design 1 provides far less precise estimates of status than any other of the designs. Further, visiting a site frequently may impact the site itself. The other temporal designs plan far fewer frequent revisits to each site; this is one of the main virtues of temporal designs like Design 5. In order for time to exert its full effect, the revisits have to be distributed across the entire life of the study; witness the limitation of Design 3. Temporal plans in Designs 4 and 5 are based on a four-year revisit cycle. There is nothing special about four years, but it has an observable effect. For many cases power starts accelerating after eight years (2×4). For a five-year cycle this would happen after 10 years, and so on. The advantage of longer cycles is that more sites are visited, and status is estimated more precisely; the disadvantage is an initial delay in detecting possibly present trend.

The temporal plans illustrated have regularly planned revisits. If an occasional site is missed, or a whole year omitted, say for budgetary reasons, little is lost, provided the missed sites are eventually revisited. This actually happened in the Oregon study used for the illustration; due to funding restrictions only new sites were visited one year – no visits were made to previously visited sites. The really critical feature for trend detection is to visit all of the sites early and late; intermediate times have far less effect. Of course, the intermediate times can be very important for other objectives of the study.

Third, the power to detect trend depends tremendously on the actual amount of trend. This should be obvious, but deserves to be noted, because no detection of trend does not mean no trend; it may mean a small trend below any detection threshold. A value like the 0.02 illustrated might be a value for which to hope. Over 20 years this would amount to a standardized change of 0.40. If the data were analyzed on the log scale, as are many ecological and environmental responses, a 0.02/year loglinear trend would amount to an increase of nearly 50% or decrease of about 33% in the response over that period.

Fourth, consider the case of trend varying across sites, a reasonable possibility. Power to detect trend has the same patterns as if this effect is absent, but the power is substantially lower. Of course, how much lower depends on the magnitude of the associated variance component. Note that if multiple slopes are present, use of the analytical model which explicitly models this variation will have lower estimates of residual variance compared to an analytical model which does not model multiple slopes, as the examples show (Tables 7.2 vs. 7.3). However, that reduction is much less than we would hope for. The decrease in power documents a reality: if the slopes vary, their average is subject to much more variation than if there is one overall slope. Namely, in addition to all of the factors which influence the estimation of a slope, the variance of an average has to be added in when the slopes vary. A major problem is that each slope must be estimated from very few observations; the one-slope model accumulates observations over sites and time and thus is based on far more observations.

On another monitoring topic, given the difficulty of accessing some sites, the frequent presence of many multiple objectives, and uncertainty about what aspects of a system might change, most monitoring studies seek to examine many responses. Different responses may indicate different index windows during which the responses should be evaluated. Because site visits often are a major component of study expenses, multiple visits to a site need to be avoided. This issue needs to be faced during the initial design stage so that response definitions and the list of responses to be evaluated is suitably modified fairly early in the planning process. It is very important that their feasibility be critically evaluated by pilot studies whose only objective is to "shake down" response definitions and field protocols.

Future research and development

Currently popular time series methodology utilizes adaptation of autoregressive models. The paper of Breidt and Fuller (1999) utilizes this methodology. Other work in progress in the UK also utilizes this model. On the surface this approach appears very different from that utilized here. On closer examination the more general mixed linear model introduced by Urquhart *et al*. (1993) has the same correlation structure as that displayed by Breidt and Fuller (1999). The major difference lies in how the residual is included in the model – multiplicatively in the autoregressive models and additively in the linear model. Probably neither is exactly correct! Both provide approximations. The advantage of the linear model approach is that it uses a slight generalization of models to which most ecologists get at least some exposure in their graduate statistics courses. Use of the autoregressive models ordinarily would require the substantial involvement of a knowledgeable statistician. An introduction to autoregressive models in a structural equation modeling context is provided in Chapter 15.

Summary

This chapter shows how trend can be detected using data from sites selected using established probability sampling methods. The approach used here relies on such a valid sampling approach, not on a statistical model someone hopes might be true. The focus is on detecting the linear component of trend, but this does not assume the underlying trend is actually linear; moreover, this approach can be extended to more complex trend models as warranted. The chapter outlined how site effects, year effects, residual variation, and potential variation in trend among sites contribute to the variance of trend estimates (i.e. how they affect power to detect trend). The data used for illustration came from a large survey of Coho salmon habitat in Oregon; the indicators or responses reflect both features of the streams and related landscape features. The results of this analysis show moderate amounts of trend can be detected if studies are continued for a substantial time using the same data-collection protocols. When trend varies among sites in the population, as was observed for many responses in the example considered in this chapter, power to detect trend is reduced; if this variation is significant, its effect can be reduced only by increasing the number of sample sites. The number of years over which monitoring has occurred, the magnitude of trend, and the site revisit plan generally have major influences on power to detect trend. In most circumstances, a panel plan should incorporate revisits to sites, even at relatively long intervals. Compared to plans that consist of either always revisiting the same sites or visiting a new sample of sites each year, other suitably chosen plans can offer a better balance between ability to detect trends and ability to estimate resource status at each point in time.

Acknowledgments

The author's involvement with the subject of trend detection was a direct consequence of his 15-year association with the US Environmental Protection Agency's (USEPA) Environmental Monitoring and Assessment Program (EMAP), mainly aquatic resources. His work was funded in part through Cooperative Agreements CR816721 and CR821738 between USEPA and Oregon State University (OSU) and CR829095 with Colorado State University. The work reported here it has not been subjected to the Agency's review and therefore does not reflect the views of the Agency, and no official endorsement should be inferred. Extensive discussions with Tony Olsen, Don Stevens, Phil Larsen, Steve Paulsen, other EMAP cooperators, and graduate students in the OSU Department of Statistics were very important in the development of ideas presented here. This writer gratefully acknowledges their collaborations. Sarah Williams started the development of the SAS code which led to Tables 7.2 and 7.3 as part of her master's project at CSU (Williams 2006). Thanks also to Frank Deviney who supplied SAS code he wrote for power evaluation. Kim Jones and Kara Anlauf, Aquatic Inventories Project, Oregon Department of Fish and Wildlife, were instrumental in providing access to and information about the data used in the numerical illustration. Two reviewers provided very useful observations and suggestions. Of course, errors of fact are mine.

Appendix 7.1. Brief description of variables used in example

See Oregon Plan Habitat Dataset 1998–2008 with Metadata (http://oregonstate.edu/ dept/ODFW/freshwater/inventory/orplan/orplnhab.htm#Dataset) for a more complete description. If a decimal point has been moved to facilitate presentation of estimates, such scaling is indicated here.

Variable	Description	Scaling
PRICHNLL	Length of primary channel in meters; excludes secondary channels, subunits, or tributaries	−2
SECCHNLL	Length of secondary channels in meters	−2
PRICHNAREA	Area of primary channel (m^2)	−3
SECCHNAREA	Area of secondary channel (m^2)	−2
PCTSCCHNLA	Percentage of the total area of the stream in the reach associated with secondary channels	−1
GRADIENT	Average of gradient (percent slope) for reach	0
VWIRCH	Valley Width Index. The ratio of the active channel to the valley floor	−1
WIDTH	Channel width in meters	0
ACW	Active or bankfull channel width in meters. This is the distance across the channel at "bankfull" flow	0
ACH	Active or bankfull channel height in meters	1
NOPOOLS	Combined count of scour and dammed pools within the reach	−1
PCTPOOLS	Combined percentage (by area) of scour and dammed pools in the reach	−1
PCTSCPOOL	Percentage of habitat units in the reach that are scour pools	−1
PCTSWPOOL	Percentage of habitat units in the reach that are slackwater pools	−1
SCRPOOLD	Average depth of scour pools in meters	1
RIFFLEDEP	Average depth of riffles and rapids in meters. A blank cell indicates no riffle or rapids	1
LRGBLDR	Count of large boulders ($\geq$ 0.5 m diameter)	−1
PCTSNDOR	Percentage of the total wetted habitat unit area consisting of sand, silt, and organics in surface substrate	−1
PCTGRAVEL	Average percentage of the total wetted habitat unit area consisting of gravel in surface substrate of all units	−3
PCTBEDROCK	Average percentage of the total wetted habitat unit area consisting of bedrock in surface substrate	−1
POOL1P_KM	Number of pools > 1 m in depth per km of total stream length	0
CWPOOL	Pool frequency or channel widths per pool: total reach length ÷ total number of pools	−1
PCTSHADE	Amount of shade provided to a stream by riparian vegetation and topography (percentage of 180 degrees)	−1
PCTEROSION	Percentage reach length of channel units with banks classified as eroding	−1
PCTUNDERC	Undercut bank unit average as percentage of unit length	−1
LWDPIECE1	Pieces of large woody debris (LWD) $\geq$ 3 m in length and 0.15 m in diameter per 100 m	−1
LWDVOL1	Volume of LWD per 100 m of primary channel length	−1
KEYLWD1	Key pieces of LWD per 100 m of primary channel length	0
RESIDPD	Average residual pool depth in meters	1
LRGBLDR1	Large boulders ($\geq$ 0.5m diam)/100 m total channel length	−2
CON_20PLUS	# of conifers $\geq$ 20 in (50 cm) dbh (diameter at breast height) per 1000 ft of stream length	−2
CON_36PLUS	# of conifers $\geq$ 36 in (90 cm) dbh (diameter at breast height) per 1000 ft of stream length	−1
BVR_DAM	Total count of beaver dams observed in each reach	0

8 Estimating variance components and related parameters when planning long-term monitoring programs

John R. Skalski

Introduction

The overall variability in observations from an ecological study generally is composed of multiple components of random error. The statistical theory of variance component estimation is well established (Searle *et al.* 1992; see also Chapters 7, 9). What is not well established is the routine use of preliminary surveys to collect information on sampling costs, expected response levels, and the magnitude of error sources (Box 8.1). Preliminary surveys should be an integrated component of every monitoring program which has as its objective more than simply long-term employment for those involved. In ecological studies, natural variation is typically too large and the sampling techniques too imprecise to leave study design to chance.

Sample-size calculations require knowledge of both the nature and magnitude of error sources. Add to this financial limitations, and the only prospect for an efficient monitoring program is often design optimization. Most monitoring programs will have sample sizes that are multidimensional – for example, a number of samples within a site and a number of sites within the landscape. Optimal allocation based on cost functions and variance component estimates can be used to determine the best allocation of survey effort. Design optimization is also useful in identifying discrepancies between desired study performance and budget that must be reconciled if a study is to be effective. All this, however, begins with variance component estimation.

The purpose of this chapter is to illustrate a general approach to variance component estimation with particular reference to long-term monitoring studies. Environmental studies have design considerations that are often not found in industrial or agricultural applications. These considerations include finite population sampling, errors in measurement, and non-additive error structures. This is particularly true when monitoring mobile populations.

In the case of demographic parameters such as animal abundance, survival, or recruitment, these responses are commonly estimated using tagging models (Chapters 18, 19), and the measurement error is estimated from the hessian matrix associated with the likelihood model (Seber 1982: 17, Edwards 1992). A parametric expression of the

Design and Analysis of Long-term Ecological Monitoring Studies, ed. R.A. Gitzen, J.J. Millspaugh, A.B. Cooper, and D.S. Licht. Published by Cambridge University Press. © Cambridge University Press 2012.

> **Box 8.1** Take-home messages for program managers
>
> Preliminary surveys are crucial in providing the necessary information on sampling uncertainty and costs needed to design effective and efficient monitoring designs. Preliminary survey data should be analyzed to estimate the magnitude of competing sources of uncertainty. These sources inevitably include natural variation and measurement error. The design of the monitoring program will determine whether this natural variation includes spatial and/or temporal variance components. The measurement error can be either design- or model-based (or both), depending on how responses are measured. Panel designs and monitoring designs based on blocking will also require preliminary survey information on correlations in responses. Well-established methods of statistical analysis exist using either finite sampling or analysis of variance to estimate those variance components. Sample-size calculations should be bracketed, taking into account the uncertainty in those variance component estimates.

measurement variance is derived as a function of demographic parameters, capture (detection) probabilities, and tag release sizes (Seber 1982). This model-based variance is then combined with designed-based variance components associated with the higher order sampling elements of the monitoring design in deriving the overall variance structure of a study.

The choice of response variable and the nature of the monitoring program will affect the overall error structure of the study and determine which variance sources are relevant. For instance, in the case of before–after, control–impact (BACI) designs used to assess site-specific impacts of development, the temporal variances and covariances between sites influence the study performance (Skalski and McKenzie 1982, Skalski and Robson 1992). In the case of accident assessment designs, the error term is composed of spatial variances and temporal covariances (Skalski and Robson 1992: 194–212). In panel designs used for long-term regional monitoring, the important sources of variability that influence study performance are also the spatial and temporal variances/covariances (Cochran 1977: 343–351; Chapters 7, 9, 10). In restoration evaluations, spatial variances and covariances drive the performance of the studies when comparing remediated sites to reference sites. However, in all cases, measurement error may be present because the response variables are typically estimated, and not directly or completely measured.

Note that other references, including other chapters in this volume, often distinguish inaccuracies in measuring the response variable at a sample unit vs. inaccuracies due to measuring only a portion of a population, commonly labelled "sampling error". Both imprecision in measurement techniques and variability from sampling finite populations are encompassed by the general variance expressions given in this chapter: measurement error, as used in this chapter, can include either or both sources of variability.

The time to consider variance component estimation is during design development (Box 8.1). As also emphasized in Chapter 2, an integral component of design development is determining how the monitoring data ultimately will be analyzed. A lack of an analysis plan at this initial stage is a good indicator of a poorly designed study. Without the statistical model for the analysis, sample-size calculations and optimal allocations are impossible. Therefore, design, analysis, and variance component structure go hand-in-hand before data collection begins. After 30 years of design experience, I recommend the 10% rule. That is, use 10% of the start-up costs for a study to collect preliminary survey data and cost information. The preliminary survey costs will invariably pay for themselves through the improved performance of the consummate design.

Determining relevant sources of variability

No automatic approach to variance component estimation exists that will safely lead the uninitiated through this complex field of quantifying sources of uncertainties. Both quantitative intuition and statistical training are necessary. A couple of examples may illustrate this point.

Example #1

Consider the case where an investigator interested in monitoring small mammal population trends sets out trap lines to collect index data. Each of the trap lines consists of the same length, number of traps and trap spacing, and produces a count n_i ($i = 1, \ldots k$). What then are the sources of variation that contribute to the observed variance in the n_i (i.e. $s_{n_i}^2$)?

Assuming for the moment that the survey effort produces a constant probability of capture (p) across sites, then

$$E\left(s_{n_i}^2\right) = p^2 \sigma_N^2 + \mu_N p(1 - p) \tag{8.1}$$

where σ_N^2 = spatial variance in abundance (N_i; $i = 1, \ldots, k$), and μ_N = mean abundance.

Equation (8.1) indicates the variance in n_i across sites will be influenced by both the spatial variance in actual animal abundance (σ_N^2) and binomial sampling variance within the sites. Therefore, if you identified at least two sources of variance, in this example, you are correct. If you allow for the probability of capture to vary independently between sites, there is a third source of variance where

$$E\left(s_{n_i}^2\right) = \bar{p}^2 \sigma_N^2 + \mu_N \bar{p}(1 - \bar{p}) + \sigma_{p_i}^2 \left(\sigma_N^2 + \mu_N^2\right). \tag{8.2}$$

The third term is associated with the variation in capture probabilities between locales (i.e. $\sigma_{p_i}^2$).

Neither Equation (8.1) nor (8.2) can be derived using traditional matrix algebraic approaches for random effects or mixed effects models (Searle *et al.* 1992). In addition, neither variance expression can be derived unless the investigator knows a priori the sources of variation to consider. While the mathematics of variance component estimation can be taught, the intuition and insight needed to set up the calculations must be acquired by experience. When a sampling design can be described by a linear model, variance component expressions can be readily, albeit sometimes arduously, derived (Searle *et al.* 1992). However, when sampling finite populations or in the case of non-additive measurement error common in monitoring mobile species, a more general approach must be used. The approach used in deriving Equations (8.1) and (8.2) will be discussed below in detail.

Example #2

Consider the case where there is interest in monitoring salmon smolt growth as an indicator of the benefits of estuary restoration efforts on fish health and survival over time. Each year, fish are captured, marked, measured, and released for subsequent recapture downriver as they out-migrate to the ocean. The response variable will be the sample mean for the difference in fish lengths between release (l_{i1}) and recovery (l_{i2}) across the n fish recaptured, i.e.

$$\bar{x} = \frac{\sum_{i=1}^{n} (l_{i,2} - l_{i,1})}{n}.$$

Stop again and consider, what are the sources of variation that must be taken into account in determining the required release size R for precision of the sample mean (x) defined as

$$P \left(\left| \frac{\bar{x} - \mu}{\mu} \right| < \varepsilon \right) = 1 - \alpha?$$

Ready?

If you identified two sources of sampling variance, you are correct. There is the variance in growth between fish (i.e. σ^2) and the binomial sampling of the fish downriver with probability p (= probability of survival multiplied by probability of capture). The required release size is the value of R that satisfies the expression (Skalski 1992)

$$\sum_{i=0}^{R} \left\{ \Phi \left(\frac{-\varepsilon\mu\sqrt{n}}{\sigma} \right) \binom{R}{n} p^n (1 - p)^{R-n} \right\} = \frac{\alpha}{2}. \tag{8.3}$$

Here, again, both biological and statistical insights are drawn upon to reformulate the problem into an objective function (8.3) that can be solved mathematically.

Regardless of whether you only thought of one source of variance or you are an ecologist with a good understanding of the relevant variance components, the best strategy for identifying and estimating relevant variance components is usually an interdisciplinary partnership with a statistician experienced in environmental sampling. There is no one-size-fits-all, cookbook approach that exists to guide the uninitiated through the complex field of variance component estimation. For investigators sampling essentially infinite populations using additive models, the book by Searle *et al.* (1992) is a good reference, and Chapters 7, 9, and 10 discuss a variance component framework associated with additive models. This chapter will introduce some of the ramifications when sampling finite populations with measurement error. In today's environment of litigation and often million-dollar studies, the best approach to study design is an interdisciplinary approach where ecologists and statisticians bring their respective strengths to bear in working with managers to design a biologically meaningful, statistically precise, and cost-effective monitoring program (see also Chapter 2).

Fundamental variance component expressions

The basic statistical formula underlying variance component analysis is the variance in stages formula:

$$\text{Var}(\hat{\theta}_i) = \text{Var}_2[E_1(\hat{\theta}_i|2)] + E_2[\text{Var}_1(\hat{\theta}_i|2)]. \tag{8.4}$$

If there are two difference variance sources contributing to the overall variance of some parameter estimate $\hat{\theta}_i$ (i.e. $\text{Var}(\hat{\theta}_i)$), then the individual contributions can be calculated in stages. The first step is to calculate the conditional expected value and variance of $\hat{\theta}_i$ over source 1, conditioning or holding source 2 momentarily constant, i.e. $E_1(\hat{\theta}_i|2)$ and $\text{Var}_1(\hat{\theta}_i|2)$. Variance and expected values are then calculated over the second error source next. This stepwise fashion of variance calculation makes the individual steps easier to conceptualize and perform. Conditioning is usually performed on the various stages of sampling.

Consider once again the small mammal trapping example and Equation (8.1). Assuming capture processes are homogeneous across sites, the overall variance in catch (n_i) can be calculated in two stages:

$$\text{Var}(n_i) = \text{Var}_2[E_1(n_i|2)] + E_2[\text{Var}_1(n_i|2)]$$

stage 1 = binomial sampling of n_i of N_i animals at site i and
stage 2 = sampling the landscape with mean abundance μ_N and spatial variance in abundance σ_N^2.

The first step in deriving Equation (8.1) is calculating the conditional moments for n_i holding the spatial sampling fixed and considering just the within-site binomial sampling process of the N_i animals:

$$\text{Var}(n_i) = \text{Var}_2[N_i p] + E_2[N_i p(1 - p)].$$

The next step is calculating the variance and expected values now, considering the spatial variance in animal abundance:

$$\text{Var}(n_i) = p^2 \sigma_N^2 + \mu_N p(1 - p).$$

In using Equation (8.4), one can condition on any convenient quantity or process. Typically, in variance component estimation, the best strategy is to condition on the various stages in the sampling design.

Multi-stage derivation

The two-stage variance expression (8.4) can be used recursively to express the variance structure in multi-stage designs. For example, in the case of a three-stage (i.e. stages 1, 2, and 3) sampling design, the total variance formula can be expressed as

$$\text{Var}(\hat{\theta}_i) = \text{Var}_3[E_2[E_1(\hat{\theta}_i|2, 3)]] + E_3[E_2[\text{Var}_1(\hat{\theta}_i|2, 3)]]$$
$$+ E_3[\text{Var}_2[E_1(\hat{\theta}_i|2, 3)]]. \tag{8.5}$$

The variance structure in Equation (8.2) was derived using the above three-stage variance expression:

$$\text{Var}(n_i) = \text{Var}_3[E_2[E_1(n_i|2, 3)]] + E_3[E_2[\text{Var}_1(n_i|2, 3)]]$$
$$+ E_3[\text{Var}_2[E_1(n_i|2, 3)]],$$

where *stage 1* = binomial sampling of N_i with probability p_i,
   stage 2 = variation in p_i between sites, and
   stage 3 = sampling the landscape with mean abundance μ_N and spatial variance in abundance σ_N^2.

Calculating the conditional moments with respect to stage 1, holding stages 2 and 3 constant for the moment,

$$\text{Var}(n_i) = \text{Var}_3[E_2[N_i p_i]] + E_3[E_2[N_i p_i(1 - p_i)]] + E_3[\text{Var}_2[N_i p_i]].$$

Next, calculating the conditional moments over stage 2, holding stage 3 constant,

$$\text{Var}(n_i) \doteq \text{Var}_3[N_i \bar{p}] + E_3[N_i \bar{p}(1 - \bar{p})] + E_3\left[N_i^2 \sigma_{p_i}^2\right].$$

Finally, calculating the moments over the third stage yields Equation (8.2), where

$$\text{Var}(n_i) \doteq \bar{p}^2 \sigma_N^2 + \mu_N \bar{p}(1 - \bar{p}) + \sigma_{p_i}^2 \left(\sigma_N^2 + \mu_N^2\right).$$

Equations (8.4) and (8.5) and their extensions to multiple error sources have wide applications. These variance-in-stages formulas are used in finite sampling to calculate sampling variances for multi-stage and nested designs (Cochran 1977). They also can be used to calculate the expected mean squares in analysis of variance (Johnson and Leone 1977). Seber (1982) used the same approach for variance calculations in many applications of animal tagging studies. Furthermore, these formulas do not require the assumptions of homogeneous variance and additive effects commonly associated in random and mixed effects models in ANOVA (Searle *et al.* 1992). However, when sampling for an infinite population can be described by an additive model, the methods in Searle *et al.* (1992) are available.

Important special case

A commonly encountered situation is when the monitoring study is estimating some parameter θ_i which is sampled over multiple locations or multiple times. An example would be when animal abundance (N_i) is estimated unbiasedly by $\hat{N}_i$ and interest is

Table 8.1 Estimates of animal abundance $(\hat{N}_i)$ and associated sampling errors $(\widehat{SE}(\hat{N}_i|N))$ for 10 spatially replicated sites.

Site	$\hat{N}_i$	$\widehat{SE}(\hat{N}_i|N)$
1	110.17	14.20
2	65.83	9.17
3	93.20	12.87
4	141.65	20.21
5	83.55	8.81
6	56.93	9.17
7	112.44	15.52
8	97.16	10.11
9	149.10	19.14
10	73.23	8.61

in mean abundance across locales or over time. In this case, where $\hat{\theta}_i$ is an unbiased estimator of θ_i, then

$$\text{Var}(\hat{\theta}_i) = \text{Var}_{\theta_i}(E(\hat{\theta}_i|\theta_i)) + E_{\theta_i}(\text{Var}(\hat{\theta}_i|\theta_i))$$
$$= \sigma_\theta^2 + \overline{\text{Var}(\hat{\theta}_i|\theta_i)},$$

where σ_θ^2 is the natural variability in θ_i, and $\overline{\text{Var}(\hat{\theta}_i|\theta_i)}$ is the average measurement error (i.e. $\text{Var}(\hat{\theta}_i|\theta_i)$ is the measurement error in $\hat{\theta}_i$). Consequently, if one calculates the empirical variance among replicate values of $\hat{\theta}_i$, it has the expectation

$$E\left(s_{\hat{\theta}_i}^2\right) = \sigma_\theta^2 + \overline{\text{Var}(\hat{\theta}_i|\theta_i)}, \tag{8.6}$$

where

$$s_{\hat{\theta}_i}^2 = \frac{\sum\limits_{i=1}^{n}(\hat{\theta}_i - \hat{\bar{\theta}})^2}{(n-1)}. \tag{8.7}$$

Using Equation (8.6) and the method of moments, an estimate of $\sigma_{\theta_i}^2$ can be calculated as

$$\hat{\sigma}_{\theta_i}^2 = s_{\hat{\theta}_i}^2 - \frac{\sum\limits_{i=1}^{n}\widehat{\text{Var}}(\hat{\theta}_i|\theta_i)}{n}. \tag{8.8}$$

For example, consider the case where animal abundance was estimated using mark–recapture methods at 10 replicate sites (Table 8.1). With model estimates of sampling variance at each site, the overall variance between the 10 estimates can be decomposed into estimates of natural variation ($\sigma_{\theta_i}^2$) and average measurement error (i.e. $\overline{\text{Var}(\hat{\theta}_i|\theta_i)}$).

The empirical variance among the replicate abundance estimates, using Equation (8.7), is $s_{\hat{N}_i}^2 = 937.7608$, and average measurement error (Table 8.1) is found to be

$$\frac{\sum\limits_{i=1}^{10} (\widehat{SE}(\hat{N}_i|N_i))^2}{10} = \frac{\sum\limits_{i=1}^{10} \widehat{Var}(\hat{N}_i|N_i)}{10} = 180.5069.$$

Therefore, an estimate of natural variation in animal abundance between locales [Equation (8.8)] is computed to be

$$\hat{\sigma}_{N_i}^2 = 937.7608 - 180.5069 = 757.2539.$$

In this example, 80.8% of the overall variance is due to spatial variance in abundance, and the remaining 19.2% is due to measurement error. Note, the average abundance estimate is an estimate of mean population abundance where

$$E_{\underset{\sim}{N}_i}[E_1(\hat{N}_i|N_i)] = E_{\underset{\sim}{N}_i}[N_i] = \mu_N, \text{ such that}$$

$$\hat{\mu}_N = \frac{\sum\limits_{i=1}^{10} \hat{N}_i}{10} = 98.326,$$

as long as the site-specific estimates $\hat{N}_i$ are from unbiased estimators of N_i $(i = 1, \ldots, n)$.

In the case of estimating the average response, say, $\hat{\bar{\theta}}$, the variance structure is

$$\text{Var}(\hat{\bar{\theta}}) = \frac{\sigma_{\theta_i}^2 + \overline{\text{Var}(\hat{\theta}_i|\theta_i)}}{n}, \tag{8.9}$$

where n is the number of replicate values of $\hat{\theta}_i$; $i = 1, \ldots, n$. As seen from Equation (8.9), the most direct way to reduce the contribution of $\sigma_{\theta_i}^2$ to the overall variance of the mean is to increase replication (n). The measurement error's contribution to the variance of the mean also decreases as a function of n but can also decline by additional within-site sampling effort and technique refinement. The magnitude of σ_N^2 can be altered in some circumstances by changing plot size (Skalski and Robson 1992: 42–46). The spatial variance in animal abundance tends to follow a negative binomial distribution. Population dispersion (σ_N^2) tends to increase as μ_N increases, but the $CV_N = \sigma_N/\mu_N$ declines as μ_N increases. Because relative precision is a function of the coefficient of variation (CV), detecting population change is typically easier in abundant rather than sparse species. Ecologists would be well advised to follow the lead of agricultural scientists who use uniformity trials (Cochran 1937, Hatheway and Williams 1958, Wight 1967) to identify optimal plot size as part of the design optimization process.

As another example, consider the case of estimating mean plant cover using replicate 1-m^2 quadrates and point sampling within the quadrate. Within each quadrate, 100 random x-,y-coordinates were randomly sampled and the number of points touching

Table 8.2 Estimates of fractional cover (p) and binomial sampling error ($\frac{p(1-p)}{m}$) at 8 replicate 1-m^2 quadrates.

Quadrate	$\hat{p}_i$	$\widehat{SE}(\hat{p}_i)$
1	0.55	0.04975
2	0.37	0.04828
3	0.63	0.04828
4	0.31	0.04625
5	0.79	0.04073
6	0.81	0.03923
7	0.46	0.04984
8	0.35	0.04770

cover recorded. A binomial sampling model can be used to estimate the within-quadrate sampling error for each of the replicate estimates of fractional cover (Table 8.2). Average fractional cover is estimated to be

$$\hat{p} = \frac{\sum\limits_{i=1}^{8} p_i}{8} = 0.5338.$$

The overall variance among replicate observations is $s_{\hat{p}_i}^2 = 0.038227$ with average mea-surement error of $\widehat{Var}(\hat{p}_i \mid p_i) = 0.00215$. The natural variation in fractional cover between replicate quadrates is then estimated to be $\hat{\sigma}_{p_i}^2 = 0.038227 - 0.00215 = 0.036077$. Using these variance component results suggests a variance expression for mean fractional cover to be

$$Var(\hat{p}) = \frac{0.036077 + \frac{1}{n}\sum\limits_{i=1}^{n}\frac{p_i(1-p_i)}{m}}{n}, \tag{8.10}$$

where n = number of replicate quadrates and m = number of point samples per quadrate. Variance expression (8.10), for sample-size purposes, can be approximated by

$$Var(\hat{p}) = \frac{0.036077 + \frac{\bar{p}(1-\bar{p})}{m}}{n}$$

and used to project the anticipated variance for various combinations of effort, m and n, for fixed $\bar{p}$. Using a cost function of the form $C_0 = C_1 n + C_2 mn$, the combination of n and m that minimizes the variance for fixed costs C_0 can be readily found. The solutions can be found algebraically or graphically as illustrated in Skalski and Robson (1992).

Fundamental covariance and correlation expressions

Rotational or panel sampling designs are commonly used for long-term monitoring (McDonald 2003). Chapters 7 and 10 examine general panel designs in the context of linear model-based estimates of trends. For design-based estimates of status and changes in status from rotational designs, the optimal fraction of sites rotated each year depends on how temporally correlated responses are at sites over time. For panel designs with simple rotation on an annual basis, the optimal rotational fraction for precision of the annually revised design-based estimates of status (Cochran 1977: 347) is equivalent to the quantity

$$\frac{1}{1 + \sqrt{1 - \rho^2}},$$ (8.11)

where ρ is the temporal correlation in annual response. For example, in the case where $\rho = 0.90$, the optimal level of rotation has proportion $= 0.70$ of the monitoring sites being replaced each year. When the correlation is only $\rho = 0.30$, the optimal level of rotation reduces to 0.51. Therefore, knowing the level of correlation to expect will have an effect on the monitoring design. The importance of correlation in blocked designs will be illustrated later in this chapter.

For any two random variables x and y, the correlation coefficient is defined as

$$\rho = \frac{\text{Cov}(x, y)}{\sqrt{\text{Var}(x) \cdot \text{Var}(y)}}.$$ (8.12)

Previously, I discussed how the variance of a random variable can be derived using the variance-in-stages approach [i.e. Equations (8.4) and (8.5)]. Attention is now turned to covariances. Analogous to variance formula (4), the covariance between two response measures $\hat{\theta}_i$ and $\hat{\theta}_j$ can also be taken in stages where

$$\text{Cov}(\hat{\theta}_i, \hat{\theta}_j) = \text{Cov}_2[E_1(\hat{\theta}_i|2), E_1(\hat{\theta}_j|2)] + E_2[\text{Cov}_1(\hat{\theta}_i, \hat{\theta}_j|2)],$$ (8.13)

and where 1 and 2 denote stages in a two-stage design. For example, $\hat{\theta}_i$ and $\hat{\theta}_j$ could be estimates of animal abundance in successive years (i.e. $\hat{N}_i$, $\hat{N}_{i+1}$). Consider the case where stage 1 is the estimate of animal abundance within a year and stage 2, the natural processes that generate the actual abundance levels over time. Then

$$\text{Cov}(\hat{N}_i, \hat{N}_{i+1}) = \text{Cov}_2[E_1(\hat{N}_i|2), E_1(\hat{N}_{i+1}|2)] + E_2[\text{Cov}_1(\hat{N}_i, \hat{N}_{i+1}|2)].$$

If we assume $\hat{N}_i$ is unbiased (i.e. $E(\hat{N}_i|N_i) = N_i$) and the estimation processes are independent, then

$$\text{Cov}(\hat{N}_i, \hat{N}_{i+1}) = \text{Cov}_2[N_i, N_{i+1}] + E_2(0)$$
$$= \text{Cov}(N_i, N_{i+1}).$$ (8.14)

This implies that the covariance between the abundance estimates is an unbiased estimate of the covariance between actual abundance levels, i.e.

$$E(\text{Cov}(\hat{N}_i, \hat{N}_{i+1})) = \text{Cov}(N_i, N_{i+1}).$$

A similar result occurs between the covariance of an abundance estimate $(\hat{N}_i)$ and a measured environmental covariate (x_i), where

$$
\begin{aligned}
\text{Cov}(\hat{N}_i, x_i) &= \text{Cov}_2[E_1(\hat{N}_i|2), x_i] + E_2[\text{Cov}_1(\hat{N}_i, x_i|2)] \\
&= \text{Cov}_2(N_i, x_i) + E_2(0) \\
&= \text{Cov}(N_i, x_i).
\end{aligned}
\tag{8.15}
$$

Equations (8.14) and (8.15) indicate that we can estimate the covariance between demographic and/or environmental responses by calculating the covariance among their estimates.

A problem emerges, however, when we re-express these estimated covariances in terms of the correlation coefficient [Equation (8.12)]. The correlation between successive annual abundance estimates has the approximate expected value

$$
\begin{aligned}
E(\hat{\rho}_{\hat{N}_i, \hat{N}_{i+1}}) &\doteq \frac{\text{Cov}(N_i, N_{i+1})}{\sqrt{\left(\sigma_N^2 + \overline{\text{Var}(\hat{N}_i|N_i)}\right)\left(\sigma_N^2 + \overline{\text{Var}(\hat{N}_i|N_i)}\right)}} \\
&\neq \rho_{N_i, N_{i+1}} = \frac{\text{Cov}(N_i, N_{i+1})}{\sqrt{\left(\sigma_N^2\right)\left(\sigma_N^2\right)}}.
\end{aligned}
\tag{8.16}
$$

Measurement error in the annual abundance estimates (i.e. $\overline{\text{Var}(\hat{N}_i|N_i)}$) will inflate the denominator of $\hat{\rho}$, resulting in a negatively biased estimate of $\rho_{N_i, N_{i+1}}$. As measurement error increases, the negative bias of $\hat{\rho}$ increases as well. For example, in the case where average measurement error equals natural variation (i.e. $\overline{\text{Var}(\hat{N}_i|N_i)} = \sigma_N^2$), then

$$
E(\hat{\rho}_{\hat{N}_i, \hat{N}_{i+1}}) \doteq \frac{1}{2}\rho_{N_i, N_{i+1}},
$$

or in other words, there is a 50% negative bias.

Properly, the correlation between $\hat{N}_i$ and $\hat{N}_{i+1}$ is the autocorrelation between successive annual abundance estimates and should not be used to infer the autocorrelation in annual population abundance. Similarly, the correlation between, say, annual abundance estimates $(\hat{N}_i)$ and an environmental covariate (x_i) has the expected value

$$
\begin{aligned}
E(\hat{\rho}_{\hat{N}_i, x_i}) &\doteq \frac{\text{Cov}(N_i, x_i)}{\sqrt{\left(\sigma_N^2 + \overline{\text{Var}(\hat{N}_i|N_i)}\right)\sigma_x^2}} \\
&\neq \rho_{N_i, x_i} = \frac{\text{Cov}(N_i, x_i)}{\sqrt{\sigma_N^2 \sigma_x^2}}.
\end{aligned}
\tag{8.17}
$$

Here, again, $\hat{\rho}_{\hat{N}_i, x_i}$ is a negatively biased estimate of ρ_{N_i, x_i}. Investigators may see no correlation between $\hat{N}_i$ and x_i because of the magnitude of the measurement error, while a correlation between N_i and x_i may actually exist. Inferences from $\hat{\rho}_{\hat{N}_i, x_i}$ should be restricted to what it is actually estimating, the correlation between annual abundance estimates and a covariate, and not the correlation between annual abundance and the covariate.

There is an additional implication of measurement error on perceived correlation. In panel designs for long-term monitoring, the precision of the study will depend, in part, on the correlation $\hat{\rho}_{\hat{N}_i, \hat{N}_{i+1}}$. Measurement error will decrease $\hat{\rho}_{\hat{N}_i, \hat{N}_{i+1}}$, also decreasing the precision and benefits of a rotational design in estimating status and trends. A lower value of $\hat{\rho}_{\hat{N}_{i_j}, \hat{N}_{i+1}}$ will also decrease the rate in which sampling stations are rotated into and out of the annual sample [Equation (8.11)].

Nested sampling designs: finite vs. infinite sampling

A hierarchical structure in nature often dictates the construction of a nested sampling frame. For example, trees in a forest, branches on a tree, and leaves on a branch form a natural hierarchy. Another example is broods of ducklings and chicks within a brood. If the intent is to make inferences to the lowest level of the hierarchical structure, it is usually easier and more efficient to start with a probabilistic sample at the highest level of organization and successively sample through the various stages rather than conduct a totally random sample at the lowest level. In the case of self-defined hierarchies, the sampling designs will not be balanced. Trees will have varying numbers of branches, and branches will have a varying number of leaves. Alternatively, hierarchical designs may be balanced by intent when the investigator defines the hierarchy – for example, sites within the landscape, quadrates within the sites, and point samples within the quadrates when estimating percentage cover.

The overall precision of a hierarchical nested sampling design will depend on the sampling effort at each level and the within-level variance components. In a balanced, three-stage sampling design where the grand mean of the population is defined as

$$\bar{\bar{Y}} = \frac{\sum_{h=1}^{N} \sum_{i=1}^{M} \sum_{j=1}^{K} y_{hij}}{NMK},$$

the sample mean is an unbiased estimator of the population mean and calculated as

$$\hat{\bar{\bar{Y}}} = \bar{\bar{y}} = \frac{\sum_{h=1}^{n} \sum_{i=1}^{m} \sum_{j=1}^{k} y_{hij}}{nmk}, \tag{8.18}$$

where n of N, m of M, and k of K sampling units are sampled at stages 1, 2, and 3, respectively.

The variance of $\hat{\bar{\bar{Y}}}$ (Cochran 1977: 286) is

$$\text{Var}(\bar{\bar{Y}}) = \frac{\left(1 - \frac{n}{N}\right) S_1^2}{n} + \frac{\left(1 - \frac{m}{M}\right) S_2^2}{m} + \frac{\left(1 - \frac{k}{K}\right) S_3^2}{k} \tag{8.19}$$

where:

$$S_1^2 = \frac{\sum\limits_{h=1}^{N}(\bar{Y}_h - \bar{\bar{Y}})^2}{N-1} = \text{between-primary-unit variance;}$$

$$S_2^2 = \frac{\sum\limits_{h=1}^{N}\sum\limits_{i=1}^{M}(\bar{Y}_{hi} - \bar{Y}_h)^2}{N(M-1)} = \frac{\sum\limits_{h=1}^{N}S_{2,h}^2}{N} = \text{within-primary and between-secondary-unit variance;}$$

$$S_3^2 = \frac{\sum\limits_{h=1}^{N}\sum\limits_{i=1}^{M}\sum\limits_{j=1}^{K}(Y_{hij} - \bar{Y}_{hi})^2}{NM(K-1)} = \frac{\sum\limits_{h=1}^{N}\sum\limits_{i=1}^{M}S_{3,hi}^2}{NM} = \text{between-tertiary, within-secondary-unit variance.}$$

Variance formula (8.19) can be used to project the anticipated sampling variance under alternative levels of sampling effort (i.e. n, m, and k) for given values of variance components S_1^2, S_2^2, and S_3^2.

Estimating the variance components

Using the data at hand from sampling the three stages, the variance of $\hat{\bar{\bar{Y}}}$ can be estimated by (Cochran 1977: 287):

$$\widehat{\text{Var}(\hat{\bar{\bar{Y}}})} = \frac{\left(1 - \frac{n}{N}\right)s_1^2}{n} + \frac{\left(\frac{n}{N}\right)\left(1 - \frac{m}{M}\right)s_2^2}{nm} + \frac{\left(\frac{n}{N}\right)\left(\frac{m}{M}\right)\left(1 - \frac{k}{K}\right)s_3^2}{nmk},$$

where

$$s_1^2 = \frac{\sum\limits_{h=1}^{n}(\bar{y}_h - \bar{\bar{y}})^2}{(n-1)},$$

$$s_2^2 = \frac{\sum\limits_{h=1}^{n}\sum\limits_{i=1}^{m}(\bar{y}_{hi} - \bar{\bar{y}}_h)^2}{n(m-1)},$$

$$s_3^2 = \frac{\sum\limits_{h=1}^{n}\sum\limits_{i=1}^{m}\sum\limits_{j=1}^{k}(y_{hij} - \bar{y}_{hi})^2}{nm(k-1)}.$$

Although s_3^2 is an unbiased estimator of S_3^2 [i.e. $E(s_3^2) = S_3^2$], s_2^2 and s_1^2 are not unbiased estimators of variance components S_2^2 and S_1^2, respectively, i.e. $E(s_2^2) \neq S_2^2$ and $E(s_1^2) \neq S_1^2$.

Because the stage 3 tertiary units were sampled and not completely measured within the secondary units, the sample variance s_2^2 has contributions of both S_2^2 and S_3^2. Similarly, because stages 2 and 3 were based on subsampling and not complete characterizations, s_1^2, the between-primary-unit variance, has contributions of not only S_1^2, but also S_2^2 and S_3^2. To use Equation (8.19) for precision and sample-size calculations, estimates of the

Table 8.3 An ANOVA table in the case of a balanced three-stage nested design with n samples at stage 1, m second-stage samples within each first stage, and k third-stage samples within each second stage. Calculations are outlined for degrees of freedom (DF), sums of squares (SS), mean squares (MS), and expected mean squares E(MS).

Source	DF	SS	MS	E(MS)
Total	nmk	$\text{SSTOT} = \sum\limits_{h=1}^{n}\sum\limits_{i=1}^{m}\sum\limits_{j=1}^{k} y_{hij}^2$		
Mean	1			
Total$_{\text{Cor}}$	$nmk - 1$	$\text{SSTOT}_{\text{Cor}} = \sum\limits_{h=1}^{n}\sum\limits_{i=1}^{m}\sum\limits_{j=1}^{k} (y_{hij} - \bar{y}\ldots)^2$		
Level 1	$n - 1$	$\text{SSL}_1 = \sum\limits_{h-1}^{n} mk(\bar{y}_{h..} - \bar{y}_{...})^2$	$\text{MSL}_1 = \dfrac{\text{SSL}_1}{n-1}$	$S_3^2 + kS_2^2 + mkS_1^2$
Level 2 within Level 1	$n(m-1)$	$\text{SSL}_2 = \sum\limits_{h-1}^{n}\sum\limits_{i=1}^{m} k(\bar{y}_{hi.} - \bar{y}_{h..})^2$	$\text{MSL}_2 = \dfrac{\text{SSL}_2}{n(m-1)}$	$S_3^2 + kS_2^2$
Level 3 within Level 2	$nm(k-1)$	$\text{SSL}_3 = \sum\limits_{h=1}^{n}\sum\limits_{i=1}^{m}\sum\limits_{j=1}^{k} (y_{hij} - \bar{y}_{hi.})^2$	$\text{MSL}_3 = \dfrac{\text{SSL}_3}{nm(k-1)}$	S_3^2

variance components S_1^2, S_2^2, and S_3^2 must be extracted from the sample variances s_1^2, s_2^2, and s_3^2.

The simplest and most direct way of estimating the variance components is to use analysis of variance (ANOVA) for a random effects model (e.g. Johnson and Leone 1977: 627–632, 743–787; see also Chapter 9). This model assumes the sampling units at each stage in a nested design are a random sample. In the case of a balanced three-stage nested design, the structure of the ANOVA is depicted in Table 8.3 [for an unbalanced design, see Johnson and Leone (1977: 628)]. Searle *et al.* (1992, Appendix F) outlines variance component estimation for one-stage, two-stage, and three-stage sampling.

Consider the expository example of sampling on the Washington State coast for polycyclic aromatic hydrocarbons (PAH) concentrations in mussels (*Mytilus* spp.). The sampling design consisted of four beaches ($n = 4$ of $N = 20$), three quadrates per beach ($m = 3$ of $M = 9$), and five mussels among hundreds per quadrate ($k = 5$ of $K = \infty$) (Table 8.4). Average PAH concentration per mussel was estimated to be $\bar{\bar{\bar{Y}}} = 15.8833$ ppb with a standard error of $\widehat{\text{SE}}(\bar{\bar{\bar{Y}}}) = 3.7749$. The nested ANOVA for the mussel data directly estimates $\hat{S}_3^2 = \text{MSL}_3 = 46.48$ (Table 8.5). Using the expected mean squares from the second and third levels of sampling (Table 8.5),

$$\hat{S}_2^2 = \frac{\text{MSL}_3 - \text{MSL}_2}{k} = \frac{99.50 - 46.48}{5} = 10.604.$$

Finally, using the expected mean squares from the first and second levels of sampling (Table 8.5),

$$\hat{S}_1^2 = \frac{\text{MSL}_1 - \text{MSL}_2}{mk} = \frac{1048.33 - 99.50}{15} = 63.255.$$

Table 8.4 Expository polycyclic aromatic hydrocarbons PAH data for mussels by beach, quadrates within beach, and individual mussels within quadrate.

Beach 1 Quadrate			Beach 2 Quadrate		
1	2	3	1	2	3
15	7	27	3	13	10
13	6	9	4	12	16
25	12	27	7	3	11
11	20	23	9	10	9
9	15	14	1	8	12

Beach 3 Quadrate			Beach 4 Quadrate		
1	2	3	1	2	3
10	21	11	31	32	39
11	15	15	14	16	27
8	15	7	47	21	28
10	8	6	11	27	36
5	17	18	14	27	45

Table 8.5 Nested ANOVA table with degrees of freedom (DF), sums of squares (SS), mean squares (MS), and expected mean squares E(MS) for Washington coast mussel data.

Source	DF	SS	MS	E(MS)
Total$_{\text{Cor}}$	59	6072.18		
Beaches	$(4-1)=3$	3144.98	1048.33	$S_3^2 + 5S_2^2 + 15S_1^2$
Quadrates within beaches	$4(3-1)=8$	796.00	99.50	$S_3^2 + 5S_2^2$
Mussels within quadrates	$4 \times 3(5-1) = 48$	2231.20	46.48	S_3^2

Combining the estimated variance components with Equation (8.19), the variance of $\hat{\hat{\hat{Y}}}$ can be expressed as

$$\text{Var}(\hat{\hat{\hat{Y}}}) = \frac{\left(1 - \frac{n}{20}\right)63.255}{n} + \frac{\left(1 - \frac{m}{9}\right)10.604}{m} + \frac{46.48}{k}$$

for alternative levels of sampling effort n, m, and k. Note in the case where $K \to \infty$, the finite population correction for the third term has the value of 1. In the case where the finite population corrections cannot be ignored (i.e. $1 - n/N$, $1 - m/M$, and $1 - k/K$), Cochran (1977: 287) provides expected values for s_1^2, s_2^2, and s_3^2, where:

$$E\left(s_3^2\right) = S_3^2$$

$$E\left(s_2^2\right) = S_2^2 + \frac{\left(1 - \frac{k}{K}\right)S_3^2}{k}$$

and
$$E\left(s_1^2\right) = S_1^2 + \frac{\left(1 - \frac{m}{M}\right)S_2^2}{m} + \frac{\left(1 - \frac{k}{K}\right)S_3^2}{mk}.$$

These expected values can be used to find the method-of-moment estimators for the variance components, taking into account the stage-specific finite population corrections.

Incorporating costs

For a cost function of the form

$$C_0 = C_1 n + C_2 nm + C_3 nmk,$$

where

$C_0 =$ total variable costs of sampling,
$C_1 =$ cost of sampling at a primary unit,
$C_2 =$ cost of sampling at a secondary unit, and
$C_3 =$ cost of sampling a tertiary unit,

Cochran (1977: 288) provides optimal allocations of effort, where

$$k_{\text{OPT}} = \frac{S_3}{\sqrt{S_2^2 - \frac{S_3^2}{K}}} \cdot \sqrt{\frac{C_2}{C_3}}, \tag{8.20}$$

$$m_{\text{OPT}} = \frac{\sqrt{S_2^2 - \frac{S_3^2}{K}}}{\sqrt{S_1^2 - \frac{S_2^2}{M}}} \cdot \sqrt{\frac{C_1}{C_2}}, \tag{8.21}$$

$$n_{\text{OPT}} = \frac{C_0}{(C_1 + C_2 m_{\text{OPT}} + C_3 m_{\text{OPT}} k_{\text{OPT}})}. \tag{8.22}$$

Multi-year data

In the case of environmental monitoring, Equation (8.19) constitutes the within-year sampling error. In this case, the measurement error is of a design-based nature which itself has multiple sources of variation. In the case of interannual monitoring, the yearly estimates of mean PAH concentration would be estimated with associated measurement error (Table 8.6). Under static environmental conditions, the variation in annual response would have expected value

$$E\left(s_{\bar{\bar{y}}}^2\right) = \sigma_{year}^2 + \widehat{\text{Var}(\hat{\bar{\bar{Y}}}|\bar{Y})},$$

composed of interannual variance (σ_{year}^2) and average measurement error of the form of Equation (8.19). Using the monitoring data in Table 8.6, average concentration over the 6 years was 15.19 ppb, with variance among yearly estimates of 11.8559 and

Table 8.6 Expository annual estimates of mean PAH concentrations in mussels with standard errors under static conditions and annual trend.

Year	Static $\hat{\bar{Y}}$	SE	Trend $\hat{\bar{Y}}$	SE
1	19.88	2.63	19.88	2.63
2	13.19	3.19	27.05	3.19
3	18.95	2.37	24.82	2.37
4	11.61	1.98	28.19	1.98
5	14.83	3.30	36.29	3.30
6	12.70	2.17	33.81	2.17

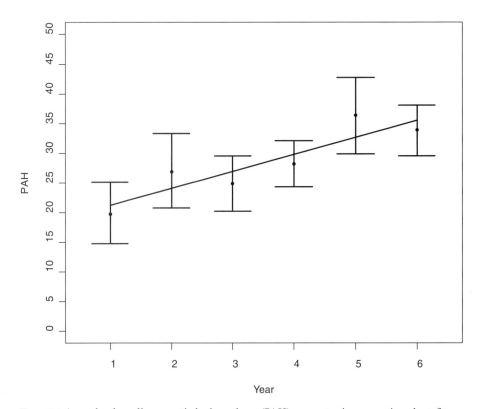

Figure 8.1 Annual polycyclic aromatic hydrocarbons (PAH) concentrations over time, best-fit linear regression, and 95% confidence intervals for individual observations.

an average sampling error $\widehat{\text{Var}}\left(\hat{\bar{Y}}|\bar{Y}\right) = 7.0382$. Therefore, interannual variance in PAH concentration is estimated to be $11.8559 - 7.0382 = \hat{\sigma}^2_{year} = 4.8177$.

In the case of a temporal trend in PAH concentrations over time (Table 8.6, Fig. 8.1), the concentration data must first be detrended in order to estimate the natural yearly variation (σ^2_{year}) superimposed on the temporal trend. Using linear regression of mean annual concentration vs. year (i.e. 1, ..., 6), the mean square error is calculated to be 8.448. Here, the MSE has expected value

$$E(\text{MSE}) = \sigma^2_{year} + \widehat{\text{Var}(\hat{\bar{Y}}|\bar{Y})}$$

and can be used to partition the unexplained variation into components of σ^2_{year}, the natural yearly variation about the time trend, and average measurement error. Hence, natural variation in PAH concentration about the trend is estimated as $\hat{\sigma}^2_{year} = 8.448 - 7.038 = 1.410$ where, again, average measurement error is estimated to be 7.038. Note that here, as throughout the chapter, variance components are estimated, leading to uncertainty that should be addressed when using them to design studies (Box 8.2).

Box 8.2 Common challenges: uncertain uncertainty

It is important to remember that the variance components are estimates which, in turn, also have uncertainty (i.e. variances). These estimates may be subject to high uncertainty and, with some estimation methods, significant bias (Chapter 9). Such uncertainty may translate into considerable uncertainty into estimates of sample sizes and length of time needed for meeting sampling objectives.

For example, in the case of the simple random sampling from a normal distribution, the variance of the sample variance (s^2) has the variance of

$$\text{Var}(s^2) = \frac{2\sigma^4}{(n-1)},$$

(Searle *et al.* 1992: 64), while the sample mean ($\bar{x}$) has the familiar and much smaller variance of

$$\text{Var}(\bar{x}) = \frac{\sigma^2}{n}.$$

An unbiased estimator of $\text{Var}(s^2)$ is

$$\widehat{\text{Var}(s^2)} = \frac{2\sigma^4}{(n+1)}.$$

Hence, the variance estimate of s^2 is much larger than that of the sample mean $\bar{x}$.

Because of this uncertainty, it behooves investigators to bracket their sample-size and optimization calculations for uncertainties in the variance estimates. Searle *et al.* (1992, Appendix F) provide variance estimates for variance components for two- and three-stage nested designs. In other situations, simulation approaches may be useful to assess potential sampling variability and bias in variance component estimators and the effects of this uncertainty on power/precision examinations (Chapter 9). Nevertheless, the available estimates of uncertainty invariably produce more effective and efficient sampling designs than those produced by ad hoc or intuitive approaches alone.

Table 8.7 The ANOVA table for a two-way design and expected values for the mean squares for the replicated small mammal trapping data. $\bar{\sigma}_{ME}^2$ denotes average measurement error.

Source	DF	SS	MS	E(MS)
Total$_{Cor}$	7	514.08		
Blocks	3	320.04	$MSL_1 = 106.68$	$2\sigma_B^2 + \sigma_{P/B}^2 + \bar{\sigma}_{ME}^2$
Plots/blocks	4	204.04	$MSL_2 = 51.01$	$\sigma_{P/B}^2 + \bar{\sigma}_{ME}^2$

The effects of blocking and intraclass correlation

Blocking in experimental designs is analogous to stratification in sampling designs. The purpose of both is to eliminate the between-block/between-stratum variance from the overall error term. The more heterogeneous the blocks are among each other, and the more homogeneous the sampling units within blocks, the more effective and efficient the blocking designs. Pairing is a special case of blocking where there are only two experimental units within each block, one for each of two treatments. The practice of pairing monitoring sites is common in restoration evaluations and impact-assessment designs. In both situations, a restored or potentially impacted site is paired with a reference or control site.

Preliminary surveys may be used to evaluate the performance of such designs compared to a completely randomized design based on estimated variance components and intraclass correlation. Skalski and Robson (1992) provide an example of such a preliminary survey on small mammal abundance. At a site in Colorado, eight 1-ha small mammal sampling grids were established in a two-stage nested design. Four 1-mile^2 sections were randomly selected within a township, and within each of those sections, two trapping grids were constructed. Small mammal abundance was estimated using the Lincoln/Petersen single mark-recapture model (see Chapter 18). Abundance estimates by section (block) and grid (plots-within-blocks) were: Block 1 = 18.25, 10.00; Block 2 = 16.50, 16.60; Block 3 = 10.67, 29.00; Block 4 = 2.00, 4.00.

A one-way ANOVA for a randomized effects model, or equivalently, a two-stage nested design, was used to analyze the estimated abundance data (Table 8.7). Johnson and Leone (1977: 606–610) provide expected mean squares for both a balanced and unbalanced one-way, randomized effect ANOVA. The data had three sources of variation: mark–recapture measurement error, plots-within-a-block variance, and between-block variance. The MSE for plots-within-block was estimated to be 51.01 with an expected value of

$$E(MSE) = \sigma_{P/B}^2 + \overline{Var(\hat{N}_i|N_i)}.$$

The mark–recapture model provided estimates of measurement error associated with each survey, (i.e. $Var(\hat{N}_i|N_i)$), from which was calculated an estimate of average measurement error of $\widehat{\mathrm{Var}(\hat{N}|N)} = 38.76$. Using this estimate and the MSL_1, the variance in abundance between plots-within-block ($\sigma^2_{P/B}$) was estimated to be

$$\hat{\sigma}^2_{P/B} = \mathrm{MSL}_1 - \widehat{\mathrm{Var}(\hat{N}|N)} = 51.01 - 38.76 = 12.25.$$

Mean abundance across all plots was $\hat{\bar{N}} = 13.38$ and is an estimate of mean abundance μ_N. Using the expected mean squares for blocks (MSL_2) and plots within a block (MSL_1), the between-block (σ^2_B) variance is estimated to be

$$\hat{\sigma}^2_B = \frac{\mathrm{MSL}_2 - \mathrm{MSL}_1}{2} = \frac{106.68 - 51.01}{2} = 27.835.$$

Hence, if a completely randomized design was implemented, the natural variance in actual annual abundance between plots would have been $\sigma^2_B + \sigma^2_{P/B}$, and estimated to be $27.835 + 12.25 = 40.085$. The variance among a simple random sample of plot-specific abundance estimates would then have the expected value

$$\sigma^2_B + \sigma^2_{P/B} + \widehat{\mathrm{Var}(\hat{N}_i|N_i)} \triangleq 27.835 + 12.25 + 38.76 = 78.845. \qquad (8.23)$$

However, if a blocked/paired design was implemented, the natural variance in actual abundance would be reduced to only the plots-within-block variance component ($\sigma^2_{P/B}$) estimated to be 12.25. Incorporating measurement error, the perceived variance for a paired design would be $12.25 + 38.76 = 51.01$. The difference between these estimates of total variation would be the estimated variance reduction due to blocking.

In the case of the single mark–recapture method (Seber 1982: 59), the measurement error can be expressed as

$$\mathrm{Var}(\hat{N}_i|N_i) = \frac{N_i(1 - p_1)(1 - p_2)}{p_1 p_2}, \qquad (8.24)$$

where

$p_1 =$ probability of capturing an animal in period 1,
$p_2 =$ probability of capturing an animal in period 2.

In the case of a replicated study, the average measurement error could be expressed as

$$\frac{\sum_{i=1}^{n} \frac{N_i(1 - p_{1i})(1 - p_{2i})}{p_{1i} p_{2i}}}{n}.$$

When designing a study a priori, this term could be approximated by

$$\frac{\mu_N(1 - p_1)(1 - p_2)}{p_1 p_2},$$

assuming common capture probabilities and mean abundance of μ_N. Combining natural and measurement sources of error, a treatment mean under a completely randomized design would have the expected variance of

$$\mathrm{Var}(\hat{\bar{N}}) = \frac{\sigma^2_B + \sigma^2_{P/B} + \frac{\mu_N(1 - p_1)(1 - p_2)}{p_1 p_2}}{n}.$$

In the case of a randomized block design, the variance of a treatment mean would be estimated to be

$$\text{Var}(\hat{N}) = \frac{\sigma^2_{\hat{P}/B} + \frac{\mu_N(1-p_1)(1-p_2)}{p_1 p_2}}{n}.$$

Study performance in either case would depend on the level of plot replication (n) and capture probabilities p_1 and p_2.

Intraclass correlation

In paired designs, the correlation between experimental units within a block determines the amount of variance reduction due to blocking. For example, in an unpaired, completely randomized design, the variance of the difference between two responses would be

$$\text{Var}(x_1 - x_2) = \text{Var}(x_1) + \text{Var}(x_2), \tag{8.25}$$

while for a paired design,

$$\begin{aligned}\text{Var}(x_1 - x_2) &= \text{Var}(x_1) + \text{Var}(x_2) - 2\,\text{Cov}(x_1, x_2)\\ &= \text{Var}(x_1) + \text{Var}(x_2) - 2\,\rho\sqrt{\text{Var}(x_1)\cdot\text{Var}(x_2)}.\end{aligned} \tag{8.26}$$

In the above small mammal example, there was no inherent labeling of the plots within the blocks by a treatment designation (e.g. control, treatment) or by order. Therefore, the standard Pearson product-moment estimator of correlation (ρ) is unsuitable. Instead, the correlation can be calculated by the intraclass correlation coefficient, defined as

$$\rho_I = \frac{\sigma^2_B}{\sigma^2_B + \sigma^2_{\hat{P}/B}}, \tag{8.27}$$

which for the small mammal example would be estimated to be

$$\rho_I = \frac{27.835}{27.835 + 12.25} = 0.694.$$

Hence, ρ_I suggests the correlation between *actual abundance levels* among plots within a block would be 0.694. In the presence of measurement error, the correlation between abundance estimates would be projected to be

$$E(\rho) = \frac{\sigma^2_B}{\sigma^2_B + \sigma^2_{\hat{P}/B} + \text{Var}(\hat{N}_i|N_i)} = \frac{27.835}{27.835 + 12.25 + 38.76} = 0.3530,$$

where, again, average measurement error was estimated to be 38.76.

To see how the intraclass correlation works within a block design, start with expressions (8.23) and (8.26). The variance for the difference between two abundance estimates within a block would have the expected value of

$$\text{Var}(\hat{N}_1 - \hat{N}_2) = \left[\sigma_B^2 + \sigma_{P/B}^2 + \text{Var}(\hat{N}|N)\right] + \left[\sigma_B^2 + \sigma_{P/B}^2 + \text{Var}(\hat{N}|N)\right]$$

$$- 2 \left(\frac{\sigma_B^2}{\sigma_B^2 + \sigma_{P/B}^2 + \text{Var}(\hat{N}|N)}\right) \sqrt{\left(\sigma_B^2 + \sigma_{P/B}^2 + \text{Var}(\hat{N}|N)\right)^2}$$

$$= \left(\sigma_{P/B}^2 + \text{Var}(\hat{N}|N)\right) + \left(\sigma_{P/B}^2 + \text{Var}(\hat{N}|N)\right), \tag{8.28}$$

which is as expected because block effects (i.e. σ_B^2) are eliminated by pairing. The intraclass correlation coefficient can be used in lieu of the Pearson product-moment correlation coefficient when necessary.

Statistical power to test for long-term trends

A basic objective of most long-term monitoring programs is to detect a temporal trend in a critical response variable (e.g. Chapters 10, 11). In this section, I provide a brief introduction to assessing power to detect a trend; Chapter 7 addresses this topic in more detail.

For endemic species, the goal is often to detect an appreciable decline in abundance. This goal translates into the null hypothesis: H_o: $\beta \geq 0$ versus the one-tailed alternative of H_a: $\beta < 0$, where β is the slope in the regression of abundance against time (Fig. 8.1). For invasive species or environmental pollutants, the opposite one-tailed hypothesis is relevant: H_o: $\beta \leq 0$ vs. H_a: $\beta > 0$.

The statistical test of slope can be based on a t-test of the form

$$t_{n-2} = \frac{\hat{\beta} - 0}{\sqrt{\dfrac{\text{MSE}}{\sum\limits_{i=1}^{n}(t_i - \bar{t})^2}}}, \tag{8.29}$$

where MSE is the mean squared error or residual mean deviance, an estimate of the unexplained variance, and t_i is the time of the ith yearly observation ($i = 1, \ldots, n$). The statistical power of the test can be calculated using a non-central F-distribution with non-centrality parameter

$$\Phi_{1,n-2} = \frac{1}{\sqrt{2}} \cdot \frac{|\beta|}{\sqrt{\dfrac{\text{MSE}}{\sum\limits_{i=1}^{n}(t_i - \bar{t})^2}}}. \tag{8.30}$$

In the case of long-term monitoring, the slope parameter can be re-expressed in terms of

$$\beta = \mu_0 \left(\frac{\Delta}{n-1}\right),$$

where Δ = fractional reduction in response at the end of n years of study, and μ_0 is the mean response at time zero (i.e. t_0). The MSE can, in turn, be expressed in terms of natural variation in response (i.e. σ^2), plus average measurement error (i.e. $\overline{\text{Var}(\hat{\theta}|\theta)}$). Hence, the non-centrality parameter [Equation (8.30)] can be rewritten as

$$\Phi_{1,n-2} = \frac{1}{\sqrt{2}} \cdot \frac{\left|\mu_0\left(\frac{\Delta}{n-1}\right)\right|}{\sqrt{\dfrac{\sigma^2 + \overline{\text{Var}(\theta|\theta)}}{\displaystyle\sum_{i=1}^{n}(t_i - \bar{t})^2}}}. \tag{8.31}$$

For example, consider a monitoring program with the statistical goal of being able to detect a 50% decline in abundance over a 10-year study, the non-centrality parameter is then

$$\Phi_{1,8} = \frac{1}{\sqrt{2}} \cdot \frac{\left|\mu_0\left(\frac{0.50}{9}\right)\right|}{\sqrt{\dfrac{\sigma^2 + \overline{\text{Var}(\hat{\theta}|\theta)}}{82.5}}}$$

for $t_i = 0, 1, \ldots, 9$. With estimates of μ_0, σ^2, and $\overline{\text{Var}(\hat{\theta}|\theta)}$, the statistical power to detect the 50% decline over 10 years can be found using the non-central F-tables of Skalski and Robson (1992, Appendices 2 and 3). Statistical software such as R (R Development Core Team 2011) can be readily used to calculate statistical power. However, when using any software to calculate power, the exact form of the non-centrality parameter should be checked because its form varies greatly between programs.

Additional issues

Expressing uncertainty on other scales

In biological systems, the variance in response usually changes with the value of the mean response. In these circumstances, the CV tends to be more stable over time and space than the variance. The reason is that environmental effects are often multiplicative and many types of biological data are approximately log-normally distributed, where

$$\text{Var}(\ln x) \doteq \frac{\text{Var}(x)}{\mu^2} = \text{CV}^2. \tag{8.32}$$

Consequently, the variance estimates and variance components may be more meaningfully expressed in terms of CVs or CVs of variance components (e.g. $\frac{\sigma_B}{\mu} = \text{CV}_B$), and precision or power calculations re-expressed accordingly. For example, expressing sampling precision in terms of relative error, i.e.

$$P\left(\left|\frac{\hat{\theta} - \theta}{\theta}\right| < \varepsilon\right) = 1 - \alpha,$$

then

$$\varepsilon \doteq \frac{Z_{1-\frac{\alpha}{2}}\sqrt{\text{Var}(\hat{\theta})}}{\theta} = Z_{1-\frac{\alpha}{2}}\text{CV}(\hat{\theta}), \tag{8.33}$$

where precision is a function of the CV of $\hat{\theta}$. Insights into the sampling process and objective function of the study can therefore be combined to provide a relatively stable platform for sample-size calculations despite limited preliminary data and natural stochasticity.

Sources of variance component information

The single best source of variance component information is from a preliminary survey. Preliminary data collection for the purpose of sample-size calculations and optimization should be an integral part of any proposed monitoring program. Design development should consider alternative measurement technologies, the benefits of blocking if appropriate, rates of rotation, and sampling costs with the objective of finding the most cost-effective and statistically powerful monitoring program (e.g. Chapters 10, 20). The best available information should be used in establishing the initial study design. Field data for the first 1 or 2 years should then be used to fine tune the consummate study design. Because the goal is to implement a long-term monitoring program, sacrificing some consistency in the first couple of years for long-term performance should not be of undue concern.

In the absence of first-hand preliminary data, investigators may look to other investigations or the scientific literature for in-kind information on anticipated measurement error and natural variability. This may require reanalyzing some other investigator's data for the explicit purposes of variance component estimation. Some investigators are reluctant to use data from other locales or time periods because they view the information inapplicable or irrelevant to the population at hand. While measured responses may very well be different and environmental fluctuations somewhat dissimilar, the multiplicative effects in biological systems often make the coefficients of variation stable and useful in sample-size calculations [Equation (8.33)] (van Belle 2008: 31–35, 88–90). Moreover, supplementing preliminary surveys with examination of existing data sets often is essential for estimating year-to-year variability, for which many years of data may be needed to produce relatively accurate estimates (Chapters 7, 9).

If all else fails and you are in the rare situation of being the first person to ever study anything remotely similar to the species or ecosystem in question, you may be able to fall back on first principles. In monitoring population abundance, spatial heterogeneity (i.e. σ_N^2) can be anticipated to be greater than or equal to the mean abundance (i.e. μ_N). In a uniform environment and in the absence of spatial competition, animal dispersion would be random and described by a Poisson process (i.e. $\sigma_N^2 = \mu_N$). Hence, a "guesstimate" of mean abundance also provides a lower bound on spatial variance. As the environment becomes more heterogeneous and animal preferences more exacting, one can expect $\sigma_N^2 \gg \mu_N$. This level of information may be adequate to begin design considerations until actual field data can be used to refine the study.

The one option to avoid is to do nothing. The crudest preliminary information is likely to produce a better design and sample allocation than the best ad hoc approach. Left to their own devices, field biologists trained to be good technicians will invariably allocate sampling effort the opposite of optimal allocation [Equations (8.20)–(8.22)]. Too much effort will invariably be exerted at reducing measurement error, and too little effort allocated to the higher levels of the sampling design that contribute the most to the overall variance. Optimization processes will usually avoid such pitfalls.

Where next?

Chapters 7 and 9 provide additional discussions of variance component estimation in a linear-model framework, with variance components corresponding to specified random effects in a general model applicable to many large-scale monitoring programs. Where to go next for more information depends on the statistical sophistication of the reader. For those of you who found reading this chapter challenging, there are two recommendations. First, a general appreciation for design and analysis issues can be found in *Statistical Rules of Thumb*, written by van Belle (2008). While the book lacks sufficient detail necessary to design a monitoring program, it does provide valuable insights into designing biological studies. It also provides an effective starting point for collaboration with a consulting statistician. In today's litigious environment and where monitoring programs can cost millions of dollars, an interdisciplinary team approach is the only responsible action. Ecologists, managers, and biologically trained statisticians should pool their respective talents and insights in developing monitoring programs that are biologically meaningful, relevant, statistically sound, and cost effective.

For the other readers interested in advanced topics and performing their own variance component analyses, you would do well to consider texts on linear models (Searle 1971, Neter *et al.* 1990), analysis of unbalanced data (Searle 1987), and variance component estimation (Searle *et al.* 1992). These approaches are most suited for the cases of additive effects, homogeneous variances, and observations without measurement error. Incorporation of measurement error can typically be handled by using the variance-in-stages formulas (8.4) and (8.5) in the last stage of error estimation. For more examples and applications of taking variances in stages, the reader is referred to Cochran (1977: 275–277, 301, 344) and Seber (1982: 9).

Recommendations for future research and development

Relevant to this chapter, the most important "development" needed is routine application of the survey-design principles and approaches outlined in this chapter and by others in this volume. Funding agencies should require, as a requisite part of any monitoring program, a rigorous statistical plan. That plan should identify the variables to be monitored, the quantitative objective functions of the investigation, and detailed sampling designs. A part of that plan should include a preliminary survey and rigorous sample-size calculations. In turn, groups charged with monitoring should expect the resources needed to develop the statistical design and analysis plans, collect the preliminary survey and cost data, and perform sample-size and optimization calculations. Recognition by all parties of the importance of statistical principles in study design will go a long way to ensuring future monitoring programs will be capable of detecting environmental changes considered important.

Summary

A monitoring program starts first and foremost with a design and analysis plan. Part of that study plan must include sample-size calculations. Inevitably, sample sizes will be multidimensional to account for both spatial and temporal replication and within-site measurement error. Variance component estimation can be used to estimate the potential magnitude of natural variation and measurement error in order to optimally distribute the sampling resources. Analysis of variance and the associated expected mean squares provide a convenient framework for variance component estimation. The complexity of the variance component analysis will depend on whether the preliminary survey data are balanced or not. However, to partition measurement error from natural variation, the measurement process must either be repeated or the techniques must provide a model-based estimate of uncertainty. For monitoring programs based on rotating panel designs or blocking, covariances or correlation coefficients must also be estimated prior to sample-size calculations. However, the perceived correlation between responses decreases as measurement error increases. In designing a biologically meaningful, relevant, statistically precise, and cost-effective monitoring program, an interdisciplinary team approach including resource managers, ecologists, and statisticians is always recommended.

Acknowledgment

I would like to thank the Bonneville Power Administration, Project 1991-051-00, for their support of this statistical review.

9 Variance components estimation for continuous and discrete data, with emphasis on cross-classified sampling designs

Brian R. Gray

Introduction

Variance components may play multiple roles (cf. Cox and Solomon 2003). First, magnitudes and relative magnitudes of the variances of random factors may have important scientific and management value in their own right. For example, variation in levels of invasive vegetation among and within lakes may suggest causal agents that operate at both spatial scales – a finding that may be important for scientific and management reasons.

Second, variance components may also be of interest when they affect precision of means and covariate coefficients. For example, variation in the effect of water depth on the probability of aquatic plant presence in a study of multiple lakes may vary by lake. This variation will affect the precision of the average depth-presence association.

Third, variance component estimates may be used when designing studies, including monitoring programs. For example, to estimate the numbers of years and of samples per year required to meet long-term monitoring goals, investigators need estimates of within- and among-year variances. Other chapters in this volume (Chapters 7, 8, and 10) as well as extensive external literature outline a framework for applying estimates of variance components to the design of monitoring efforts. For example, a series of papers with an ecological monitoring theme examined the relative importance of multiple sources of variation, including variation in means among sites, years, and site-years, for the purposes of temporal trend detection and estimation (Larsen *et al.* 2004, and references therein).

Due to the scientific and management value of variance components and their role in study design, the statistical properties of variance component estimators should be investigated. Specifically, investigators may explore bias, precision, and other properties of estimators using variance component estimates derived from, for example, simulated data sets. Such evaluations may be undertaken to compare the performance of alternative estimators in general, or to assess the performance of one or more estimators under the constraints of a specific study or monitoring design (e.g. Chapter 10). Note that

Design and Analysis of Long-term Ecological Monitoring Studies, ed. R.A. Gitzen, J.J. Millspaugh, A.B. Cooper, and D.S. Licht. Published by Cambridge University Press. © Cambridge University Press 2012.

understanding the properties of variance component estimators represents an important aspect of wisely using variance component estimates for designing monitoring programs.

What, specifically, are variance components? Consider each of multiple measurements, $y_{ij}, i = 1, \ldots, n$, from each of multiple units ("sites"), $j = 1, \ldots, m$, and presume that y_{ij} derives from the model $y_{ij} = \beta_{00} + u_{0j} + e_{ij}$, where β_{00} denotes the expected value (grand mean) of y_{ij}, $E[y_{ij}]$, u_{0j} a site-specific effect on y_{ij}, and e_{ij} residual variation. Presume further that the u_{0j}'s are independent and identically distributed (iid) with mean zero and variance σ_{site}^2, that the e_{ij}'s are iid (given the u_{0j}'s) with mean zero and variance σ_e^2, and that the e_{ij}'s are uncorrelated with the u_{0j}'s. Then it may be shown that the variance of y_{ij}, σ_y^2, is $\sigma_{\text{site}}^2 + \sigma_e^2$. Hence, we see that σ_{site}^2 and σ_e^2 are components of σ_y^2, the variance of y (Searle *et al.* 1992).

This chapter aims to help readers estimate variances associated with monitoring designs, to think critically about the value of variance estimates from small samples, and to emphasize that estimator performance may affect study design conclusions. The previous chapter in this volume (Chapter 8) provides a broad introduction to identifying components of variation and applying general, fundamental variance expressions to specify the variance structure relevant to any particular ecological study. Chapters 7, 8, and 10 discuss the use of variance estimates in designing monitoring surveys. In this chapter, I focus more specifically on estimation of variance components, using not only analysis of variance (ANOVA) but also maximum likelihood (ML) – whether full (FML) or restricted (REML), partial or quasi-likelihood – and Bayesian approaches. More attention is paid to ML estimation because ANOVA is traditionally restricted to use with linear models, use of ANOVA is demonstrated in Chapters 7 and 8, and variance component estimation has been addressed in the literature more extensively using likelihood than Bayesian methods.

This chapter addresses variance components estimation from clustered continuous, categorical, and count data (Box 9.1). Variance component estimation for categorical and count data involves considerations and approaches that may seem highly technical to some of this volume's readers. However, advanced readers may find this coverage useful, given that categorical and count data are commonly encountered in ecological studies. Throughout the chapter, general comments from the literature are augmented by results from simulations. Foci include estimation from small samples and from data from cross-classified sampling designs.

A cross-classified model

The ecological and temporal foci of this volume suggest the consideration of two-way cross-classified random effects sampling designs. Such designs permit multiple observations from, for example, multiple sites in each of multiple years. A candidate model equation may be represented by

$$y_{ijk} = \beta_{00} + u_{0j} + u_{0k} + u_{0jk} + \varepsilon_{ijk} \tag{9.1}$$

where y_{ijk} denotes the ith observation, $i = 1, \ldots, n_{jk}$, within the jkth site-year, $j = 1, \ldots, m_{\text{site}}, k = 1, \ldots, m_{\text{year}}$; β_{00} denotes the expected value (grand mean) of y_{ijk}; u_{0j},

Box 9.1 Take-home messages for program managers

Managers often rely on estimates from grouped or clustered data. Examples include estimates of mean animal abundance or pollutant concentration from multiple observations from each of multiple lakes, streams, or years. Such data should typically be presumed correlated within clusters, with the correlation arising because data from one cluster will typically be more like other data from that same cluster (and less like data from other clusters). Failure to address correlation in clustered data may yield invalid conclusions.

Correlation *within* clusters may be viewed as the flip side of variation *among* clusters. For example, the average or mean abundance of a species may vary from year to year, a finding that implies that data from a single year are relatively similar. As discussed in the previous paragraph, this similarity ensures abundance observations are correlated within years. Consequently, variation of means among years implies correlation within years.

This information – correlation and variation among year or cluster means – will often lead, from a planning perspective, to treating the number of clusters as more important than the number of data points within clusters. Since data from clusters are typically correlated, the information associated with those data is partially shared with other observations from the same cluster. This sharing of information means that data from the same cluster contain less information than is contained in an otherwise equivalent set of uncorrelated observations. Therefore, the contribution of observations within clusters to *effective* sample size is smaller than their contribution to the stated or nominal sample size. For example, 100 observations derived as 20 clusters of 5 observations might contain information equivalent to only 70 independent observations.

For designing future studies, investigators need estimates of variation among and within clusters. Given study goals, these estimates would then be used to select combinations of numbers of clusters and of observations per cluster (see Chapter 8). Determining the ratio between numbers of clusters and observations per cluster will typically also take into account the costs of sampling within clusters and of traveling between clusters. In the context of long-term monitoring, where years may often be viewed as clusters, there may be logistical/budgetary benefits associated with sampling less frequently but more intensively (e.g. every kth year).

Estimates of variation among clusters may also be used for scientific purposes. For example, variation in levels of invasive vegetation among lakes may suggest a causal agent that operates for entire lakes (e.g. lake substrate or lake management practice).

This chapter reviews variance estimation for types of data and situations that are commonly encountered in ecological studies. Major points include that estimation of among-cluster variances will often involve bias and low precision unless the number of clusters is moderate to large (20 to possibly as many as 100). The accuracy and precision concerns described in this chapter should be considered when designing monitoring programs (and pilot studies to guide further selection of monitoring survey designs) and when interpreting variance components for scientific and management purposes.

u_{0k}, and u_{0jk} denote random site, year, and site × year effects (respectively), and ε_{ijk} denotes residual error. The random terms are presumed distributed with means zero, and variances σ^2_{site}, σ^2_{year}, $\sigma^2_{\text{site-year}}$, and σ^2_{ε}, respectively. Note that the model is a random effects model (covariates are absent) and that, from a multilevel (hierarchical) modeling perspective, the three group-level terms are at the same level while the observations occur at a lower level, the observation level. For further information about modeling outcomes with cross-classified random group effects, see Meyers and Beretvas (2006) and the multilevel modeling texts by Raudenbush and Bryk (2002), Goldstein (2003), and Hox (2010).

Simulation models

The above cross-classified model and the specified variance components provide a general framework relevant to many long-term monitoring efforts and other ecological studies. For example, this model essentially is of the same form as the models specified in Chapter 7 and applied in Chapters 7 and 10. However, the majority of studies in the literature assess estimator performance for designs less relevant to the focus of this volume. Further, many of those studies presume sample sizes that exceed those typically associated with ecological studies. Specifically, pilot studies designed to quantify variance components for use in designing monitoring studies will often be associated with cost and time constraints that yield both few observations per site-year (n) and few years (m_{year}), and often relatively few sites (m_{site}). For these reasons, I augmented findings from the literature with those from data simulated under Equation (9.1) with relatively small sample-size assumptions.

For most simulation scenarios conducted for this chapter, site, year, site × year, and residual effects were distributed as normal random variates, with means 0 and variances as defined under (9.1). For a few scenarios, however, random effects were generated from non-normal distributions to explore the effects of violating normal distributional assumptions common to most estimation methods. In other scenarios, I altered the relative magnitudes of variance components, sample sizes (m_{site}, m_{year}, or n), and other factors relevant to specific data types (e.g. a naïve Poisson distributional assumption for conditional, negative binomial-distributed count data). With the exception of those scenarios specified as unbalanced, numbers of observations per site-year were constant (i.e. $n_{jk} = n$ for all jk) and small ($n = 5$ or 10). As a practical matter, note that when using the model associated with Equation (9.1) without supplemental information, it is necessary to have multiple visits to at least some sites within some years (i.e. $n > 1$ for some site-year groups) to separate year-specific variation among sites ($\sigma^2_{\text{site-year}}$) from residual variation.

In all simulations, site and year effects were independent, and site-year (u_{0jk}) effects and residual errors (ε_{jk}) conditionally so. In practice, real data may be more complex – site effects may be spatially correlated, annual effects may be temporally correlated, and conditional site-year effects may be spatio-temporally correlated. Given small sample sizes, however, such concerns will not easily be addressed. The appropriateness of these simplifying assumptions does not affect the conclusion that small sample sizes may jeopardize accurate and precise estimation of variance components. Owing to the temporal focus of this volume and because variation in year effects may most directly affect temporal inferences (Larsen *et al.* 2004), I provide results primarily for σ^2_{year} estimators; results for other estimators may be obtained using code supplied as an online supplement at this volume's web site.

Variance components estimation and linear models

Variance component estimation using linear models with random effects is a relatively developed topic, dating to at least 1861 (Airy 1861, Scheffé 1956), and is the focus of many texts (e.g. Searle *et al.* 1992, Rao 1997, Cox and Solomon 2003). Variance components have traditionally been estimated using ANOVA and, beginning in the 1960s and 1970s, increasingly with ML and REML (Searle *et al.* 1992). More recently, computational advances have stimulated interest in Bayesian methods of estimating variance components (Draper 2008). See Appendix 9.1 for description of ANOVA, ML, and Bayesian methods of estimating variance components using linear models. Chapters 7 and 8 provide additional discussion and demonstration of ANOVA estimation.

The maturity of variance components estimation using linear models, combined with the common problem for ecological studies of small sample sizes, motivated this chapter's focus on issues associated with sample size. For variance component models, the most relevant sample size is typically defined relative to the variable of interest – years if among-year variance is of interest, sites if among-site variance is of interest, and so on. This perspective often leads to greater attention being paid to the estimation of group-level than to observation-level components: sample sizes at group levels are typically much smaller than sample sizes at the observation level, although exceptions may occur when σ^2_e is estimated by group or by other data subset.

Bias of variance component estimators

Bias is commonly considered when evaluating the performance of estimators. However, the importance of bias as a criterion for evaluating estimators is often treated as less important for variance component than for fixed-effect estimators. For ANOVA, this reflects that the favorable property of the unbiasedness of ANOVA variance estimators comes at the expense of accepting negative variance estimates (e.g. see Chapter 7) while the common solution of setting those negative estimates to zero yields positive bias. However, FML and REML estimators of group-level variance components are often biased (Searle *et al.* 1992). Another concern is that bias reflects the idea of multiple repetitions of the same study (Searle *et al.* 1992, section 2.3; see also Chapter 3). Of course, variance components are often estimated from observational data – where replication may be empirically unattainable.

Despite the above issues, bias of variance component estimators remains an important, albeit qualified, concern. For example, Browne and Draper (2000, 2006) documented negligible bias with REML for group-level variance components, but relatively large negative biases with FML at 6 and 12 groups (∼–21% and –11%, respectively); biases associated with Bayesian estimation using Gibbs sampling depended on prior and posterior summary (mean, median, mode), but were generally greater in magnitude than those documented for ML. Biases decreased in magnitude for both ML and Bayesian methods when numbers of groups increased to 24 and 48. Raudenbush (2008) argued that, in a likelihood setting, small numbers of observations per group may yield group-level variance component estimates that are biased low; by contrast, the residual error variance, σ_ε^2, which is estimated from all observations from all groups, will generally be accurately estimated (and, for this reason, estimation of σ_ε^2 will represent a minor focus of this chapter).

The above findings from the literature are congruent with those from the simulations conducted for this chapter using small sample sizes and cross-classified random effects designs (Table 9.1). Relative biases associated with ANOVA and REML estimators of σ_{year}^2 were ignorable when $m_{year} = 10$ and 20, and modest when $m_{year} = 3$ and 5 (scenarios 1–4; ANOVA, REML). The unexpected finding of lower bias for the ANOVA and REML estimators when years $= 3$ is an artifact of the treatment of negative values: negative ANOVA estimates were, as is commonly done, set to zero while REML (and FML) estimation ensures nonnegative estimates by setting negative solutions to zero. (For all estimation methods, the proportion of non-positive variance estimates under the years $= 3$ scenario was $\geq 10\%$.) In contrast to the ANOVA and REML cases, biases for FML were severe at small sample sizes (scenarios 1 and 2). As noted in Appendix 9.1, the bias of the FML estimator is approximately $-k$/sample size (provided the proportion of zero estimates is small), where k denotes the number of linearly independent fixed covariate terms. For this and other reasons, REML is typically preferred over FML for estimating variance components. See McCulloch and Searle (2001) for a fuller treatment of this topic.

The effects of manipulating attributes unrelated to year on bias of $\widehat{\sigma_{year}^2}$ were typically minor. Relative to that of the $m_{year} = 5$ reference (scenario 2), bias improved modestly when n was doubled but was not improved by doubling the number of sites (scenarios 5, 6). Bias improved trivially when σ_ε^2 was substantially decreased (scenario 9); such decreases may occur with sampling protocol refinement or adjustment for covariates that vary at the observation scale. As also described by Cools et al. (2009), the effects on bias were not severe when datasets were unbalanced (scenarios 7 and 8).

Precision

The topics of precision and of precision comparisons across estimators of group-level variances, particularly when numbers of groups are few, have received limited attention. Browne and Draper (2006) documented improvements in the precision of group-level variance component estimators as numbers of groups increased, when data sets were balanced, and as the ratio of among-group variance to within-group variance (i.e. the intraclass correlation coefficient) increased. While Monte Carlo standard deviations for scenarios with few groups ($m = 6$, 12) were typically smaller for FML than for REML and for both relative to Bayesian estimators, among-method differences narrowed or disappeared when number of groups reached 24. Coverage by Bayesian credible intervals approximated nominal coverage with as few as 12 groups. See Browne and Draper (2000) for a related study, and Gelman and Hill (2007, section 19.6) for comments on selection of priors for estimating group-level variances in studies with few groups.

Table 9.1 Estimates of among-year variances, σ^2_{year}, from random effects linear models by estimation method. Unless otherwise indicated, numbers of sites, years, and observations per site-year (n) = 10, 5, and 5, respectively, $\sigma^2_\varepsilon = 1$, $\sigma^2_{site} = \sigma^2_{year} = 0.3$, and $\sigma^2_{site-year} = 0.15$.[a]

		$\widehat{\sigma^2_{year}}$ by estimation method (Monte Carlo standard deviation), % zero estimates[b]		
Scenario	Attribute varied	ANOVA	FML	REML
Sample size modifications				
1	Years = 3	0.303 (0.345), 0.10	0.207 (0.235), 0.13	0.303 (0.345), 0.10
2	Years = 5 (reference)	0.285 (0.239), 0.02	0.236 (0.195), 0.02	0.285 (0.239), 0.02
3	Years = 10	0.298 (0.159), 0.00	0.278 (0.146), 0.00	0.298 (0.159), 0.00
4	Years = 20	0.297 (0.103), 0.00	0.290 (0.100), 0.00	0.297 (0.103), 0.00
5	$n = 10$	0.292 (0.236), 0.01	0.244 (0.193), 0.02	0.292 (0.236), 0.01
6	Sites = 20	0.284 (0.221), 0.01	0.233 (0.179), 0.01	0.284 (0.221), 0.01
Unbalanced data sets				
7	Unbalanced-1[c]	0.284 (0.308), 0.06	0.228 (0.205), 0.06	0.279 (0.251), 0.05
8	Unbalanced-2[d]	0.300 (0.322), 0.06	0.231 (0.221), 0.11	0.293 (0.267), 0.08
Variance component modifications				
9	$\sigma^2_\varepsilon = 0.5$	0.286 (0.233), 0.02	0.239 (0.191), 0.02	0.286 (0.233), 0.02
10	$\sigma^2_{site} = 1$	0.285 (0.239), 0.02	0.251 (0.204), 0.02	0.285 (0.239), 0.02
11	$\sigma^2_{year} = 1$	0.954 (0.740), 0.00	0.777 (0.595), 0.00	0.954 (0.740), 0.00
Distributional violations				
12	Uniform[e]	0.302 (0.201), 0.02	0.251 (0.165), 0.02	0.302 (0.201), 0.02
13	Uniform, years = 20	0.302 (0.079), 0.00	0.295 (0.077), 0.00	0.302 (0.079), 0.00

[a] Estimates represent means of simulation-specific estimates; simulations per scenario = 500; 'ANOVA,' 'FML', and 'REML' denote analysis of variance (Type III estimation), full maximum likelihood, and residual maximum likelihood, respectively (see Littell *et al.* 2006 for computational details); negative variance estimates (ANOVA only) were set to zero; model convergence proportions for ML and REML = 1.00. Results were generated using SAS' linear mixed modeling procedure (PROC MIXED; SAS 2008).
[b] $\leq$ 1E-4 units; includes negative estimates (ANOVA only).
[c] All cells (site-year groups) had observations, but in three of five years, 70% of sites had only one observation.
[d] Some cells empty; in three of five years, 70% of sites had no observations.
[e] Random effects generated as random uniform random variates with nominal variance.

The precisions of σ^2_{year} estimators under the small sample-cross-classified design simulation study of this chapter were poor, with Monte Carlo standard deviations approaching or exceeding in value those of the corresponding variance component point estimates (Table 9.1). Substantial improvements in precision occurred only with increased numbers of years, chiefly when years reached 10 and 20 (scenarios 3, 4). As may be expected, precision was largely unaffected by increases in numbers of sites sampled, numbers of observations sampled per site-year, and changes in σ^2_{site} and σ^2_ε (scenarios 5, 6, 9, 10). Precision deteriorated modestly when data sets were unbalanced, with less deterioration when cells contained one rather than no observations (scenarios 7, 8; cf., Cools *et al.* 2009).

> **Box 9.2** Common challenges[1]: dealing with inaccuracy
>
> As this chapter documents, variance components estimated from small sample sizes will often be characterized by low precision and – depending on the data type and estimator – substantial levels of bias. Investigators need to consider these concerns when using variance estimates from small samples for study design. Often, a simulation approach is useful for evaluating the potential magnitude of available estimates, following the same approach demonstrated in this chapter but with sample sizes and other details tailored to the specific situation. The bias/precision observed in such simulations can be useful for determining a range of plausible values for each variance component; study design options may then be evaluated using multiple plausible values rather than that of a single point estimate (see also Chapter 8). This problem may also be addressed through full integration of uncertainty in variance component estimates (Sims *et al.* 2007). In some cases, although absolute results of such design studies may change greatly across the range of plausible values (e.g. expected power for trend detection or sample size needed for a specified precision), the comparative trade-offs among alternative design options may be relatively robust to uncertainties in the variance estimates. In the absence of such a finding, further data collection may be required.
>
> The concern with poorly estimated variance components is especially problematic when monitoring program personnel wish to estimate the minimum number of years until power to detect a given trend reaches a given level. The value of σ^2_{year} will often have an overriding influence on power to detect temporal trends (e.g. Urquhart *et al.* 1998), and bias and imprecision associated with estimators of σ^2_{year} may lead to biased and imprecise estimates of power to detect trends (Sims *et al.* 2007). In these cases, the simulation approach described in the previous paragraph will be useful. Also, in some cases, suitable long-term data sets conducive to meaningful estimation of σ^2_{year} will be available, either from the system under investigation or from systems with patterns of variation hypothesized to be similar.

In practice, precision concerns when numbers of groups are small may be addressed by treating inferences as provisional pending more years of monitoring (presuming variance component magnitudes remain approximately constant) and by decreasing variances by adjusting for covariates. When planning for new studies, investigators should consider that available or pilot variance component estimates derived from few groups may not only be biased but also be estimated imprecisely. The importance of these concerns may be addressed using simulations (Box 9.2).

Mean squared error

Bias and precision may be quantified in combination using a number of methods, including that of root mean squared error (RMSE $= \sqrt{\text{bias}^2 + \text{variance}}$). For the simulation

---

[1] With Robert Gitzen.

study and as may be inferred from the preceding discussion, relative RMSEs for $\widehat{\sigma^2_{\text{year}}}$ decreased over the 3, 5, 10, and 20 year series (0.35, 0.24, 0.16, and 0.10 units, respectively), with the largest decreases occurring when numbers of years were relatively small (ANOVA/REML, Table 9.1).

Distributional assumptions

Group-level random effects such as those associated with Equation (9.1) are typically assumed to be normally distributed (although ANOVA estimation does not strictly require this assumption; Appendix 9.1, Chapter 7). Maas and Hox (2004) addressed the reasonableness of this assumption by intentionally treating skewed (chi-squared with one degree of freedom), heavy-tailed (uniform), and light-tailed (Laplace) group effects as normally distributed. The effects of these distributional violations on variance component point estimates from REML were modest, with the largest reported relative bias (12%) associated with their study's smallest sample size scenario (30 groups with five observations per group).

Modest effects of distribution violations were also seen when group effects in this chapter's simulation study were random uniform but were naïvely modeled under a normal distributional assumption (scenarios 12, 13, Table 9.1). At years $= 5$ and relative to the balanced case, mean variance point estimates were larger while Monte Carlo standard deviations decreased (scenarios 2, 12). Increasing the number of years to 20 narrowed differences among point estimates but widened relative differences among precision estimates (scenarios 3, 13).

The possibility that distributional assumptions have not been essentially met should lead investigators to treat variance component estimates from small samples with additional caution. Snijders and Berkhof (2008) address diagnostic checks for model residuals at group and observation scales. Unfortunately, the reasonable satisfaction of distribution assumptions may be unclear when numbers of groups are small; in these cases, the practical importance of distributional-assumption failure may be addressed using simulations.

Variance partition coefficients

Group-level variance components are often reported as proportions of total variation. Such proportions are termed variance partition coefficients (VPCs), and may be used for both study planning and scientific purposes (Kincaid *et al.* 2004; Browne *et al.* 2005; Hox 2010, Chapter 12). For linear models without covariates with random coefficients, VPCs are equivalent to the intracluster or intraclass correlation coefficient (ICC) familiar from survey statistics (see also Chapter 8). For these models (and given a single group-level variance term), ICC $=$ VPC $= 0$ denotes independent observations while ICC $=$ 1 denotes an absence of variation within groups; most studies with grouped data yield ICC estimates between 0 and 1.

Table 9.2 Among-year variance partition coefficient (VPC$_{year}$) estimates from random effects linear models by estimation method. Unless otherwise indicated, numbers of sites, years, and observations per site-year (n) = 10, 5, and 5, respectively, $\sigma_\varepsilon^2 = 1$, $\sigma_{site}^2 = \sigma_{year}^2 = 0.3$, and $\sigma_{site-year}^2 = 0.15$.[a]

		$\widehat{VPC}_{year}$ by estimation method (VPC = 0.17) [Mean (Monte Carlo standard deviation), median]		
Scenario	Attribute varied	ANOVA	FML	REML
Sample size modifications				
1	Years = 3	0.15 (0.13), 0.12	0.11 (0.10), 0.09	0.15 (0.13), 0.12
2	Years = 5 (reference)	0.17 (0.09), 0.13	0.16 (0.08), 0.12	0.17 (0.09), 0.13
3	Years = 10	0.17 (0.07), 0.16	0.16 (0.07), 0.15	0.17 (0.07), 0.16
4	Years = 20	0.17 (0.05), 0.17	0.17 (0.05), 0.17	0.17 (0.05), 0.17
5	$n = 10$	0.16 (0.10), 0.14	0.14 (0.09), 0.12	0.16 (0.10), 0.14
6	Sites = 20	0.15 (0.09), 0.14	0.13 (0.08), 0.12	0.15 (0.09), 0.14
Unbalanced data sets				
7	Unbalanced-1[b]	0.15 (0.12), 0.12	0.13 (0.10), 0.11	0.15 (0.11), 0.13
8	Unbalanced-2[c]	0.15 (0.12), 0.13	0.13 (0.10), 0.11	0.15 (0.12), 0.13

[a] See footnote "a", Table 9.1, for analytical details.
[b] All cells (site-year groups) had observations, but in 3 of 5 years, 70% of sites had only 1 observation.
[c] Some cells empty; in 3 of 5 years, 70% of sites had no observations.

The two-way cross-classified random effects model associated with Equation (9.1) yields VPCs that correspond to correlation between two outcomes from the same site and year, same site but different years, and same year but different sites, respectively (cf. Raudenbush and Bryk 2002):

$$VPC_{site-year} = \frac{\sigma_{site}^2 + \sigma_{year}^2 + \sigma_{site-year}^2}{\sigma_{site}^2 + \sigma_{year}^2 + \sigma_{site-year}^2 + \sigma_\varepsilon^2}$$

$$VPC_{site} = \frac{\sigma_{site}^2}{\sigma_{site}^2 + \sigma_{year}^2 + \sigma_{site-year}^2 + \sigma_\varepsilon^2} \quad (9.2)$$

$$VPC_{year} = \frac{\sigma_{year}^2}{\sigma_{site}^2 + \sigma_{year}^2 + \sigma_{site-year}^2 + \sigma_\varepsilon^2}$$

The effects of sample size and balance on VPC estimation are illustrated using $\widehat{VPC}_{year}$, as derived from simulation study estimates. For scenarios 1–8, biases in mean $\widehat{VPC}_{year}$ from ANOVA and REML estimation were <15% (Table 9.2). However, biases in the medians of the typically right-skewed $\widehat{VPC}_{year}$ exceeded |20%| when years ≤ 5 and, for ML, reached −32% and −47% when years = 5 and 3, respectively. The properties of VPC estimators for linear models with multiple random effects are poorly addressed in the literature; the results from the current study suggest caution when estimating VPCs using ML estimates from few groups.

Variance component estimation and categorical data

Binomial data

The binomial distribution derives from the idea of a Bernoulli trial, which is defined as an experiment with only two exclusive outcomes (e.g. female or male, hatched or failed to hatch). Let p denote the probability of success (e.g. female, hatched). A binomial experiment is defined as a sequence of identical and independent Bernoulli trials, each with success probability p; the binomial outcome, a count, represents the number of successes. Data that might be treated as binomial include counts of the number of females in a litter or of the number of hatched eggs within nests.

Models of binomial data typically address fundamental characteristics of those data, including that expected responses (probabilities) are non-negative and do not exceed 1 and that sampling variances are functions of the expected responses. These characteristics are typically viewed as precluding the use of standard linear models with binomial data. Instead, binomial data are typically modeled via the inverses of cumulative distribution functions, including those of the standard logistic, standard normal, and standard extreme value distributions. Use of these functions yield logistic (logit), probit, and complementary log-log models, respectively (McCullagh and Nelder 1989).

Binomial data are commonly modeled using generalized linear models (GLMs). GLMs generalize linear regression by incorporating a linear predictor while also permitting linear and nonlinear associations between response means and predictors. The combination of a linear predictor and model nonlinearity is permitted by use of a function that links the expected value of the response with the linear predictor. Link functions commonly used for binomial data include the logit, probit, and cumulative log-log functions mentioned above. The addition of random effects to a GLM yields a generalized linear mixed model (GLMM). Further descriptions of GLMs and GLMMs are provided by Lindsey (1997) and McCulloch and Searle (2001).

A two-way random effects cross-classified GLMM equation for the expected value of a binomial probability, p_{jk}, with logit or log odds link is:

$$\text{logit}(p_{jk}) = \log\left(\frac{p_{jk}}{1 - p_{jk}}\right) = \beta_{00} + u_{0j} + u_{0k} + u_{0jk}. \tag{9.3}$$

The terms in the linear component of Equation (9.3) correspond to those in Equation (9.1): β_{00} denotes a grand intercept, and u_{0j}, u_{0k}, and u_{0jk} denote, respectively, random site, year, and site × year effects. Analogous to the case with Equation (9.1), the random terms are presumed distributed (albeit now on the logit scale) with means zero and variances σ^2_{site}, σ^2_{year}, and $\sigma^2_{\text{site-year}}$, respectively. The omission of index "i" [cf. Equation (9.1)] arises from treating grouped Bernoulli outcomes as binomial counts, with binomial index n; the error term, denoted ε_{ijk} in Equation (9.1), is subsumed in the nominal distributional assumption (e.g. logistic).

An important change from the linear models case is that variances for binomial data may be estimated on both link and measurement scales, whereas for linear models these scales are equivalent. As with linear models, variance estimates on the link scale may be important for evaluating properties of variance component estimators and for study planning purposes. On the other hand, scientific interest will often focus on the scale at which the response occurs and is measured. For standard models of binomial data, we may then speak about variances on both the link (e.g. logit) and measurement (or probability) scales; both are addressed in the following sections. Also, I focus principally on logit normal models. The use of link functions other than the logit for GLMMs of binomial data is comparatively rare. From a variance components estimation perspective, Callens and Croux (2005) evaluated the complementary log-log function, while Browne *et al.* (2007) provided variance component models using mixed effects probit models. Although I focus here on binomial data from hierarchical designs, the analysis of multinomial data – whether ordered or nominal – from hierarchical designs may be treated similarly to the case with binomial data; this topic is addressed in detail by Fielding (2003), Hedeker (2008), and Hox (2010).

Variance component estimation on the link scale

GLMMs of binomial data may be fitted using multiple methods, including linear approximations. Linearization methods employ Taylor series expansions and a linear model of the resulting data approximations. These methods are broadly divided into those that use Taylor series expansions about the expected marginal value of random effects (i.e. zero) and those that expand about subject-specific predictions; these methods are often defined as marginal quasi-likelihood (MQL) and penalized or predictive quasi-likelihood (PQL), respectively. Linearizations may be first- or second-order (MQL1 and MQL2, and PQL1 and PQL2, respectively); linearized "data" may also be evaluated using residual maximum likelihood (RMQL, RPQL).

While linearization methods are commonly used, they may yield negatively biased variance component estimates when used with binomial outcomes with small numbers of observations per group, n, or when among-cluster variances are large (> 0.5 on the logit scale; Goldstein and Rasbash 1996, Pinheiro and Chao 2006). Other studies confirm these concerns, and also demonstrate superior variance component estimation by PQL1 relative to that of MQL1 and of PQL2 and MQL2 relative to PQL1 and MQL1, respectively (Rodriguez and Goldman 1995, 2001; Goldstein and Rasbash 1996; Guo and Zhao 2000; Callens and Croux 2005; Browne and Draper 2006; Diaz 2007).

GLMMs may also be fitted by approximating the marginal likelihood using numerical approximations. These methods, which include Laplace estimation and Gaussian quadrature, have yielded variance component biases that are smaller in magnitude than those seen with MQL and PQL (see below; Pinheiro and Chao 2006, Diaz 2007, Moineddin *et al.* 2007). Further, these methods yield likelihood estimates and so permit the calculation of likelihood-based information criteria and likelihood ratio tests. The downsides are that these methods are more computationally intensive than their MQL and PQL counterparts, and may not be suitable for designs with multiple random components.

In particular, Gaussian quadrature is precluded for models with crossed random factors (McCulloch and Searle 2001), including the model associated with Equation (9.3). Another consideration is that, due to improved relative precision of the PQL estimator, PQL may outperform Laplace estimation of variance components under a mean squared error criterion (Callens and Croux 2005, Diaz 2007).

The relative performance of Bayesian estimators of variance components from hierarchical models of binomial data has seen relatively little study. Findings to date suggest the potential comparability or superiority of Bayesian estimators relative to Gaussian quadrature estimators, and the superiority of both relative to MQL and PQL estimators (Rodriguez and Goldman 2001, Browne and Draper 2006). Fitting hierarchical models to binomial data using Bayesian methods is described by Gelman and Hill (2007).

We earlier saw with linear models that violation of distributional assumptions for group-level effects may be associated with biased variance component estimators (Table 9.1). Whether such biases might be non-ignorable under reasonable assumptions for GLMMs has been poorly studied. Moineddin et al. (2007) described distribution-related biases that were resolved with increased n when group-level effects were random uniform but not when those effects were $t_{(df=3)}$-distributed. Unfortunately, the Moineddin et al. study employed models that were relatively complex given many of the considered sample sizes. The topic of distributional assumptions in the context of variance component estimation from binomial data requires further research.

Results from the simulation scenarios conducted with binomial data for this chapter confirm the expectation of bias for variance components estimated from few groups (scenarios 1–3; Table 9.3). At years $= 5$, biases were typically substantial in relative magnitude (range: –43% to –17%) and also varied substantially among estimation methods. At years $= 20$, however, biases were relatively modest ($\leq |17\%|$) and varied less among methods (–17% to –7%). Variation in the magnitude of bias among estimation methods was generally in the order MQL1 > PQL1 > Laplace > RPQL1 > Bayesian estimation using Markov chain Monte Carlo (MCMC). Second-order linearization methods were not evaluated in this simulation study.

As with the linear model simulations, improvements in $\widehat{\sigma^2_{year}}$ biases were typically no more than modest when attributes other than number of years were varied (scenarios 4–8). In particular, relative biases were similar regardless of whether σ^2_{year} was moderate (0.3 units) or large (1 unit; scenario 8).

Precisions of σ^2_{year} estimators were generally poor, with most improvement seen when numbers of years were increased (compare Monte Carlo standard deviations, Table 9.3). For example, Monte Carlo standard deviations decreased by $\sim$40% as numbers of years increased from 5 to 20 (scenarios 1, 3) but by roughly half that amount when median p increased from 0.12 ($\beta_{00} = -2$; scenario 1) to 0.5 ($\beta_{00} = 0$; scenario 6). Note that ranking of precisions by estimation method generally followed $\widehat{\sigma^2_{year}}$ and, hence, typically appeared best for MQL1 and PQL1, the estimators with the greatest negative biases. For PQL1 and from an MSE perspective, the improved relative precision typically trumped the importance of increased bias – leading to lower MSEs for this method than for other likelihood-based methods. This latter finding is consistent with those reported by Callens and Croux (2005) and Diaz (2007).

Table 9.3 Estimates of among-year variances σ^2_{year}, *on the logit scale* from random effects logistic models by estimation method. Unless otherwise indicated, number of sites = 20, number of years = 5, number of Bernoulli observations per site-year (n) = 5, mean probability on the logit scale (β_{00}) = −2 (i.e. median p_{jk} = 0.12), σ^2_{site} = σ^2_{year} = 0.3, and $\sigma^2_{site\text{-}year}$ = 0.15.[a]

		Estimates of σ^2_{year} on logit scale by method (Monte Carlo standard deviation)				
Scenario	Attribute varied	MQL1	PQL1	RPQL1	Laplace	MCMC
Sample size modifications						
1	Years = 5 (reference)	0.17 (0.17)	0.19 (0.19)	0.25 (0.24)	0.21 (0.22)	0.33 (0.41)
2	Years = 10	0.22 (0.16)	0.24 (0.15)	0.27 (0.17)	0.26 (0.17)	0.31 (0.24)
3	Years = 20	0.25 (0.12)	0.26 (0.11)	0.27 (0.12)	0.28 (0.12)	0.30 (0.14)
4	n = 10	0.18 (0.16)	0.21 (0.20)	0.26 (0.25)	0.22 (0.22)	0.31 (0.30)
5	Sites = 40	0.19 (0.15)	0.21 (0.17)	0.27 (0.22)	0.21 (0.19)	0.31 (0.30)
Mean and variance component modifications						
6	β_{00} = 0	0.17 (0.13)	0.20 (0.17)	0.26 (0.21)	0.23 (0.19)	0.36 (0.35)
7	σ^2_{site} = 1	0.16 (0.14)	0.22 (0.21)	0.26 (0.26)	0.23 (0.24)	0.31 (0.34)
8	σ^2_{year} = 1	0.53 (0.44)	0.65 (0.55)	0.84 (0.69)	0.76 (0.68)	1.19 (1.09)

[a] Estimates represent means of simulation-specific estimates; methods include marginal quasi-likelihood (MQL1), penalized quasi-likelihood (PQL1), restricted PQL (RPQL1), Laplacian estimation (Laplace), and Bayesian analysis using Markov chain Monte Carlo (MCMC). MQL, PQL, RPQL, and Laplacian estimation was performed using SAS' GLIMMIX procedure (SAS 2008) while MCMC estimation was implemented in WinBUGS via the R package "R2WinBUGS" (Lunn *et al.* 2000, Sturtz *et al.* 2005, R Development Core Team 2011). For the Bayesian analyses, priors for all variance components (σ^2_{site}, σ^2_{year}, $\sigma^2_{site\text{-}year}$) were gamma(0.001, 0.001), with means 1 and variances 1000. Each Bayesian estimate was obtained using 4000 posterior samples taken from a total of 22 000 iterations (the first 2000 iterations were used as burn in; samples were obtained every fifth iteration from the remaining 20 000 samples). Quasi-likelihood methods employ first-order Taylor series linearizations; percent model convergence per scenario ≥ 0.92 for QL methods, and 1.00 for Laplacian estimation; replicates per scenario = 200. R, SAS, and WinBUGS code is available as an online supplement at the volume's website. All models presumed p varied as a logit-normal random variable.

The above evidence suggests that investigators should treat estimates of group-level variances from binomial outcomes with few groups as imprecise and, for likelihood-based estimators, as negatively biased. The practical importance of these concerns may be addressed using simulation studies prior to the use of those estimates for study planning.

Variance component estimation on the probability scale

As already mentioned, variance components from GLMMs will typically be more meaningful to scientists and managers when reported on measurement scales. For example, consider a hypothetical example where hatch success (eggs hatched/eggs laid) of a songbird is measured at each of j sites during each of k years. For ecological interpretation, probability-scale estimates (e.g. of year-to-year variation in mean probability of success) would typically be more useful than logit-scale variance estimates. Methods for

Table 9.4 Estimates of among-year variances, σ^2_{year}, *on the probability scale* from random effects logistic models by estimation method. Unless otherwise indicated, number of sites $= 20$, number of years $= 5$, number of Bernoulli observations per site-year $(n) = 5$, mean probability on the logit scale $(\beta_{00}) = -2$ (i.e. median $p_{jk} = 0.12$), variances on the logit-scale of $\sigma^2_{site} = \sigma^2_{year} = 0.3$ and $\sigma^2_{site-year} = 0.15$.[a]

		Estimates of σ^2_{year} on probability scale by method (Monte Carlo standard deviation)			
Scenario	Attribute varied	MQL1	PQL1	RPQL1	Laplace
Sample size modifications					
1	Years $= 5$ (reference)	0.0029 (0.0033)	0.0029 (0.0033)	0.0036 (0.0036)	0.0027 (0.0035)
2	Years $= 10$	0.0039 (0.0032)	0.0036 (0.0028)	0.0039 (0.0030)	0.0036 (0.0029)
3	Years $= 20$	0.0042 (0.0023)	0.0036 (0.0017)	0.0037 (0.0020)	0.0038 (0.0023)
4	$n = 10$	0.0032 (0.0033)	0.0031 (0.0031)	0.0036 (0.0036)	0.0030 (0.0034)
5	Sites $= 40$	0.0033 (0.0031)	0.0031 (0.0029)	0.0038 (0.0033)	0.0029 (0.0029)
Mean and variance component modifications					
6	$\beta_{00} = 0$	0.0083 (0.0061)	0.0089 (0.0067)	0.0110 (0.0083)	0.0099 (0.0077)
7	$\sigma^2_{site} = 1$	0.0030 (0.0032)	0.0032 (0.0032)	0.0039 (0.0042)	0.0030 (0.0031)
8	$\sigma^2_{year} = 1$	0.0113 (0.0013)	0.0099 (0.0087)	0.0130 (0.0115)	0.0110 (0.0112)

[a] See footnote "a", Table 9.3, for estimation and design details. Estimates derive from the alternate or difference method described in Appendix 9.2.

estimating variance components on measurement scales include simulations and, for categorical outcomes, a latent variable method (Goldstein *et al.* 2002). See Appendix 9.2 for an overview of these approaches. As variance component estimation methods on measurement scales have seen little study and appear to have rarely been employed with ecological data (see Li *et al.* 2008 for an exception), comments will center on results from this chapter's simulations.

Concerns associated with estimation of variance components on the probability scale include that expected values may be method dependent, and that investigators have not settled on a protocol for method selection in all cases. For cross-classified models, a further concern is that methods have not been evaluated in the peer-reviewed literature. For these reasons, I infer bias in $\widehat{\sigma^2_{year}}$ primarily by comparison with estimates from the scenario with the largest number of years.

From the simulation study, we see that σ^2_{year} on the probability scale increased with number of years (scenarios 1–3, Table 9.4). As explained above, this finding is concordant with an assumption of declining bias in $\widehat{\sigma^2_{year}}$ with increasing number of years, a finding that also parallels that seen with the logit scale estimates (Table 9.3). As also seen with the logit scale and linear model estimates, biases in $\widehat{\sigma^2_{year}}$ were largely unaffected by changes in n, sites, and σ^2_{site} (scenarios 4, 5, 7; Table 9.4).

Variance component estimates on the probability scale for the simulation study were generally small (Table 9.4). This partly arose from the constraint on variability associated with the small median probability [i.e. median $p_{jk} = $ antilogit(-2) $= 0.12$]. As evidence, note that, while holding variance components on the logit scale constant, the

σ_{year}^2 estimate on the probability scale more than doubled when the median probability increased to ~ 0.50 (scenario 6). As this example demonstrates, comparing variance components on the probability scale will be challenging unless those components have means or medians that are comparable (i.e. that are roughly equal distances from $p = 0.5$).

Estimating variances on inverse link scales becomes more challenging when, as may often be the case for ecologists, covariates are present (Goldstein *et al.* 2002, Li *et al.* 2008). For example, Browne *et al.* (2005) used multi-covariate, multi-level logistic models of literacy within states and districts within states in India to infer the importance of addressing literacy needs of females in rural areas and of whole states with low literacy rates (rather than districts with low rates). This seemingly promising approach has seen essentially no use in the ecological literature.

Variance partition coefficients for binomial models

While VPCs from GLMMs may be estimated on link and measurement scales, VPCs on the link scale seem most useful for study planning purposes (cf. Gray and Burlew 2007). A challenge to estimating VPCs on the link scale is that of defining the variation of an observation on that scale. Gray and Burlew (2007) addressed this issue for count models using the delta method while, for categorical outcomes, the latent variable approach mentioned above may also be used.

VPC estimates on the probability scale may vary by estimation approach, with VPCs calculated using the simulation approach often yielding smaller estimates than those estimated using the latent variable approach (Goldstein *et al.* 2002, Browne *et al.* 2005, Li *et al.* 2008). These findings were also seen with this chapter's simulation study (Table 9.5). A major difference between the simulation and latent variable approaches is made clear by the behavior of VPC estimates as β_{00} increased from -2 to 0 (scenarios 1, 6). VPCs estimated using the latent variable approach varied little among β_{00} values because $\widehat{\sigma_{\text{year}}^2}$ varied little *on the logit scale* with β_{00}. By contrast, VPCs estimated using the simulation approach varied substantially with β_{00} because $\widehat{\sigma_{\text{year}}^2}$ varied substantially *on the probability scale* with β_{00}. The simulation approach will often be preferred over the latent variable approach (Goldstein *et al.* 2002, Li *et al.* 2008).

Categorical data and classification errors

Ecological data that are categorical often incorporate classification errors, a concern that is relevant to variance components estimation when classification error probabilities are heterogeneous. For species detection/non-detection data, a misclassification occurs when a species that is present is not detected. Classification errors may also occur for multi-category outcomes such as frog calling indices. Failure to address classification errors may yield erroneous inferences for both dichotomous and ordered multinomial outcomes (Royle and Link 2005, Mackenzie *et al.* 2006, Holland *et al.* 2010, Holland and Gray 2011; see also Chapter 18).

Table 9.5 Among-year variance partition coefficient, VPC_{year}, estimates on the probability scale from random effects logistic models by estimation method and VPC estimation approach.[a]

Scenario	Attribute varied	VPC$_{year}$ estimates on the probability scale by method (Monte Carlo standard deviation)			
		MQL1	PQL1	RPQL1	Laplace
Simulation approach					
1	Years $= 5$ (reference)	0.021 (0.021)	0.021 (0.022)	0.027 (0.025)	0.021 (0.024)
2	Years $= 10$	0.027 (0.020)	0.027 (0.019)	0.029 (0.020)	0.028 (0.020)
3	Years $= 20$	0.030 (0.016)	0.028 (0.012)	0.029 (0.014)	0.030 (0.016)
4	$n = 10$	0.022 (0.020)	0.023 (0.021)	0.027 (0.024)	0.023 (0.023)
5	Sites $= 40$	0.023 (0.020)	0.023 (0.020)	0.029 (0.024)	0.022 (0.021)
6	$\beta_{00} = 0$	0.034 (0.025)	0.036 (0.027)	0.045 (0.034)	0.040 (0.031)
7	$\sigma^2_{site} = 1$	0.018 (0.018)	0.022 (0.020)	0.026 (0.025)	0.021 (0.020)
8	$\sigma^2_{year} = 1$	0.067 (0.058)	0.068 (0.053)	0.086 (0.065)	0.076 (0.066)
Latent variable approach					
1	Years $= 5$ (reference)	0.043 (0.040)	0.047 (0.044)	0.059 (0.050)	0.048 (0.048)
2	Years $= 10$	0.056 (0.036)	0.060 (0.036)	0.065 (0.039)	0.063 (0.039)
3	Years $= 20$	0.064 (0.030)	0.064 (0.026)	0.066 (0.028)	0.069 (0.031)
4	$n = 10$	0.046 (0.038)	0.052 (0.042)	0.061 (0.048)	0.053 (0.046)
5	Sites $= 40$	0.048 (0.038)	0.052 (0.041)	0.064 (0.048)	0.051 (0.044)
6	$\beta_{00} = 0$	0.045 (0.033)	0.050 (0.037)	0.062 (0.047)	0.056 (0.044)
7	$\sigma^2_{site} = 1$	0.037 (0.032)	0.047 (0.040)	0.055 (0.047)	0.046 (0.041)
8	$\sigma^2_{year} = 1$	0.121 (0.087)	0.136 (0.093)	0.166 (0.105)	0.151 (0.108)

[a] See footnote "a", Table 9.3, for estimation and design details. Estimates under "Simulation approach" derive from the alternate or difference method described in Appendix 9.2.

Addressing classification errors becomes challenging when the probabilities of those errors vary among sampling units. For dichotomous outcomes, failure to address such variation typically will yield biased estimators of both the probability of detection and the probability of the true state (Royle and Dorazio 2008). The analogue of this problem has also been demonstrated for multinomial models (Holland and Gray 2011).

In many settings, variation in a parameter would typically be addressed by allowing the parameter to vary according to a mixing distribution (e.g. the logit-normal distribution employed above). For two-category models like those commonly used to estimate occupancy, however, the probability of the state variable of interest (e.g. site occupancy) is not identifiable across mixing distributions, with the practical importance of this concern increasing as the mean detection probability decreases and as unexplained among-site variation in the classification error increases (Royle 2006). It seems reasonable to expect that the same problem will apply to multinomial abundance models. The result is that, while variation in misclassification parameters is estimable, such estimation may not always qualitatively improve inferences on the variable of interest. This highlights the importance of achieving low overall probabilities of classification errors (e.g. high detection probabilities) to reduce effects of variation in these probabilities (Chapter 18).

Variance component estimation and count data

As with their categorical counterparts, count data are typically modeled using GLMs and, given random group effects, using GLMMs. GLMs and GLMMs of counts typically presume counts are distributed as Poisson or negative binomial (NB) random variates (conditional on any fixed or random effects). An NB distributional assumption allows for the common case where sampling variation exceeds that expected under a Poisson assumption. GLMMs of counts are typically formulated using a log link, a link that conveys the advantage of ensuring predicted means are non-negative.

The estimation problems described above for modeling clustered categorical data have not generally been addressed for count models. Consequently, few studies have compared variance component estimators for count models as a function of estimation method. A modest exception is found in an adaptive Gaussian quadrature study by Pinheiro and Chao (2006). As part of that study, the authors evaluated RPQL, Laplace, and adaptive Gaussian quadrature estimators of fixed effects and variance components under a single simulation scenario. Data arose from many groups (300) of 2 conditional Poisson outcomes each (i.e. $n = 2$); the median group mean was large (~ 17) while the among-cluster variance on the log scale was low (0.09 units). Given this setting, fixed effects and variance components were, as might be expected, estimated with ignorable bias ($<1.7\%$) by all three methods. For comparisons among GLMMs and a GLMM elaboration, see Lee and Nelder (2001) and references therein.

Simulations of count data for this chapter suggest wide differences among estimators in performance when outcomes were NB distributed and, to lesser degree, when outcomes were Poisson distributed (Table 9.6). Simulations were conducted under the log-link analog of model equation (9.3):

$$\log(\mu_{jk}) = \beta_{00} + u_{0j} + u_{0k} + u_{0jk} . \tag{9.4}$$

Convergence rates for PQL1 and RPQL1 models were poor when outcomes were NB-distributed (Table 9.6: upper panel) and often little better when outcomes were Poisson distributed (Table 9.6: lower panel). Restricting subsequent attention to MQL1 and Laplace estimators, bias and precision were typically best under Laplace estimation. Note that the MQL1 estimators of the NB dispersion parameter and of $\sigma^2_{\text{site-year}}$ were typically positively biased and estimated imprecisely. For scenario NB3, for example, the MQL1 estimator (with Monte Carlo standard deviations) yielded NB dispersion parameter and $\sigma^2_{\text{site-year}}$ estimates of 1.00 (0.24) and 0.37 (0.17) – with true values 0.5 and 0.15, respectively. By contrast, the corresponding Laplace estimates were 0.50 (0.03) and 0.15 (0.02), respectively. Evidence of positive bias in $\sigma^2_{\text{site-year}}$ was also seen with MQL1 models of Poisson outcomes. As also seen with the binomial models, varying attributes other than number of years typically led to only minor effects on the bias and precision of σ^2_{year} (Table 9.6).

Naïvely treating NB-distributed outcomes as Poisson-distributed yielded generally minor changes in σ^2_{year} estimates (compare scenarios NB1 vs. NB10; Table 9.6). Instead, the effects of misspecifying the conditional generating distribution was seen in inflated

Table 9.6 Among-year variance, σ^2_{year}, estimates on the log scale from random effects count models by estimation method. Unless otherwise indicated, number of sites $= 20$, number of years $= 5$, number of counts per site-year (n) $= 5$, median count mean (λ_{jk}) $= 5$, $\sigma^2_{site} = \sigma^2_{year} = 0.3$, and $\sigma^2_{site\text{-}year} = 0.15$.[a]

Scenario	Attribute varied	σ^2_{year} estimates on the log scale by method (Monte Carlo standard deviation), % convergence			
		MQL1	PQL1	RPQL1	Laplace
Negative binomial outcomes (dispersion parameter $= 0.5$)					
NB1	Years $= 5$ (reference)	0.21 (0.16), 96	0.21 (0.13), 49	0.21 (0.16), 46	0.23 (0.16), 100
NB2	Years $= 10$	0.27 (0.16), 96	0.27 (0.10), 14	0.26 (0.11), 14	0.27 (0.12), 100
NB3	Years $= 20$	0.33 (0.20), 93	0.27 (0.09), 29	0.29 (0.09), 27	0.29 (0.09), 100
NB4	$n = 10$	0.21 (0.16), 93	0.25 (0.18), 22	0.27 (0.22), 29	0.23 (0.16), 100
NB5	Sites $= 40$	0.23 (0.18), 92	0.17 (0.12), 14	0.23 (0.14), 15	0.23 (0.15), 100
NB6	$\lambda_{jk} = 2$	0.20 (0.15), 90	0.19 (0.13), 46	0.24 (0.19), 42	0.23 (0.16), 100
NB7	$\lambda_{jk} = 20$	0.21 (0.16), 98	0.23 (0.16), 47	0.25 (0.17), 49	0.23 (0.16), 100
NB8	$\sigma^2_{site} = 1$	0.21 (0.16), 88	0.25 (0.17), 37	0.27 (0.19), 33	0.24 (0.16), 92
NB9	$\sigma^2_{year} = 1$	0.60 (0.46), 87	0.66 (0.47), 34	0.77 (0.53), 31	0.62 (0.35), 90
NB10	Poisson assumption	0.20 (0.15), 87	0.22 (0.14), 77	0.27 (0.20), 73	0.23 (0.16), 100
Poisson outcomes					
P1	Years $= 5$ (reference)	0.21 (0.17), 99	0.22 (0.15), 71	0.24 (0.17), 67	0.22 (0.15), 100
P2	Years $= 10$	0.29 (0.24), 78	0.26 (0.12), 83	0.28 (0.13), 85	0.27 (0.12), 100
P3	Years $= 20$	0.32 (0.23), 72	0.28 (0.09), 48	0.29 (0.10), 46	0.29 (0.09), 100
P4	$n = 10$	0.22 (0.16), 90	0.23 (0.15), 72	0.27 (0.18), 72	0.23 (0.15), 100
P5	Sites $= 40$	0.21 (0.15), 84	0.23 (0.15), 78	0.27 (0.18), 82	0.23 (0.15), 100
P6	$\lambda_{jk} = 2$	0.22 (0.17), 94	0.22 (0.15), 66	0.27 (0.19), 64	0.23 (0.16), 100
P7	$\lambda_{jk} = 20$	0.22 (0.16), 98	0.23 (0.15), 92	0.27 (0.18), 94	0.23 (0.15), 100
P8	$\sigma^2_{site} = 1$	0.22 (0.17), 94	0.25 (0.17), 60	0.28 (0.19), 61	0.24 (0.16), 100
P9	$\sigma^2_{year} = 1$	0.63 (0.48), 96	0.70 (0.51), 65	0.81 (0.56), 66	0.74 (0.50), 100

[a] Estimates represent means of simulation-specific estimates; replicates per scenario $= 200$. Methods include marginal quasi-likelihood (MQL1), penalized quasi-likelihood (PQL1), restricted PQL1 (RPQL1), and Laplacian estimation (Laplace). Quasi-likelihood methods employ first-order Taylor series linearizations.

$\sigma^2_{site\text{-}year}$ estimates; these estimates were, on average, biased high by 59% and 68% for MQL and Laplace estimation, respectively (relative to those derived under the conditional NB assumption). These findings reflect the wisdom of addressing extra-Poisson variation in models of count outcomes.

Variance component estimation for clustered count data requires further examination. While the results described in this section provide modest evidence in favor of Laplace estimation, readers interested in VC estimation from count data should also consider MCMC and, for fully nested models, adaptive Gaussian quadrature.

The contamination of count data with structural zeroes has been a common concern for modelers and ecologists (Cameron and Trivedi 1998, Gray 2005). This topic of zero inflated count data has been addressed from a model specification rather than estimation method perspective for grouped count data by Min and Agresti (2005), Lee et al. (2006), Moghimbeigi et al. (2008), and Gray et al. (2010).

Variance components from count models on the measurement scale

Interest in variance components (and in VPCs in particular) on the measurement scale for count outcomes has lagged that for binomial outcomes. An implicit exception is Goldstein *et al.* (2002). While Goldstein *et al.* focused on binomial outcomes, those authors noted that three of the four methods they proposed – model linearizations using Taylor series approximations, a normal distributional approximation, and simulations – might be considered for use with other nonlinear models. Stryhn *et al.* (2008) considered the above three methods from a count perspective, and also considered exact formulae (calculated using integration formulae for exponential functions). They found the simulation and integration methods yielded results that were not only similar but also superior to those derived by the linearization and normal approximation methods. The Stryhn *et al.* paper is sparse on methodology and results (but see presentation at http://people.upei.ca/hstryhn/iccpoisson.ppt); a more detailed follow-up paper is expected (H. Stryhn, personal communication, 25 August 2009).

Estimating VPCs on the measurement scale for cross-classified random effects count models apparently has not been addressed in the literature. One approach that seems promising but which has apparently not been evaluated in the published literature would adapt the provisional method supplied in Appendix 9.2 for counts by substituting the exponential for the logistic function in step 1 and by modifying $\mathrm{var}(y_{ijk})$ in step 5 to reflect the assumed count distribution.

Study design

Given the focus of this book, many readers may be interested in variance component estimation for the purposes of designing new monitoring efforts and other ecological studies. Other chapters in this volume address study design at length, but here I briefly mention a few tools useful for study design for the types of data considered in this chapter.

The design of such studies may be explored using study-specific simulations or using a number of specialized freeware packages. For example, MLPowSim estimates statistical power using either R or the multilevel software package MLwiN, and using ML, REML, or MCMC; MLPowSim may be used with continuous, binary, and count data (Browne *et al.* 2009). Another package, PINT, calculates approximate standard errors for estimates of fixed effects, as well as optimal sample sizes, for linear models with two levels (Snijders and Bosker 1993; PINT is available at http://stat.gamma.rug.nl/multilevel.htm). The program Optimal Design calculates sample size, statistical power, and optimal allocation of resources for multilevel studies with continuous and binary outcomes (Spybrook *et al.* 2009). Note that Wang and Gelfand (2002) address sample-size determination from a Bayesian perspective.

Methods for addressing study design for clustered count outcomes are underdeveloped. Models of trends among grouped counts with a design focus have presumed linearity (Gibbs *et al.* 1998, Urquhart *et al.* 1998, Kincaid *et al.* 2004) or that counts were lognormally distributed (Gerrodette 1987, 1991). Exceptions include Purcell *et al.* (2005) and Gray and Burlew (2007). Purcell *et al.* (2005) estimated statistical power to detect trends in a hierarchical count setting using Monte Carlo methods under an assumption that data were Poisson, conditional on random observer and/or route effects. Gray and Burlew (2007) provided algorithms for estimating precision of and power to detect trends in a single population using GLMMs with conditional Poisson and NB distributional assumptions.

Future research and development

Sampling designs often incorporate variable selection probabilities, a complication not otherwise considered in this chapter. For example, sampling units associated with heterogeneous habitats may be oversampled relative to those from homogeneous habitats. Variation in sampling probabilities may be addressed using survey statistical methods through the use of sampling weights (Chapters 5, 6, 14). Unfortunately, the improper use of sampling weights using models (as distinct from design-based methods) may lead to biased estimates of among-group variation. Addressing this shortcoming through so-called design-adjusted models represents an area of ongoing research. Statistical modeling packages that accommodate sampling weights include Mplus, MLwiN, and Stata (Muthén and Muthén 1998–2010, Rasbash *et al.* 2009, StataCorp 2009). Further discussion of this topic is provided by Rabe-Hesketh and Skrondal (2006) and Carle (2009).

Summary

Variance components may be estimated for scientific, management, and study planning purposes. Scientific purposes include, for example, examining whether nitrate concentrations in lakes vary more among than within lakes, and whether either component of variation might be associated with putatively causal agents (e.g. agricultural runoff). Variance component estimates are used for study planning when, for example, an investigator wishes to select the number of groups (e.g. lakes) and the number of observations within each group for a future study. A major concern is that variance component estimators may be biased and yield imprecise estimates when the number of groups is small. This concern appears especially relevant for ecologists who, for logistic or cost reasons, may design studies with few sites and/or few years, such as in pilot studies conducted as part of development of a monitoring survey design.

This chapter reviews the estimation of variance components and variance partition coefficients (VPCs) for continuous, categorical, and count data that are clustered, and with emphases on studies with small sample sizes and crossed random effects. Variance components estimated from few groups using linear models of continuous outcomes may exhibit only modest bias when estimated using ANOVA or REML, but may be substantially biased when estimated using FML. For all three estimation methods, however, precision is expected to be poor unless the number of groups is modest to large (e.g. more than 10, and possibly as many as 100).

Variance components estimated from GLMMs of categorical data may be expected to be both biased and imprecise when number of groups are few (e.g. <20 to as high as <100, depending on estimation method). The performance of Bayesian estimators of variance components from categorical data appears promising, but has received relatively little attention in the literature.

GLMM estimators of variance components from count data have received less attention than have their categorical counterparts. Information supplied in this chapter suggests that, for cross-classified random effects models of count data, the Laplace estimator should be preferred over first-order quasi-likelihood (QL) variance component estimators; the QL estimators suffered from poor convergence rates (PQL and RPQL) or substantial bias associated with $\sigma^2_{\text{site-year}}$ (MQL). Readers interested in VC estimation from count data should also consider Markov chain Monte Carlo and, for fully nested models, adaptive Gaussian quadrature.

The estimation of VPCs has received relatively little attention in the ecological literature. This is particularly the case for VPCs from categorical and count data (for which methods appear to have first been published in 2002 and 2008, respectively; Goldstein *et al.* 2002, Stryhn *et al.* 2008). For these discrete outcomes, VPCs may be estimated on both measurement and modeling or link scales. A method for estimating VPCs for binary outcomes on the measurement scale from two-way cross-classified random effects designs is proposed in Appendix 9.2.

Acknowledgments

I thank Sherwin Toribio for providing the Bayesian estimates in Table 9.3, David Afshartous, Jialiang Li, and Chuck Rose for helpful reviews, Bob Gitzen for helpful editorial suggestions, and Bethany Bell for reviewing the SAS code used for many of the simulations. This study was partially funded by the Upper Mississippi River's Long Term Resource Monitoring Program. References to proprietary software do not imply endorsement by the US government.

Appendix 9.1. Methods of variance components estimation using linear models

Analysis of variance

Analysis of variance (ANOVA) may be used to estimate variance components by equating sums of squares to expected values. Chapter 8 provides an overview and example of ANOVA, while Chapter 7 illustrates another application of ANOVA estimation of variance components for a cross-classified model very similar to that associated with

Equation (9.1). Here, I briefly review general properties of ANOVA estimators of variance components.

These estimators possess a number of advantages. One is that the ANOVA estimators make no distributional assumptions (other than that random effects have means zero and finite variances; e.g. Chapter 7). Another is that they are unbiased. (However, note that Searle *et al*. 1992, section 2.3, question the importance of unbiasedness for variance component estimators.) A third is that, given balanced data sets and normality, ANOVA variance component estimators are "best unbiased" in the sense that, among unbiased estimators of variance components, they have uniformly smallest variance. A final and heuristic advantage is that variance components may often be defined in closed form – an advantage seen when variance component concepts are motivated using ANOVA arguments even when alternatives to ANOVA are recommended (Searle *et al*. 1992, Snijders and Bosker 1999).

However, ANOVA estimators of variance components also possess a number of negative attributes. First, their unbiasedness comes at the cost of allowing group-level variance estimates that are negative (variances, by definition, are non-negative). For example, negative estimates were produced for several responses in the example data set analyzed in Chapter 7. Unfortunately, the usual solution of setting negative variance estimates to zero eliminates the unbiasedness property of these estimators. Second, the minimum variance properties mentioned above as a positive attribute for balanced data do not apply under unbalanced data assumptions. Third, unbalanced data – the common case for observational data – lead to estimators that are not only more complex but also which are not unique. Further details associated with ANOVA estimation of variance components for one-way and multi-way classifications are provided by Searle *et al*. (1992), Cox and Solomon (2003), and, with a less technical approach, Muller (2009).

Maximum likelihood

Maximum likelihood (ML) requires assuming an underlying probability distribution for a given set of data. ML may then be used to estimate values of the parameters associated with that distribution. For ML, the estimates are those that are considered most likely – given the data and the selected distribution.

ML estimators (MLEs) possess a number of favorable properties. These include those properties that are asymptotic – that are approached as sample size goes to infinity. These latter properties include: consistency (under fairly weak assumptions, MLEs converge to the value being estimated), normality, and minimum variance (among asymptotically unbiased estimators, and given commonly attained conditions). Another favorable property is that MLEs must be within the allowed range of the given parameter (thereby eliminating the negative variance estimates that were possible under ANOVA). Likelihood-based methods also accommodate data that are missing at random (e.g. Chapter 15). Last, the distributional assumption may be tailored to the process in question. For example, count data may be presumed to follow one or more of a number of potential count distributions.

Disadvantages associated with using ML include some of those associated with the favorable properties listed above. First, the favorable properties that are associated with large samples may not be reasonable for small samples. For a given study, investigators may infer that a sample is "large enough" based on experience, including that associated with Monte Carlo simulations (e.g. Table 9.1). Another concern relates to the possibility of unreasonable distributional assumptions. Given small sample sizes, for example, it may be difficult to determine on statistical grounds which distributional assumption is most reasonable. A rejoinder is that scientific theory and information from previous studies may (and arguably should) be used to select distributional assumptions.

An important concern related to the use of full maximum likelihood (FML – as distinct from REML, described below) to estimate variance components is that FML does not adjust for degrees of freedom associated with fixed effects. For data without clustering, for example, the FML variance estimator is $\sum (y_i - \mu)^2/n$ rather than $\sum (y_i - \mu)^2/(n - k)$, where k denotes the number of linear independent predictors. Therefore, the bias of the FML estimator is approximately $-k$/sample size (provided the proportion of zero estimates is small). See McCulloch and Searle (2001) for a fuller treatment of this topic. Further information about MLEs is provided by Casella and Berger (1990) and Searle et al. (1992).

Restricted maximum likelihood

Restricted or residual maximum likelihood (REML) represents ML on a function of the data, specifically that function of the data from which fixed effects have been removed. REML confers the twin advantages of yielding estimators that are invariant to fixed effects, and of eliminating the variance component bias related to degrees of freedom described for FML estimators in the previous paragraph. For balanced data, REML estimators of variance components equal the expected value of the ANOVA estimates – provided that negative ANOVA estimates are set to zero. REML estimators of variance components are typically preferred over their ML counterparts (McCulloch and Searle 2001). See Chapter 10 for an example of the use of REML estimators of variance components.

Bayesian estimation

In Bayesian statistics, each parameter is treated as a random variable, with variation described by a probability distribution. This distribution, which is assigned without reference to the data in question, is termed a prior distribution. The prior distribution is then updated using information from sample data, thereby yielding a posterior distribution for the parameter in question. This updating of the prior distribution occurs via Bayes theorem (see Chapter 4). The updated, posterior distribution is used for making inferences on the parameter in question. Introductions to Bayesian analysis with an ecological flavor are provided by Link et al. (2002) and Link and Barker (2010); more detailed treatments are provided by, for example, Gelman and Hill (2007) and

Draper (2008). Chapters 4, 12, 19, and 20 demonstrate general applications of Bayesian estimation.

Bayesian methods are valid with small samples, will not yield negative variance estimates, and do not require normality assumptions when variances are estimated. On the other hand, Bayesian estimators of group-level variances are not unbiased and have received relatively little attention in the literature; estimating variances using Bayesian methods also may be more computationally intensive than is the case under classical methods. Fitting models of grouped data using Bayesian methods is addressed in Goldstein (2003, section 2.13), Browne and Draper (2006), Gelman and Hill (2007), Draper (2008), and in Hox (2010, section 11.4); the Goldstein and Hox references are less technical.

Appendix 9.2. Estimating variance components and variance partition coefficients for two-way cross-classified random effects designs on the measurement scale using simulations

Overview

Variance components on the inverse link scale may best be estimated using simulations (Li *et al.* 2008). In this case, estimation begins by treating parameter and variance estimates on the link scale as population values. The user then generates a large number of means (say, $m = 5000$) under the population assumptions, and then transforms those means using an inverse link transformation. The variance of the means on the inverse link scale is then estimated by method of moments; variance at the measurement scale (given the model) is estimated under the sampling distribution assumption and across all simulated means.

For example, consider a random effects logistic regression model with a single group-level random effect, and with estimated grand intercept and variance on the logit scale of -1 units and 0.7 units, respectively. Presume group effects on the logit scale are, as is typically assumed, normally distributed. Then generate a large number (e.g. 5000) of random normal variates and treat these (after adding a grand intercept) as the population of group means. Trivially, presume two random normal variates of 0.5 and -0.5 are generated. Given the grand intercept of -1, the corresponding means on the logit scale are then $(-1 + 0.5) = -0.5$ and $(-1 - 0.5) = -1.5$, respectively. These means may be transformed to probabilities using the inverse logit transformation

$$p_j = \frac{\exp(z_j)}{1 + \exp(z_j)} = \frac{1}{1 + \exp(-z_j)}$$

where z_j and p_j denote means on logit and probability scales, respectively. By this transform the two means on the measurement scale yield $p_j = 0.38$ and 0.18, respectively. The variance of these means, by method of moments, is

$$\frac{1}{n-1} \sum_{j=1}^{2} (p_j - \widehat{p})^2 = 0.02$$

where $\widehat{p}$ denotes the sample mean. The variance at the measurement scale is estimated by the mean variance of the putative Bernoulli observations

$$\frac{1}{n}\sum_{j=1}^{2}p_j(1-p_j) = [0.38(1-0.38)+0.18(1-0.18)]/2 = 0.1916.$$

The variance at the measurement scale for binomial models is typically presumed that of a Bernoulli outcome – because the number of trials per binomial count may vary and because covariates may vary across binary observations (Goldstein *et al.* 2002). Of course and as already emphasized, the choice of $m = 2$ groups for this example represented a heuristic device; calculations should routinely be performed with much larger m. An elaboration of this method for two-way cross-classified random effects models is provided below.

Other methods for estimating variance components on the inverse link scale include Taylor series linearizations and, for categorical outcomes, a latent variable method (Goldstein *et al.* 2002). The latter method is commonly employed but appears appropriate only when the outcome of interest might reasonably derive from a continuous distribution (Snijders and Bosker 1999, chapter 14; Goldstein *et al.* 2002). For example, the probability that an organism may succumb to a toxicant may be postulated to derive from a standard logistic or standard normal cumulative distribution function. In these cases, the outcome may be treated as arising from a threshold model, and with variance that of a standard logistic or Gaussian outcome (i.e. $\pi^2/3$ or 1, respectively). For the example above, variance components at the group and measurement scales using the latent variable method would be the group-level variance (0.7 units) and $\pi^2/3$, respectively.

As with variance components, VPCs appear more informative from scientific and management perspectives when expressed on measurement scales. For the example above, VPCs calculated using the simulation and latent variable approaches are $0.02/(0.02 + 0.1916) = 0.09$ and $0.07/(0.07 + \pi^2/3) = 0.18$, respectively. Note that the latent variable approach yields the same VPC estimates for both link and measurement scales.

Simulation method for the cross-classified model

As described above, the simulation method of estimating variance components and VPCs on the measurement or probability scale represents a reconstruction of the data generation process by computer simulations paired with a recording of the observed variation. Here, I propose a simulation method for use with the two-way cross-classified random effects model associated with Equation (9.3). This method was adapted from the corresponding methodology for fully nested models (Goldstein *et al.* 2002, Browne *et al.* 2005, Li *et al.* 2008); I thank Bill Browne for reviewing an early draft of the proposed method. Note that the method simplifies when, as may often be the case, variation in random interaction effects, var(p_{jk}), is treated as either inestimable or as identically zero. The method is as follows:

1. From the model [e.g. as associated with Equation (9.3)], simulate a large number M1 (say 5000) of main effects $u_{0j}, j = 1, \ldots, \text{M1}$, using the corresponding sample estimate of variance (e.g. $\widehat{\sigma^2_{\text{site}}}$). For each j, simulate M2 (say 5000) of main effects $u_{0k}, k = 1, \ldots, \text{M2}$, using the corresponding sample estimate of variance (e.g. $\widehat{\sigma^2_{\text{year}}}$). For each unique combination, jk, simulate M3 (say, 30) interaction effects $u_{0jkl}, l = 1, \ldots,$ M3, using the corresponding sample estimate of variance (e.g. $\widehat{\sigma^2_{\text{site-year}}}$). Calculate the $M = \text{M1} \times \text{M2} \times \text{M3} \ p_{jkl}$'s as

$$p_{jkl} = \frac{1}{1 + \exp(-(\widehat{\beta_{00}} + u_{0j} + u_{0k} + u_{0jkl}))}.$$

where $\widehat{\beta_{00}}$ denotes the grand intercept estimate.

2. Calculate the uncorrected (marginal) variation of the p_j's within k by method of moments. It may be convenient to use a single l replicate per jk:

$$\text{var}(p_j)_m = \frac{1}{\text{M2}} \sum_{k=1}^{\text{M2}} \text{var}(p_{jkl}|k, l = 1) \approx \text{var}(p_j) + \text{var}(p_{jk}).$$

3. Similarly, calculate the uncorrected (marginal) variation of the p_k's within j by method of moments. It may be convenient to use a single l replicate per jk:

$$\text{var}(p_k)_m = \frac{1}{\text{M1}} \sum_{j=1}^{\text{M1}} \text{var}(p_{jkl}|j, l = 1) \approx \text{var}(p_k) + \text{var}(p_{jk}).$$

4. Calculate the variation of the p_{jk}'s by method of moments:

$$\text{var}(p_{jk}) = \frac{1}{\text{M1M2}} \sum_{j=1}^{\text{M1}} \sum_{k=1}^{\text{M2}} \text{var}(p_{jkl}).$$

5. Then,

$$\text{var}(p_j) \approx \text{var}(p_j)_m - \text{var}(p_{jk})$$
$$\text{var}(p_k) \approx \text{var}(p_k)_m - \text{var}(p_{jk})$$
$$\text{var}(y_{ijk}) = \frac{1}{\text{M1M2}} \sum_{j=1}^{\text{M1}} \sum_{k=1}^{\text{M2}} p_{jkl}(1 - p_{jkl})|l = 1. \tag{9B.1}$$

6. VPCs may be estimated from Equation (9B.1) using Equation (9.2).

The method ignores sampling variation in $\widehat{\beta_{00}}$ and in the variance estimates, and presumes linearity on the probability scale. The first concern may be addressed by nesting the method within a larger Monte Carlo simulation. The second concern remains unaddressed (Li *et al.* 2008). This method may be adapted to accommodate covariates after Goldstein *et al.* (2002), Browne *et al.* (2005), and Li *et al.* (2008).

The above method is somewhat demanding computationally. An approximation to this method which is less demanding but which elaborates the linearity assumption relies on differencing rather than on replicating on jk to estimate var(p_{jk}). Under the model defined by Equation (9.3) and the Table 9.3 legend, this alternative "differencing" method yields σ_{year}^2 estimates on the probability scale that are smaller than those estimated using the above-described method by 17%, 6%, 28%, and 15% (for scenarios 1 to 5, 6, 7, and 8, respectively). Corresponding differences for VPC_{year} estimates were similar: -16%, -5%, -28%, and -14%, respectively. For this alternative method, steps 1–3 above are followed with the caveat that the interaction effects are not replicated (i.e. $M3 = 1$). Steps 4 and 5 become:

Alternate step 4. Calculate the uncorrected (marginal) variation of the p_{jk}'s by method of moments:

$$\text{var}(p_{jk})_\text{m} \approx \text{var}(p_j) + \text{var}(p_k) + \text{var}(p_{jk}).$$

Alternate step 5. Then,

$$\text{var}(p_j) \approx \text{var}(p_{jk})_\text{m} - \text{var}(p_k)_\text{m}$$
$$\text{var}(p_k) \approx \text{var}(p_{jk})_\text{m} - \text{var}(p_j)_\text{m}$$
$$\text{var}(p_{jk}) \approx \text{var}(p_j)_\text{m} + \text{var}(p_k)_\text{m} - \text{var}(p_{jk})_\text{m} \qquad \text{(9.B1(alt))}$$

$$\text{var}(y_{ijk}) = \frac{1}{M1M2} \sum_{j=1}^{M1} \sum_{k=1}^{M2} p_{jk}(1 - p_{jk}).$$

The properties of these two methods have not been rigorously investigated (cf. Li *et al.* 2008).

10 Simulating future uncertainty to guide the selection of survey designs for long-term monitoring

Steven L. Garman, E. William Schweiger, and Daniel J. Manier

Introduction

A goal of environmental monitoring is to provide sound information on the status and trends of natural resources (Messer *et al.* 1991, Theobald *et al.* 2007, Fancy *et al.* 2009). When monitoring observations are acquired by measuring a subset of the population of interest, probability sampling as part of a well-constructed survey design provides the most reliable and legally defensible approach to achieve this goal (Cochran 1977, Olsen *et al.* 1999, Schreuder *et al.* 2004; see Chapters 2, 5, 6, 7). Previous works have described the fundamentals of sample surveys (e.g. Hansen *et al.* 1953, Kish 1965). Interest in survey designs and monitoring over the past 15 years has led to extensive evaluations and new developments of sample selection methods (Stevens and Olsen 2004), of strategies for allocating sample units in space and time (Urquhart *et al.* 1993, Overton and Stehman 1996, Urquhart and Kincaid 1999), and of estimation (Lesser and Overton 1994, Overton and Stehman 1995) and variance properties (Larsen *et al.* 1995, Stevens and Olsen 2003) of survey designs. Carefully planned, "scientific" (Chapter 5) survey designs have become a standard in contemporary monitoring of natural resources.

Based on our experience with the long-term monitoring program of the US National Park Service (NPS; Fancy *et al.* 2009; Chapters 16, 22), operational survey designs tend to be selected using the following procedures. For a monitoring indicator (i.e. variable or response), a minimum detectable trend requirement is specified, based on the minimum level of change that would result in meaningful change (e.g. degradation). A probability of detecting this trend (statistical power) and an acceptable level of uncertainty (Type I error; see Chapter 2) within a specified time frame (e.g. 10 years) are specified to ensure timely detection. Explicit statements of the minimum detectable trend, the time frame for detecting the minimum trend, power, and acceptable probability of Type I error (α) collectively form the quantitative sampling objective.

The values specified in this sampling objective affect the required sampling effort. A smaller minimum detectable trend requirement, higher power, a shorter time frame, and a lower acceptable Type I error rate generally increase the effort required to achieve the sampling objective. In addition, the spatial and temporal variability of an indicator

Design and Analysis of Long-term Ecological Monitoring Studies, ed. R.A. Gitzen, J.J. Millspaugh, A.B. Cooper, and D.S. Licht. Published by Cambridge University Press. © Cambridge University Press 2012.

influences sampling effort; higher variability increases the required effort. Estimates of indicator variability are acquired in pilot monitoring or related studies, and used in prospective statistical power analyses to determine the sample size and other aspects of the survey design needed to satisfy the sampling objective (Chapters 2, 7, 8). Budgetary limitations generally dictate a parsimonious annual sampling effort. In many cases rotating sampling effort among groups of sites (called panels) over time, with eventual revisitation of sites, can provide high power for trend detection while also providing suitable precision for status estimation (McDonald 2003; Chapter 7).

Determining sample sizes and re-visitation schedules that satisfy a sampling objective within the limits of allocated programmatic resources requires a concerted effort prior to selection of a final survey design and implementation (Box 10.1). In many situations, the power and precision of candidate survey designs can be assessed analytically, even with complex panel designs (Chapters 7, 8). Often, however, a simulation approach can offer higher flexibility, particularly (i) when analytical power/precision relationships are not well defined for a planned trend/status analysis approach; (ii) for examining how violation of key assumptions affects the performance of the planned trend-analysis approach; (iii) in situations where spatially explicit aspects of the monitoring scenarios need to be examined; or (iv) for examining effects of other factors not easily accounted for using analytical power/precision equations (e.g. Chapters 17, 19, 20). For example, simulation modeling has been used to determine optimal sampling designs for varying spatial patterns of an indicator (Pooler and Smith 2005, Heywood and DeBacker 2007, Morrison et al. 2008), to evaluate design variance and bias (McDonald et al. 2001), to assess sampling design sensitivity to disturbances (Edgar and Burk 2006), and to assess power to detect long-term trends and properties of sampling strategies (e.g. Eng 2004, Stevens and Olsen 2004, Field et al. 2005). In some cases, analytical power calculations may be feasible once suitable estimates of variance are available and sampling objectives have been specified, but simulation modeling of patterns and dynamics of the population under various levels of change may be important for helping examine variance scenarios and even defining what minimum level of trend should be of concern.

Although even basic power/precision examinations lead to more informed choices of survey designs (Chapter 8), ecologists and statisticians also should consider complications likely to arise in most monitoring situations. Prospective power analyses with estimates of historical and current variances assume that future variance components will be of similar magnitudes. In ecological systems, however, changes over time in the spatial and temporal variance of an indicator are inevitable due to stochastic processes such as disturbances and climatic fluctuations, as well as human-mediated impacts such as anthropogenic climate change. Failure to account for potential future changes in indicator variance can lead to over-optimistic assessments of expected design performance. Inflating indicator variance in an analytical evaluation can compensate for future increases, but this does not explicitly consider how agents of change may modify the spatial and temporal variance structure of an indicator. Adaptive monitoring is often invoked to safe-guard against underestimating future performance of a design, where repeated evaluation of accumulating monitoring data and the sampling objective determine when and how to modify the survey design (Ringold et al. 1996, Lindenmayer and

---

Box 10.1 Take-home messages for program managers

Quantitative evaluation of alternative monitoring survey designs is essential for choosing a design that efficiently provides useful information. "Useful" needs to be defined explicitly in the form of a carefully considered sampling objective. For example, a program focused partly on trend detection needs to define what forms and levels of change would be of concern and how quickly such a change needs to be detected to enable managers to take action. In our example in this chapter, the sampling objective is to have an 80% probability of detecting a 3.5% annual decline in grass cover in our arid-system park within 10 years if such degradation occurs, with no more than a 10% chance of falsely concluding there is a trend if degradation does not occur. A program also has to specify a general trend- or status-analysis approach that will be used once monitoring data are collected; this serves as the framework for examining power and precision.

Examination of power, precision, sample sizes, etc., also requires estimates of relevant sources of variation that will affect the accuracy of trend and status estimates once monitoring is operational. Estimates of current variability based on pilot studies and historical data are an essential starting point, but future changes in the spatial and temporal variance of a monitoring indicator are inevitable due to disturbances, climatic fluctuation, and anthropogenic stresses such as climate change. Failure to account for plausible future spatial and temporal changes in indicator variance can lead to overestimation of design performance during the planning phase of monitoring. More importantly, in newly implemented monitoring efforts, the inability of the implemented sampling design to provide the desired precision of status and trend information due to changes in spatial or temporal variance may not be recognized before an indicator has fallen below a critical level. We recommend directly considering the effects of changes in indicator variance resulting from plausible future disturbances and stressors (change agents) when evaluating survey designs for operational implementation. Simulation modeling, as demonstrated in this chapter, is a highly flexible approach for such examinations. Comparison of multiple survey designs under plausible future scenarios helps to inform the selection of a survey design most robust to near-term uncertainty.

Likens 2009; Chapter 20). Based on our experience, available programmatic resources are largely expended on the initial implementation of operational monitoring. Once operational monitoring has commenced, expanding sampling efforts becomes very difficult. Additionally, trend detection usually requires an extended period of time. In the short term, deficiencies of a sampling design may go undetected before an indicator has fallen below a critical level (see also Chapter 2).

In this chapter, we illustrate the importance of considering future uncertainty and demonstrate the use of a simulation approach in the design of monitoring surveys. We develop a simulation strategy that emulates the effects of future agents of change on the annual observations of a spatially explicit indicator population, and that has the ability to apply and compare the performance of multiple survey designs. This approach does not employ complex ecosystem modeling or perceptions of future variance structure of a population. Instead, it emphasizes modeling, in a relatively simple manner, the salient effects of key agents of change on the spatial and temporal dynamics of an indicator. Future variance of the indicator population, and ultimately of survey designs, derives from these modeled effects. We show that this simulation-based comparison of competing designs under plausible future variance helps to inform the selection of a survey design most robust to future uncertainty. To introduce this simulation strategy, we provide an overview of our recommended approach, followed by an extended example focusing on comparison of alternative survey designs for monitoring grasslands in Canyonlands National Park, Utah, USA.

Design and use of simulations of future uncertainty in assessments of survey designs

We first describe several aspects of the strategy we recommend in developing a simulation strategy for comparing monitoring designs while accounting for future uncertainty.

(i) Any evaluation of candidate survey designs for long-term monitoring should be seen as an integrated step in development and implementation of a monitoring study (Chapter 2). In our context, we assume that overall monitoring objectives, target population, general analytical approach to be used for assessing status/trend once monitoring commences, response design (measurement approach at each site), sample frame, and quantitative sampling objectives have been carefully defined, and that suitable estimates of current status and variance components have been obtained (Chapters 2, 7, 8, 9).

(ii) With our approach, the population to be monitored is represented in a spatially explicit manner. The minimum spatial unit represents the size of units in the sample frame, and therefore the size of sites (plots) to be used in the actual monitoring effort. This is not a general requirement of simulation approaches in monitoring design. However, when agents of change are expected to have disproportionately higher effects on some portions of the target population than others but these subpopulations are not partitioned in separate strata, spatially explicit simulations tailored to the population of interest will be most informative.

(iii) The dynamics of a population are simulated using estimates of initial status, annual variability, and trend. These parameters determine the future spatial pattern of an indicator population given contemporary ecological processes, and are referred to as baseline parameters. Parameters are distributed across each frame element and used to simulate annual observations at the frame-element level.

(iv) Key agents of change are selected for simulation assessments on the basis of their potential to have a large effect on an indicator, or because they are a high concern to management (Box 10.2). Simulated properties of change agents include effect, frequency, extent, and pattern. Effects are approximated by modifying the

Box 10.2 Common challenges: simulation choices

In the context of evaluating survey designs based on current and potential future levels of variability, there are three key components of simulation assessments that can be problematic. These include the specification of the initial spatial pattern of an indicator, baseline population parameters which represent the current population of an indicator, and key change agents for assessments. Detailed information on the initial status of an indicator over the target population often is lacking. Results from pilot monitoring or from relevant research studies may provide site-level estimates of an indicator, which may be spatially interpolated to the target frame with kriging or other statistical methods. In some cases, remotely sensed information may be used to spatially delineate relative condition of an indicator (e.g. low, medium, high percent cover), and relationships between these conditions and indicator values used to assign initial status values. Where data are lacking, initial status may be estimated based on biophysical properties of the sample frame given understanding of how these properties influence the indicator. Similarly, baseline population parameters may be estimated from pilot monitoring and other research efforts, and aligned with biophysical properties of the target area. Iterative evaluation of population parameters tends to be necessary. A future projection over the time interval of an assessment using initial estimates is first scrutinized to determine reasonability of spatial patterns, based on historical patterns or trends. In the absence of historical data, professional judgment alone determines reasonability. As necessary, parameters are modified and subsequent projections are used to determine further modifications.

Considering all possible future change agents and all possible permutations of their properties (effect, frequency, extent, and pattern) adds considerable time and cost to a simulation assessment. Also, results from extensive permutations can be difficult to organize, and comparisons among the different combinations can be tedious and confusing. Efficient use of a simulation approach requires a parsimonious number of simulation runs. Informed judgment is thus critical to limit assessments to change agents with the greatest potential to influence future indicator properties, and to salient properties of change agents. Monitoring programs typically use conceptual models of disturbance and stressor dynamics to synthesize current empirical understanding, to aid in understanding potential system change, to incorporate management goals and objectives, and to inform indicator selection (e.g. Britten *et al.* 2007; Chapters 2, 22). These models are a valuable source for selecting the key agents for simulation assessments. Ideally, models are developed with stakeholder input, but where this is not the case, stakeholders should have the opportunity to identify change agents of greatest concern. Also, conceptual models that illustrate causal linkages can inform the types and value ranges of change-agent properties to include in an assessment. Focusing on a limited number of change agents and varying properties facilitates interpretation of results, and expedites the survey-design assessment process.

variability of annual values of an indicator or overall trend. The effects, temporal frequency, aerial extent, and pattern (aggregated, dispersed, random) are varied over a gradient of assumed possibilities to account for future uncertainty. This gradient may represent perceptions of nominal to worst-case or severe conditions. Future projections of an indicator population are then simulated.

(v) Survey designs sample a simulated population similarly to real-world monitoring. Survey designs specify the number of sample sites, the revisitation schedule of sites, and the spatial locations of sites (relative to the sampling frame), and are used to extract annual observations from the (simulated) target population. In our example, the measurement design and effort for each site visit was fixed, but alternative measurement approaches (e.g. producing higher or lower residual variability due to measurement error) could be easily incorporated into the simulation approach. Multiple designs are applied to the same simulated population for comparisons of performance relative to the specified sampling objective.

(vi) Parsimonious summaries of major patterns observed in these comparisons are produced to facilitate discussion with managers and administrators about advantages and trade-offs of alternative designs. Again, this is a general consideration for any quantitative examination of design alternatives.

The key advantage of a simulation approach is its flexibility. In our context, this advantage is the ability to explicitly represent patterns of future possible changes based on current system understanding and to assess survey-design performance in light of these potential changes. Simulating possible changes in a spatially explicit manner provides estimates of potential future indicator variability as well as realistic assessments of the accuracy of sample variance estimates under assumed change scenarios. This aids in assessing the adequacy of designs to sample patterns of future change, and overall, to satisfy the sampling objective. Survey designs selected from this procedure are likely to be more robust to future changes relative to designs based solely on historical estimates of variance.

Simulation assessment example

In the following example, we demonstrate how simulation assessments can be used to assess which survey designs will best meet sampling objectives in the face of future uncertainty. Our case study focuses on monitoring in Canyonlands National Park (CANY), which encompasses 136 610 ha in southeast Utah (Fig. 10.1). Given the potential impacts of social trailing on soil erosion and degradation of herbaceous and shrubland ecosystems in this dryland system, the NPS Northern Colorado Plateau Inventory and Monitoring program (NCPN) selected grassland and shrubland ecological site types for monitoring (O'Dell *et al.* 2005). Our example focuses on the evaluation of survey designs for perennial grass cover in the combined Desert and Semidesert Sandy Loam (four-wing saltbush) ecological site types, which have similar soil and vegetation properties and represent the majority of native grasslands in CANY (Fig. 10.1).

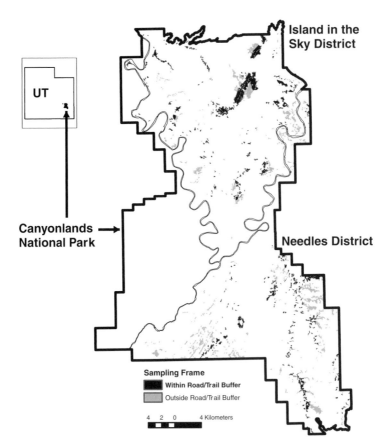

Figure 10.1 Location of Canyonlands National Park, Utah, USA, and the sample frame for the example presented in this chapter. Frame elements are additionally delineated by distance from roads and trails ($\leq$ 500 m, > 500 m).

The target population consists of 5064 ha in these site types accessible by foot and within 4 km of a road or trail. The sample frame was derived using procedures developed by Garman *et al.* (2010). The target population was gridded using an element (i.e. cell) size of 0.25 ha, which equates to the size of an NCPN monitoring site. Using transportation network data layers, sample-frame elements within 500 m of roads and trails (ca. 48% of the sample frame) were delineated to identify regions with different potential for human disturbance.

Simulation system and initialization

We developed a customized system to simulate the annual observations of a single indicator (e.g. grass cover) across a user-provided gridded sample frame, and to simulate impacts of future change on an indicator. Annual observations are generated from estimates (mean and standard deviation) of initial status, annual fractional change (log-linear slope), and variability around a log-linear trend line (Root Mean Squared

Error – RMSE). Means and standard deviations define distributions which are then sampled to derive parameter values for each frame element.

For this study, we estimated population parameters for percent grass cover using pilot monitoring data from Schelz (2002) and Witwicki (2010), and distributed parameter values to mirror patterns generated from kriging contemporary samples of herbaceous cover. Baseline parameters were mean = 15 (SD = 8.5) for initial status (% perennial grass cover); 0.0 (0.01) for log-linear slope; and 0.30 (0.13) for RMSE. Baseline parameters of individual frame elements were then modified to emulate impacts of change agents. Any number of survey designs can be applied to extract annual observations from a simulated population. Extracted sample observations are used to determine agreement between sampled and population variance, and in analyses of status and trend.

Our system is stochastic, in that the variability of annual observations and the implementation of change-agent properties involve a random component. Multiple replicates of a simulated population are generated using different random number seeds to provide a range of possible trajectories. Similarly, a survey design is repeatedly simulated using multiple sets of locations generated with different random number seeds to assess the average performance and sample-to-sample consistency of a design.

Survey designs

Proposed design

Besides informing choice of survey design when a monitoring effort is being developed, power/precision examinations can be used to assess benefits of modifying an operational design (see *Discussion*). Our example emphasizes the use of a simulation approach to enhance a design selected using historical variances. We used the survey design proposed by the NCPN for monitoring grasslands in CANY as a starting point. This design was based on programmatic resources and power-for-trend assessments using variance estimates acquired from a pilot monitoring effort (Witwicki 2010). The proposed design includes two groups of panels. One group consists of 7 panels with 3 sites per panel, each of which are visited for 2 consecutive years then revisited after 5 years. The second group consists of 7 panels, each containing 6 sites, which are visited once and revisited after 6 years (e.g. X = 3 and Y = 6 in Fig. 10.2). Across both groups of panels, 12 sites are visited annually and the total number of sites is 63; this proposed designed is referred to as 2PS_12 (Table 10.1). This type of design, a split-panel design, performs well, compared to alternatives, for both estimation of trend (emphasized by panels 1–7 in Fig. 10.2) and status (emphasized by panels 8–14 in Fig. 10.2) (Urquhart *et al.* 1998, Breidt and Fuller 1999, Urquhart and Kincaid 1999, McDonald 2003; Chapter 7). Note that in the first group of panels, two panels are visited each year, resulting in an overlapping pattern of visits to panels. This pattern is one approach for ensuring that the revisit design is "connected" in an experimental design sense (see Chapter 7).

Alternative designs

We generated seven additional survey designs that met the following practical constraints and programmatic goals of the NCPN. First, only designs with ≤15 sample sites per

Table 10.1 Split-panel survey designs evaluated in the case study. Notation for Design Code is described in Fig. 10.2. Short-hand notation of Revisit Design (i.e. panel plan) was adapted from McDonald (2003), and is defined as follows: The sets of paired parentheses designate the temporal sub-design for each of the two or three groups of panels in each split-panel design. For each group, the first number inside the paired parentheses is the number of consecutive years a panel is sampled and the second number is the number of years between revisits; the superscript is the number of panels in that group and the subscript is the number of sites per panel.

Design code	Revisit design	Sites per year	Total unique sites
2PS_12	$[(2-5)^7_3, (1-6)^7_6]$	12	63
2PS_13	$[(2-5)^7_3,(1-6)^7_7]$	13	70
2PS_14	$[(2-5)^7_4,(1-6)^7_6]$	14	70
2PS_15	$[(2-5)^7_4,(1-6)^7_7]$	15	77
3PS_12	$[(2-5)^7_2,(1-6)^7_6, (3-3)^2_2]$	12	60
3PS_13	$[(2-5)^7_2,(1-6)^7_7, (3-3)^2_2]$	13	67
3PS_14	$[(2-5)^7_3,(1-6)^7_6, (3-3)^2_2]$	14	67
3PS_15	$[(2-5)^7_3,(1-6)^7_6, (3-3)^2_3]$	15	69

Panel	1	2	3	4	5	6	7	8	9	10	11	12	13	14	15	16	17	18	19	20
1	X	X						X	X						X	X				
2		X	X						X	X						X	X			
3			X	X						X	X						X	X		
4				X	X						X	X						X	X	
5					X	X						X	X						X	X
6						X	X						X	X						X
7	X						X	X						X	X					
8	Y							Y							Y					
9		Y							Y							Y				
10			Y							Y							Y			
11				Y							Y							Y		
12					Y							Y							Y	
13						Y							Y							Y
14							Y							Y						
15	Z	Z	Z				Z	Z	Z				Z	Z	Z				Z	Z
16				Z	Z	Z				Z	Z	Z				Z	Z	Z		

Figure 10.2 Structure of the two types of split-panel designs evaluated in this study. The first type consists of seven panels (X) with sites in each panel visited two years in a row and revisited after five years and seven panels (Y) with sites in each panel visited for one year and revisited after six years. Survey designs with this panel structure are designated for the case study as 2PS_##, where the "2PS" indicates two types of panel-revisit structures, and ## is the number of plots visited annually (i.e. 12–15). The second type, designated as 3PS_##, has the same two sets of seven panels, but also has a third panel-revisit structure (Z) in which two additional panels are visited for three consecutive years and revisited after three years. More generally, adapting notation from McDonald (2003), the structure of the first design can be indicated as $[(2-5)^7, (1-6)^7]$, while that of the second design can be indicated as $[(2-5)^7, (1-6)^7, (3-3)^2]$ (Table 10.1).

year were feasible given NCPN budgetary limitations. Second, annual visitation was not desirable because plot fatigue due to constant soil disturbance by monitoring crews is a serious concern in the dryland systems of the Colorado Plateau. Third, under the scenario that population variance for percentage grass cover would remain at current estimated values, the designs had to satisfy a sampling objective defined by the NCPN to provide early detection of possibly 'important' trends (Witwicki 2010). This objective is to detect an annual decline of 3.5% within 10 years with power $\geq$ 80% and $\alpha = 0.10$.

Three of the additional designs were generated by adding one, two, and three sample sites annually to the NCPN proposed design (Table 10.1). Adding one additional site to panels 8–14 emphasized status estimation (2PS_13). Adding one additional site to the first seven panels emphasized trend estimation (2PS_14). Adding one additional site to each panel of the proposed design increased the annual sampling effort by three, and resulted in a design with the highest total number of sample sites of all designs considered (2PS_15).

The other four alternative designs were a variant of the proposed split-panel design that included a third panel-revisitation structure (Fig. 10.2). This structure consisted of sites visited in three consecutive years and revisited after three years. Two panels of this structure were added. Starting with a 12 site per year design (3PS_12, Table 10.1), additional sites were added, with a maximum of 15 sites visited per year (3PS_15). The revisitation schedule of the third panel structure was employed because increasing revisitation frequency increases power to detect trends (Urquhart et al. 1998, McDonald 2003). For the same annual sampling effort, these designs had fewer total number of sample sites than designs with only the two types of panel structures (Table 10.1).

All survey designs incorporated a Generalized Random Tessellation Stratified (GRTS) spatially balanced probability design (Stevens and Olsen 2004; Chapter 6). We generated samples and panel assignments with equal probability of selection, using R package "spsurvey" (Kincaid et al. 2009, ver. 2.1; see Chapters 6, 14). For each survey design, we used 100 replicates of simulated samples in each scenario.

Modeling future scenarios

Disturbance scenarios

Climate variability and human-mediated impacts to soil stability are key processes likely to influence future conditions of grasslands in CANY. Soil compaction from social trailing, off-road vehicles, and even occasional trespass livestock grazing promotes soil erosion and loss of native grass cover (Miller 2005). We modeled reduction in grass cover due to trampling by first randomly selecting a sample frame element within 500 m of a road or trail, then expanding to a 1-ha patch which could form beyond this distance. Disturbed elements within the 500 m buffer were assigned a log-linear slope of –0.14; otherwise elements were assigned a slope of –0.10. A higher rate of decline near roads and trails reflects a tendency for social trails to initiate along transportation networks and to be repeatedly used. These slope values represent a severe loss of grass cover in a relatively short period of time. We simulated six levels of disturbance (0.5, 1, 10, 20, 30, and 40% of the sampling frame) over a 20-year period. This gradient provided a context for evaluating design performance from nominal to extreme occurrence of disturbance. Impacted frame elements were selected, and slope values were modified in the first 5 years of a simulation run. Twenty replicates were generated for each of the six disturbance scenarios.

Climate scenarios

Impacts of climate variability over the next 20 years were modeled using Global Circulation Model (GCM) projections. We used 12-km downscaled climate projections from the NOAA Geophysical Fluid Dynamics Labs' GFDL CM2.1 model, A2 emission scenario (http://cascade.wr.usgs.gov/data/Task1-climate/index.shtm), to generate a future climate signal. We converted the downscaled daily temperature and precipitation values to annual measures, and used the WebWIMP model (The Web-based, Water-Budget, Interactive Modeling Program – http://climate.geog.udel.edu/~wimp) to generate an annual drought index (Willmott and Feddema's Moisture Index; Willmott and Feddema 1992) for each of the next 20 years for each GFDL grid cell. We derived the proportional difference between a future annual index and the historical long-term mean (i.e. yearly deviations) and input this information to our survey simulator as a geospatial data layer.

Within the simulator, the 12-km GCM grid cell covering a sample frame element was determined, and the corresponding climate time series was used in the derivation of annual observations. To emulate effects of climate extremes, we imposed coherence of observations whenever the yearly deviation in the drought index exceeded a threshold level; otherwise, annual observations varied independent of climate. When the threshold was exceeded, a scaled variant of the difference was added to the annual observation of all affected frame elements. Based on exploratory assessments, we selected two thresholds to emulate grass cover response to climate extremes. Annual observations were affected when the absolute proportional difference between the annual and long-term drought index was >0.30 and >0.20, where the lower threshold value represents high sensitivity. These thresholds correspond to about 20% and 33% of the 20 years experiencing a climate-extreme signal, respectively. Twenty replicates were simulated for each threshold level.

Changes in variance structure and accuracy of variance estimates from alternative designs

Assessing variance components

Site, year, and residual variance components influence status and trend estimation in the linear mixed-effects model used as the framework for our calculations (Larsen *et al.* 1995, 2001; Urquhart *et al.* 1998; Urquhart and Kincaid 1999; Kincaid *et al.* 2004; Chapter 7). As also specified in Chapters 7 and 9, site variance reflects inherent variation among sites that is consistent across years (i.e. persistent site effects). Coherent year to year variance reflects population-wide year effects, excluding systematic effects of a linear trend. Residual variance includes unexplained process variance (e.g. unexplained variation in trend among sites, Chapter 7) and measurement variance.

A linear mixed-effects model was used to derive population variance components of simulated populations to determine the effects of modeled change agents on the underlying variance structure, and to derive estimates from each replicated simulated survey to assess level of agreement between survey design estimates and actual population variance. The latter determines a design's ability to capture the actual population variance.

The linear mixed-effects model used for estimation of variance components mirrored the "one trend" model specified in Chapter 7 [Equation (7.7)], with "time" as a numeric fixed effect, and "site" and "year" as random effects. We used the following mixed-effects model ("PROC MIXED", SAS Institute 2001) to estimate the three variance components.

```
Proc Mixed Method=REML;
      class site year;
      Model LnCover = time/DDFM=KENWARDROGER;
      random site year;
      Run;
```

The SAS commands listed above and in Chapter 7 are very similar. However, we used Restricted Maximum Likelihood (REML) estimation (see Chapter 9), under assumed normality of random effects. Due to unbalanced designs (each sample site did not have observations in each year), the Kenward–Roger method was used to derive degrees of freedom (Spilke *et al.* 2005). Natural-log transformation of percent cover (LnCover) enhanced normality and constancy of variance. "Time" was included to de-trend observations because the coherent year variance component is tied to random year effects apart from systematic trend (see also Chapters 7, 9).

Comparison of survey designs

In disturbance scenarios, population-wide site and residual variance components increased with increasing disturbance level (Table 10.2). In general, sampled estimates of variances were within 5–8% of population values for disturbance levels $\leq 10\%$. Agreement with population site variance increased with increasing sampling effort per year. For simplicity, we present results only for designs with the lowest (12 sites/year) and highest (15 sites/year) annual efforts throughout the rest of the chapter. Using 10% disturbance as a worst case, all designs reliably sampled the population variance, with best performance from designs measuring 15 sites per year.

Under climate scenarios, the coherent year variance of simulated populations increased with increasing sensitivity of annual observations to a climate signal, while mean site variance decreased slightly with increasing sensitivity (Table 10.2). Similar to the disturbance scenarios, mean sample estimates of variance components were within 5–10% of population values. For both threshold levels of sensitivity to climate change, sample means for site and year variances decreased with increasing annual sampling effort.

Table 10.2 Mean (standard error) estimates of variance components from simulated populations (20 replicates per scenario) for the disturbance and climate scenarios. Disturbance levels are expressed as percentage of the sample frame. Climate thresholds are threshold levels to invoke a climate effect on simulated annual observations. Values were derived from a log-linear, mixed-effects model with site and year as random effects, and time as a fixed effect to de-trend observations. Each replicate contained 20 years of simulated, annual values of percent grass cover for each of the 20 256 sample frame elements.

	Variance component		
Scenarios	Site	Year	Residual
Disturbance level			
0.5	0.601 (0.001)	0.000 (0.000)	0.102 (0.000)
1	0.610 (0.001)	0.000 (0.000)	0.105 (0.000)
10	0.730 (0.012)	0.000 (0.000)	0.158 (0.001)
20	0.810 (0.020)	0.000 (0.000)	0.200 (0.001)
30	0.852 (0.028)	0.000 (0.000)	0.229 (0.001)
40	0.870 (0.036)	0.000 (0.000)	0.245 (0.002)
Climate threshold			
> 0.30	0.572 (0.002)	0.003 (0.000)	0.100 (0.000)
> 0.20	0.561 (0.001)	0.016 (0.000)	0.100 (0.000)

In both disturbance and climate scenarios, all survey designs estimated variance components fairly accurately. This was observed despite the strong additional population-level variability produced in disturbance scenarios, in which trampling disturbance tended to be aggregated along transportation networks. Without explicitly addressing this spatial pattern, the sampling approach adequately captured population variability. This may be a result of the dispersion of transportation networks throughout the target population (Fig. 10.1) as well as the efficiency of spatially balanced sampling. In addition, the total sample sizes relevant to each variance component were relatively high (based on results in Chapter 9), with relevant sample sizes of > 60 sample sites for capturing site variance, 20 years for capturing year variance, and 240–300 total site-visits for capturing residual variance.

Assessing performance in relation to the sampling objective

Assessing power and precision

In our example, we used simulations to project the current estimated population forward and to assess multiple realizations of each survey design, while we calculated power and precision of status estimates following the analytical framework of Urquhart *et al.* (1993, 1998) and Larsen *et al.* (1995, 2001). The assumed analytical model is the general linear mixed-effects model described above. Chapter 7 summarizes how the year, site, and residual variance components contribute to the variance of the estimate of the linear trend, and therefore affect power to detect trend. With this analytical model, estimates of the slope and intercept of the linear trend line can be used to calculate a model-based estimate of expected status in each year. The standard error (SE) of this status estimate is derived from the variance-covariance structure of the linear regression coefficients.

We used a set of functions developed in the R language by Tom Kincaid, US Environmental Protection Agency, Corvallis, Oregon, to calculate power and precision of status estimates for each year of simulated monitoring for each scenario, survey design, and replicate. Site, year, and residual variance components were specified as described above. Use of log-scale variance estimates was appropriate given our focus on a log-linear trend. The general variance structure for the mixed-model also includes site and year correlation components, which were set to 1 and 0, respectively [see Larsen *et al.* (1995) and Chapter 7].

To simplify comparisons among designs, we used the minimum year to achieve a power of ≥ 80% for detecting an underlying annual −3.5% trend, with $\alpha = 0.10$ as a measure of performance. This measure reflects the desire to achieve at least 80% power to detect this level of change within 10 years to provide early warning of undesirable change (Witwicki 2010). We generated a mean (and standard error) for the minimum year at which 80% power was achieved by averaging values across the replicates of a design-scenario combination (100 replicates of a survey design × 20 replicates of a scenario). For each set of extracted observations, we derived the average standard error of status over the 19-year period, and then derived a mean average across the replicates of a design-scenario combination. We used an ANOVA of ranked values and Scheffe's multiple contrast for statistical comparison of mean minimum year to achieve the target power, and of mean average standard error of status among the eight survey designs.

Power and precision of alternative designs

In disturbance scenarios, the mean minimum time to achieve the target power (Fig. 10.3a) and mean average model-based standard error (Fig. 10.3b) increased with increasing population variability. Within each disturbance level, mean minimum time to detect the −3.5% trend only differed at most by ca. 1 year and mean average standard error differed by 0.017 among designs. Although these were minor differences in mean values among designs, both the minimum time and standard error means were significantly different ($P < 0.05$) among designs with different annual sampling effort, with mean values decreasing with increasing number of annual samples. That is, with higher annual effort, desired power was obtained more quickly, and status estimated more precisely. Of the designs evaluated, the 15 sites per year designs appear to be the most robust to trampling effects. Although the differences were slight, only designs with 15 sites per year consistently achieved the target power level within 10 years for ≤ 10% disturbance levels (Figs. 10.3a). Above 10% disturbance, all designs failed to achieve the target power within a decade.

The survey-design assessments for disturbance scenarios illustrated the importance of the power and time-frame components of the sampling objective. We used a 10-year time frame for achieving 80% power, which was based on ecological and management considerations (Witwicki 2010). If 80% power within 12 years was deemed sufficient, all designs we evaluated would satisfy this objective for all simulated disturbance levels

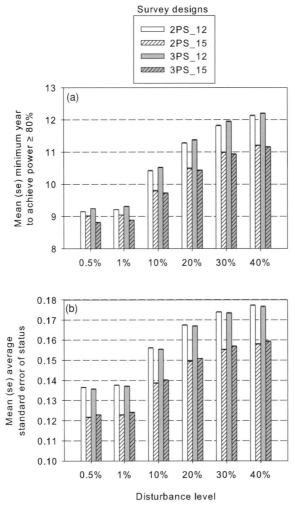

Figure 10.3 (a) Mean (standard error) minimum year to achieve a power of $\geq 80\%$; (b) 19-year mean (standard error) average standard error of status for survey designs for the disturbance scenarios. Values for designs were generally linear between the 12 and 15 plots per year designs. For visual clarity, only results for designs with 12 and 15 plots per year are shown. Power and standard error of status were based on a -3.5% trend and $\alpha = 0.10$. Means were derived from 100 replicates of sample locations for each survey design × 20 replicates of a disturbance scenario ($n = 2000$).

(Fig. 10.4). If 90% power was the target level, all designs would reach this target within 14 years.

Precision of status estimates responded primarily to differences in annual sampling effort among designs (Fig. 10.3b), but the type of split-panel design had minor effects. Although the 19-year mean averages were similar, the revisitation schedule of the 3PS_15 design provided higher precision of status (lower mean values for standard errors) in the first 5 years compared to the 2PS_15 design (Fig. 10.5). After year 8, the

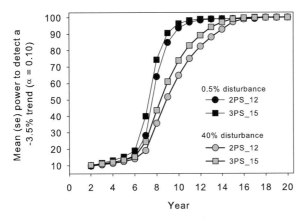

Figure 10.4 Mean power for two survey designs for the nominal and extreme disturbance scenarios. Designs bracket the minimum (2PS_12) and maximum (3PS_15) power across all designs. Means were based on $n = 2000$. Standard error bars of means are too small to be visible.

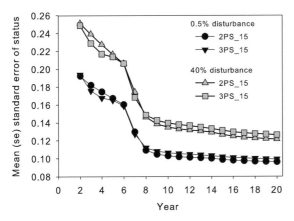

Figure 10.5 Mean standard error of status for survey designs with 15 plots per year for the nominal and extreme disturbance scenarios. Curves illustrate temporal differences between designs, where means for the 3PS designs were lower in the first 7 years then higher in the remaining years compared to 2PS designs. Means were based on $n = 2000$. Standard error bars of means are too small to be visible.

larger overall sample size of the 2PS_15 design (Table 10.1) resulted in slightly higher precision.

For climate scenarios, comparative power and precision showed patterns similar to those for the disturbance scenarios. The minimum year to achieve the target power (Fig. 10.6a) and mean average standard error of status (Fig. 10.6b) increased with increasing population variance for all survey designs. These measures for the two designs with 15 sites per year were not significantly different ($P > 0.05$), but were significantly less ($P < 0.05$) than means of other designs. With the lower sensitivity threshold, only the designs with 15 sites per year satisfied the sampling objective by year 10 (Fig. 10.6a). At the higher level of climate sensitivity, no designs satisfied the sampling objective.

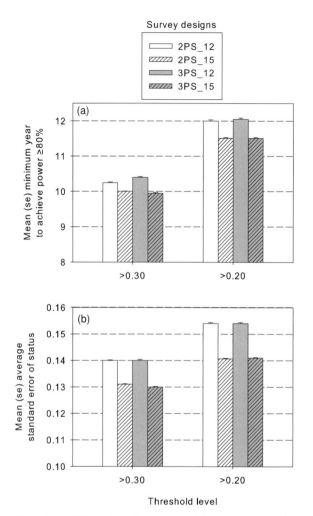

Figure 10.6 (a) Mean (standard error) minimum year to achieve a power of $\geq 80\%$, and (b) 19-year mean (standard error) average standard error of status of survey designs for the two threshold levels used to invoke a climate effect on simulated observations. Values for designs were generally linear between the 12 and 15 plots per year designs. For visual clarity, only results for designs with 12 and 15 plots per year are shown. Power and standard error of status were based on a -3.5% trend and $\alpha = 0.10$. Means were derived from 100 replicates of sample locations for a survey design $\times$ 20 replicates of a climate scenario ($n = 2000$).

In addition to examining precision of status estimates based on a linear-model framework, for all scenarios we also examined confidence-interval coverage for a design-based estimator of status. We calculated Horvitz–Thompson estimates of population means and corresponding 95% confidence intervals using the function *total.est*() in the "spsurvey" package, with confidence intervals calculated based on the local neighborhood variance estimator (Stevens and Olsen 2003; see Chapters 6, 7, and 14). Coverage of confidence intervals was calculated at years 10 and 20. For all survey designs and scenarios, >95% of the 95% sample confidence intervals contained the true population mean.

Implications for monitoring at Canyonlands National Park

In this example we limited our assessments to two key change agents, and only evaluated alternative survey designs that would be operationally viable by the NCPN, based on current programmatic limits. The designs evaluated were not substantially different in terms of annual and total sampling effort, and as expected, did not widely differ in terms of absolute differences of the power and status metrics we evaluated.

However, a key finding was that only the two designs with 15 sites per year consistently achieved the target power level within 10 years for ≤10% disturbance levels and with the lower climate sensitivity threshold. This increase in sampling effort of three sites per year also produced the highest precision. These results provide incentive to consider one of the two designs with 15 sites per year to ensure that the sampling objective is satisfied over plausible, near-term changes in percent grass cover. Differences in temporal precision, and plot fatigue potential of the three-year visitation scheme would factor into the choice between the 2PS_15 and 3PS_15 design for implementation. From a programmatic perspective, the added costs of these designs must be weighed against the benefit of achieving the sampling objective only ca. one year sooner than the proposed design for worst-case conditions. The time frame specified in a sampling objective is ideally selected to ensure early warning of degrading conditions. Not exceeding this time frame may be essential to avoid threshold behaviors (i.e. a rapid, nonlinear decline in an indicator) and costly restoration.

The spatially balanced GRTS design used for selecting sample sites in our example accommodates the addition of sites to the design without compromising spatial balance (Chapters 5, 6). Logistically, expanding an existing monitoring effort by three sample sites per year may appear trivial. In practice, monitoring programs are likely to monitor numerous target populations, and budgetary limitations may require convincing evidence for even nominal expansion of any sampling design. For instance, there are five target populations of terrestrial vegetation in Canyonlands NP and different survey designs for each, and Canyonlands NP is just one of 16 park units in the NCPN program. In addition to terrestrial vegetation monitoring, 12 other monitoring protocols are funded by the NCPN (see http://science.nature.nps.gov/im/units/ncpn). Simply adding three sites per year to each design may help safeguard against future uncertainty, but the additional workload and cost of even a nominal increase for each target population can be surprising large. Comparable "trivial" increases across all target populations and protocols in the NCPN would far exceed programmatic resources. Results from examinations of current designs based on results to date and from simulation assessments of future uncertainty may be a necessary incentive to expand even a few existing designs by one to three sites per year.

Discussion

Flexibility is a key advantage of a simulation approach for developing or modifying sampling designs. In our example, a simulation approach allowed us to explicitly represent patterns of future possible changes based on current system understanding and to assess survey-design performance in light of these potential changes. Adjustments of current variance estimates by professional judgment to account for future uncertainty generally are ad hoc, and lack consideration of patterns of future change. Spatially explicit simulations potentially provide more insight into potential variability under assumed change scenarios. Survey designs selected from this procedure are likely to be more robust to future changes relative to designs based solely on historical estimates of variance. A simulation approach to assess potential future variability is especially important in situations where systems are highly susceptible to known or suspected agents of change, there is high probability of future change due to these agents, and delayed detection of trends estimates can lead to substantive loss of ecological integrity or high restoration costs.

Assumptions of how future change agents may affect an indicator population can be crafted to evaluate different levels of future uncertainty. Such assumptions are necessary to account for uncertainty in the effects, frequency, extent, and pattern of future change agents. Expert opinion, professional judgment, stakeholder concerns, and objectives of an assessment determine the range of assumptions employed in an assessment. Assumptions may reflect perceptions of reasonable change based on judgment, historical rates, or proposed land-management plans with the potential to increase disturbance (e.g. increased human disturbance around proposed campgrounds, roads, and trails). They may also represent an upper level of change that may or may not be realistic, but ensure the selection of an efficient survey design for a wide spectrum of future change. For example, in our work in grasslands we have incorporated a range of potential trends to emulate trampling effects; however, the two values (10 and 14% annual decline) we currently use and reported on in our example are considered upper limits on plausible change. Trampling effects over >10% of the target population are likely unrealistic, but assessments using these extreme disturbance levels help to identify limitations of survey designs. Using 10% disturbance as an upper level for design assessments provides a stringent but not excessive criterion, and ensures that the selected design is likely to satisfy sampling objectives up to plausible levels of future change, based on current assumptions of what is "plausible". Varying each change-agent property across a range of values deemed reasonable in the future, and simulating all permutations of these values provides a gradient of plausible future outcomes. A focus on the most severe plausible impacts of change agents may be used to understand the implications of underestimating a worst case scenario, which may encourage selection of a highly robust survey design.

More generally, analytical and simulation assessments to inform selection of a monitoring survey design are ideally performed in the initial planning phase. For each target population, comprehensive assessments could employ a wide range of designs unconstrained in terms of annual and total sampling effort, but potentially reflecting other practical constraints. For instance, only revisitation schedules with low potential for plot fatigue (effects on a site's status or trend due to frequent disturbance by monitoring personnel) may be included. Varied spatial properties of candidate designs can help determine the need for stratification or the implications of using unequal probabilities in sample site selection, given current variability and future potential changes in indicator variance. For each target population, assessments can help establish optimal survey designs.

These assessments also may be instrumental in determining the feasibility of monitoring certain indicators given estimates of current and future uncertainty. Where sampling effort of an optimal design exceeds programmatic resources or requires an inordinate amount of effort relative to programmatic goals, the indicator may be dropped from consideration. Alternatively, attributes of the sampling objective (e.g. precision, power) may be modified – i.e. expectations may be lowered if this does not compromise the underlying monitoring objectives.

When operational monitoring is already in place, power/precision/effort assessments can still provide important evidence for modifying designs. As in our example, the range of alternative designs considered may be restricted to those that are moderate modifications of the current design. Broad-scale monitoring programs frequently need to monitor numerous indicators and target populations, but this need must be balanced with the need to collect meaningful information for each indicator and population. Assessments across a suite of existing designs can identify those in greatest need of enhanced sampling effort. Providing evidence for adding even a nominal number of samples to a few existing designs may be critical given budgetary limitations.

Accounting for future possible changes in indicator patterns and variance aids in the selection of a survey design that is robust to future uncertainty, but attention to actual design performance over time is still critical. Future dynamics of an indicator may result in spatial and temporal variance more extreme than anticipated or what was perceived to be plausible. Over the long term, adaptive monitoring will always be critical to ensure a sampling strategy continues to satisfy the sampling objective (Ringold *et al.* 1996, Lindenmayer and Likens 2009). Quantitative power/precision examinations can help ensure selection of a survey design that will be effective based on current variation, while simulation-based sampling of projected future populations helps to safeguard against plausible, near-term (10+ years) changes in indicator variance. In practice, these examinations ensure acquisition of reliable status and trend information while providing sufficient time to determine necessary adaptive changes to a sampling design.

Future research and development

We are not aware of an existing user-friendly program or package that can simulate spatially explicit dynamics of an indicator population, that allows a user to implement and simulate a range of change-agent properties, and that has the ability to implement a range of survey designs. Even our customized simulation system is not generalized; implementation of different change-agent properties requires modifying C code. Results from extensive simulations can be difficult to organize, and analyses and comparisons of multiple survey designs can be tedious. Automating performance analyses is thus equally important to facilitate use of simulation approaches. Monitoring programs currently seeking to use simulation assessments must develop their own customized system. This requires access to expertise in programming languages or in commercial packages with simulation capabilities. Development of a generalized and flexible package for wide-spread use would help many monitoring programs more efficiently examine candidate survey designs under a range of scenarios, to justify the selected design, and to re-examine the operational design as more information is obtained. Developing such a system in the freely available R language (R Development Core Team 2011) is perhaps the most viable option, particularly given the available packages in R for spatial sampling and analysis (e.g. Chapters 6, 14) and for spatio-temporal modeling.

Summary

A simulation approach is a highly flexible tool for assessing comparative performance of alternative sampling designs to address factors not easily incorporated into analytical power and precision examinations. We demonstrated this flexibility in the context of assessing design performance under potential future changes in population variances. Selection of an optimal sample survey design for operational monitoring tends to be based on historical or contemporary estimates of indicator variance components, with the assumption that variances will not change in the future. Even relatively near-term future changes in the spatial and temporal variance of a monitoring indicator are inevitable due to disturbances, climatic fluctuations, and human-mediated stressors such as climate change. Failure to account for future changes in indicator variance can lead to over-estimation of potential survey design performance. We recommend using simulation approaches to model plausible near-term (1–2 decades) changes in indicator variance, such as those due to disturbances or climate-change impacts, when evaluating survey designs for operational implementation, as well as to enhance existing designs using historical variances. This simulation approach allows the user to assess the potential effects of change agents on the variance, status, and trend of an indicator population across a designated sample frame, and to apply multiple survey designs to sample the affected population. Comparison of these competing designs helps to inform the selection of a survey design most robust to near-term future uncertainty. Over the long term, unpredictable changes in indicator conditions may dictate more substantive changes in a sampling strategy. However, operational survey designs selected on the basis of current variation and under plausible change scenarios are likely to provide reliable status and trend information while providing sufficient time to determine the need for more significant changes to a sampling design.

Acknowledgments

Data for the delineation of grass-cover patterns were provided by the US National Park Service Northern Colorado Plateau Inventory & Monitoring Network (Moab, UT), and

Mark Miller (USGS, Flagstaff, AZ). We are indebted to Scott Urquhart for simulator-design recommendations, and assistance with the analysis and interpretation of survey-design samples. We also thank Lee Ann Starcevich for assistance with variance estimation of unbalanced designs, and Tom Kincaid for the power-for-trend R program. We thank Robert Gitzen and two anonymous reviewers for constructive comments that improved this chapter.

Section III

Data analysis

11 Analysis options for estimating status and trends in long-term monitoring

Jonathan Bart and Hawthorne L. Beyer

Introduction

This chapter describes methods for estimating long-term trends in ecological parameters. Other chapters in this volume discuss more advanced methods for analyzing monitoring data, but these methods may be relatively inaccessible to some readers. Therefore, this chapter provides an introduction to trend analysis for managers and biologists while also discussing general issues relevant to trend assessment in any long-term monitoring program.

For simplicity, we focus on temporal trends in population size across years. We refer to the survey results for each year as the "annual means" (e.g. mean per transect, per plot, per time period). The methods apply with little or no modification, however, to formal estimates of population size, other temporal units (e.g. a month), to spatial or other dimensions such as elevation or a north–south gradient, and to other quantities such as chemical or geological parameters. The chapter primarily discusses methods for estimating population-wide parameters rather than studying variation in trend within the population, which can be examined using methods presented in other chapters (e.g. Chapters 7, 12, 20). We begin by reviewing key concepts related to trend analysis. We then describe how to evaluate potential bias in trend estimates. An overview of the statistical models used to quantify trends is then presented. We conclude by showing ways to estimate trends using simple methods that can be implemented with spreadsheets.

Although some authors have criticized the efficiency and utility of monitoring focused on detecting trends (e.g. Chapter 18), trend assessment is a primary objective of many programs (Chapters 3, 7). The importance of estimating trends accurately is widely recognized (Balmford *et al.* 2003, Gregory *et al.* 2005). Readers seeking to gain more in-depth knowledge of statistical methods for trend analysis should be aware of the extensive literature on this topic. Trend estimation methodology has a long history with numerous statistical textbooks and monographs (Chatfield 1989, Diggle *et al.* 1994, Brockwell and Davis 1996). Even with this long history, trend estimation is an active field of research in many disciplines including economics (Van Campenhout 2006, Chang *et al.* 2009, Shimotsu 2009), ecology (Soldaat *et al.* 2007, Zhang *et al.* 2007,

Design and Analysis of Long-term Ecological Monitoring Studies, ed. R.A. Gitzen, J.J. Millspaugh, A.B. Cooper, and D.S. Licht. Published by Cambridge University Press. © Cambridge University Press 2012.

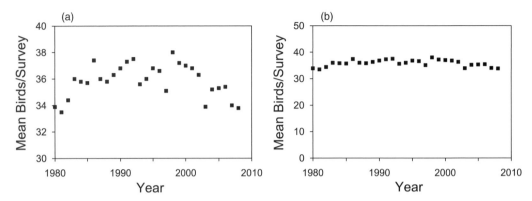

Figure 11.1 The importance of defining scale for the *y*-axis. Each panel shows the same data for American robins from the Breeding Bird Survey (BBS).

Kéry and Schmidt 2008, Thompson and La Sorte 2008, Adamowski *et al.* 2009, Beck *et al.* 2009, Dauwalter and Rahel 2009, Humbert *et al.* 2009, Kéry *et al.* 2009, McPherson and Myers 2009), physiology (Afshinpour *et al.* 2008), and statistics (Mazzetta *et al.* 2007, Burman and Shumway 2009, Shao 2009). Several books on quantitative methods in ecology contain chapters on trend estimation (Skalski *et al.* 2005, Bart *et al.* 1998, Thompson *et al.* 1998). Analysts should become familiar with some of this literature and its existence should be recognized by anyone contemplating development of a "new" method in trend estimation.

Describing "the trend"

In statistical analysis, it is usually helpful to begin by defining the parameter of interest. This is particularly important in trend estimation because "the trend" often depends on the definition (Chapter 3). As an example, consider several examples using data from the North American Breeding Bird Survey (Sauer *et al.* 2008), an annual count of birds during the breeding season carried out widely across the USA and southern Canada. A large sample of routes is surveyed each year. For common species, analyzed at the range-wide level, precision is high [Coefficients of Variation (CVs) < 0.03].

A single data set for American robins (*Turdus migratorius*) illustrates a first principle in defining trends (Fig. 11.1). In Fig. 11.1a, the *y*-axis starts at 30. The scatter plot clearly shows an increasing, and then decreasing, trend. In Fig. 11.1b, the same data are plotted but with the *y*-axis starting at 0 so that the scatter plot emphasizes *proportional* change. Most observers would probably summarize Fig. 11.1b by saying that little change had occurred in population size during the survey period. The fundamental issue is whether proportional or absolute change is of greatest interest. In some applications (e.g. change in body temperature for humans) absolute change is clearly of greater interest. In estimating trend in population size, however, proportional change is nearly always of greater interest. In a well-planned monitoring program, this question will have

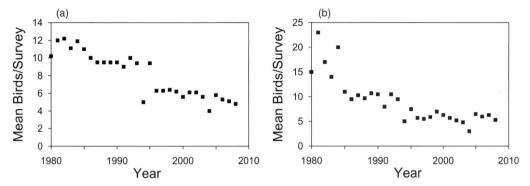

Figure 11.2 Population sizes usually do not all fall on a smooth curve. Shown are mean numbers of chipping sparrows recorded per BBS survey in Maryland (a) statewide, and (b) on 10 well-surveyed routes.

been considered when monitoring objectives are specified during initial program design (Chapter 2).

Another important point about defining trends is that real ecological variables usually do not change in a smooth manner, as illustrated with BBS data for chipping sparrows (*Spizella passerine*) in Maryland (Fig. 11.2). Sample sizes (16 routes per degree block) for this data set are large and we assume variation about the trend indicates real variation. At the statewide level (Fig. 11.2a), the trend is generally downward but annual means (means/survey) well below the trend were reported in 1980, 1994, and 2004. For 10 well-surveyed routes, the trend was also generally downward but mean counts/survey increased substantially in several years (e.g. 1981, 1983, 2005; Fig. 11.2b). The significance of this observation is that the equations used to describe trends never fit the real numbers perfectly. This, in turn, means that the choice between equations used to describe trends is often not obvious nor can one choice necessarily be viewed as "wrong" and another "right". Conversely, as we will see later, a statistical model used for trend analysis incorporates both an equation for describing the systematic change over time (trend) and unexplained deviations from this systematic pattern due to natural variation and survey error; the systematic portion of a trend model should not be expected to provide perfect fit.

A corollary of the point above is that different descriptions of trends may be equally legitimate; as emphasized in Chapter 3, "trend" is an amorphous term. Consider BBS data for northern cardinals (*Cardinalis cardinalis*) in New York State (Fig. 11.3). "The trend" in this scatter plot might be described with equal accuracy as (a) generally increasing, (b) increasing until the early 1990s but then nearly stable, or (c) increasing but with declines in the mid 1980s, mid 1990s, and mid 2000s followed by recoveries. It is probably best to question which of these descriptions is most *useful*, rather than which of them is most *valid*. Answering the former question is often difficult, but it is important to consider the conservation, management, or research issue that the data are being used to address.

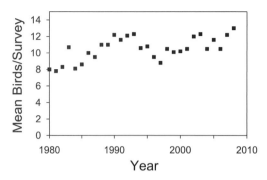

Figure 11.3 "The trend" can often be described in different but equally valid ways (BBS data for northern cardinals in New York State).

Trends are usually described by choosing a model, or set of models (e.g. linear, exponential, polynomial), and fitting the model to data. Trend, or change in population size, is then often defined as the change in the fitted values of the model during a stated interval, such as from the start of the monitoring program to the last year of data or some subset of interest during this period. The exponential model is widely used because it changes at a constant proportional rate, and this provides a relevant and convenient way to describe change in population size. The equation for the exponential curve may be written as

$$E(y) = \beta_0 e^{\beta_1 t} = \beta_0 R^t \tag{11.1}$$

where $E(y)$ is the model-estimated value at time t (specified as the number of years or other time steps since the start of monitoring); β_0 and β_1 are parameters; and R is the annual proportional rate of change in the model. If our estimate of $R = 1.01$, we say the population, on average, is estimated to be increasing at a rate of 1% per year. If $R = 0.98$, we say it is decreasing at an average rate of 2% per year.

The point that different models may be used to describe the trend in population size can be illustrated with the nationwide BBS results for Brewer's sparrows (*Spizella breweri*) during 1980–2008 (Fig. 11.4). Assuming that the survey results accurately index population trends, population size declined during this period. An exponential curve has been fit to the scatter plot. The annual proportional change is 0.9925. During the 29 years, the total change was $0.9925^{29} - 1 = -0.20$, a 20% decline. In this example, the exponential curve resembles a straight line (Fig. 11.5a). While the exponential model seems to fit the scatter plot fairly well, note that the survey results early and late in the period tend to lie above the curve whereas results in the middle part of the period tend to lie below the curve. A second-order polynomial appears to fit the scatter plot better than the exponential or linear curves (Fig. 11.5b) in the sense that, throughout the duration of the study, depicted on the x-axis, observed values occur both above and below the curve.

Linear and polynomial curves do not change at a constant proportional rate, so describing the annual or total change is more complicated than with an exponential curve. One approach is to describe the overall (proportional) change between times as

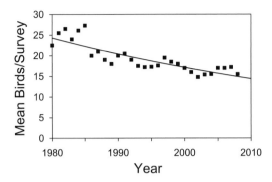

Figure 11.4 Example of using an exponential curve to describe trend. Data are US nationwide BBS summaries for Brewer's sparrows.

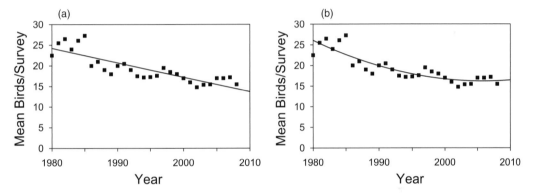

Figure 11.5 Other ways to describe trend in Brewer's sparrow data set: (a) linear model, and (b) second-order polynomial model.

(value at time 2/value at time 1) − 1. In many cases, including the Brewer's sparrow example above, trend estimates based on the exponential, linear, and polynomial models are similar, so the exponential model may be preferred because the results are easy to understand and describe. Again, "trend" encompasses multiple alternative definitions even for this data set; the choice of a model is driven partly by what aspect of change is of interest (Chapters 2, 3). In contentious situations, decisions about how best to describe, analyze, and interpret trends in any data set can lead to significant controversy (Box 11.1).

Examining scatter plots for outliers and influential points is always worthwhile. However, with the long series of data needed to estimate trends, single points usually do not have enormous influence. For example, BBS survey results for Carolina wrens (*Thryothorus ludovicianus*) in New York State during 1998–2003 (Fig. 11.6a) show an increasing trend except for a high value in 2002. It might be thought that the high value would dominate the trend. In fact the annual rate of increase estimated from an exponential model is 8.4% with the high value included and 6.2% with it excluded. In either analysis, we would conclude that population size increased substantially.

Box 11.1 Take-home messages for program managers

Analysis of data from ecological monitoring programs is often complex, partly because sampling plans may be poorly defined or not implemented as planned, and partly because controversies over how best to conduct the analysis are unfortunately common in the field. As a result, analyses often take longer than expected and obtaining conclusions that will be universally accepted is sometimes nearly impossible. When the outcome of the analysis is important, and is expected to be controversial, a substantial effort may be needed to fortify the conclusions against opponents who may be more interested in discrediting the result than in discovering the truth. As shown by the debates over spotted owls, salmon, wolves, grizzly bears, and many other species, such efforts can unfortunately be quite time-consuming, complex, and expensive. Program managers should thus be careful to make realistic estimates of how much time and money will be needed for analysis.

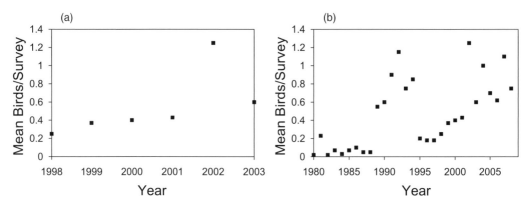

Figure 11.6 Single points usually have little influence but short-term trends often do not indicate long-term trends. Data are BBS summaries for Carolina wrens in New York State for (a) 1998–2003, and (b) 1980–2008.

A much more serious problem, however, is trying to predict long-term trends from short-term data. For the Carolina wren in New York, several years of data starting, for example, in the late 1980s (sharp increase), early 1990s (sharp decrease), or early 2000s (sharp increase) would suggest very different long-term trends than the ones that actually occurred (Fig. 11.6b). This reality indicates that extrapolating trends into the future, especially when the trends are based on <10 years of surveys, may lead to serious errors. It also reemphasizes that different interpretations of trend may be equally valid depending on the time period considered (Chapter 3).

Perhaps the most serious problem of all, however, in defining trends, is doing so without examining a scatter plot of the data. Consider a data set for Lazuli buntings (*Passerina amoena*). The exponential curve is slightly negative (proportional rate of change $R = 0.99$); it declines 17% over the survey interval (Fig. 11.7a). The scatter

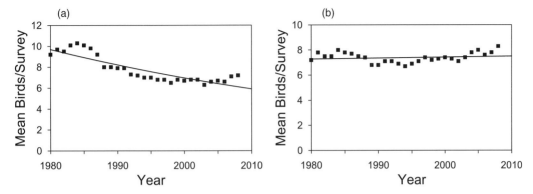

Figure 11.7 Exponential curves often do not describe trends well. Data sets and fitted exponential curves for (a) Lazuli bunting; (b) Yellow-breasted chat.

plot shows, however, that population indices increased for 4–5 years, then decreased for about a decade, and then increased for several years. Most biologists, after inspecting the scatter plot, would not be comfortable summarizing the trend by saying the species was stable or slightly declining. A similar problem occurs with yellow-breasted chats (*Icteria virens*; Fig. 11.7b). The curve is almost flat (20-year change = +2%) but the scatter plot shows rapid increases and decreases followed by a long upward trend. The exponential model thus does not provide a very satisfactory description of the trend for either species. In part, this is because wildlife populations are rarely stable. With a long series of surveys, the population is likely to switch from an increasing to a decreasing period, or vice versa. However, the exponential curve does not change direction so it often gives misleading results. The larger point, however, is that one should never use any trend estimate without examining a scatter plot of the data. Doing so may lead to absurd (and potentially embarrassing) conclusions.

Bias in trend estimates

Sources of bias

Previous chapters have outlined some of the important potential sources of error – random or systematic – that must be considered in survey design and analyses (Chapters 2, 3, 5, 6). Here, we briefly review potential sources of bias in the context of trend estimation.

Bias caused by frame errors

Bias may be caused by difference (if any) between the "population of interest" ("target universe", Chapter 2) and the "sampled population". Both phrases refer to the statistical units being selected. The population of interest is the study area and study period, for example, a wildlife refuge during specified months of winter for a winter waterfowl survey. The sampled population is the set of locations (and the time frame) that can enter the sample – the sample frame. Ideally, the population of interest and sampled population are the same in which case frame bias, i.e. bias caused by frame error, is zero. However, if some locations are inaccessible, for example due to ownership or safety issues, bias is present if the trend in the accessible area is not the same as the trend in the area of interest (Chapter 3).

Selection bias

Selection bias results from variation in selection probabilities that are not acknowledged in the sampling plan and analysis. It usually results from an undefined sampling plan or a sampling plan that is not followed. One example that arises frequently involves stratified sampling based on habitats. Strata are often given names based on habitats like "conifer" or "shrub". The habitat maps used to delineate strata, however, are not perfect. In addition, habitat boundaries may be dynamic over time, even though the strata boundaries are fixed. As a result, for example, the stratum called "conifer" may actually contain a substantial amount of other habitat. When surveyors arrive at a plot in the "conifer" stratum, and find that the plot is some other habitat type, they may be inclined to move to the nearest patch of conifers and conduct the survey there, failing to recognize that the "conifer stratum" refers to all potential sample units within the delineated stratum boundary, not just to conifer patches within this boundary (Chapter 3). This procedure can cause substantial errors. Consider the following example of estimating density for a particular species. Suppose the "conifer" stratum is two-thirds conifer and one-third other habitat and suppose the species of interest is restricted to conifers. Replacing some of the randomly selected non-conifer sample units in the "conifer" stratum causes the sample to include a higher proportion of conifer plots than are in the stratum so density will be overestimated. Problems can also occur in trend estimation if the species of interest does occur in non-conifer habitats, but its trend there is different than in conifers. The problem probably arises from confusion over the objectives of (i) estimating density or trend in a specific habitat vs. (ii) estimating region-wide density or trend through the use of stratification, with strata boundaries overlapping imperfectly with habitat-type boundaries. It should be added, however, that even if the goal is to estimate density or trend in a specific habitat, selecting a point and then moving to the nearest conifer plot is not a very good sampling plan because it gives conifers on the edge of other habitats a higher (and essentially unknowable) selection probability than plots in the interior of conifer patches. The effects of selection bias may be minor, but it can be very difficult to demonstrate this convincingly. Note also that if the goal is to estimate long-term trend in population size by habitat types, stratification based on non-stable habitat boundaries may be problematic even in the absence of selection bias (Chapters 3, 5, 22).

Measurement bias

Measurement bias results from measurement error that is systematic – consistently recording either too high or too low of a value for the attribute of interest. Consider estimating trend in the abundance of some organism using plot counts on which an effort is made to record all individuals present at the time of the survey. In bird surveys, it is well known that much variation exists in how many birds different observers record (Sauer *et al.* 1994, Kendall *et al.* 1996, Diefenbach *et al.* 2003). If observer skill changes gradually so that at the start of the survey period 20% of the animals are missed on average (i.e. average detectability = 0.80), but by the end of the period only 5% are missed, then a trend in the detection ratio (described below) is present and the sample results will tend to increase even if the population is actually stable. Numerous statistical methods have been developed to correct counts for imperfect and non-constant detectability (Butcher and McCulloch 1990; Link and Sauer 1997, 1998; Mackenzie and Kendall 2002; Etterson *et al.* 2009; Kéry *et al.* 2009; Chapters 18, 19). This is a topic of active research. The analyst should consider the available methods and make use of them as appropriate, while recognizing that the data need to have been collected in a way that meets the assumptions of the analytical method (Chapter 18).

Model bias

Model bias refers to using an analytic model that is inappropriate for the sampling plan used to collect the data. The most common form of model bias is probably due to pseudoreplication in which lack of independence is not properly acknowledged in the analysis. Deciding whether counts in different years can be considered independent is sometimes difficult. Here is an example in which the survey results can be considered independent though this might not initially be obvious. The red knot (*Calidris canutus*) is a shorebird that breeds in the Arctic and winters at the southern tip of South America. Each spring red knots congregate in the Delaware Bay to feed on horseshoe crab eggs. Weekly aerial surveys are made of all beaches they use to monitor the numbers present (Niles *et al.* 2009). It might seem that the counts are not independent because the same beaches are surveyed each year. However, all beaches used (which comprise the entire target universe) are surveyed, so there is no sampling in space. Sampling error – the random deviations of a sample estimate from the expected value for the entire target universe resulting when only a subset (a sample) of the population is measured – is not an issue here, nor is selection bias. Assuming that bias from other sources is either absent, constant, or varies randomly in magnitude from year to year, results in one year have no effect on results the next year. Stated more explicitly, suppose we knew that the mean number recorded in one year was below its expected value. This information would give us no reason to predict that the survey result the next year would be either less than or more than its expected value. Results from this survey are therefore independent with respect to measurement uncertainty. In general, surveys that include sampling in space and repeated surveys of the same locations are seldom independent, and must use appropriate analytical methods that account for dependencies (e.g. Chapter 7), whereas surveys that do not include sampling in space, or in which a new sample of locations is selected each year, often are independent.

Evaluating bias: a simple rule

In ecological surveys, the sampled population and population of interest are often different, the sampling plan is often not fully planned, and counts may be incomplete

---

Box 11.2 Common challenges: salvage jobs

The sampling plans in many environmental surveys are unfortunately not very well designed (Anderson 2001). As a result, analysts often have difficulty deciding which analytic methods (all of which require assumptions about the sampling plan followed) to use and disagreements about the best "salvage" methods are common. Systematic consideration of the potential magnitude and impact of biases produced by inadequacies of the sampling design or by uncontrollable factors may help analysts arrive at meaningful conclusions in some cases. Expensive supplemental surveys may be needed to assess key assumptions (e.g. similarity of trends near vs. far from roadsides). However, it may not be feasible to retrospectively address errors, and major objectives of the monitoring effort may have to be abandoned.

When detection rates vary, as is common in many wildlife surveys, then doubts may arise as to whether the average detection rate changes in a directional manner during the survey. For example, average surveyor skill may increase as the survey becomes better known and funded and attracts better observers. If detection rates increase, this will cause an increase in population indices even if the population is stable. This problem can be difficult to detect, and it may be nearly impossible to demonstrate convincingly that it did not occur, if detection rates are not estimated as part of the survey. Many methods have been proposed for estimating detection rates, especially in bird surveys, but there often is little agreement, and a good deal of acrimony, over which methods to use in specific surveys. Furthermore, there is an ongoing need for rigorous field evaluations of methods designed to estimate detection rates.

---

and may be affected by numerous factors such as observer skill, time of day and season, and weather conditions. These issues can be caused by poor planning when the survey was designed (Box 11.2; Chapters 2, 5), unforeseen or unexpected events (e.g. budget reductions, changes in spatial extent of the target population as discussed in Chapter 3), and simply the realities of dealing with highly variable and non-cooperative natural populations. In the face of so many factors that might cause bias, knowing how (and even whether) to proceed with an analysis can be difficult. A simple rule, however, can sometimes be useful by helping the analyst focus attention on the most critical question concerning potential bias:

Bias in a trend estimate equals the trend in the "detection ratio" estimated using the same method as used to estimate the trend in population size.

The "detection ratio" (Bart *et al.* 1998) for a given year is defined as the ratio of the expected value of the survey result to population size. For example, with BBS data, the detection ratio is the ratio of the expected count on BBS routes to the number of birds present during the survey period throughout the survey area. Note that "detection ratio" is a very general term; it includes but is not restricted to issues of detectability. For example, measurement error, non-random selection of sample locations, and incorrect measurement of stratum sizes in stratified sampling all affect the detection ratio. The numerator and denominator in the detection ratio can both be expressed in any units (e.g. the numerator could be mean birds per survey or per each of the 50 locations per survey route; the denominator could be defined as continental population size, as density, or as the mean number within, say, 100 m of the BBS survey route). However, whatever units are used for the numerator and denominator, they must be the same throughout the survey period.

An example, again using the BBS, may make the utility of this result clearer. Counts are made solely along roadsides, surveys on routes with few birds may be discontinued, and the counts are affected by numerous factors. It is therefore easy to imagine that bias in trend estimates calculated from BBS data could be large. The critical question, however, is whether the ratio, count/population size, is likely to show substantial change during the survey. This issue can be investigated through literature reviews, analyses, and careful consideration of how the data were collected. Bias caused by restricting counts to roadsides has been studied by several authors (Bart *et al.* 1995, Rotenberry and Knick 1995, Keller and Scallan 1999, Lawler and O'Connor 2004, Betts *et al.* 2007, Harris and Haskell 2007). Analysts can therefore study these results to estimate how serious the restriction of counts to roadsides is in their area. They can also investigate how many birds are recorded on routes later discarded to judge how much discontinued routes affect the trend estimate. If detection rates are changing through time, then many species would probably be affected in similar ways. For example, if surveyors, on average, are becoming less capable (perhaps due to hearing loss associated with advancing age), then many species (especially those hard to hear) should show declining trends. Analysts can therefore assess how similar trends are within groups of species likely to be similarly affected. This rule regarding the detection ratio thus helps the analyst focus on the most important issues in assessing potential bias.

For example, a good deal of debate has occurred over the reliability of counts from surveys in which bias may occur. Some authors have argued that bias may be substantial and even that surveys using index methods rarely provide useful information (Anderson 2001, 2003; Ellingson and Lukacs 2003; Chapters 4, 18, 19). Others have responded with equal intensity that this view over-states the problem and ignores the difficulty of carrying out surveys in which the design alone insures that no bias exists (Hutto and Young 2003, Engeman 2003, Johnson 2008). One fact to keep in mind, because it is still largely overlooked in these debates, is that variation in detection rates does not automatically cause bias in trend estimates. Even if < 20% of the animals present are detected, and the percentage varies widely between years, no bias occurs in the estimator of population trends unless there is a trend in the detection ratio: i.e. there is no bias if the average percentage of animals detected does not show a long-term change. Furthermore, the bias exactly equals the trend in detection ratio and thus may be small enough (relative to the change in population size) to ignore. The argument, however, cuts both ways. Unless one can demonstrate convincingly that an observed change in survey results could not plausibly have been caused by a change in the detection ratio, then no change in population size can be inferred. It is for this reason that most people working on index methods agree that detectability biases – and all other sources of bias – should be avoided whenever possible (Bart *et al.* 1995, Williams *et al.* 2002).

Examples of considering bias

We conclude this section with two examples illustrating the questions an analyst might ask while trying to evaluate potential bias in trend estimation. Both come from real consulting experiences.

Bias example 1

Annual aerial surveys were made for a species of special concern during a 20-year period. The results showed a sharp increase and had high precision. The survey covered the entire area of interest with a systematic sample of transects that were surveyed once each year. The analyst initially assumed that there could not be frame errors or selection bias because the entire area was covered by the survey and transect lines were selected under a well-defined plan and not moved. Further investigation, however, revealed concerns that some birds might be arriving after the beginning of the survey and others – whose nests failed early – might have already left by the end of the survey. The biological population of interest was clearly the number of birds that attempted breeding so if some of them were not present in the study area during the survey then bias was present due to frame error in the temporal dimension (Chapter 2). Furthermore, it is well known that phenology for many species is changing, probably in response to climate change (La Sorte and Thompson 2007, Zuckerberg *et al.* 2009), and there was concern that habitat quality might be declining in the study area, which could readily cause an increase in nest failure and early departure of the adults. Thus, it was difficult to rule out the hypothesis of change in fraction of the population present at the time of the survey, which would have tended to produce a negative trend in estimated population size as more and more birds left the study area prior to the end of the survey.

A much more serious problem, however, came to light when the analyst investigated measurement error. Observations were made by the pilot who conducted the survey during the entire 20-year survey and by a second observer who was replaced every several years. The survey requires great skill, and the analyst became concerned that each secondary observer's success at detecting birds might not be constant but rather might improve over that observer's years of participation. Total counts for each survey did not provide meaningful evidence for or against this possibility; the pilot (being more experienced) recorded most of the birds so the effect of learning by the second observer was not obvious. When the data for the primary and secondary observer were plotted separately, it became clear that the second observers tended to record many fewer birds during their first few years than in later years. The data had initially been analyzed using observer as a covariate, since different observers clearly recorded different numbers of birds. The output of the model was thus very sensitive to the increase in number recorded within observers. When the data were re-analyzed, either by ignoring the second observers' data or by adding covariates for years of experience by the second observer, the estimated trend was substantially lower.

Bias example 2

The second example concerns counts of migrating shorebirds at a US National Wildlife Refuge on a large lake. A carefully designed 5-year survey had just been completed in part because much concern has been expressed that shorebird populations may have declined seriously during the past few decades (Stroud 2003, Bart *et al.* 2007). Five years is usually not long enough to estimate long-term trends so the biologists leading the project were delighted to learn that a series of surveys had been conducted on the Refuge during the mid-1950s. The numbers counted in each survey were available, and major differences in counts during the two periods were obvious, but there was no information on where the earlier surveys occurred. There was thus much opportunity for both frame error and selection bias. However, a more serious problem became apparent when the analyst considered measurement error. Water levels vary on the lake and in the impoundments managed by the Refuge. When conditions are favorable vegetation may grow to heights > 1 m. Eventually, a high-water event kills the vegetation and it begins to grow again. This vegetation has an enormous affect on visibility and detection rates of shorebirds. Furthermore, the effects vary among species (depending on size and foraging habitats) and in different parts of the refuge (depending on when the last high-water event occurred). After much investigation, discussion, and soul-searching it was decided that there was no way to determine whether differences in numbers between the two periods were due to change in number present or simply to measurement bias. The effort to estimate the 50-year change in numbers present was therefore reluctantly abandoned.

We wish that these examples were unusual and that in most analyses potential bias could be briefly considered and then ignored. In many ecological surveys, however, the estimation of trends, means, or other parameters along with associated confidence intervals is straightforward; it is the evaluation of potential bias that often proves difficult, controversial, and far less conclusive than wished by either the analysts or those they are advising (Boxes 11.1 and 11.2).

Model specification

Many classes of models exist for analyzing population change. Choosing a model (model specification) is a non-trivial process that must be based on the characteristics of the data, the question being addressed, understanding of the system, intended uses of the analysis, and the scope of inference of the analysis. The choice of analysis method can have a profound influence on inferences regarding population change (Thomas and Martin 1996); therefore, model specification and selection is a critical aspect of the scientific process (see Chapter 12). In this section, we outline major general types of models frequently used for trend analyses; subsequent chapters build on this introduction.

Models can be characterized in several ways. One useful distinction is whether the model is intended to be descriptive/explanatory or predictive. Descriptive models focus primarily on how well the data fit the model at the possible expense of predictive power. An example of a question that would motivate a descriptive model might be "has the population increased or decreased over the survey period, and by how much?" Descriptive models are often simple, phenomenological, statistical models (e.g. linear regression) that require the estimation of few parameters, and that do not explicitly model the biological process that generates the observed pattern. Conversely, predictive models would be motivated by questions such as "what is the population size likely to be next year?" or "how long is this population likely to persist?" Predictive models are often more complex, mechanistic, and conceptual/heuristic, require the estimation of more parameters than descriptive models, and are based on explicit hypotheses about the underlying biological process. Thus, it is important to be clear on what questions are being addressed with the model and to understand which modeling frameworks are likely to be most suitable for addressing those questions.

One must also consider whether the assumptions of the modeling approach are reasonably met by a particular dataset. For example, simple linear models assume that there is a linear relationship between the response and predictor variables, and that the samples are independent. Violations of assumptions may undermine the validity of statistical inferences based on the model, thus an understanding of the sampling design (randomization, control, stratification) and possible groupings or structure in the data is essential. A conceptual understanding of the biological process that generated the data is useful in determining whether the assumption of linearity is reasonable. If, however, we know very little about the system then simple linear models may be an excellent starting point for learning about a system.

Several trade-offs should be considered in the model specification process. First, more complex models often require more data, or more types of data, to reliably estimate parameters. This can be a key issue in biological systems, where sample sizes are often small. Confidence intervals for parameter estimates are inversely related to sample size, so if sample sizes are small there may be limited ability to estimate parameters and evaluate competing models. Second, while a more complex model may improve the fit of the model to the data, there is a risk of overfitting a model, i.e. selecting a more complex model than is warranted by the data with the result that the model describes the random error rather than the underlying process. Overfitting is a serious problem as it can lead to incorrect inferences and poor predictive power. Although this problem is not a concern for most trend analyses using simple linear models with time as the only covariate, it becomes more relevant as nonlinear terms (e.g. polynomial terms) and other covariates are incorporated or with more complex models. Careful judgment is required to determine an appropriate level of model complexity.

Biological knowledge of the system is invaluable in the model specification process. Biological understanding allows us to develop hypotheses, and is the basis for judging whether a model makes sense and is useful and appropriate for a specified purpose. Modeling is often an iterative process that requires careful judgment at all stages of the model specification, fitting, and evaluation process. Here we provide an overview of some of the common modeling approaches for monitoring data, and discuss the strengths and weaknesses of each.

Linear models

The simple linear model describes the relationship between a "response variable", y (in our case, annual means), and a single "predictor variable", x (year or whatever temporal unit is being used). The equation is

$$y_i = \beta_0 + \beta_1 x_i + \varepsilon_i \tag{11.2}$$

where β_0 (the intercept) is the expected value of y when $x = 0$, β_1 is the slope of the line, ε_i is the error term, and the subscript i is an index on the sample point $(1, \ldots, n,$ where n is the number of years in the survey period). The error represents both process error (variation unexplained by the model) and measurement error. Often, the error term is implied and not explicitly written. The simple linear model can be easily extended to include multiple predictor variables:

$$y = \beta_0 + \beta_1 x_1 + \beta_2 x_2 + \cdots + \beta_p x_p + \varepsilon \tag{11.3}$$

where x_p refers to the pth predictor variable. Note that the subscript i as used in expression (11.2) is often omitted for simplicity. This equation can be more succinctly written using the matrix form:

$$y = X\beta + \varepsilon \tag{11.4}$$

where y and ε are $n \times 1$ vectors, X is a $n \times p$ matrix of predictor variables, and β is a $p \times 1$ vector of estimated coefficients. The intercept term is included by making the first column of X all 1s.

In the context of population trend analysis the response variable corresponds to an estimate or index of population abundance (annual means in our examples). If time (e.g. year) is the single predictor variable, the model estimates a constant rate of population change over the study period. Multiple regression can be used to model the relationship between annual means of population counts and several predictor variables we hypothesize might influence population trends. For instance, the objective may be to investigate whether habitat loss, habitat fragmentation, and increased anthropogenic disturbance (e.g. road and housing density) are correlated with changes in population numbers. Or we may wish to determine if there is evidence for a declining population trend above and beyond the effects of other factors such as weather. In estimating temporal trends, some covariates could reduce unexplained year-to-year variability apart from any systematic trend and therefore increase the precision of trend estimates (Chapter 7). We can begin to examine the relative importance of these effects by including them as covariates in the linear model provided that the covariates have been measured at the same temporal and spatial scale as the population data. More complex models, such as hierarchical or mixed-effects models, may be required to accommodate covariate data measured at different spatial or temporal scales.

The assumption of these linear models is that the response variable changes as a linear function of the predictor variables, and that the error terms (residuals) are independent and identically distributed (normally distributed with mean 0 and standard deviation σ^2 in the case of the linear model). Polynomial expressions, such as $x + x^2$ (a second-order polynomial) can be included if we believe this may be a better reflection of the relationship between the predictor and response variables (e.g. Fig. 11.5b).

The assumptions of the linear model are often not met in population trend data for several reasons. First, when the response variable is based on count data the errors are often not normally distributed. Counts (and densities) cannot be negative so 0 defines a lower boundary on the response variable, and this can cause a departure from normality in the residuals. Second, errors are not identically distributed if the variance of the errors is not constant throughout the survey period (this is called heteroscedasticity). There are many possible causes of heteroscedasticity including, for instance, measurement error that is positively correlated with population size, or when a model is a good fit to the data at high population densities but not at low population densities. Third, counts may be correlated in space and time: sample locations close together may be more similar than distant sample locations (spatial autocorrelation), and population counts one year may be correlated with population counts in previous years (temporal autocorrelation). Autocorrelation violates the assumption of independence of residuals and results in an overestimation of the effective sample size (Hoeting 2009; Chapter 9). The variance of the parameter estimates (β) is therefore underestimated, which has important implications for inference because predictor variables may be incorrectly considered to be statistically significant when the data do not, in fact, support that conclusion (Hoeting et al. 2006; Chapter 20). Lack of independence among residuals can also lead to incorrect model selection and higher prediction errors (Hoeting et al. 2006).

Generalized linear models

The linear models previously discussed are a special case of generalized linear models (GLMs) that extend the linear model by allowing for response distributions other than the normal distribution, and for a degree of nonlinearity in the model structure (Chapter 9). GLMs are appropriate for data to which a linearizing transformation can be applied such that

$$g(u) = X\beta \tag{11.5}$$

where g is the "link function" that transforms the u – the expected values of the response variable y – such that there is a linear relationship between the transformed expected value of y and $X\beta$. An error probability distribution is specified for y. There are a number of link functions and families of error distributions that can be used within the GLM framework (Nelder & Wedderburn 1972, Faraway 2006). The ones used most commonly (see Chapter 9) are the binomial distribution with logit link for binary data (i.e. logistic regression), and the Poisson distribution with log link for count data (i.e. Poisson regression). Thus, GLMs can resolve the problem of assuming normally distributed errors in linear models when that is not a valid assumption. For instance, Both *et al.* (2010) used Poisson regression to examine evidence for a link between climate change and population trends of migrant bird species in the Netherlands, and Trimble and van Aarde (2011) used a quasi-Poisson GLM that included a detection probability term to assess population trends of 37 South African bird species.

GLMs provide a powerful framework for analyzing trends, but there are several circumstances in which other approaches are required. As with linear models, GLMs are sensitive to spatial or temporal correlations among residuals, and they assume a linear relationship, in this case between the expected value of the response variable on the "link scale" and the predictors X [Equation (11.5)].

Splines

Splines are a flexible device for fitting complex curves to data. They can be useful for describing the relationship between two variables without making prior assumptions about the form of that relationship. There are many forms of splines, which can be generally characterized as either interpolating splines (the curve is forced to pass through the data points), or smoothing splines. The cubic spline is a common form of interpolating spline, for instance, in which the spline is constructed from segments of piecewise, cubic polynomial curves $(a + bx + cx^2 + dx^3)$ based on each consecutive pair of data points. For smoothing splines, the level of smoothing is controlled by the degrees of freedom (df). Few df corresponds to little smoothing (df $= 1$ is a straight line). The choice of df is somewhat subjective and depends on the purpose of the model, and this may have important effects on inferences. Splines are useful for visualizing and exploring data, but can also be incorporated into other models to facilitate more rigorous analysis (e.g. Gurrin *et al.* 2005, Witherington *et al.* 2009). We discuss one example of this in the context of *Generalized Additive Models* next.

Generalized additive models

Generalized additive models (GAMs) are an extension of GLMs in which the relationship between the predictor variables and response can be nonlinear and non-monotonic (*monotonic* means always increasing or always decreasing). In contrast to GLMs, where the value of a parameter β is a constant that describes a linear, monotonic relationship between x and the transformed (link scale) expected value of y, in GAMs the linear predictor is replaced by an *additive predictor* that allows the link function of the mean response variable to follow a smooth and potentially complex curve as a function of the predictor variables (Hastie & Tibshirani 1986, 1990). The general form of a GAM model can be expressed as:

$$g(\mu) = X\beta + f(x_1) + f(x_2) + f(x_3, x_4) + \cdots \tag{11.6}$$

where $g(\mu)$ is the transformation of the expected value of the response variable, $X\beta$ are the strictly parametric model components, and the $f()$ functions are smooth functions of the covariates x_i (Wood 2006). The smoothing functions can take a variety of forms, although splines are often used. GAMs are very flexible and, depending on the link and smoothing functions, can encompass a wide range of other models including linear models and GLMs.

Fewster *et al.* (2000), for example, apply GAMs to the analysis of population trends of farmland birds using data from the Common Bird Census with the goal of modeling both site effects and nonlinear trends. Their model was:

$$log(\mu_{i,t}) = a_i + s(t) \qquad (11.7)$$

where $\mu_{i,t}$ is the expected count in site i at time t, α_i is the site effect, and $s(t)$ is the smoothing function (piecewise cubic polynomial splines). Defining the appropriate level of smoothing (controlled by the df) in the spline curves is somewhat subjective and depends on the objectives of the analysis. An approximate rule of thumb is that the df should be one-third the length of the time series, although Fewster *et al.* (2000) strongly recommend plotting smoothed curves with a variety of df prior to selecting one.

One potential issue with GAMs is that it becomes difficult to decide if an important or significant decline or increase has occurred. A possible solution is to use bootstrap sampling to estimate the 95% confidence intervals on the GAM model, which can be used for approximate hypothesis tests of statistically significant change (Fewster *et al.* 2000). For instance, Hewson and Noble (2009) used this method as a basis for determining if the first and last years in a time series were different. This approach may be sensitive to anomalous years at the start or end of the time series, and the confidence intervals are also sensitive to the level of smoothing in the spline (less smoothing results in wider confidence intervals). Furthermore, the reliability of the fit of GAM models is lower at the start and end of the time series where less data are available for smoothing, resulting in wider confidence intervals.

Mixed models

Groupings in data are common and result in correlations among observations. In the context of monitoring, the grouping usually arises as a result of the sampling design (e.g. in designs with repeated observations of sites across years and multiple sites observed within each year, Chapters 7, 9; or stratified sampling, which groups observations within strata) but groupings may also arise as a result of social structure (e.g. herds, family groups), spatial structure (e.g. subpopulations in a metapopulation), environmental factors (e.g. habitat type), and so on. Groupings can be represented as a factor (a categorical variable) that describes the membership of observations in a group, and enter the model as either a fixed effect or a random effect. A factor is modeled as a fixed effect when it contains all levels of the variable about which we want to make inference (Pinheiro and Bates 2000); as we are interested in their effect on the *mean* of the response variable they are termed *informative* (Crawley 2007). The models discussed so far have all been fixed-effects models. In contrast, random effects are used when the grouping can be considered a random subset of the total population of groups (e.g. such as sites randomly selected from a larger area). Mixed-effects models (mixed-models or hierarchical models) incorporate both fixed and random effects (e.g. Chapter 7).

The distinction between fixed and random effects is not always obvious. A factor can be considered random when the experimenter has not explicitly controlled for levels of the variable in the experimental design, but was randomly sampled from the levels within a population. For example, if a sample of lakes was randomly selected from a population of lakes, sampling occurred within lakes in each year, and the purpose was to estimate trend in the population of lakes, then it would be appropriate to classify lake as a random effect. In contrast, if inferences were restricted to the lakes selected then the lakes would be considered fixed effects (because no inferences would be made beyond the lakes studied).

In trend analyses, random effects are often used to control for correlations among observations, and unbalanced designs in the numbers of observation among levels of a factor (VanLeeuwen et al. 1996, Piepho and Ogutu 2002). The Breeding Bird Survey is an example of a survey design where correlations among observations may arise as a result of site (route), observer, and year effects, and sample sizes are frequently unbalanced whereby remote areas are sampled less frequently than areas near population centers (Thomas 1996, Sauer et al. 2004). The advantage of mixed models is, therefore, that they can accommodate complex datasets in which there may be multiple levels of correlations in the data that need to be accounted for analytically. Mixed models also facilitate analysis of within- and between-factor variation, which can lead to further insight into the system (Laird and Ware 1982, Gelman and Hill 2007; Chapter 9). GLMs and GAMs that include random effects are referred to as generalized linear mixed models (GLMM; Pinheiro and Bates 2000, Bolker et al. 2009) and generalized additive mixed models (GAMM; Wood 2006), respectively.

Although mixed-effects models are more complex than fixed-effects models and are, therefore, more difficult to fit and evaluate (Bolker et al. 2009), random effects provide a rigorous method of addressing structure in data sets that it is statistically inappropriate to ignore. For example, mixed models can be used to estimate population trends when surveys are conducted at multiple sites across years, even with complex panel designs (e.g. Piepho and Ogutu 2002; Chapter 7). Sauer et al. (2008) incorporated strata, years, and observers as random effects in a log-linear hierarchical model to estimate population trends for American woodcock (*Scolopax minor*). They conclude that the hierarchical modeling approach is generally more efficient than alternative modeling frameworks and results in smaller credible intervals (the Bayesian equivalent of confidence intervals). Similarly, Rittenhouse et al. (2010) used random effects to account for observer and site effects in an analysis of Breeding Bird Survey data. Chapters 12, 19, and 20 illustrate applications of Bayesian hierarchical modeling for analyzing monitoring data, with the latter two chapters using forms of GLMs with Poisson or negative binomial error distributions and log links.

Process models

Earlier we made the distinction between exploratory or descriptive models and heuristic or conceptual models. Arguably, all statistical models imply a conceptual model of the process that generated the observed data, although this is often not explicitly defined or discussed. Linear models, GLMs, and GAMs are statistical models that help us to explore the relationship between response and predictor variables. Alternatively, we may wish to explore how well different *process models* fit the observed data, thereby developing a more mechanistic understanding of the system.

A prerequisite for fitting a process model is one or more hypotheses describing the process. In the case of population trends, we may have a reasonable understanding of the biology of the species that allows us to build competing models of reproduction and population growth. These population models could include age and sex effects, density dependence (e.g. a carrying capacity), Allee effects, winter mortality, habitat effects, and so on. Population models can be more powerful for detecting treatment effects than non-mechanistic models which do not incorporate realistic biological processes or dynamics (de Valpine 2003). Moreover, population models can also be used to address complex ecological questions such as "does a population exhibit density dependence, and what is the form of the density dependence?", "what are the relative influences of exogenous and endogenous factors in driving population dynamics?", "is there spatial or temporal synchrony among populations?", or "do populations interact?" For instance, van de Pol *et al.* (2010) use population dynamics models to investigate the mechanisms (changes to vital rates) by which climatic variability affects population viability.

Process models are able to accommodate data related through time and space, and a wide variety of model structures (de Valpine 2002). Process models are typically non-linear, and cannot be fit using the techniques previously discussed. Hidden process models, or state-space models, provide one framework for fitting these sorts of models (Buckland *et al.* 2004, Newman *et al.* 2006) in a Bayesian framework. Importantly, state-space models can be used to explicitly model the effects of both measurement error and process error on parameter estimation. However, process models often require advanced statistical software and a good understanding of how to specify a model in terms of probability distributions. This is a barrier to many ecologists (Bolker 2009), although a number of good reference texts are available (McCarthy 2007, Gelman & Hill 2007, Clark 2007).

Trend estimation with spreadsheet software

To estimate trends, quantitative ecologists and statisticians generally would use statistical/mathematical packages such as SAS (SAS Institute 2008) or R (R Development Core Team 2011), or use other programming languages. However, many project leaders and others for whom this chapter is written are not trained to use such software. For these people, spreadsheet packages such as Microsoft Excel can be a useful alternative. Therefore, in this section we describe the estimation of trends with spreadsheets using Microsoft Excel as an example.

Many simple but widely used trend models can be implemented using standard curve-fitting procedures in Excel, which are quite flexible and easy to use. The curves include linear, logarithmic, polynomial, power, exponential, and moving average with various periods. The exponential and polynomial are the most useful in trend estimation. In fitting the power and exponential curves, Excel calculates the natural logarithm of each y value so these curves are unavailable if any values are 0. A common solution is to replace zero values with small positive values, although the sensitivity of the results to the choice of the value added needs to be considered. The equation and r^2 may be displayed, but no other measures of precision or significance are given, although they can be added fairly easily with built-in Excel commands. The "Data Analysis" add-in program, which is included with Excel but must be loaded by the user, carries out a standard multiple regression analysis. It is easily used to fit a linear regression.

Estimates of precision calculated from Excel's linear regression routine will be valid only if estimates from different years are independent. If instead a sample of locations is revisited so that counts in different years are not independent, then a more complex method must be used. As noted above, many approaches have been described but most are not suitable for use in Excel, except perhaps for people who can program in Visual Basic. Bart *et al.* (2003) proposed a very simple method for repeated surveys at multiple locations that can easily be carried out in Excel. Sauer *et al.* (2004) expressed concern about the method, and Bart *et al.* (2004) responded that they did not consider the concerns well founded. As with other controversies in the trend estimation field, analysts will have to decide for themselves whether the method is appropriate in their situation. It relies on the fact that with relatively small annual changes ($0.9 < R < 1.1$) in the rates of change of a linear curve, R_{lin}, and an exponential curve, R_{exp}, are nearly identical (see table 1 in Bart *et al.* 2003). This fact justifies estimating R_{lin} as a useful approximation for R_{exp} if the rate of change is within 0.10 of 1.0. Let

$$R_{lin} = \frac{B^*}{Y^*_{mid}} + 1 \tag{11.8}$$

where B^* is the slope of the linear regression of the estimates of the population means (based on the sampling design used) against years and Y^*_{mid} is the value of the regression line at the midpoint of the study interval. For any complete array of numbers with n rows and T columns, it is easy to show that

$$B^* = \bar{B} = \frac{1}{n}\sum_{i=1}^{n} B_i \tag{11.9}$$

and

$$Y^*_{mid} = \bar{Y}_{mid} = \frac{1}{n}\sum_{i=1}^{n} Y_{mid,i} \tag{11.10}$$

where B_i = the slope of the ith row of data regressed against year and $Y_{mid,i}$ = the value of the regression line for the ith row at its midpoint. Without loss of generality, we may therefore write

$$R_{lin} = \frac{B}{Y_{mid}} + 1. \tag{11.11}$$

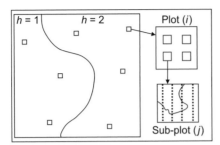

Figure 11.8 Hypothetical sampling plan for trend-estimation example. The study area is divided into two strata; plots are randomly selected within each stratum; a systematic sample of subplots in each plot is established; and data are collected on four transects in each subplot.

The utility of expression (11.11) is that the B_i and $Y_{mid,i}$ are independent, even though the observations within rows are not independent, so they are easy to estimate. The usual estimators for the trend and its standard error, based on standard survey sampling theory (Cochran 1977, Bart *et al.* 2003), are

$$r_{lin} = \frac{\bar{b}}{\bar{y}_{mid}} + 1 \tag{11.12}$$

and

$$se(r_{lin}) = \frac{1}{\bar{y}_{mid}} \left[v(\bar{b}) + r_{lin}^2 v(\bar{y}_{mid}) - 2r_{lin} cov(\bar{b}, \bar{y}_{mid}) \right]^{0.5}. \tag{11.13}$$

Expressions (11.12) and (11.13) are general, but the specific formulae for the terms in them depend on the sampling plan. However, the specific formulae used in the following example will apply to the sampling plans used most commonly in environmental surveys.

Example: spreadsheet trend estimation

Great concern exists in the intermountain west of North America about the disappearance of sagebrush (*Artemisia tridentata* and others; Knick *et al.* 2003). To determine trend in the amount of sagebrush, let us imagine that a large study area is subdivided into $L = 2$ strata, locations for plots are randomly selected, and each plot is visited for 10 years. In each plot, four subplots are established systematically (Fig. 11.8). Each subplot is further stratified into cover types. Four lines are then established systematically within each subplot and the proportion of the lines in each cover type covered by sagebrush is recorded. In succeeding years, the same plots are visited. Overall, this design incorporates many design elements discussed in Chapter 5; stratification, random sampling of plots within each stratum, cluster sampling, and repeated visits to survey locations are all part of the sampling plan. Further, the number of lines per subplot could change over time due to budgetary considerations. The following calculations require that clusters (plots) be selected with equal probabilities and that results be equally weighted; however these practices are the norm in ecological surveys. In a given year, let:

n_h = number of plots in stratum h (3 in stratum 1, 4 in stratum 2)

n_{hi} = number of subplots in plot i of stratum h (4 in all plots)

L_{hij} = number of cover types in subplot j of plot i in stratum h (2 in the example but variable among subplots)

p_{hijk} = proportion of the lines in cover type k of subplot j of plot i of stratum h covered by sagebrush

W_{hijk} = proportion of subplot j covered by cover type k in plot i of stratum h

W_h = proportion of the overall study area in stratum h (i.e. stratum weight)

$$p_{hij} = \sum_{k=1}^{L_{hij}} W_{hijk} p_{hijk} = \text{estimated proportion of subplot } j \text{ in plot } i \text{ of stratum } h \text{ covered}$$
by sagebrush

$$p_{hi} = \frac{1}{n_{hi}} \sum_{j=1}^{n_{hi}} p_{hij} = \text{estimated proportion of plot } i \text{ in stratum } h \text{ covered by sagebrush}$$

$$p = \sum_{h=1}^{L} W_h \frac{1}{n_h} \sum_{i=1}^{n_h} p_{hi} = \text{estimated proportion of the study area covered by sagebrush.}$$

Note that p can be considered a design-based estimate of status, where 'status' in this example is the current proportion of the study area covered by sagebrush.

This sampling plan involves returning to the same plots each year. The annual esti-mated mean proportions are thus not independent even though the sampling plan within plots may vary between years. An analytic method that acknowledges this dependency must therefore be used. Note, too, that the variation between years in the sampling plan within plots in no way compromises the trend estimate because in each year an unbiased estimate is produced. Finally, note that while the sampling plan within plots is complex, the trend estimation method we present only uses means per plot. The analyst therefore calculates means per plot (in this case, mean proportions; p_{hi}) prior to starting the trend analysis and enters only these means into the spreadsheet. If cluster sampling had not been used and the plot-level data consisted of just one observation rather than a cluster of observations, then p_{hi} would be equivalent to the single observation for plot i in stratum j that year.

A simplified form of the necessary spreadsheet is displayed in Fig. 11.9. The upper part of the figure displays the data from the seven plots and 10 years. Below the data for each year the means for each stratum (p_h) are calculated along with the grand annual mean (p). In calculating the grand mean we assume that stratum 1 occupies 40% of the study area and stratum 2 occupies 60%. The plotted data are overall proportions in each year with an exponential curve fitted to the data.

To the right of the 10th year in each row, standard Excel commands are used to obtain the intercept and slope of the regression of the proportions on year for each plot. These are then used to calculate the mid-point of the regression line [here, by plugging in the value $t_{mid} = 5.5 = 1 + (10 - 1)/2 = $ start year + (end year – start year)/2 into each regression equation], producing:

b_{hi} = slope of plot-specific regression line for plot i in stratum h

$y_{mid,hi} = intercept_{hi} + (b_{hi})(t_{mid}) = $ midpoint value of the plot-specific trend line for plot i in stratum h

Strata	Plot	Year										Intcpt	Slope	y
		1	2	3	4	5	6	7	8	9	10			
1	1	0.52	0.35	0.41	0.87	0.49	0.33	0.24	0.11	0.38	0.57	0.52	-0.02	0.43
1	2	0.00	0.15	0.00	0.23	0.00	0.21	0.11	0.32	0.15	0.00	0.07	0.01	0.12
1	3	0.43	0.56	0.47	0.54	0.63	0.42	0.32	0.11	0.23	0.43	0.58	-0.03	0.41
2	1	0.63	0.73	0.45	0.10	0.21	0.10	0.00	0.00	0.00	0.00	0.67	-0.08	0.22
2	2	0.21	0.22	0.13	0.30	0.29	0.36	0.33	0.36	0.37	0.29	0.18	0.02	0.29
2	3	0.41	0.45	0.49	0.32	0.33	0.36	0.37	0.33	0.40	0.45	0.41	0.00	0.39
2	4	0.53	0.55	0.45	0.40	0.43	0.39	0.35	0.32	0.26	0.28	0.57	-0.03	0.40
Strata														
1	Mean	0.32	0.35	0.29	0.55	0.37	0.32	0.22	0.18	0.25	0.33	0.39	-0.01	0.32
2	Mean	0.45	0.49	0.38	0.28	0.32	0.30	0.26	0.25	0.26	0.26	0.46	-0.02	0.32
Grand mean		0.39	0.43	0.35	0.39	0.34	0.31	0.25	0.22	0.26	0.29	0.43	-0.02	0.32

Strata	Stat	Slope	y
1	Var	0.00013	0.01025
2	Var	0.00047	0.00179
1	Cov	-0.00109	
2	Cov	0.00042	

Study area	Slope	y
Variance	0.00019	0.00229
Covariance	-0.00002	
r	0.94	
se(r)	0.15	
cv(r)	0.16	

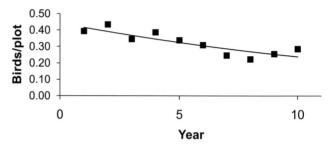

Figure 11.9 Spreadsheet to analyze example data collected according to sampling plan in Fig. 11.8.

$s^2(b_{hi})$ and $s^2(y_{mid,hi})$ = empirical variance estimates of the n_h slope and midpoint estimates in stratum h.

The mean regression slope, midpoint, and their variances for the entire study area are then estimated using the following (resulting values are shown to the left of the data plot in Fig. 11.9):

$$\bar{b} = \sum_{h=1}^{L} W_h \frac{1}{n_h} \sum_{i=1}^{n_h} b_{hi} \qquad (11.14)$$

$$\bar{y}_{mid} = \sum_{h=1}^{L} W_h \frac{1}{n_h} \sum_{i=1}^{n_h} y_{mid,hi} \qquad (11.15)$$

$$v(\bar{b}) = \sum_{h=1}^{L} W_h^2 \frac{s^2(b_{hi})}{n_h} \qquad (11.16)$$

$$v(\bar{y}_{mid}) = \sum_{h=1}^{L} W_h^2 \frac{s^2(y_{mid,hi})}{n_h}. \qquad (11.17)$$

Finally, the estimated annual change in the trend line (r) and its standard error are calculated using (11.12) and (11.13). The covariance term in expression (11.13) may be calculated as

$$cov(\bar{b}, \bar{y}_{mid}) = \sum_{h=1}^{L} W_h^2 \frac{cov(b_{hi}, y_{mid,hi})}{n_h} \qquad (11.18)$$

where

$$cov(b_h, y_{mid,h}) = \frac{\sum_{i=1}^{n_h} (b_{hi} - \bar{b}_h)(y_{mid,hi} - \bar{y}_{mid,h})}{(n_h - 1)}.$$

The estimated trend is 0.94, a decline of 6% per year. The SE, however, is 0.15 and the associated P-value (for a test of the null hypothesis that $R = 1$) is 0.69 indicating that the estimate is far from significant. If the dependence between years due to repeated visits to the same plots had not been acknowledged, the trend would have been highly significant. For example, the significance of the a linear trend line fit to the annual means, calculated using Excel's Data Analysis extension, is 0.001. This shows the importance of acknowledging between-year dependence.

Future research and development

Ideally, design and analysis in any monitoring program would be done as a partnership with a statistical expert. A rich array of sophisticated trend estimation methods can be implemented in SAS, the programming language R, and other statistical/mathematical packages and programming languages. However, many environmental biologists lack access to these programs and/or the background to use them yet still need to summarize trends as part of their work. For example, for the type of data set described in the previous example, many environmental biologists without statistical support likely would have difficulty implementing analytical approaches that account for the dependence in the data set produced by repeated visits to the same sample plots over time, as well as features of the example design such as stratification with unequal strata weights (W_h). If these biologists were unable to collaborate with someone who had the relevant expertise, they might be tempted to simply apply a trend model to the annual means or even ignore the stratification, potentially producing misleading results. Making suitable analytical tools accessible to these practitioners is a continued need. For example, implementing the method described above may help them avoid some analytical mistakes as long as they carefully assess whether the method is appropriate in their situation. More broadly, these practitioners need user-friendly, statistically robust, free, Windows-driven trend estimation programs that they can use for the majority of their analytic needs. A program has been designed for this purpose and can be obtained free, along with a users' manual, by contacting the senior author.

Summary

This chapter provides an introduction to trend analysis in the environmental field. Advanced analytical topics are presented in other chapters. As in Chapter 3, we stress that "the trend" is not a well-defined parameter like a population mean or total, and that the choice between different definitions of trend should be compared mainly on the basis of utility rather than by applying statistical principles or results. Trends are calculated by fitting a model (e.g. exponential, polynomial) to monitoring data. Analysts should always examine a scatter plot of the data as part of choosing the model.

Error in trend estimates may be partitioned into sampling errors (random differences between the estimated and true value that arise from observing only a subset of the population) and non-sampling error (i.e. bias or all other error). Estimating the effects of sampling error is straightforward if new survey locations (or times) are selected each year. In this case, standard regression methods may be used to estimate trends. When locations are revisited, more complex methods, that acknowledge the lack of independence between years, must be used.

Major sources of bias from non-sampling errors include bias caused by frame errors, selection bias, and measurement bias. All three are common in environmental studies. Investigating their potential effects is often difficult but always important. While many factors can cause bias, their combined effect can be summarized in a simple rule: *the bias in a trend estimate is equal to the trend in the detection ratio*. The detection ratio is the ratio of the expected survey result to true population size (or other parameter of interest) in a monitoring program. The rule helps clarify sources of bias. It shows, for example, that even substantial variation in detection rates does not automatically cause any bias in estimates of trends in abundance.

We provide a brief overview of several types of models used in trend estimation, including splines, linear models, generalized linear models, generalized additive models, mixed models, and process models. We discuss the key assumptions and relative merits of these models, as well as some of the issues that must be considered when specifying a model. We conclude the chapter with suggestions for estimating trends using widely available software. A simple approach that can be implemented in Excel is described. Analysts able to use SAS or R, however, have many more options.

Acknowledgments

Kathryn M. Irvine and an anonymous reviewer made helpful suggestions on a draft of the chapter. The US Geological Survey supported J. Bart during preparation of the manuscript. Any use of trade, product, or firm names is for descriptive purposes only and does not imply endorsement by the US Government.

12 Analytical options for estimating ecological thresholds – statistical considerations

Song S. Qian

Introduction

The concept of ecological thresholds increasingly is attracting the attention of ecologists and managers of natural resources. As defined by Groffman *et al.* (2006), "an ecological threshold is the point at which there is an abrupt change in an ecosystem quality, property or phenomenon, or where small changes in an environmental driver produce large responses in the ecosystem." Because of the complicated nonlinear dynamic of a threshold change and multiple factors that can affect the same ecosystem, detection and quantification of ecological thresholds is a challenging task.

Related to the threshold concept in ecology is the general category of changepoint models in statistics. This category encompasses a large collection of statistical models. For example, a recent review of the topic resulted in a 230-page annotated bibliography (Khodadadi and Asgharian 2008). The term "change point" has evolved in the past 50 years. In the 1950s, we find descriptions such as "a change in parameter occurring at an unknown time" (Page 1955) and "a linear regression system obeying two separate regimes" (Quandt 1958). More general terms such as "threshold regression" (Dagenais 1969), "segmented strait lines" (Bellman and Roth 1969), and "piecewise regression" (McGee and Carleton 1970), and more abstract terms such as "change point" [which probably first appeared in Hinkley (1970) and Hinkley and Hinkley (1970)] appeared in the late 1960s and early 1970s.

These earlier studies focused on two general types of models. Initially, a changepoint model was a model describing a step change, in that distributional parameters stay the same along the covariate until the change point where a distributional parameter shifts. A piecewise regression describes the problem where a regression line's slope changes at an unknown point. Because of very different computational needs, these two types of problems usually did not mix until Smith (1975, 1980), who used "changepoint" to describe both types of problems because they can be lumped into the same general class.

The relationship between the step change and piecewise regression problems is especially apparent under a Bayesian framework, under which even the computational method is unified. In the 1990s, with the advent of the Markov chain Monte Carlo (MCMC)

Design and Analysis of Long-term Ecological Monitoring Studies, ed. R.A. Gitzen, J.J. Millspaugh, A.B. Cooper, and D.S. Licht. Published by Cambridge University Press. © Cambridge University Press 2012.

simulation, changepoint computation for more complicated problems became accessible. Stephens (1994) showed how the Gibbs sampler can be used for estimating parameters of a series of changepoint models. However, applying the Gibbs sampler required intensive computer programming, often from scratch. As the right outcome of a Gibbs sampler cannot be known, software validation became a serious problem (Cook *et al.* 2006). The software package WinBUGS (Lunn *et al.* 2000) and later JAGS (http://www-ice.iarc.fr/~martyn/software/jags/) removed the programming hurdle for most problems.

In ecological applications, various changepoint methods have been used for detecting and quantifying ecological thresholds. For example, Andersen *et al.* (2008) discussed several, mostly classical statistical, methods for step function threshold models. Brenden *et al.* (2008) discussed the step function as well as the piecewise linear model (also known as the hockey stick model). Qian *et al.* (2003) suggested use of the Bayesian changepoint method for normal and binary response variables and Qian *et al.* (2004) presented the Bayesian changepoint model for a multinomial response variable for analyzing changes in species composition along a disturbance gradient.

The focus of these ecological applications usually has been the location of the change point. These applications are retrospective in nature, that is, they estimate a threshold after it has been crossed. The estimated threshold values are then used for supporting environmental management decisions, such as setting environmental standards (Box 12.1). The methods discussed in this chapter generally are not used for detecting changes in real time.

Most existing applications of changepoint models use environmental stressors (e.g. phosphorus concentration in the Everglades wetlands) as the predictor, instead of using "time" as in analyses focused on temporal trend estimation (e.g. Chapters 7, 11). When applied to long-term monitoring data for temporal trend analyses, these methods can be used to detect whether a change has already happened, but not for predicting when a change will happen in the future. For example, section 6.1.2 of Qian (2010) used a hockey stick model to detect changes in the first bloom dates of lilacs in the US, using long term records of bloom dates in several locations in the US.

This chapter provides a framework for the use of statistical modeling to address changepoint problems. The focus of this chapter is the question of finding the most appropriate model for the problem in hand. This question is not a simple statistical question of model comparison using various information criteria. Rather, it is a scientific question first. When a specific model is proposed, we are posing a specific interpretation. Chamberlin (1890) suggested that multiple working hypotheses should be proposed when explaining new phenomena. In ecological modeling, we should also propose multiple alternative models, each representing a different interpretation of the data. These models should be fit to the same data and proper comparisons should be made.

This modeling approach is outlined in the context of linear (or generalized linear) threshold models, which are the changepoint model types used most commonly. Models in this class include the step function changepoint models discussed in Qian *et al.* (2003), the piecewise linear model as in Qian and Richardson (1997), and a general model encompassing both the step function and piecewise linear models. An examination of

Box 12.1 Take-home messages for program managers

When used properly, statistical changepoint models can be effective in detecting changes in an ecosystem in response to changes in the environment. There are many types of changepoint models and each represents a different underlying pattern of change. In a long-term monitoring context, threshold analysis may be used for assessing patterns of change over time. More commonly, threshold analysis is often associated with a specific management problem in which some ecological attribute of concern is affected by the level of a stressor. For example, a changepoint model was used by a state management agency in Ohio for setting nutrient criteria for small rivers and streams (Miltner 2010). Recognizing that an estimated "threshold" is associated with a specific model is important because different models imply different assumptions about the pattern of change and may lead to different threshold values. A justification of the chosen model should be always part of assessing an ecological threshold.

When addressing a management problem, the concept of a threshold response should be discussed in the context of the management endpoint of concern. For example, EPT [Ephemeroptera (mayfly) + Plecoptera (stonefly) + Trichoptera (caddisfly)] taxa richness is often used as an indicator of stream water quality. In Ohio, a good water body is defined as one that supports at least 10 EPT taxa. When setting a criterion for phosphorus, the criterion is the maximum level of phosphorus that can occur in the watershed while keeping EPT taxa richness $\geq$ 10. The statistical change point and the management endpoint can be the same if the response of EPT taxa richness to phosphorus can be approximated by the step function; i.e. if EPT

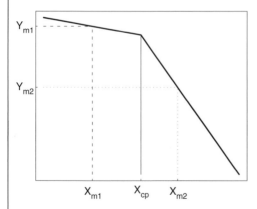

A hypothetical hockey stick model (thick solid line segments) illustrates the difference between a mathematical change point and management thresholds. A management threshold should ensure, with a certain degree of confidence, that the ecological indicator or attribute of interest does not go past some specified endpoint. The two dashed lines are two alternative hypothetical endpoints (Y_{m1}, Y_{m2}) and the corresponding respective thresholds for the predictor variable (X_{m1}, X_{m2}). A change point is the value of the predictor variable (X_{cp}) at which there is a change in the relationship between the predictor variable and underlying resource attribute of interest.

taxa richness remains constant at some value ≥ 10 as long as phosphorus remains below some threshold, but richness drops to some alternate constant value < 10 as soon as the phosphorus level goes and stays above the threshold.

However, the mathematical change point can differ from the management threshold if the underlying model is something other than the step function, such as a piecewise linear model (see figure on page 281). The statistical change point (X_{cp}) would not be protective if managers sought to maintain the ecological attribute above Y_{m1} but would be overly protective if the management threshold was Y_{m2}. In other words, a clear understanding of the management objective is needed in order to correctly interpret the significance of the change point relative to management thresholds.

these three linear threshold models using four synthetic data sets, each representing a different underlying model, is used to illustrate the importance of identifying the correct model form. A graphical diagnostic method is proposed for testing the existence of a threshold and helping identify its form. Models discussed in this chapter assume normality of the response variable, but can be extended to non-normal situations easily.

Statistical inference and threshold modeling

Statistical inference about an estimated quantity such as the estimated change point is mostly focused on the uncertainty of the estimation. The question whether or not a change point exists can be addressed using the uncertainty information. For a Bayesian analysis, Smith (1975) recommended that the estimated posterior distribution of the change point can be used for such inference, as will be demonstrated later in the chapter. Authors using classical methods are usually less concerned about the question of whether or not a change point exists. However, the confidence interval of the estimated change point can be used for addressing this question.

Although quantitative methods are inevitably statistical in nature, almost all discussions about threshold identification and quantification neglect the underlying probabilistic assumptions. A statistical model is a tool for facilitating inductive reasoning through hypothetical deduction. Using R.A. Fisher's terms, a statistical modeling problem consists of three questions: the question of distribution (of which distribution are the data random samples?), the question of estimation, and the question of interpretation (Fisher 1922). These questions are consistent with modern statistical practices. In a statistical modeling problem, the first question is always which model to use. To answer this question, we not only must have the necessary statistical knowledge, but more importantly, must know the subject matter well so that we can propose a reasonable response-variable distribution. Once the distributional assumption is made, the appropriate type of model can be derived. In a typical statistical class, model formulation is often not emphasized as the task is always more of a scientific problem than a statistical one. Instead, estimation of model parameters represents the bulk of what we learn from taking a statistics course.

A model is a collection of assumptions. The role of modeling is to assess evidence in the data either for or against the constituent assumptions. The general linear threshold model represents a class of models, reflecting a general assumption that the underlying ecosystem has undergone an abrupt change along a gradient of some related covariate, such as a stressor. To test the existence of a change point, we should contrast the class of changepoint models against a model without a change point. When the existence of a change point is confirmed, we identify the nature of the threshold and quantify the threshold. This chapter will not discuss methods of model selection and model averaging based on the Bayes factor as the author believes that such an approach is contradictory to the spirit of scientific research, of which the search of the underlying model is the main objective.

Overview of threshold modeling

Statistical models

In a statistical modeling problem, we assume that the response variable of interest y follows a probabilistic distribution characterized by a parameter representing the mean and, potentially, a second parameter representing the scale (variance). The objective of a model is to describe the relationship between the mean and relevant predictor variables. A threshold problem is defined by assuming that the parameters of the probability distribution model are a function of one or more environmental stressors (x), and that a threshold exists when the distribution parameter changes as the environmental stressor crossing a specific value (ϕ):

$$y_i \sim \begin{cases} \pi\left(\theta_1\left(x\right)\right) & \text{if } x_i < \phi \\ \pi\left(\theta_2\left(x\right)\right) & \text{if } x_i \geq \phi \end{cases} \tag{12.1}$$

where π represents a generic probability distribution function with parameter vector $\theta\left(x\right)$, and changes in model parameters occur when the predictor variable (covariate) value crosses the threshold (ϕ).

Most (if not all) existing threshold models in the literature can be summarized in terms of Equation (12.1). Quantitative options for estimating threshold lie in the selection of the response-variable distribution (π) and the dependency of the distribution mean variable on one or more predictor variables (x). When using a classical statistics approach, the above model (12.1) is modified to suit a specific response variable distribution. Different distributions often have very different computational methods, leading to numerous models in the literature. Consequently, when selecting a model, it is important to know the assumptions and conditions of a method. When using a Bayesian approach, identifying the probability distribution function π is always the first step. Once π is selected, computation using MCMC simulation is straightforward.

In either case, threshold modeling involves model selection and parameter estimation. Model selection is a scientific problem (what is the underlying relationship between the response and environmental stressor?), and parameter estimation is a mathematical problem (given the model and data, what are the likely model parameter values?). The scientific question is not related to whether we are Bayesian or not. From a practical view point, once the scientific question is settled, both Bayesian and classical approaches can be used for parameter estimation. The problem of choosing among quantitative options arises because we do not know for certain what is the underlying model. In statistics, we learn how to estimate model parameters *given* the underlying probabilistic model. Different quantitative methods are needed when π changes from one distribution to another, or when the relation between model parameters θ and environmental stressors x changes from linear to nonlinear. In a scientific question, we must first hypothesize the likely (alternative) model forms. These alternative model forms should be presented in terms of different π or different functional relations between θ and x. Observed data should be used to fit these alternative models and evaluate the likelihood of each to be the true underlying model that generated the data.

To illustrate this process, we will focus on the simplest situation where π is a normal distribution. A normal distribution has two parameters ($\theta = \{\mu, \sigma\}$). When the response variable distribution can be approximated by the normal distribution, the changepoint model assumes that either the mean or standard deviation (or both) changes when the predictor crosses a specific value (or threshold).

The step function model

The Bayesian changepoint model in Qian *et al.* (2003) assumes the change can be approximated by the step function:

$$y_i \sim \begin{cases} N(\mu_1, \sigma_1) & \text{if } x_i < \phi \\ N(\mu_2, \sigma_2) & \text{if } x_i \geq \phi \end{cases} \tag{12.2}$$

The step function assumes that the mean and variance of the response variable do not change along the covariate axis (e.g. disturbance gradient or time) until the covariate passes a threshold. The changes in mean and/or variance are abrupt.

Piecewise linear model

The model presented in Qian and Richardson (1997) is a piecewise linear model with the response variable assumed to be from a normal distribution:

$$y_i \sim \begin{cases} N(\alpha_0 + \alpha_1 x_i, \sigma_1) & \text{if } x_i < \phi \\ N(\beta_0 + \beta_1 x_i, \sigma_2) & \text{if } x_i \geq \phi \end{cases} \tag{12.3}$$

and assuming that the two line segments meet at $x = \phi$:

$$\alpha_0 + \alpha_1 \phi = \beta_0 + \beta_1 \phi$$

which defines the continuity constraint. When assuming equal variance, Equation (12.3) can be simplified to be:

$$y_i \sim N(\beta_0 + (\beta_1 + \delta I(x_i - \phi))(x_i - \phi), \sigma) \tag{12.4}$$

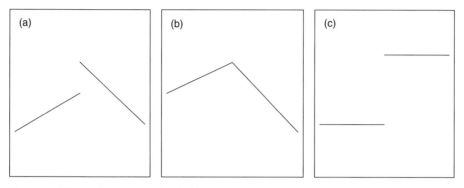

Figure 12.1 The family of linear threshold models: (a) the general model; (b) the hockey stick model; and (c) the step function model.

where $I(a)$ is the unit step function [$I(a) = 0$ when $a < 0$ and $I(a) = 1$ when $a \geq 0$]. The piecewise linear model is also known as the hockey stick model. This model assumes that change in mean is continuous and the rate of change (slope) over time or along the disturbance gradient is constant until the covariate crosses the threshold. The change in the rate is abrupt.

The general model

The family of linear threshold models with a normal response variable can be summarized by Equation (12.3) without the continuity constraint (Fig. 12.1a). This model is referred to as the general model. The piecewise linear model with the continuity constraint (Fig. 12.1b) is a special case of the general model, as is the step function model (Fig. 12.1c), which assumes the slopes of the two line segments are both 0.

Modeling the existence and form of threshold change

Statistical methods are designed for specific models. When scientific information can identify the correct model form, we can select the appropriate method for parameter estimation. However, the objective of a scientific inquiry is often to identify the underlying model. As a result, the first question is often whether a threshold model is appropriate. If the answer is yes, we need to identify the appropriate model form before parameter estimation because an estimated threshold is associated with a specific model (e.g. Fig. 12.1). Obviously, when a wrong model is selected, the resulting threshold estimate may be meaningless, and, in a management context, lead to inappropriate or suboptimal management outcomes (Box 12.2). Therefore, we need a systematic means for identifying the likely model form. The three models described above are used as examples of alternative models to illustrate the process of model selection and evaluation in the context of a threshold modeling problem.

Box 12.2 Common challenges: crossing the threshold

As discussed earlier, a statistical change point is associated with a specific model and a change point is not necessarily the management threshold of interest. Communication of the uncertainty associated with results of the threshold analysis is often the most important aspect of a study, but often the least emphasized. A difficulty in communicating the uncertainty is the lack of definite statistical test for differentiating a threshold model (representing an abrupt change) from its smooth alternative (representing a gradual change). In fact, such a test may not exist. Chiu *et al.* (2006) suggested that statistical evidence is usually unavailable to separate an abrupt-change model from a gradual-change model. Yet distinguishing these two types of responses is essential in setting an appropriate environmental management criterion. The figure below shows two examples of abrupt-change models versus gradual-change models, with the ecological indicator of interest on the *y*-axis and the stressor level on the *x*-axis. A changepoint model, representing an abrupt change, will report ϕ as the threshold, while the point of interest in a management problem is often $\phi - \gamma$ (where a gradual change starts). That is, if the gradual change model best describes the underlying ecological relationship, then assuming an abrupt-change model may lead to establishment of threshold values for the stressor of concern that offer lower protection than expected. These problems are not widely recognized.

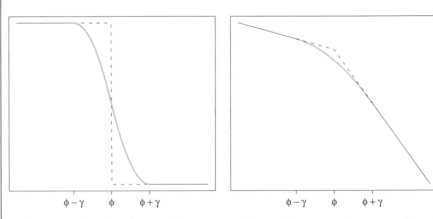

An hypothetical step function model has two possible transition patterns between two "steady states" (thin black solid lines, left panel) and an hypothetical piecewise linear model has two possible transition patterns between linear regimes (thin solid line segments, right panel): an abrupt change (thin dashed line segments) and a smooth transition model (thick gray line segment).

Under a Bayesian setting, a threshold is treated as a random variable. The range of the threshold is conveniently limited to be the observed range of the predictor variable. If the estimated threshold value is at the lower or upper bound of the range, we conclude that a threshold does not exist. Whether a threshold exists or not is determined by comparing the probability of no threshold and the probability that a threshold exists. When a wrong model is used, the likelihood of no threshold is likely to increase and the model's fit to the data will be poor. For the three linear threshold models described in Fig. 12.1, we can estimate the thresholds for each model and compare the estimated threshold distribution and models' fit using deviance or other measures such as the Bayes factor (Kass and Raftery 1995). The commonly used information criteria (AIC, BIC, and DIC) which penalize models with more parameters are less relevant for this problem because the objective is to select the most appropriate model to represent the underlying pattern rather than to select a parsimonious predictive model.

Parameters of the three threshold models can be estimated using classical statistics methods such as the maximum likelihood estimator. However, the classical MLE is often case-specific. Different threshold models may have different likelihood functions and the likelihood function and the first order derivative of the likelihood function are discontinuous. As a result, special computational methods are needed for different types of models. The R (R Development Core Team 2011) package "segmented" (Muggeo 2003, 2008) and the hockey stick model described in Qian (2010) can be used for the piecewise linear model. Qian *et al.* (2003) described two statistical methods for the step function model. One uses a classical method similar to the classification and regression tree (CART) model; the threshold identified by this "nonparametric" deviance reduction method is the first split of a CART model using the time or the environmental stressor as the only predictor. The other method is the analytical solution of the Bayesian posterior distribution of the change point.

Under a Bayesian framework, the changepoint computation strategy is unified for all model types (Smith 1980). In most cases, we have no or little information on the location of the change point and a flat prior is usually used. As a result, the difference between a Bayesian estimator and an MLE is largely in the definition of a random variable. An unknown quantity can be assigned a probability distribution under the Bayesian framework. When a flat prior distribution is used, the Bayesian posterior distribution is proportional to the classical likelihood function. Under a classical framework, we find the parameter value that maximizes the likelihood function and use the likelihood function profile to estimate the confidence interval. Under the Bayesian framework, we use either the mode (the same as the MLE) or the posterior distribution mean as the estimate of the unknown quantity. In other words, there is no practical difference between a Bayesian estimate and the MLE if a flat prior is used.

Traditionally, computation of a threshold problem was difficult both for the Bayesian estimate and for MLE. Such difficulty disappeared when MCMC simulation was introduced. Examples of implementing the Bayesian changepoint model using MCMC can be found in the BUGS/JAGS (Lunn *et al.* 2000) examples (e.g. example "stagnant" in the BUGS examples collection, volume 2; http://www.mrc-bsu.cam.ac.uk/bugs/winbugs/Vol2.pdf). The BUGS code of the three models (with a normally distributed response

Table 12.1 Estimated mean threshold and its 95% confidence limits derived from the Bayesian posterior distribution (general, hockey stick, and step function models) or bootstrap sampling distribution (CART).

Estimation model	Estimate	2.5%	97.5%
Data set 1 (generated from general model)			
General	0.500	0.488	0.512
Hockey stick	0.339	0.282	0.398
Step function	0.500	0.488	0.512
CART	0.488	0.443	0.488
Data set 2 (generated from hockey stick model)			
General	0.406	0.298	0.548
Hockey stick	0.412	0.336	0.502
Step function	0.679	0.602	0.766
CART	0.733	0.583	0.779
Data set 3 (generated from step function model)			
General	0.500	0.488	0.511
Hockey stick	0.726	0.156	0.993
Step function	0.500	0.488	0.511
CART	0.488	0.443	0.488
Data set 4 (generated from linear model)			
General	0.686	0.023	0.992
Hockey stick	0.871	0.059	0.993
Step function	0.420	0.396	0.579
CART	0.395	0.288	0.579

variable) is included in this chapter's online supplement, as are the R scripts for running these models. Readers can modify the R scripts and BUGS models if their response variable has a distribution other than the normal distribution. The online code returns plots of the posterior distributions of the estimated change point under the three models and the deviance of each model.

Assessing the performance of alternative models

The threshold modeling process is illustrated by using four simulated data sets: three of them (Fig. 12.2a–c) are generated based on the models in Fig. 12.1 and the fourth is a simulated data set without a change point (Fig. 12.2d). The results are compared to the "nonparametric" deviance reduction/CART approach (Table 12.1) described in Qian *et al.* (2003). The term "nonparametric" is somewhat misleading, as the model represented by the method is the step function threshold model (discussed above). Details of the simulated data sets are provided in the online supplement.

For each simulated data set, each of the three models represented in Fig. 12.1 was applied to estimate the existence and location of a threshold. The estimated threshold distributions from each estimation model are compared graphically (Figs. 12.3 –12.6). When a model used for estimation represents the underlying data generating process

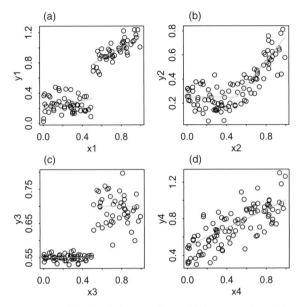

Figure 12.2 Simulated data sets from: (a) the general model; (b) the hockey stick model; (c) the step function model; and (d) linear (no threshold) model. Sample size = 100 observations for each data set.

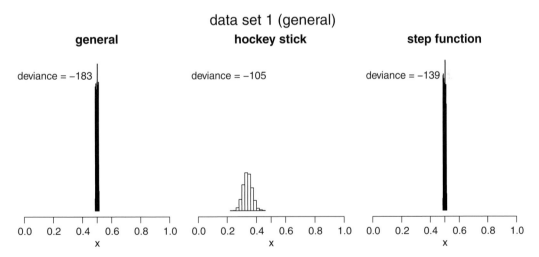

Figure 12.3 Posterior distributions of the estimated threshold parameter from the three threshold models applied to the data set generated based on the general model.

(i.e. when the model used for analyzing the data set was of the same type as the model from which the data set was generated), the estimated threshold should be concentrated around the true changepoint value (e.g. left panel of Fig. 12.3). When the model used is not representative of the underlying process, the estimated threshold distribution often (but not always) shows recognizable patterns. In general, we can divide misuse

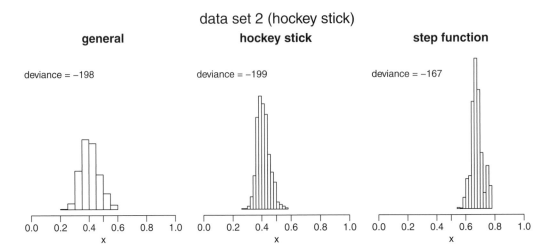

Figure 12.4 Posterior distributions of the estimated threshold parameter from the three threshold models applied to the data set generated based on the hockey stick model.

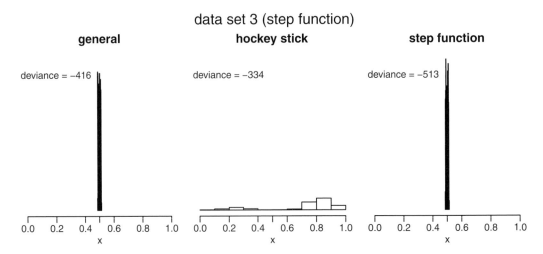

Figure 12.5 Posterior distributions of the estimated threshold parameter from the three threshold models applied to the data set generated based on the step function model.

of a changepoint model into two situations: (i) a wrong changepoint model is used to estimate the threshold, and (ii) a changepoint model is applied to data without a change point. Two of the three panels in Figs. 12.3–12.5 represent the first situation. Results shown in Fig. 12.6 represent the second situation.

When a change point does not exist (Fig. 12.6), ideally, the model used for estimation will identify both ends of the predictor variable range as the change point (effectively lumping all data points into one group), leading to a "U"-shaped threshold distribution (e.g. Fig. 12.6 left and middle panels). The estimated threshold distribution can also have a large variance, covering a wide range of the predictor variable range. The step function

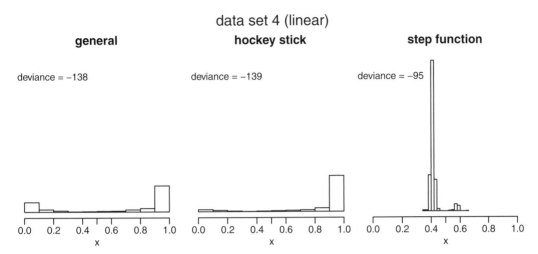

Figure 12.6 Posterior distributions of the estimated threshold parameter from the three threshold models applied to the data set generated based on the linear model.

model is likely to produce a unimodal threshold distribution when the underlying relation between the response and the predictor is linear (Fig. 12.6) or more generally, monotonic, because the data can always be divided into two groups with different response variable means.

When a wrong model was used (e.g. applying the step function model to data generated by the hockey stick model, right panel of Fig. 12.4), the estimated threshold distribution cannot be used as the sole diagnostic tool. In most cases, a wrong model may lead to a wide spread in the threshold distribution (e.g. left panel of Fig. 12.4). However, a wrong model often cannot be easily discerned from the correct one by comparing the estimated threshold distributions alone. If the changepoint signal is strong enough, two or three models may perform equally well in terms of estimating the underlying change point. Additional diagnostic information is necessary to properly separate the "good" from the "bad."

As in a simple linear regression problem, model residual plots are often the most informative tool for model assessment. For the three models discussed here, model residuals should be normally distributed with mean 0. We allow the variance to vary in the step function model (12.2), and used a constant residual variance for the other two models. A difficulty in using residuals is how to obtain model residuals when the model parameters are not well-defined. One solution is to treat each residual as a random variable. Numerically, the residual distribution is obtained by using the MCMC samples of model coefficients; that is, calculating one set of residuals for each set of MCMC samples of model coefficients. As a result, the residual for each observation is represented by a distribution (summarized by the mean and, for example, the 95% credible interval).

Figures 12.7 –12.10 show the residual versus predictor plots for the fitted models. Model residuals were obtained through posterior simulations with the R package "rv"

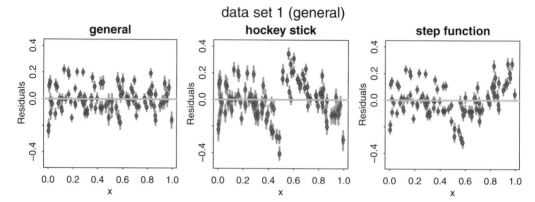

Figure 12.7 Residual distributions from the three threshold models applied to the data based on the general model (open circle = mean; light gray = 95% credible interval; darker gray where visible = 50% credible interval).

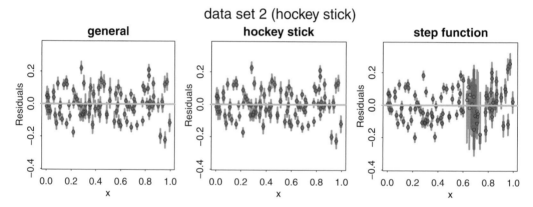

Figure 12.8 Residual distributions from the three threshold models applied to the data based on the hockey stick model (open circle = mean; light gray = 95% credible interval; darker gray where visible = 50% credible interval).

(Kerman and Gelman 2007; see online supplement). These plots aid in identifying the most appropriate model from those that produced similar threshold distributions. In summary:

• When the underlying model was the general model (Fig. 12.1a), the general model and the step function model both correctly estimated the threshold (Fig. 12.3, left and right panels). The residuals of the general model (Fig. 12.7, left panel) showed the typical random scatter, while those for the step function model did not (Fig. 12.7, right panel, showing a linear pattern between $x = 0.5$ and $x = 1$). The same information is also reflected in that the general model had a much lower deviance value (Fig. 12.3). Residuals from the hockey stick model show a pattern of two line segments, separated by the change point at $x = 0.5$ (Fig. 12.7, middle panel). The middle and right panels

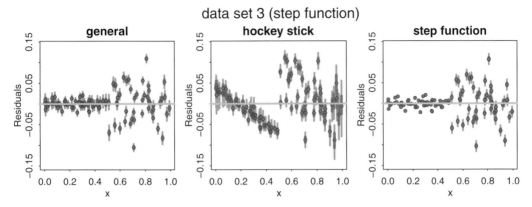

Figure 12.9 Residual distributions from the three threshold models applied to the data based on the step function model (open circle = mean; light gray = 95% credible interval; darker gray where visible = 50% credible interval).

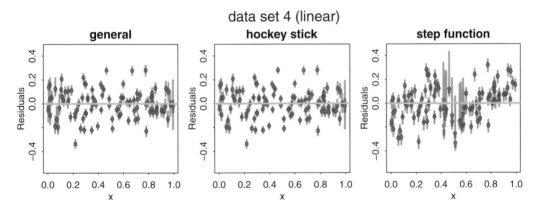

Figure 12.10 Residual distributions from the three threshold models applied to the data based on the linear model (open circle = mean; light gray = 95% credible interval; darker gray where visible = 50% credible interval).

of Fig. 12.7 show a typical residual pattern of an inappropriate model (regions of non-random patterns).

- When the underlying model was the hockey stick model, both the general model and the hockey stick model produced estimated threshold distributions that covered the true value of 0.5, while the estimate from the step function model did not (Fig. 12.4). By the estimated threshold distribution alone, we would have concluded that the step function model produced a more certain estimate. When the additional information (residual deviance and residual distributions in Fig. 12.8) is examined, we see an increasing trend in the step function model residual plot after $x = 0.6$, suggesting that the step function model is not appropriate.
- When the underlying model was the step function with different variances above and below the threshold, both the general model and the step function model performed

well in that they both captured the true underlying threshold, while the hockey stick model failed to detect a threshold (Fig. 12.5). Because the general model assumes a common residual variance and the step function model allows different variances, the step function model has the lower deviance value.

- When the underlying model was the linear model, both the general model and the hockey stick model correctly suggested that a change point does not exist, but the step function model produced a fairly concentrated threshold distribution (Fig. 12.6). The misperception of a threshold presented by the step function model is easily corrected when additional information, especially the residual plots (Fig. 12.10), are examined.
- In all four cases, the step function model produced a concentrated changepoint distribution. This result is not surprising, as the step function is designed to detect changes in the mean. As long as the response variable changes monotonically along the predictor gradient, the step function will always find a "change point" that results in a "change" in the mean of the response. Showing the residual plot for a step function model is an important step to guard against misleading conclusions.

Discussion

Statistical modeling is a tool for inductive reasoning. The goal is to find the model that likely represents the underlying data-generating process. As the true model is always elusive, the model fitting and model selection processes require careful sleuthing. The starting point is always the assumptions associated with a model. In this chapter, all three models assume that the response variable follows a normal distribution. The threshold response is represented by the mean function. Which model to use in an application depends on our knowledge of the underlying process. For example, the piecewise linear model used in Qian and Richardson (1997) was based on the understanding that a wetland has a limited capacity of assimilating phosphorus. Consequently, a continuous response model was scientifically defensible (Richardson and Qian 1999).

In the absence of such knowledge, the problem becomes identifying the alternative models and choosing which one fits the data the best. In this chapter, this question is addressed in two steps by using three alternative linear changepoint models. The first step is to determine whether a threshold response exists, using the posterior threshold distributions from these alternative models. The estimated threshold distributions can provide basic information on the appropriateness of each threshold model. When a specific threshold response model is not appropriate, we often see a widespread (maybe U-shaped) threshold distribution. When a concentrated threshold distribution such as the right panel of Fig. 12.6 is obtained, residual plots should be used to further evaluate whether the fitted model is appropriate as the second step of model checking and evaluation. For example, the residual plot (Fig. 12.10) for the model shown in the right panel of Fig. 12.6 suggests that a step function is not appropriate for this data set.

As long as the underlying assumption of the model is properly evaluated, whether model parameters are estimated using a Bayesian method or a classical statistics method will make no practical difference [compare results from Bayesian methods with those from the "nonparametric" deviance reduction method (Qian *et al.* 2003) – CART – in Table 12.1]. If threshold modeling uses a classical statistics method, the confidence interval of the estimated change point can be used for statistical inference. The null hypothesis of no change point is rejected at a 0.05 significance level if the 95% confidence interval does not include either end of the covariate gradient. (Note, however, that the conventional hypothesis testing procedure is often criticized for its emphasis on avoiding Type I error at the expense of statistical power.) However, this approach is not appropriate for the step function type of change point when the confidence interval is estimated using the bootstrap method (Bühlmann and Yu 2002, Banerjee and McKeague 2007) – the bootstrapping confidence interval is always too narrow. Banerjee and McKeague (2007) proposed a classical confidence interval for step function changepoint problems with a normal response variable.

However, parameter estimation usually is not the major challenge in threshold modeling. Many alternative models have been proposed for identifying ecological and environmental thresholds without adequate discussion about their respective underlying probabilistic models. In general, most models in the literature can be represented by the basic changepoint model Equation (12.1), even those claimed to be "nonparametric". For example, Brenden *et al.* (2008) reported that the "quantile regression" model is often the most robust threshold identification method, and they described the model as a "nonparametric" regression model. A careful examination of the case studies where the quantile regression model was used reveals that most of these cases can be summarized as having a log-normal or Poisson (or "overdispersed Poisson") response variable. These distributions have variances proportional to their means. When the response variables are log-transformed, a quantile regression changepoint problem is equivalent to the hockey stick model. The claim that quantile regression is most robust is expected because most environmental concentration variables can be approximated by the log-normal distribution, and the standard deviation of a log-normal distribution is proportional to its mean. If the response variable mean changes along an environmental gradient, the standard deviation (and a given quantile) of the distribution will change. When a step function or a hockey stick model is used to describe the changes, we can simplify the problem by parameterizing the changes in terms of log-mean. The term "nonparametric" is often misleading because the term often means "non-normal." A specific probability distribution is always assumed, but not explicitly stated. Discussing the outcome without knowing the exact assumptions is often unhelpful. A nonparametric method, therefore, is most appropriate for exploratory data analysis aimed at generating hypothesis.

In contrast, the need to specify probability distributions in applying a Bayesian method forces researchers to layout assumptions explicitly. This process is often difficult, but necessary. Because the Bayesian method is a generalization of the classical counterpart (Gelman *et al.* 2004), it is almost always more flexible than classical methods.

This chapter is not intended to "reinvent" statistical methods already in the literature. In fact, most published methods papers in ecological literature cannot claim originality about statistical methods. What is new in this chapter is the way that these statistical methods are used for threshold quantification. Even this line of thinking is not new, as Fisher had already defined the logic of probabilistic inference and Tukey provided exploratory tools for better implementing this logic (Lenhard 2006).

Future research and development

The flexibility of the Bayesian approach (e.g. Chapters 19, 20) allows us, for example, to expand the models discussed in this chapter into a hierarchical structure to pool data from multiple sites or over multiple time periods. For example, Qian (2010) described a hockey stick model of the first-bloom date of lilacs as a function of year for modeling the effect of climate change. A hockey stick model assumes that the first bloom date stayed relatively stable until the climate started to change, after which, lilacs in the United States started to bloom earlier each year in response to the warmer temperature. This model is of the form of the piecewise linear model with constant variance (12.4). The parameters of interest are ϕ, the timing of the climate change felt by the species, and δ, the magnitude of the effect. Because the first bloom date data were collected from multiple locations, the timing and magnitude of the climate change effect may vary. To study the among-location variation, we can use a hierarchical modeling approach to pool data from all stations together. This approach is conceptually straightforward. First, we use the same model form for all locations:

$$y_{ij} \sim N(\beta_{0j} + (\beta_{1j} + \delta_j I(x_{ij} - \phi_j))(x_{ij} - \phi_j), \sigma_j) \tag{12.5}$$

where j is the index of locations where lilac first-bloom dates were recorded. Model coefficients are location-specific. To study the response of all locations together and still allow location-specific features to persist in the model, we assume that model coefficients from different locations have the same prior distributions:

$$
\begin{aligned}
\beta_{0j} &\sim N(\beta_0, \sigma_{\beta_0}) \\
\beta_{1j} &\sim N(\beta_1, \sigma_{\beta_1}) \\
\delta_j &\sim N(\Delta, \sigma_\delta) \\
\phi_j &\sim N(\Phi, \sigma_\phi)
\end{aligned}
\tag{12.6}
$$

Often, flat (or vague) prior distributions are used. By assuming that model coefficients for different locations are from a common prior distribution, data from all sites are pooled only partially (Gelman et al. 2004). Partial pooling of data from multiple locations will allow us to study the among-location variation and explore factors affecting the timing and magnitude of the climate change effect.

The models discussed in this chapter are focused on a univariate response variable. From an ecological perspective, such models are limited to study the changes of an ecosystem one aspect at a time. A more realistic and more complicated question in ecological studies is how to define the threshold response at an ecosystem level. Using the Bayesian hierarchical modeling approach, this question can be addressed by partially pooling data from multiple ecological metrics (or species) and fitting a model for them together. The model setup is the same as in Equation (12.5), with the location index j

representing the index of ecological metrics. The hyper-distribution of the change point, $N(\Phi, \sigma_\phi)$, is now the distribution of all change points representing variation among "species". This hyper-distribution is used in ecotoxicology literature as the species sensitivity distribution (Posthuma *et al.* 2002). Using a Bayesian approach (especially the Bayesian hierarchical modeling approach), the estimated posterior distribution of ϕ_j represents a much improved estimate than the empirical distribution approach used in the ecotoxicology literature.

Summary

The concept of the ecological threshold is increasingly used in ecological studies and in environmental management. Various quantitative methods for estimating ecological thresholds are often specific to the particular response variables of interest. The number of quantitative methods not only reflects the numerous types of ecological responses to various disturbances (and numerous types of potential temporal trends), but also indicates the varying characteristics of different ecological response variables, which must be modeled by different probabilistic distributions. In this chapter, a unified definition of a threshold problem is presented and the emphasis of the chapter is on the process of model identification through examining multiple alternative models. A set of simulated datasets are used to demonstrate two types of plots that can be used to examine (i) whether a threshold exists, and (ii) whether the underlying model is appropriate, respectively. Plotting the fitted model and data is especially important when the step function model is used, because a step function model will return a statistically significant change point no matter whether a change point exists or not, as long as the response and stressor relationship is monotonic. Unfortunately, many names are used in the literature to describe the step function model, so care is needed to recognize when a step function model is being applied. The chapter concludes with a discussion of the Bayesian hierarchical modeling approach for synthesizing threshold modeling results on multiple ecological metrics from the same ecosystem, and/or results from multiple sites representing different natural conditions.

The emphasis of this chapter is on the general steps of model formulation, parameter estimation, and model interpretation in the context of quantifying ecological thresholds. The detailed analysis of three linear threshold models illustrates a general procedure for identifying the most likely model form. An annotated R script file is available as an online supplement.

13 The treatment of missing data in long-term monitoring programs

Douglas H. Johnson and Michael B. Soma

Introduction

Monitoring environmental resources, such as populations of animals and plants and their habitats, is essential for determining their general status (e.g. Likens 1983, Thompson *et al*. 1998). The objective of a monitoring program typically is to track a target variable over time to detect changes, either with periodic (e.g. annual) measures of an attribute such as population size or with estimates of overall trend during that time. In the case of monitoring focused on animals or plants (e.g. Ralph and Scott 1981, Menges and Gordon 1996, Elzinga *et al*. 1998), often a fixed set of sites within a species' range are visited periodically and the number of individuals at each site counted. Our discussion is in terms of counts of animals or plants but applies more generally to other natural resources. The sites may have been chosen on a systematic basis, may have been randomly selected, or may encompass the entirety of the species' known range. If all sites are monitored on all occasions, analysis of resulting data is relatively straightforward; sampling variability, however, may come into play if only a sample of the possible known locations are monitored on each occasion.

It can happen that not all sites are examined on every scheduled occasion. There are contexts in which missing data are not a critical problem, depending on why data are missing (Chapters 7, 15). In programs such as the North American Breeding Bird Survey, certain results are based on mean counts of sites (routes) within each stratum. Missing counts – routes not surveyed – will reduce the precision of an estimator but not bias it unless a relation exists between the abundance of a species on a route and the probability that the route is not surveyed (e.g. if some routes are less likely to be surveyed because they have too few "interesting" birds to appeal to observers). For the types of monitoring programs that are the focus of this chapter, however, missing data can be problematic.

In this paper we consider a variety of alternatives for handling missing data. For convenience, we will assume that sampling is conducted periodically for T occasions (e.g. years), indexed by t, $t = 1$ to T. Sites are indexed by i, $i = 1$ to I. Denote the count of individuals at Site i on occasion t as C_{it}. If all sites are examined on all occasions, then the sum $C_{.t} = \Sigma_i C_{it}$ is the total population count for occasion t. If the sites are a random

Design and Analysis of Long-term Ecological Monitoring Studies, ed. R.A. Gitzen, J.J. Millspaugh, A.B. Cooper, and D.S. Licht. Published by Cambridge University Press. © Cambridge University Press 2012.

Box 13.1 Take-home messages for program managers

In many long-term monitoring programs, it can happen that not all monitoring sites are examined on every scheduled occasion. Missing data may not be problematic in some situations; e.g. some analytic approaches that address trend do not require values for each occasion. However, missing observations often can complicate treatment of the data – for example, when the total count from all sample sites visited on an occasion is used as an index to population size on that occasion.

Although replacing missing counts by zeroes sounds unreasonable, it actually is commonly but implicitly used; methods that present the total count of organisms observed on each occasion often fall into this category. The assumption that missing counts are truly zeros can be realistic, if it is known that a zero would have resulted from a survey at the site. Otherwise, this method is not acceptable. Other relatively simple methods of imputing (replacing) missing data use other data from that site or all sites. For example, a missing observation may be replaced by the overall mean of the observed values at the site, by the previous observation from that site, or more generally by replacing the missing value with a weighted average of previous and possibly subsequent observations at that site. These methods are better alternatives than simply ignoring all data from occasions with any missing data or (in most cases) replacing non-zero missing observations with zeros. However, the accuracy of each method depends on the pattern of missing data (e.g. mostly early vs. late in the monitoring program), the magnitude of variability in the target variable, and on the type of systematic temporal pattern in the population (e.g. no trend vs. a linear trend vs. cyclical dynamics)

These factors also affect the performance of model-based imputation, which replaces missing values of the target variable with values generated from a specified model that relates the target variable to covariates. The model can be developed from observations without missing values. The covariates can be simply the site identification and occasion number, but might include measured covariates. When predictability from covariates is high, model-based estimators can perform well.

In general, imputation based on a model with good predictability is recommended. If such a model is not available, the selection of an imputation method should be based on an examination of the behavior of the target variable. More broadly, how missing data should be treated depends markedly on how they came to be missing. When visits to a site need to be skipped due to logistics or budgetary limitations, a monitoring program should document the reasons data are not collected.

---

or representative sample of all possible sites and are surveyed every occasion, then the sum of the counts over all sites on occasion t, $C_{.t}$, is an index to the population. $C_{.t}$ will be an index also if the counts themselves represent only a consistent fraction of the actual number. We are concerned with situations in which certain sites are not surveyed on some occasions, that is, some C_{it} are missing. Our objective is to review methods available for treating such missing data, discuss situations in which each is applicable, and offer some recommendations (Box 13.1). Our coverage is aimed at biologists and natural resource managers seeking straightforward methods for addressing problems caused by missing data.

The statistical literature on the general missing-data problem is rich, with hundreds of scientific papers published and entire books devoted to the topic. Tools developed by statisticians to handle the general missing-data problem are appropriate in some monitoring applications, but not all of them. Many of the general statistical situations involve estimating the effect of one or more covariates on a response variable. Monitoring programs of the type considered here, in contrast, focus on the count itself, which in this context corresponds to the response variable. Such programs may not measure any covariates associated with the response. Further, the default method used by most statistical software to deal with missing data is to exclude from analysis every observation for which any variable is missing. That approach would defeat what we are assuming in this chapter to be the primary objective of monitoring, which is to have an estimate of, or index to, the population size on each occasion.

Some definitions

- *Monitoring program*: a series of periodic surveys at fixed *sites* on prescribed *occasions* to measure or estimate specified variables. A combination of a site and occasion is a monitoring *site visit*. [Note that in some cases a single "site visit" may actually encompass multiple data-collection visits during an occasion; e.g. in a breeding-bird monitoring program, replicate point counts might be conducted at each site during each year (occasion).] We illustrate methods for which all sites are intended to be surveyed on each occasion, although the methods we describe are more general.
- *Target variable*: the variable of primary interest, usually a count of some organism.
- *Covariates*: variables measured or estimated, along with the target variable. *Local covariates* are those measured at the same sites on the same occasions as the target variable; these might encompass vegetation metrics at the site, for example. *Global covariates* are those available even without a visit to a site, such as temperature at a nearby weather-recording station. *Site-visit covariates* are the site identification (i) and occasion number (t); these covariates are always available.
- *Imputation*: the replacement of missing values by other values.

Typically, the primary objective of most monitoring programs is to track the target variable over time to detect changes (Chapter 3). This objective could involve estimating a "trend" during a period of time, or determining values that reflect the target variable on each occasion. Secondary objectives might include examining spatial variation in the target variable, comparing trajectories (temporal variation) by location, and estimating the effects of certain covariates on the target variable.

Analysis of monitoring data

Methods to analyze monitoring data differ depending in part on whether the primary objective is to estimate a temporal trend in the target variable or to obtain an index to the target variable on each occasion. We focus on the latter objective; many methods addressing trends require indices for each occasion. Conversely, indices can be generated from the results of certain trend estimation procedures. Nonetheless, some approaches that address trend do not require values for each occasion, but estimate a trend parameter directly under some assumed model of the target variable (e.g. Clark and Bjørnstad 2004, Humbert *et al.* 2009). However useful these methods are, the concept of "trend" is not well-defined and is strongly dependent either on an assumed model or on the start and end points for the period over which trend is estimated (Chapters 3, 11).

The importance and effects of missing data depend on how the monitoring data are analyzed. Several methods have been employed historically. For example, the chain index method (Ogilvie 1967) uses the ratio of counts on successive occasions but only for those sites that were surveyed on both occasions; therefore, missing data on one occasion reduce the value of data on the previous and subsequent occasions. These annual ratios are "chained" together from some arbitrary base year to provide a series of annual indices. The chain index method does not use all the available data and further is subject to generating spurious trends in the time series (reviewed by ter Braak *et al.* 1994).

Mountford (1982) generalized the chain index method to employ ratios of counts on non-consecutive, as well as consecutive, occasions. This method assumes changes among occasions are similar for all sites. That is, change depends on occasion but not site. This assumption is a drawback of the Mountford method, as are its computational difficulties and unsuitability for incorporating effects of covariates, such as observers, habitat, etc. (ter Braak *et al.* 1994).

Route regression, developed by Geissler and Noon (1981), focuses directly on estimating a trend, from which annual indices can be generated. For each site (North American Breeding Bird Survey routes, in their original application), log(Count) is regressed on Occasion, and trend is estimated as the regression coefficient. Route regression allows trends to vary among individual sites. Changes depend on site but not occasion, in contrast to the Mountford method. Estimates of trend at higher spatial levels (strata) are made by aggregating trends at sites within each stratum. These estimates are weighted by factors involving the abundance at a site and the variance of the estimated trends at sites within the stratum. The model underlying route regression for a particular site i is:

$$\text{Count}_{it} = S_i \times (b_i)^t \times \text{error}_{it} \tag{13.1}$$

so that

$$\log(\text{Count}_{it}) = \log(\text{Site}_i \text{ effect}) + \log(\text{Trend}_i) \times \text{Occasion} + \log(\text{error}_{it})$$

Poisson regression involves a model similar to Mountford's:

$$\log(\text{Expected Count}) = \text{Site effect} \times \text{Occasion effect},$$

but now this mean function is treated as the parameter of a Poisson distribution (e.g. Underhill and Prys-Jones 1994). Other covariates can be included in the model readily.

Generalized linear models (Chapter 11) can be used to estimate the parameters. As with the Mountford method, the assumption of consistency of Occasion effects among all sites is restrictive.

A more recent approach involves the use of hierarchical models (Link and Sauer 2002; Sauer and Link 2002, 2011; Chapters 12, 19, 20). These models allow Occasion effects, as well as Stratum and Observer effects, to be governed by parameters that themselves are random variables, hence the hierarchy. Regional summaries are defined as functions of the underlying regional abundance and population change parameters. For the North American Breeding Bird Survey, Sauer and Link (2011) modeled population change with an overdispersed Poisson regression with fixed and random effects. Counts $Y_{i,t}$ are viewed as independent Poisson random variables with means that are described by loglinear functions of explanatory variables: stratum-specific intercepts and slopes, effects for observer/route combination and year, and overdispersion.

Other chapters in this volume provide more extensive discussion of approaches for modeling temporal changes using forms or extensions of the types of models discussed above (Chapters 7, 11, 12, 19, 20). Again, the remainder of our chapter focuses on situations in which the objective is to obtain an index to the target variable on each occasion.

Mechanisms for missingness

How missing data should be treated depends in part on how they came to be missing (Rubin 1976, Harrell 2001, Gelman and Hill 2007; see also Chapter 15). The most straightforward mechanism resulting in missing data is that values are missing completely at random (MCAR). That is, each site visit (combination of site and occasion) has the same probability of having a missing value. The MCAR mechanism seems unlikely to occur in typical monitoring programs. The second class, missing at random (MAR), has a similar name but different mechanism. Here the probability that a site visit has a missing value depends on other variables that were measured. Perhaps difficulty of access is measured for all sites, and on certain occasions the more difficult sites are less likely to be accessed than more accessible sites. Data sets with MCAR and MAR features are said to have ignorable missingness (or non-response, a term used initially when applied to surveys of humans). In the third class, the probability that a value is missing depends on variables that were not measured. This situation more commonly occurs in monitoring programs; some data are missing, but the reason, which may have been known to the observer at the time, was not recorded for posterity. The fourth class involves situations in which the probability that a value is missing depends on that value itself. A plausible example of this situation is if certain sites are not surveyed because it is somehow known that they are unoccupied. These latter two classes constitute data sets with non-ignorable non-response and are the most difficult to address. A fifth class that arises in monitoring occurs when sites are not surveyed on one occasion because no organisms had been present at the site on one or more previous occasions (Johnson *et al.*

2011). In this situation the probability that a count is missing depends on a measured variable (the count on the previous occasion); modeling the missing value in terms of that variable seems like a promising avenue for research.

Methods for treating missing values

We next describe several methods used to account for missing data, with an emphasis on their applicability to monitoring situations. The first method simply deletes an observation with any missing value; all the others impute values for the missing values. The second method replaces missing values by zeros. The next four methods replace a missing value by a value based on other values of the target variable. The remaining methods exploit the relations between the target variable and any covariates. For more details about these methods, see Rubin (1976), Allison (2001), or Little and Rubin (2002).

1. Complete-case method

When an observation of the target variable is missing for any sites on a scheduled occasion, one possible approach is to disregard *all* data for that occasion. This approach is called the complete-case method and is the default method used by most statistical software. The method is simple but can result in misleading conclusions if the data that are missing follow a pattern different from the observed data. Also, it can be wasteful of information and is singularly inappropriate for monitoring. For example, with a large number of sample sites and fluctuating budgets, there may be many occasions in which some sites were not visited; disregarding these occasions might mean discarding most of the data set. We do not discuss this method any further.

2. Treat missing values as zero

Despite the unreasonable-sounding label we give it, this method actually is commonly – albeit unknowingly – used; methods that present the total count of organisms observed often fall into this category. The assumption being made, often implicitly, is that all sites where the organism occurs are surveyed. Any time a monitoring program fails to survey one or more sites, and the estimate for the occasion is the sum of the observed counts, the implicit assumption is that sites that were not surveyed had no organisms. This assumption can be realistic if there is a justifiable belief that a zero would have resulted from a survey of the site. For example, if in a survey of diving ducks, a site is known to be a wetland that has gone dry, it is reasonable to assume a count of zero for that site. If such an assumption is incorrect, however, the method results in a biased low estimate of the total of the target variable on each occasion a value is missing.

3. Mean-substitution method

When values of target variables are missing at random, one solution is to substitute for the missing data the overall mean of the observed values at the site. This method is unlikely to bias an estimator as much as does the treat-as-zero method. However, for the primary monitoring objective, the main drawback arises if temporal variation is high, if there is some trend and missing values are more frequent early or late in the period, or if the overall mean is based on few observations, so that it is unlikely to approximate the unobserved value.

4. Last-measured (and next-measured) method

Another simple imputation method involves substituting for a missing value the value from the just-previous occasion. If monitoring data are analyzed after subsequent occasions, one might replace the missing value with the average of the previous observed (last measured) value and the subsequent value. Either replacement would be expected to do well if counts at the site vary little from occasion to occasion. The latter replacement also should be very good if the trajectory was steady, especially linearly.

5. Random-imputation method

This method selects a value at random from all the observed values of the missing variable. It is a relatively simple method that introduces appropriate variability into the data set to avoid artificially decreasing the estimated variation. One could impute from previous occasions at the same site, if variation among occasions within a site was minor, or impute from other sites on same occasion, if sites varied little among occasions. In general, though, the method neglects any other information that could be used to improve estimates.

6. General weighted-average method

The previous three methods can be considered special cases of an approach in which a missing value is replaced by a weighted average of the counts on, say, k previous occasions. (Data from subsequent occasions can be used as well, if the analysis is conducted after further data collection.) The mean-substitution method uses weights $(k^{-1}, \ldots, k^{-1})$, that is, equal weights for all preceding occasions, the number of which is k. The "last-observed" method has a weight of one for the just-previous occasion and zero for all others. The method that uses the mean of the just-previous count and the subsequent count has weights one-half for values from each of those two occasions and zero for others. The random-imputation method has a weight of one for the randomly selected occasion and weight of zero for all others.

A more general weighted-average approach would use weights that decline with the number of occasions between an observed value and the occasion with the missing value. Two reasonable selections of weights are those that decline exponentially with the number of occasions involved in the average and weights that decline linearly with the number of occasions involved. If one wanted to use counts from, say, k previous occasions, exponentially declining (by $1 - \alpha$), weights would be

$$w_j = (\alpha^{j-1})(1 - \alpha)(1 - \alpha^k)^{-1} \qquad (13.2)$$

where $j = (1, \ldots, k)$ is the number of occasions before the one with the missing value. For example, with a proportionality value of $\alpha = 0.5$ and $k = 4$, weights would be 8/15, 4/15, 2/15, and 1/15 for $t - 1$, $t - 2$, $t - 3$, and $t - 4$, respectively. This approach requires the selection of α as well as k. Linearly decreasing weights could be of the form

$$w_j = 2(k + 1 - j)(k^{-1})(k + 1)^{-1}. \qquad (13.3)$$

For example, with $k = 4$, weights would be 4/10, 3/10, 2/10, and 1/10 for $t - 1$, $t - 2$, $t - 3$, and $t - 4$, respectively. This approach requires only the selection of k. In either approach k need not necessarily be the total number of previous occasions.

7. Imputation from models with covariates

Model-based imputation replaces missing values of the target variable with values generated from a specified model that relates the target variable to covariates. The other imputation methods involve models, too, but typically they are not explicitly stated. The model can be developed and regression coefficients estimated from observations without missing values. The covariate could be simply the site-visit covariates – site identification and occasion number – as well as local or regional covariates. Note, however, that in monitoring programs, any local covariates such as proximate vegetation features typically are measured concurrently with the target organism and at the same site. If the count of the target variable is missing, measures of such covariates likely would be missing as well.

This imputation procedure can be taken another step by replacing missing values of the target variable with values predicted from the model just described, then refitting the model now using the full data set. Then the initial predicted values would be replaced by predictions from the second model. This process could be repeated until values converge. Whether the second or subsequent steps offer any improvements is questionable. These methods make full use of all data and should perform well if the target variable can be closely predicted from the covariates.

8. Homogeneous-groups method

If the count at a site is related both to previous counts at that site and to counts at other sites on the same occasion, the value of a missing count can be estimated from those previous counts and counts at other sites. Although such patterns may not hold true generally for all sites, it may be feasible to divide the sites into relatively homogeneous groups, that is, groups of sites that closely covary over occasions. For example, piping plovers (*Charadrius melodus*) in the Great Plains nest both in riverine situations (e.g. sandbars) and lacustrine shorelines (Elliott-Smith and Haig 2004). When river levels are high, little nesting habitat is available along the rivers and some birds shift from riverine sites to lacustrine sites. Because of that shift, numbers of plovers at different riverine sites tend to vary together, and numbers at many lacustrine sites tend to vary together. It then becomes feasible to model plover counts at a site on an occasion as a function of previous counts at that site and the current count at other sites, all within a relatively homogeneous group.

9. Iterative imputation from models

Iterative imputation makes full use of data in which some values of both the target variable and the covariates are missing. Separate models are needed for each variable that has some missing values. This procedure begins with successively fitting models with each such variable as the dependent variable. One can use either the data set with only complete observations or replace the missing values by, say, means. Next, each model is refit using these substitute values. This procedure can be repeated, that is, missing values now replaced by values predicted from the models, until estimates of missing values converge. Estimates could be poor, however, if a large fraction of the observations have missing values.

Multiple imputation

The single-imputation methods described above substitute a value for a missing value once and then proceed as if that substituted value was the one actually observed. Single imputation will not reflect variation typical of a real sample. Multiple imputation is a way to acknowledge the uncertainty introduced with the imputed value (Rubin 1987). Under this approach, values for the missing observation are imputed several times, and the analysis performed with each imputation. Averages and other summary measures arising from the multiplicity of imputations are calculated. The variability in the results reflects uncertainty caused by a value being unknown.

Evaluating options for treating missing values

Simulation scenarios

To evaluate some methods for handling missing data, we simulated data from a variety of monitoring designs. Our simulations were not intended to provide a comprehensive array of possible scenarios, but to illustrate a variety of situations with features typical of actual monitoring data. For all designs we set the number of occasions at 30, but varied the number of sites, patterns in the expected values of the target variable, predictability from covariates, variability, percentage of missing values, and mechanisms that caused values to be missing. The number of sites was either one or 20. We considered six general patterns in the expected value of the target variable:

  (i) *Stationary.* The expected value of the target variable was constant, that is, the variable was stationary, with no temporal trend.

(ii) *Linear increasing.* The expected value of the target variable was linearly increasing, at a fixed rate per occasion equal to 3% of the value of the target variable measured on the first occasion at each site.

(iii) *Logarithmic increasing.* The expected value of the target variable increased logarithmically, at a rate that decreased with the occasion number. Specifically, the expected value on occasion t ($t > 1$) was the expected value on occasion $t - 1$ multiplied by $[1 + 0.3/(t - 1)]$.

(iv) *Exponential decreasing.* The expected value of the target variable was exponentially decreasing, at a rate per occasion of 3% of the initial value. That is, the expected value on occasion t was 0.97 multiplied by the expected value on occasion $t - 1$.

(v) *Cyclic.* The expected value of the target variable was stationary but cyclical. That is, the expected value on occasion t was $X_1 + 4$ (8 in the high-variance scenarios) multiplied by the square root of the initial value multiplied by $\sin[(t - 1)/3]$.

(vi) *Stationary and linear increasing.* For the 20-site situations we also looked at a mixture with 8 sites following pattern (i) above (stationary) and 12 sites following pattern (ii) (linearly increasing).

For the single-site scenario, we obtained the initial value of the target variable from a Normal distribution with mean $= 100$ and variance $= 400$. For the 20-site scenario, each initial value came independently from a Normal distribution with mean $= 50$ and variance $= 400$.

After generating initial values of the target variable, subsequent counts were generated from Normal distributions with means as described for each pattern. We simulated data with both low and high variability around the expected values of the target variable, as follows. For the stationary pattern, (i) above, subsequent counts were generated with standard deviation either 10% or 30% of the initial count. For the linearly increasing pattern, (ii), and the logarithmically increasing pattern, (iii), standard deviations were either two or five times the square root of the value of the target variable on the initial occasion. For the exponentially decreasing pattern, (iv), standard deviations were either one or three times the square root of the value of the target variable on the initial occasion. For the cyclic pattern, (v) above, standard deviations were either three or five times the square root of the value of the target variable on the initial occasion.

We examined situations in which either 10% or 30% of target variable values were missing. Three mechanisms for missingness were simulated: (a) at random (equal probability for all site visits); (b) with probability proportional to the value of target variable; and (c) with probability inversely proportional to the value of target variable.

The criterion we used to evaluate the methods for handling missing values was simply the average of absolute differences between the known (simulated) values and imputed values. Other criteria we could have used include similarity between the estimated and the actual trend (but trend itself is a slippery concept), or the correlation coefficient between estimated and actual counts. We selected the criterion we used to reflect our interest in having good estimates of the count on each occasion.

Table 13.1 Performance of several imputation methods from simulations under different patterns of variation in the target variable and extent of variability, averaged over percentage of observations missing, and mechanism for missingness: single-site scenario. Table entries are averages of absolute differences between the known (simulated) values and imputed values, scaled to average zero across each row. Largest negative values, in bold, indicate best-performing methods for each combination of pattern and level of variability.

		Imputation method			
Pattern	Variability	Mean	Just previous	Random	Weighted average
Stationary	High	**−0.67**	0.51	0.66	−0.50
Stationary	Low	**−0.05**	−0.01	0.11	−0.06
Linear increasing	High	0.12	**−0.74**	1.32	−0.70
Linear increasing	Low	0.62	**−1.63**	1.61	−0.60
Logarithmic increasing	High	1.19	−0.63	1.08	**−1.64**
Logarithmic increasing	Low	3.39	**−5.99**	3.29	−0.69
Exponential decreasing	High	0.29	−0.51	1.03	**−0.81**
Exponential decreasing	Low	0.53	**−1.85**	1.50	−0.19
Cyclic	High	**−1.36**	−1.27	3.41	−0.80
Cyclic	Low	**−0.24**	−0.12	0.61	−0.25

Imputation methods

We applied methods 3–6 to all simulated data sets, and also methods 2 and 7 to the 20-site scenarios. For the general weighted-average (method 6), we replaced a missing value with a weighted average of values for that site on all previous occasions, with weights declining linearly. For the imputation-from-models method (method 7), we considered four cases: (a) only the site-visit variables (Site and Occasion) available; (b) no predictability from covariates, $R^2 = 0$; (c) medium predictability from covariates, $R^2 = 0.40$; and (d) high predictability from covariates $R^2 = 0.80$.

Simulation results

Results for each method include the following patterns:

- The method of replacing missing counts by zeros was not applicable to single-site scenarios and, as expected, it performed uniformly poorly in all 20-site scenarios (Table 13.2). We do not consider it further.
- Replacing each missing value by the site mean of the variable was the best or at least a competitive strategy for both single-site (Table 13.1) and 20-site scenarios (Table 13.2) with a stationary pattern and for cyclic patterns except for the 20-site scenario with high variance. It also was competitive for the mixed stationary and linearly increasing pattern.
- Replacing each missing value by the value of the variable on the just-previous occasion was effective for monotonic patterns (either increasing or decreasing), especially when

Table 13.2 Performance of several imputation methods from simulations under different patterns of variation in the target variable and extent of variability, averaged over percentage of observations missing, and mechanism for missingness: 20-site scenario. Table entries are averages of absolute differences between the known (simulated) values and imputed values, scaled to average zero across each row. Largest negative values, in bold, indicate best-performing methods for each combination of pattern and level of variability.

		Imputation method						Linear model		
Pattern	Variability	Zero	Mean	Just previous	Random	Weighted average	Site-visit covariates	Low predictability	Medium predictability	High predictability
Stationary	High	5.50	−1.25	−0.42	−0.39	−1.19	−1.07	0.73	−0.20	**−1.72**
Stationary	Low	6.62	**−1.63**	−1.35	−1.36	−1.61	−1.56	1.20	0.44	−0.75
Linear increasing	High	8.01	−1.29	−1.10	−0.39	−1.52	−1.57	0.96	−0.44	**−2.65**
Linear increasing	Low	9.01	−1.40	**−2.61**	−0.90	−1.96	−2.18	1.24	0.32	−1.53
Logarithmic increasing	High	12.96	−1.50	−2.47	−0.47	−2.72	−2.50	1.45	−0.55	**−4.20**
Logarithmic increasing	Low	14.37	−1.17	**−4.65**	−0.92	−3.03	−3.65	1.87	0.23	−3.05
Exponential decreasing	High	2.37	−0.25	−0.06	0.47	−0.40	−0.60	0.89	−0.50	**−1.91**
Exponential decreasing	Low	3.46	−0.44	−1.00	−0.05	−0.80	−0.92	0.81	0.01	**−1.07**
Cyclic	High	1.02	0.28	−1.17	3.02	0.25	1.23	1.67	−1.29	**−5.00**
Cyclic	Low	3.31	−0.46	−0.83	0.98	−0.43	−0.19	1.08	−0.59	**−2.88**
Stationary & Linear increasing	High	6.25	−1.31	−0.76	−0.58	−1.43	−1.10	1.10	−0.11	**−2.03**
Stationary & Linear increasing	Low	7.47	**−1.65**	−2.09	−1.52	−1.90	−1.49	1.67	0.60	−1.09

variability was low. The method was less successful when variability was high, as would be expected, but it still was at least competitive in many single-site scenarios.
- Imputation by replacing a missing observation by a randomly selected value was competitive only for the 20-site scenario with stationary pattern and low variance.
- Replacing a missing value by a weighted average of previous values was most effective with stationary and the mixed stationary and linearly increasing patterns, especially the low-variance situations. This imputation method was best or competitive when the variability was high under all but the exponential decreasing and cyclic patterns with 20 sites.
- Model-based estimators where predictability was high ($R^2 = 0.80$) performed the best for all patterns when variability was high and for exponentially decreasing and cyclic patterns when variability was low. Model-based estimators were competitive when predictability was medium only under cyclic patterns and never when predictability was negligible. Models involving only site-visit covariates never were the best but were reasonably competitive in many situations.

Recommendations

Missing data are a frequent occurrence in many monitoring programs. Imputing values for missing values is, in general, recommended by statisticians (e.g. Harrell 2001), but seems especially important in monitoring programs (unless only an estimate of some "trend" parameter is sought; see *Analysis of Monitoring Data*, above). Also important is identifying the reason why data are missing. In some situations an absence of data arises because a site actually was surveyed, but no organisms were present and the observer neglected to record a zero. In other situations data are missing because no one visited the site, but organisms may or may not have been present.

Assuming that data truly are missing and not unrecorded zeros, values should be imputed if estimates are needed for each sample site on each occasion. The choice of an imputation method depends on what is known about the pattern of values that have occurred, the variability of the pattern, and the extent to which values can be predicted from other information. For a single-site program (Table 13.1), our simulations suggest that replacing a missing value by the value recorded on the previous occasion was effective for monotonic patterns (increasing or decreasing). For analyses conducted after subsequent data had been collected, the average of the values on the previous and the next occasions likely would be even better. For non-monotonic patterns, the mean of all non-missing values, or a weighted average of previous values, performed best.

For multi-site programs (Table 13.2), imputation by model-based values often performed best, as long as missing values could be predicted well by other variables that were recorded. If not, then a weighted average of non-missing values, or just the value on the occasion prior to the missing value, is recommended.

If a sufficient body of data is available, an analyst could explore the data in search of relationships that could be used to predict missing values for that particular set of data (Box 13.2). In addition, it may be feasible to conduct an analysis similar to those we did

> **Box 13.2** Common challenges: imputation choices
>
> The most appropriate imputation strategy in any situation depends on the predictive ability of potential imputation models, as well as on the underlying magnitude of variability and systematic temporal patterns in the target variable. Yet, the choice of an imputation method must be made despite incomplete information about these factors. For a monitoring program with substantial historical data, it would be worthwhile to explore those data to identify any possible strong relationships between the target variable and other variables. Subsequently, some values of the target variable could be treated as missing. Imputed values for missing ones could be generated from the models just mentioned, as well as with any of the other methods discussed in this chapter. These imputed values could then be compared with the actual known values. That analysis could form a strong basis for imputing actual missing values.
>
> The choice of an imputation strategy must also consider the role played by the reason data are missing in deciding how missing values should be imputed. For example, a site might not be visited on a particular occasion because a zero was recorded there on the previous occasion. Should a value missing for that reason be treated the same as a value that is missing for other reasons, such as inaccessibility, poor weather conditions, or simply being overlooked? Identifying reasons for missing data is critical to understanding how they should be treated. Additional problems arise when the reason why data are missing has not been recorded.

on simulated data by replacing certain values by missing, using a variety of methods to impute the missing values, and determining which method yields values closest to the actual ones.

Future research and development

Although the general statistical problem of missing data has been well-studied, issues associated with this problem in monitoring programs have not. Especially problematic are situations in which data are missing because of properties of the data themselves (non-ignorable non-response, as mentioned earlier). Sometimes monitoring sites are not visited on a particular occasion because a count of zero was recorded on the previous occasion. One avenue of study might be the distribution of counts on an occasion following a zero count and how that information could influence a decision to visit such a site.

Another promising area of research involved Bayesian statistics. Unlike the frequentist approach to statistics, the Bayesian paradigm does not distinguish conceptually between variables and parameters; it instead distinguishes between known and unknown quantities. Hence in a Bayesian treatment, missing values can be thought of as unknown quantities, as opposed to the known, non-missing values. Then the missing values can be estimated, along with other unknown quantities of interest (parameters). Another useful aspect of the Bayesian approach is that, instead of providing a point estimate along with confidence intervals for an unknown parameter, as a frequentist analysis might generate, a Bayesian approach yields an entire probability distribution for the unknown quantity. This (posterior) probability distribution reflects both what was known before data were gathered (expressed by the prior distribution) and what the data reveal about the unknown quantity (expressed by the likelihood function). For introductions to Bayesian analysis, see Carlin and Louis (2008), Gelman *et al.* (2004), or Link and Barker (2010).

Summary

Monitoring programs generally are intended to detect changes in some feature, such as population size of an animal in a specified area. Most methods of analyzing results from such programs either estimate some model-based "trend" or seek to estimate some index to population size on each monitoring occasion. We focus on the latter situation, for which it is problematic if data are missing for certain monitoring sites on certain occasions. In this context, imputing reasonable values to replace missing values is highly recommended. We discuss several methods of treating missing data and evaluate some of them with a small simulation study. In general, the recommended method is to replace missing values by values estimated from a model that fit well the set of non-missing data. If such a model cannot be created, selecting the optimal replacement will require an examination of the behavior of the target variable.

Acknowledgments

We are grateful to Paul H. Geissler, John R. Sauer, Terry L. Shaffer, and two anonymous referees for valuable suggestions that improved our presentation.

14 Survey analysis in natural resource monitoring programs with a focus on cumulative distribution functions

Thomas M. Kincaid and Anthony R. Olsen

Introduction

Typical objectives for environmental resource monitoring programs include estimating the current status of the resource, estimating change in status between two time periods, and estimating trends in status over time. For most monitoring programs, status is estimated using multiple indicators (monitoring variables) determined for each sample site that are derived from measurements of biological, chemical, and physical attributes obtained at the site. What is meant by estimating current status? First, the estimate applies to a specified portion of the region included in the monitoring program, typically the entire region. Second, a specific summary measure must be chosen. Common summary measures are estimates of the population mean, the percentage of the population that is less than a particular value (e.g. percentage of water bodies meeting a water quality standard or percentage of a dry forest region that has tree densities $< x$ trees/ha), the population median, the percentage of the population occurring in categories, or the population standard deviation for an indicator.

An additional summary measure is an estimate of the population cumulative distribution for the indicator. The population cumulative distribution is simply the percentage of the population that is less than or equal to each possible value of an indicator. The population cumulative distribution provides complete information about the indicator distribution (Box 14.1). It has the advantage that it can be used for both categorical and continuous data. It is common in statistics to call the cumulative distribution the cumulative distribution function (CDF) and we will adopt that convention in this chapter. Measurements for indicators obtained from a probability survey can be used to estimate these summary measures of status. For complex survey designs that employ stratification or unequal probability of selection, estimation of current status for any of these summary measures must use weights that are derived from the stratification or unequal probability of selection used in the design.

The material presented in this chapter is applicable to natural resources that can be classified as either finite (discrete) or continuous. A finite resource consists of a

Design and Analysis of Long-term Ecological Monitoring Studies, ed. R.A. Gitzen, J.J. Millspaugh, A.B. Cooper, and D.S. Licht. Published by Cambridge University Press. © Cambridge University Press 2012.

Box 14.1 Take-home messages for program managers

Implementing an environmental resource monitoring program based on a spatial survey design not only includes choosing an appropriate survey design to meet a program's monitoring objectives but also includes ensuring that the statistical analyses are appropriate for the survey design. This chapter focuses on the utility of the population cumulative distribution as a summary measure for a monitoring variable. The population cumulative distribution is simply the percentage of the population that is less than or equal to each of all possible values of a variable. As a measure of resource status, the population cumulative distribution provides complete information about the distribution of values of a variable, which can be especially useful in management contexts. A natural resource agency may have a management-threshold goal, e.g. percent of the stream miles with water temperature below 22°C. While the threshold goal is 22°C, how the percentage would change if the threshold was different is valuable to know as well. The cumulative distribution provides this information. Estimating the cumulative distribution based on sample data requires knowing the properties of the survey design, e.g. if it was stratified or used unequal probability of selection. These more complex survey designs require a more complex statistical analysis. For example, with more complex designs it is no longer appropriate to simply tabulate the number of sites with water temperature below 22°C and divide by the number of sites to estimate the proportion of stream miles below 22°C. A weight for each site, based on the survey design, must be used to calculate a weighted mean. Software (a free R package) is available for appropriate estimation of CDFs and comparisons of CDFs (e.g. at two time points or for two different subregions).

---

collection of distinct units, such as lakes or stream reaches, within a study region or spatial domain. An example of a finite resource is the set of lakes in California, USA, with a surface area > 1 ha. Each lake will have a single value for an indicator such as a fish assemblage index of diversity. An example of a continuous resource is the collection of perennial streams in Oregon. Conceptually, the value of an indicator varies continuously throughout the stream linear network, e.g. pH of the water or density of fish per 100 m. To define the continuous resource population in this example, streams are modeled as having no width. Another example of a continuous resource is the land area within a national forest. An indicator of interest may be an index of vegetation diversity which would vary continuously throughout the national forest. Another aquatic example of a continuous resource is the collection of estuaries for the State of Oregon. In this case it is assumed that the indicators (e.g. dissolved oxygen concentration) vary not only across the estuaries but also within an estuary. The latter two examples are examples of an areal continuous resource. In geographic information systems (GIS) terminology, a finite resource is a point GIS layer, a linear network is a linear layer, and an areal resource is a polygon layer (see also Chapter 6). For all types of environmental resources, we will use the term "extent" to reference the size of the resource, i.e. the number of discrete units for a finite resource, total length for a linear resource, and total area for an areal resource.

This chapter is divided into five sections. The first section introduces population summary measures and procedures for estimation with a focus on the CDF. The second section focuses on using the sample CDF to estimate population percentiles. The third section introduces deconvolution as a way to remove the effect that measurement error associated with indicators has on CDF estimates. The fourth section introduces hypothesis testing for comparing two CDFs. The fifth section briefly discusses unresolved issues in need of further research or development.

Population summary measures and their estimation

The Horvitz–Thompson theorem (Horvitz and Thompson 1952) and its generalization is the statistical basis for estimation of population summary measures. For a function z defined on a finite resource U, the theorem provides a procedure for estimating the population total for z,

$$T_Z = \sum_{x \in U} z(x). \tag{14.1}$$

Note that the term "population", as used in this chapter, references the set of values of z (either the indicator value directly or a function of the indicator) defined on the resource U (the target universe). An example may make this more concrete. Let the universe be the set of all lakes in Oregon and x be the indicator lake pH, and assume the function z identifies lakes with pH less than 5 ($z = 1$ if pH < 5 and $z = 0$ otherwise). The population total is then simply the number of lakes with pH < 5. If the total number of lakes in the population is known, then the percentage of lakes with pH < 5 is simply the population total divided by the total number of lakes. If the function z is simply equal to x, then an estimate for the mean pH is the population total divided by the total number of lakes.

For a sample S selected from U such that, for every element $x \in U$, there is a positive probability, $\pi(x)$, that the element x is included in S, an estimate of the total is

$$\hat{T}_Z = \sum_{x \in S} \frac{z(x)}{\pi(x)}. \tag{14.2}$$

The quantity $\pi(x)$ is the inclusion probability for an element, and a sample S that is selected with a positive inclusion probability for every element is called a probability sample (see also Chapters 5 and 6). The inverse of the inclusion probability is called the survey design weight. It can be seen that $\hat{T}_Z$ is a weighted sum of the sample values of the function z, where weights are the survey design weights. Horvitz and Thompson (1952) also provided a formula for the variance of $\hat{T}_Z$. The variance formula requires the pairwise inclusion probability, $\pi(x_i, x_j)$, which is the probability that both elements x_i and x_j are included in the sample S. (Note that Chapter 5 provides a function for estimating these pairwise inclusion probabilities.) Under the condition that $\pi(x_i, x_j)$ is positive for every pair of elements x_i and x_j in U, Horvitz and Thompson (1952) provided a formula for an unbiased estimator of the variance of $\hat{T}_Z$.

Cordy (1993) developed an extension of the Horvitz–Thompson theorem to sampling from a continuous resource. For this case, the population total is given by

$$T_Z = \int_U z(x)dx, \qquad (14.3)$$

where the summation used previously is replaced by integration. Cordy (1993) defined an inclusion density function for a continuous resource that is analogous to the inclusion probability for a finite resource. He also established that the estimator $\hat{T}_Z$ is applicable when sampling from a continuous resource with $\pi(x)$ referring to the inclusion density instead of the inclusion probability.

For a target universe corresponding to an environmental resource U, the cumulative distribution function, CDF, for a random variable z defined on U is given by $F_Z(t) = \Pr(z \le t)$ for all t. For a probability sample S, an estimate of the CDF is given by

$$\hat{F}_Z(t) = \frac{\sum\limits_{x \in S} \frac{I(z(x) \le t)}{\pi(x)}}{\sum\limits_{x \in S} \frac{1}{\pi(x)}}, \qquad (14.4)$$

where the indicator function, I, is defined as

$$I(z(x) \le t) = \begin{cases} 1, & z(x) \le t \\ 0, & \text{otherwise} \end{cases}.$$

The expressions in both the numerator and denominator of $\hat{F}_Z(t)$, Equation (14.4), are Horvitz–Thompson estimators. The numerator expression estimates the extent (i.e. size) of the spatial domain for which values of the function z are less than or equal to the value t. The denominator expression, which is the sum of the survey design weights for the sample, estimates the entire resource extent (i.e. size). If an estimate of the mean of an indicator is desired, then the estimate is given by

$$\hat{\bar{x}} = \frac{\sum\limits_{x \in S} \frac{x}{\pi(x)}}{\sum\limits_{x \in S} \frac{1}{\pi(x)}} \qquad (14.5)$$

where the numerator is the weighted sum of the indicator values and the denominator is the sum of the weights. Diaz-Ramos et al. (1996) give detailed examples on the estimation of CDFs and other summary measures of status.

As an introduction to CDF estimation, consider the CDFs displayed in Fig. 14.1. The lakes used to create the figure were extracted from the sample frame that was employed for the National Lakes Assessment (USEPA 2009b) in the USA. Specifically, the 818 lakes that were included in the National Eutrophication Survey, NES (USEPA 1975), were retained. This set of lakes will be used as the resource of interest. An index of biotic integrity (IBI) for lakes was used as the response variable z for constructing the CDFs. For illustration purposes, use of data with known properties is helpful; therefore, we simulated IBI data for all lakes from a Normal distribution with mean 50 and standard

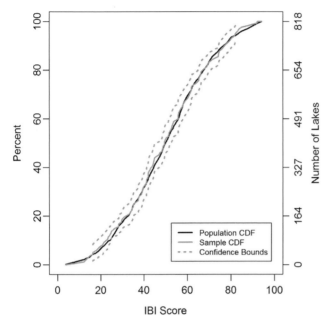

Figure 14.1 Simulated index of biotic integrity (IBI) CDFs for lakes in the US National Eutrophication Survey that were included in the survey frame for the National Lakes Assessment. Shown are the known (simulated) CDF for all of the lakes in the survey (the population CDF) and the CDF and its associated confidence bounds estimated from a sample of lakes (identified as the sample CDF).

deviation 20. The IBI CDF for all of the NES lakes (i.e. the population CDF) is therefore known in this case. A sample of lakes was selected from the population, and the IBI CDF and associated confidence bounds were estimated (Fig. 14.1). Because they tend not to be very informative, confidence bounds for the tails of the distribution are not displayed. Note that the confidence bounds enclose the population CDF.

Estimation of the CDF requires knowledge of the survey design used to select the sample of lakes and matching the survey analysis to that design, which can be challenging in many monitoring situations (Box 14.2). For our example, we used an unequal probability survey design where the unequal probability of selection was based on four lake area size classes with 25 lakes selected in each for a total sample size of 100 lakes. A spatially balanced sample was selected using the generalized random tessellation stratified (GRTS) design (Stevens and Olsen 2004; see Chapter 6). We implemented the design using the *grts* function in the "spsurvey" package for R (R Development Core Team 2011) demonstrated in Chapter 6, although other options are available. The "spsurvey" package is available on the R website (http://www.r-project.org/) and the US Environmental Protection Agency's Aquatic Resource Monitoring (ARM) website (http://www.epa.gov/nheerl/arm/).

The IBI values for the 100 selected lakes were then used to estimate the population CDF using the *cdf.est()* function in the "spsurvey" package:

Box 14.2 Common challenges: design and analysis issues

Estimating summary measures, such as the population CDF, and their standard errors or confidence intervals is computationally easily accomplished with available software. Small sample sizes, especially associated with a complex design, can result in large standard errors and consequently wide confidence intervals. If the weights associated with the design are positively correlated with the values of the indicator, i.e. sites with large weights have large indicator values, then the standard errors will be small. For example, this would occur if stratification was used to separate the population into homogeneous strata with the goal of reducing the variance. Estimating the population CDF can also be impacted by one or two observations that are extreme and also have large weights. A general rule of thumb is to have the weights vary by less than a factor of 100, which can be difficult to accomplish in some cases. Increasing the sample size and reducing the range of the weights will result in lower standard errors and shorter confidence intervals.

Monitoring programs may implement a monitoring design where it takes multiple years (5–10) before accumulating enough sites for any meaningful CDF or other summary measure estimate can be made. Accumulating enough sites can be accomplished by using any number of panel designs (see Chapters 7, 8, and 10). Estimating the CDF requires specifying a temporal period, e.g. 2005–2009, that defines the population in time. Having data from multiple years adds another source of variation to all the sources of variation included in measurement error, potentially increasing bias in CDF estimates. Some panel designs include revisits to the same sites within the temporal period. When this is the case, a question arises on how the data from these multiple visits is used in the estimation process. One approach is to average the values and then use the average in the estimation of the CDF. This results in the sites with averages having smaller measurement errors than sites that have no revisits. Consequently, the CDF estimate is impacted since the convolution of the measurement error with the distribution of interest differs for values that are averaged and those that are not. An alternate solution is to use only one of the revisit values and ignore the remaining revisits. More complex analyses based on linear models could be used to include all the data without averaging.

```
CDF_Sample <- cdf.est(z=NESsample$IBI,
       wgt=NESsample$wgt,
       x=NESsample$xcoord, y=NESsample$ycoord)
```

where "z" is the IBI data, "wgt" is the survey design weights for the lakes, and "x" and "y" are the lake location coordinates. The *cdf.est()* function implements the Horvitz–Thompson estimator $\hat{F}_Z(t)$ to calculate the CDF estimate. This estimator produces values that range between 0 and 1. Typically, the estimate is multiplied by 100 to express the estimate as a percentage. Since *cdf.est* by default uses the local mean variance estimator (Stevens and Olsen 2003; see also Chapters 6 and 7) to calculate standard error (se) of the CDF estimate, it is necessary to provide the lake location information. An option also exists to use the standard Horvitz–Thompson variance estimator. Confidence bounds, $\hat{F}_Z(t) \pm z^* \cdot se$, are calculated for each unique value in the sample and connected pointwise to display the confidence bounds (Fig. 14.1).

The population CDF can also be estimated in terms of the resource extent, i.e. in this example in terms of the number of lakes in the population with indicator values $\leq t$. For a linear or areal resource the estimate would be in terms of length or area instead of number of finite units. Two options exist when estimating the population CDF in terms of the resource extent. First, if the resource extent is known, then the CDF estimate is calculated using a ratio estimator: the product of the known resource extent and the ratio of Horvitz–Thompson estimators [Equation (14.4)] used to calculate the CDF estimate for proportion of a resource. This first CDF estimate in terms of resource extent will range between zero and the resource extent. If the resource extent is not known, the CDF estimate in terms of the resource extent is calculated using the Horvitz–Thompson estimator in the numerator of Equation (14.4). This second type of CDF estimate for total of a resource will range between zero and the sum of the survey design weights, which is an estimate of the resource extent. The y-axis displayed on the right side of Fig. 14.1 expresses the CDF as the resource extent. Note that the maximum value is 818 lakes, the known resource extent.

Percentile estimation

The $(100p)$th percentile for a random variable z defined on an environmental resource U is given by the smallest value ξ_p such that $F_Z(\xi_p) \geq p$. Note that ξ_p will belong to the range of values allowed for the response variable z. Percentile estimates are calculated by inverting the estimated sample CDF, $\hat{F}_Z(t)$. Specifically, the percentile estimate, $\hat{\xi}_p$, is provided by the value t such that $\hat{F}_Z(t) = p$. Note that t must be one of the response values at which the CDF was estimated. When no value t is available that satisfies the equality, linear interpolation is used to calculate $\hat{\xi}_p$. Confidence bounds for $\hat{\xi}_p$ are calculated by inverting the confidence bounds for $\hat{F}_Z(t)$. Note that the upper bound for $\hat{F}_Z(t)$ provides the lower bound for $\hat{\xi}_p$ and conversely regarding the lower bound of $\hat{F}_Z(t)$. As a result of the procedure used to calculate confidence bounds, standard error estimates for $\hat{\xi}_p$ are not calculated.

Measurement error and CDF deconvolution

We want the estimate for the population CDF to be an unbiased estimate of the true population CDF. The true population CDF reflects the frequency distribution of an indicator of interest for an environmental resource when the indicator is measured without error. This frequency distribution reflects the variation of the indicator across the resource and is the distribution that is of interest. For convenience, we refer to this variation as site-to-site variation. The survey design, measurement process, and components of natural variation introduce additional variation that can lead to biased estimation of the CDF. Overton (1989) classified the additional variance as measurement errors. Kincaid *et al.* (2004) used the term "extraneous variance" to reference these "measurement errors". We adopt the term "measurement error" where it includes all the sources of variation involved in obtaining a value for an indicator at a single sample site, comparable to the use of this term in previous chapters. Although our interest is in the distribution of an indicator "*x*", we actually observe "*y*" where $y = x + \varepsilon$ and ε is the measurement error. The observed indicator values convolute the measurement error variation with the true population variation resulting in the observed distribution having greater variation than the underlying true population. This produces a bias in the CDF estimate, which means that the CDF estimate will have a greater proportion of the estimate in the tails of the CDF compared to an unbiased CDF estimate.

Urquhart *et al.* (1998) and Larsen *et al.* (2001) described a variance model that addresses major sources of variation affecting status estimation and trend detection in monitoring surveys (see Chapter 7). This model identified four variance components: (i) site variance, (ii) year variance, (iii) site-by-year interaction variance, and (iv) residual variance. Site variance is site-to-site (spatial) variation in the population. Year variance is the year-to-year variation affecting all sites equally. Site-by-year interaction variance is year-to-year variation affecting individual sites that is not accounted for by year variance. Residual variance is the remaining variation not accounted for by the other variance components. See Larsen *et al.* (1995) for discussion regarding the components of residual variance, i.e. measurement variance.

The relative magnitude of measurement variance to the true population variance determines the amount of bias in the CDF estimate. The relationship between the amount of measurement variance and bias in the CDF was considered by Overton (1989). Bias resulting from measurement variance is illustrated in Fig. 14.2 using a simulated species richness variable. The upper portion of Fig. 14.2 displays CDF estimates, and the lower portion displays corresponding probability density estimates. Densities are included since it is easier to see that the effect of measurement variance is to increase the variation in the observed distribution. Density estimates were calculated using an R function that implements the average shifted histogram (ASH) algorithm (Scott 1985). Data for the species richness variable were simulated using the Normal distribution with mean 20 and standard deviation 5. The other variables displayed in the figure were created by adding 25%, 50%, and 100% measurement variance to the original variable. As an example, to create the variable with 100% measurement variance, data simulated using the Normal distribution with mean 0 and standard deviation 5 were added to the simulated species richness variable.

As can be observed, measurement error causes the CDF and density estimates to occur across a greater range of species richness values in comparison to the CDF estimate with no measurement error. In addition, bias in the CDF and density estimates increases as measurement variance increases. Note also that bias in the CDF is not constant but is greatest approximately midway between the median value (50th percentile) and the tails of the distribution. For species richness equal to 15, the CDF estimate for the original variable is 11.1%, which increases to 14.2%, 15.9%, and 20.4% for 25%, 50%, and 100% measurement variance, respectively.

Convolution is a term that refers to a variable that is a mixture of two or more distributions. In the context of this chapter, a variable that contains measurement error

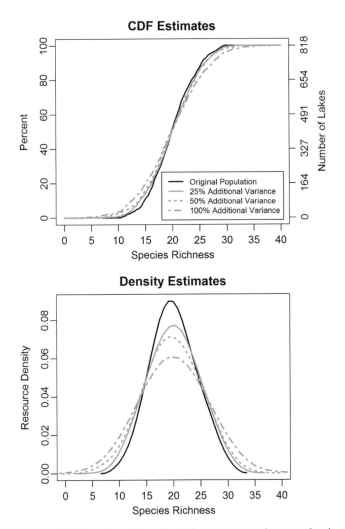

Figure 14.2 Illustration of the effect of extraneous variance on the simulated population CDF and population probability density function for IBI score for lakes in the National Eutrophication Survey that were included in the survey frame for the National Lakes Assessment.

is the convolution of the distribution of the variable of interest and a distribution that reflects the measurement error. Deconvolution is the name for a process that removes from the CDF estimate the bias caused by measurement error. Although deconvolution will remove bias from the CDF estimate, the cost associated with deconvolution is increased width of confidence bounds for the deconvoluted CDF estimate. Note that the removal of bias and the increased confidence interval width result in confidence intervals that better reflect their stated confidence level. Kincaid *et al.* (2004) discussed the cost associated with deconvolution and assessed the cost using two measures: (i) increase in the width of CDF confidence bounds, and (ii) increase in the sample size required to achieve confidence bounds equivalent to presence of no measurement error.

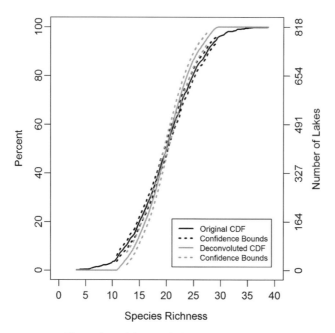

Figure 14.3 Illustration of deconvolution using simulated species richness CDFs for lakes in the National Eutrophication Survey that were included in the survey frame for the National Lakes Assessment. Shown are the CDF for species richness with extraneous variance (identified as the original CDF) and its associated confidence bounds, and the deconvoluted CDF and its associated confidence bounds (identified as the deconvoluted CDF).

Estimated CDFs using deconvolution are illustrated in Fig. 14.3. The simulated species richness variable to which 100% measurement variance was added will be used as the response variable. The *cdf.est()* function in "spsurvey" was used to calculate the CDF for the original variable with added variance. The *cdf.decon()* function in "spsurvey" was then used to calculate the deconvoluted CDF. The deconvolution process implemented in the *cdf.decon* procedure is based on Stefanski and Bay (1996). As discussed previously, tails of the deconvoluted CDF are shifted towards the distribution center. Increased width of confidence bounds for the deconvoluted CDF in comparison to the original CDF can be observed in Fig. 14.3.

Hypothesis testing for CDFs

Although extensive methodology exists for inference about CDFs in the context of simple random sampling, there is relatively little literature that addresses inference about the CDF for sampling designs involving stratification and unequal probability of selection or more complex designs. For simple random sampling the Kolmogorov–Smirnov (KS) and Cramér–von Mises (CvM) statistics typically are used for inference about CDFs. Use of the KS and CvM statistics for testing CDFs is discussed in numerous books, e.g. Conover (1980).

As a means for comparing CDFs from complex sample surveys, Krieger and Pfeffermann (1997) use an approach based on grouping the data into a fixed number of intervals and using the standard Pearson chi-squared test for categorical data analysis (χ^2) to conduct inference. The χ^2 test assumes multinomial sampling, i.e. simple random sampling. When applied to a complex survey design, χ^2 does not perform adequately. For complex survey designs, the Wald (1943) statistic provides a mechanism for incorporating features of the design into categorical data analysis. Rao and Scott (1981) and Scott and Rao (1981) examined the impact of stratification and clustering on performance of χ^2 for testing goodness of fit and independence in two-way contingency tables and developed corrections to the Pearson statistic. Rao and Scott (1981) developed first-order and second-order corrections to χ^2. Rao and Thomas (1988, 1989) present detailed descriptions of the Wald and Rao–Scott test statistics. The Wald and Rao–Scott test statistics are referenced to the chi-squared distribution. Thomas *et al.* (1996) developed modifications to the Wald and Rao–Scott tests that are referenced to the F distribution.

Kincaid (2000) conducted simulations on the performance of the Wald and Rao–Scott test statistics to hypothesis testing for statistical difference between CDFs. To utilize the tests, the CDFs were classified into a set of non-overlapping classes, and the statistics were calculated. Simulation results presented in Kincaid (2000) indicated a strong tendency for statistical power to increase as the number of classes was decreased. In subsequent simulations (T. Kincaid, unpublished data) the F distribution versions of the statistics show greater power in comparison to versions using the chi-squared distribution. Overall, the conclusion is that the F-based version of the Wald statistic is the best choice to use for inference regarding CDFs.

The *cdf.test*() function in the "spsurvey" package for R calculates the chi-squared distribution and F distribution versions of the Wald and Rao–Scott test statistics. The *cdf.test*() function was used to test for differences among IBI score CDFs for subpopulations defined by the lake size classes used for the survey design described previously. In order to provide an illustration of the test, random noise from the Uniform distribution was added to the IBI scores for the smallest lake size class. The noise served to shift the distribution for the smallest lakes toward larger values. As expected (Table 14.1), the smallest lake size class was significantly different from the other lake size classes for the standard CDFs. None of the other lake size classes was significantly different.

Future research and development

Although procedures for estimating CDFs are well developed, some aspects of estimating confidence intervals and testing of hypotheses need additional work. Estimating confidence limits at either end of the cumulative distribution, e.g. say less than 5% or greater than 95%, by relying on $\hat{F}_Z(t) \pm z^* \cdot se$ is known to be problematic in giving confidence intervals with the expected confidence level. A general solution to the problem is not currently available.

Although the procedures for testing for differences between CDFs that were discussed provide a mechanism for accommodating complex survey designs, discretizing the CDFs

Table 14.1 Results of example hypothesis test for differences between simulated IBI CDFs for subpopulations defined by lake size classes for lakes in the National Eutrophication Survey that were included in the survey frame for the National Lakes Assessment.

Lake size class 1 (ha)	Lake size class 2 (ha)	Wald statistic	P-value
0–200	200–1000	4.570	0.002
0–200	1000–5000	3.120	0.027
0–200	> 5000	3.463	0.018
200–1000	1000–5000	0.517	0.566
200–1000	> 5000	0.871	0.401
1000–5000	> 5000	0.005	0.995

by dividing them into a set of classes is less than ideal. Developing test statistics that treat the CDF as a whole is a future research objective.

Summary

This chapter focuses on the utility of the population cumulative distribution as a summary measure for variables measured by a long-term monitoring program. The population cumulative distribution provides complete information about the variable distribution. Estimating the cumulative distribution based on sample data requires knowing the properties of the survey design, e.g. if it was stratified or used unequal probability of selection. These more complex survey designs require a more complex statistical analysis.

In this chapter we define the population cumulative distribution, provide estimators of the population cumulative distribution and associated standard errors and confidence intervals, and introduce tests to determine if the cumulative distributions from two time periods or two subregions are different. While calculating the estimates and tests is more complex, open source software is available to complete the calculations. The chapter also discusses the impact (potential bias) of measurement error on the estimated cumulative distribution and presents a solution (deconvolution) to remove the bias.

Acknowledgments

The information in this document has been funded by the US Environmental Protection Agency (USEPA). This manuscript has been subjected to review by the National Health and Environmental Effects Research Laboratory's Western Ecology Division and approved for publication. Approval does not signify that the contents reflect the views of the Agency, nor does mention of trade names or commercial products constitute endorsement or recommendation for use.

15 Structural equation modeling and the analysis of long-term monitoring data

James B. Grace, Jon E. Keeley, Darren J. Johnson, and Kenneth A. Bollen

Introduction

The analysis of long-term monitoring data is increasingly important; not only for the discovery and documentation of changes in environmental systems, but also as an enterprise whose fruits validate the allocation of effort and scarce funds to monitoring. In simple terms, we may distinguish between the detection of change in some ecosystem attribute versus the investigation of causes and consequences associated with that change. The statistical framework known as structural equation modeling (SEM) can contribute to both detection of changes and the search for causes. This chapter summarizes some of the capabilities of SEM and shows a few ways it can be used to model temporal change. Because of its ability to test hypotheses about whether rates of change are zero or nonzero, it can be used for change detection with repeated-measures data. As more of the capabilities of SEM are presented, its capacity for evaluating causal networks is highlighted. Here is where its potential for making a unique contribution to the analysis of long-term monitoring data is revealed. Thus, one's primary motivation for using SEM with monitoring data will be to investigate hypotheses about what factors may be driving change (Box 15.1).

In this chapter it will be necessary to first introduce notation to describe the elements of structural equation models (SE models) so as to permit an unambiguous presentation of their various forms. The first part of the chapter works through the fundamental features of models of increasing complexity, while the second part of the chapter illustrates several of these possibilities using a real example.

The elements of structural equation models

Structural equation models can often be most easily introduced by contrasting their form with the familiar univariate model

$$y = \beta \mathbf{x} + \varepsilon \tag{15.1}$$

where y is a single response variable, $\mathbf{x}$ is a vector of predictor variables, β is a vector of prediction coefficients, and ε is an error variable that represents the effects of all

Design and Analysis of Long-term Ecological Monitoring Studies, ed. R.A. Gitzen, J.J. Millspaugh, A.B. Cooper, and D.S. Licht. Published by Cambridge University Press. © Cambridge University Press 2012.

Box 15.1 Take-home messages for program managers

Monitoring programs seek to detect change if it is occurring, to infer the causes behind observed changes, and to evaluate the implications of these changes. The methodology presented in this chapter, structural equation modeling (SEM), has characteristics that make it useful for all three of these goals in the face of common difficulties such as short or incomplete monitoring records and the challenge of discerning the relative importance of multiple factors operating simultaneously. SEM is a flexible methodology for building and testing models that seek to describe processes and relationships that regulate the complex behavior of an ecological system. SEM applications range from simple ones readily implemented with off-the-shelf software to more complex ones that require special programming and experience. For example, using a relatively simple latent trajectory model to consider how multiple sites change over time, it is possible to estimate linear trends with as few as three time periods and nonlinear trends with four or more. It is also possible to consider the divergent trends that may occur for different sites and the network of contributing factors that cause the divergence. For example, in this chapter we demonstrated the use of SEM to examine post-fire temporal trajectories in plant species richness for 88 study plots in a shrub ecosystem, incorporating a network of direct and indirect relationships among plot-level richness, cover, soil characteristic, and fire severity as well as area-wide yearly precipitation. Groups of sites can be formally compared (e.g. those exposed to some factor such as a management treatment and those not exposed) and strategies for handling missing data are available. The SEM process should be guided by careful thought to identify specific questions about potentially important processes and relationships in the system being monitored. The investment in SEM can lead to valuable insights into how and why changes may be occurring and how various factors are affecting the system – information that is essential for guiding how managers respond to these changes.

---

influences on y other than those in x (see Chapter 11 for further review). This model can be generalized to the multivariate response case (e.g. multivariate regression or MANOVA) by allowing y and ε to be vectors $\mathbf{y}$ and $\mathbf{\varepsilon}$ of variables and error terms.

For SEMs, in contrast, the equation for relations among observed variables is

$$\mathbf{y} = \mathbf{B}\mathbf{y} + \mathbf{\Gamma}\mathbf{x} + \zeta, \tag{15.2}$$

where $\mathbf{y}$ is a vector of endogenous variables, $\mathbf{x}$ a vector of exogenous variables, $\mathbf{B}$ and $\mathbf{\Gamma}$ are coefficient matrices, and ζ is a vector of error variables with non-zero elements for each non-zero element in the $\mathbf{y}$ vector on the left side of the equation. The term "endogenous" refers to variables in a network of relations that have arrows pointing to them (i.e. they are response variables somewhere in the model – e.g. variables b, c, and d in Fig. 15.1a–c are all endogenous). The term "exogenous" refers to variables in a network that have no arrows pointing to them (i.e. they only serve as predictors and are not explained by other variables in the model – e.g. variable "a" in Fig. 15.1 is

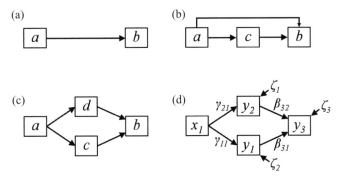

Figure 15.1 Some graphical models discussed in the text: (a) simple regression, (b) path model with direct and indirect pathways explaining the covariance between a and b, (c) path model where total relationship between a and b is explained by the mediating factors c and d, and (d) statistical notation for model in C.

exogenous). The model in Equation (15.2) is sometimes referred to as the econometric model. It is worth noting that SEMs need not be either linear or Gaussian, although often such assumptions are employed.

The inclusion of the term **By** in Equation (15.2) has rather profound implications from the standpoint of the statistical models and scientific hypotheses that can be specified and evaluated. In the univariate model [Equation (15.1)] or its multivariate extensions, the reliance on βx to convey the processes influencing y permits only a consideration of direct effects and strongly constrains possible causal interpretations. The combination of terms **By** + **Γx** in the SEM permits the representation of a network of direct and indirect effects. This equational framework for representing (potentially complex) networks of causal relations creates a need for a graphical means of representation, which is why graphical modeling and path modeling were invented simultaneously, though the term "graphical modeling" is a recent development.

Wright (1921) first illustrated the decomposition of the net relationship between a cause and an effect into multiple causal pathways including indirect pathways mediated by intervening variables. Wright immediately recognized the need for a graphical representation to accompany the analysis of networks of relationships and developed a system for graphical modeling. As Pearl (1998) has argued, the use of a graphical modeling system to address causal networks comes not only from the need to represent the implications of systems of equations, but also from the fact that the "=" sign in equations only describes mathematical equivalence and is potentially ambiguous with regard to the flow of causation, while the elements of graphical mathematics, such as →, permit an explicit statement of causal direction (see also Shipley 2000).

Figure 15.1 provides the graphical representation of some simple models. In Fig. 15.1a, we see a directional relationship between two variables, a and b. The single-headed arrow from a to b implies a flow of causation, although some will prefer to think of this as conditional dependence. With an error term for b implied but not shown, the model in Fig. 15.1a corresponds with a simple regression. From a Wrightian perspective, we might ask the scientific question, "To what degree can the dependence of b on a be explained by the effect of a on an intermediary cause c?" This question can be represented in graphical modeling terms as shown in Fig. 15.1b. Here we have specified the possibility that part of the net effect of a on b is propagated through c, while there also exists an independent mechanism whereby a influences b independent of c (the direct path). We may proceed further in our investigation of the causes behind the dependence of b on a and ask whether the remaining direct effect in Fig. 15.1b can be explained by a second intermediary cause d. Such a result could be represented by Fig. 15.1c, which is a model in which variables a and b are considered to be "conditionally independent given c and d" ($a \perp b | c, d$). Mechanistically, we would infer that the effects of a on b are mediated by effects that a has on c and d. Representing variables a–d in terms of x's and y's (Fig. 15.1d) permits us to present our example causal network as a series of equations in matrix form corresponding to our general representation in Equation (15.2):

$$\begin{bmatrix} y_1 \\ y_2 \\ y_3 \end{bmatrix} = \begin{bmatrix} 0 & 0 & 0 \\ 0 & 0 & 0 \\ \beta_{31} & \beta_{32} & 0 \end{bmatrix} \begin{bmatrix} y_1 \\ y_2 \\ y_3 \end{bmatrix} + \begin{bmatrix} \gamma_{11} \\ \gamma_{21} \\ 0 \end{bmatrix} [x_1] + \begin{bmatrix} \zeta_1 \\ \zeta_2 \\ \zeta_3 \end{bmatrix}. \tag{15.3}$$

Note that additional matrices can allow us to model the variances and covariances among exogenous variables as well as covariances among our ζ_i variables.

The enterprise of SEM, because it is concerned with underlying causes, has long utilized latent variables in models. Latent variables can be thought of as hypothesized influences, ones for which we do not have direct measurements but whose influences on measured parts of the system are conspicuous. The classic early example of a latent variable was human intelligence, which was, and remains, an attribute whose properties can only be assessed indirectly, through correlated responses. There are now numerous examples of latent variable ecological models (Grace 2006) involving concepts such as animal body size and life history. In this chapter we also use latent variables to represent random slopes and intercepts for the trajectories over time of individual observation units in sample populations. As we outline in the next paragraph, latent variables can be represented in structural equation models by generalizing the equational framework. Including latent variables in models creates a number of requirements for the estimation process, though ones that can generally be satisfied using procedures developed for this purpose. A further discussion of latent variables and how they are used in SEM can be found in Bollen (2002). Chapters 19 and 20 in this volume also incorporate latent variables as part of their hierarchical modeling approaches.

The latent-variable formulation of SE models is represented by an elaboration of Equation (15.2) into three equations. The relationships among latent variables are specified as

$$\eta = B\eta + \Gamma\xi + \zeta, \tag{15.4}$$

where η is a vector of latent responses, ξ is a vector of latent predictors, B and Γ are now matrices containing latent variable covariances (as well as their variances) and ζ is a vector of errors for the endogenous latent variables. Analogous to the observed-variable

SE model, correlations among the errors of latent endogenous and exogenous variables are specified by matrices (respectively labeled Ψ and Φ). The two remaining equations

$$\mathbf{x} = \Lambda_x \xi + \delta \tag{15.5}$$

and

$$\mathbf{y} = \Lambda_y \eta + \varepsilon \tag{15.6}$$

allow the specification of relationships between latent and observed variables, where the Λs are loadings connecting latent with observed variables. We will see in the next section some of the ways that latent variables can be used in SEMs when we use them to help us model longitudinal data.

Before moving to a discussion of how SEMs can be used to represent the processes associated with temporal dynamics, it is important we mention that the SEM enterprise places the specification and estimation of SEMs into a workflow process that is designed to allow us to learn about the causal processes operating in complex systems. The broad enterprise of SEM includes (i) theory formalization, (ii) development of general hypotheses, (iii) specification of theory-based statistical models, (iv) estimation of parameters, (v) evaluation of models, and (vi) synthesis of results. Descriptions of this broader enterprise as it relates to ecological systems can be found in Grace (2006) and Grace *et al.* (2010), but are also briefly presented later in this chapter.

Modeling temporal dynamics using structural equation models

Modern SEM owes much of its development to its adoption and use in the human sciences including psychology, sociology, economics, education, and marketing. While the environmental sciences are now initiating and implementing many large monitoring programs, the human sciences have been collecting such data for a great deal longer and with a much greater commitment of resources. As a result, there is likely much we can learn about analyzing and modeling long-term environmental data from the SEM tradition in the human sciences.

SEMs have been applied to long-term ecological data in a few cases. A simple approach sometimes used has been to summarize net system responses. For example, Clark *et al.* (2007) examined factors controlling net diversity responses to experimental fertilization treatments in herbaceous ecosystems. Larson and Grace (2004) examined the factors affecting net proportional changes in exotic plant populations subject to biocontrol additions. These studies model temporal change by looking at spatial variations in net changes over time; this simple approach can be useful when interest in temporal change can be reduced to some summary measures (final time minus initial time). As this chapter illustrates, SEM can also be used to represent much more complex temporal models.

(a)

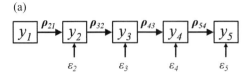

(b)

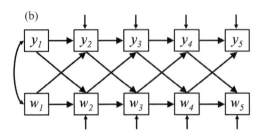

Figure 15.2 Two examples of autoregressive models. In both cases, the subscripts refer to time intervals (i.e. y_3 represents the measurement of variable y at time 3): (a) a simple autoregressive chain, and (b) an autoregressive cross-lagged model involving a response y and a covariate w.

The autoregressive model (ARM)

For the analysis of temporal dynamics using structural equations, two of the main approaches that have been used are (i) autoregressive models (ARMs), and (ii) latent trajectory models (also known as latent growth curve models). The first of these, the autoregressive model, utilizes a time-step equation of the form

$$y_{it} = \alpha_t + \rho_{t,t-1}y_{i,t-1} + \varepsilon_{it}, \tag{15.7}$$

where y_{it} and $y_{i,t-1}$ represent responses by the ith individual (e.g. an organism or sample site) at time t and $t - 1$ respectively, α_t is the time-specific intercept, $\rho_{t,t-1}$ is the autoregressive coefficient, and ε_{it} is the error for individual i at time t (assumed to have mean of zero, to be free of autocorrelation, and to be uncorrelated with y_{it}). For the autoregressive model, the values of y observed at any time are predictable from its values the previous time (or some earlier time periods) plus coefficients predicting changes between times based on either intrinsic or extrinsic factors. Such a model approach can be used to estimate, for example, the fidelity of a population of animals to a particular site over time. Larson and Grace (2004) used an autoregressive SE model to estimate the fidelity from one year to the next of biocontrol insects (*Aphthona* spp., flea beetles) living on their exotic host, leafy spurge (*Euphorbia esula* L.).

An equation that specifies how y_{it} depends on covariates is needed if a general model for temporal dynamics is to be developed from the autoregressive equation. A simple hypothesis that can be evaluated is that $\rho_{t,t-1}$ is constant for all time periods. A more elaborate model would be that $\rho_{t,t-1}$ is a function that drives a general shape to the time trajectory, such as a geometric or asymptotic trend. A graphical representation of a simple autoregressive model is shown in Fig. 15.2a. For such models we typically assume errors are uncorrelated; however, autocorrelation among errors is permitted when appropriate. We will illustrate the incorporation of an autoregressive relationship into an SEM later in the chapter.

Autoregressive models can be elaborated to represent interacting trajectories of responses. One such elaboration is the autoregressive cross-lagged model (ARCL) shown in Fig. 15.2b. Such models and their variants can allow us to partition several kinds of interactive processes between ecological entities, such as fidelity from time to time, competitive or facilitative interactions, or interactions between organisms and their resources. Again, in such models various kinds of correlated error structures are permitted when appropriate. The ecological interaction between the plant leafy spurge and its biocontrol insects, mentioned above, is one example where a cross-lagged autoregressive model was used to represent interactions playing out over time. In this case, it was possible to assess whether effects on plants from the insects caused lag responses from the previous year or only immediate impacts from the current year.

The latent trajectory model (LTM)

A second approach to time-change modeling describes a curve or trajectory that can be fit through the repeated observations of individuals. This also is referred to as latent curve or growth curve modeling. The first level equation for the latent trajectory model is

$$y_{it} = \alpha_i + \lambda_t \beta_i + \varepsilon_{it}, \qquad (15.8)$$

where y_{it} represents individual responses, α_i describes the across-time intercept for the trajectory of y_i, and ε_{it} is again the error for individual i at time t. In this equation, β_i is the random slope for individual i and λ_t is a device for coding time (and can be either linear or nonlinear depending on the coding). Following the convention of Bollen and Curran (2006), for the linear model the time periods are numbered $0, 1, \ldots, T-1$; therefore, $\lambda_1 = 0$, $\lambda_2 = 1$, and $\lambda_t = T-1$, where $T =$ total number of time periods in the data set.

Because the intercept and slope terms are random effects in Equation (15.8), it holds that there is a second level of equations

$$\alpha_i = \mu_\alpha + \zeta_{\alpha i} \qquad (15.9)$$

and

$$\beta_i = \mu_\beta + \zeta_{\beta i} \qquad (15.10)$$

where μ_α and μ_β are means for the intercept and slope, respectively, and the ζs are error terms. The combination of these three equations represents a standard hierarchical or multilevel model of the sort commonly used in statistical analysis (e.g. Chapters 7, 11, 12, 19, 20). In this case, Equation (15.8) is referred to as a level 1 or lower-level equation while Equations (15.9) and (15.10) are level 2 or upper level equations (μ_α and μ_β are sometimes referred to as hyperparameters, although this is a limiting terminology where there are more than two levels in the model). One difference between the use of this multilevel modeling feature in SEM versus conventional mixed models (e.g. Pinheiro and Bates 2000) is that in SEM covariate effects at both levels can be organized in networks, as illustrated later in the chapter. Other major differences have to do with (i) the fact that the covariance procedures in SEM can lead to rejection of our models and the search for specifications more consistent with the data, (ii) the precise control of assumptions about parameters (e.g. variance inequalities, error correlations) in SEM, and (iii) the universal capability for using latent variables to address measurement questions.

There has been significant interest in the use of latent trajectory models within the SEM framework. Meredith and Tisak (1990) first showed that the SEM latent variable modeling system [Equations (15.4)–(15.6)] could be used to represent random-effects trajectory models by setting the loadings for a factor model to a set of time steps. To elaborate, in factor models one or more latent variables are hypothesized to explain the covariances among a set of observed variables. As an example, Sewall Wright (1918) developed a factor model that hypothesized a set of animal body-size developmental factors (latent variables) to explain the patterns of correlations among bone lengths. In typical factor models, the loadings (regression weights linking observed to latent variables) are freely estimated (although sometimes one loading will be set to a value of 1.0 to specify the scale of measurement for the latent variable). This "factor-analytic" architecture is easily represented using the linear structural relations (LISREL) equations (Jöreskog 1973, equations 4–6a) and commonly included in SEMs. The invention of Meredith and Tisak (1990) was to conceptualize the slope of a time trajectory as a latent variable and then set the loadings for its relationship to a sequence of measurements to the time-step values (one can peek ahead to Fig. 15.7 to see a graphical representation of this model showing the loadings). A number of authors have elaborated on this framework and there are now some excellent treatments of latent trajectory modeling from an SEM perspective (Bollen and Curran 2006, Duncan *et al.* 2006).

The unconditional latent trajectory model, unconditional in the sense that it ignores covariates and only describes the shape of the trajectory, can be specified using the measurement model [Equation (15.6)]:

$$\mathbf{y} = \mathbf{\Lambda_y}\mathbf{\eta} + \mathbf{\varepsilon}. \tag{15.11}$$

Here $\mathbf{y}$ describes the vector of repeated measures ($y_0 \ldots y_{T-1}$) while $\mathbf{\eta}$ represents the latent factors describing the trajectories. In the simplest case, $\mathbf{\eta}$ has two elements

$$\mathbf{\eta} = \begin{pmatrix} \alpha_i \\ \beta_i \end{pmatrix}, \tag{15.12}$$

the random intercepts (α_i) and slopes (β_i) for each individual in the sample. In this simple case, we require a minimum of three time periods of measurement to be able to identify the estimated slope. For trajectories with more complex shapes, $\mathbf{\eta}$ can have more terms and we have the general requirement that there be at least one more time period of measurement than slope parameters (e.g. a quadratic model would require four time periods).

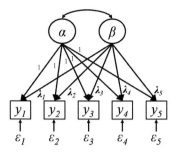

Figure 15.3 Simple latent trajectory model (LTM). In this model the trajectory described by observed measurements of response variable y over five time periods can be explained by an intercept α and slope β. For the linear model, the values for $\lambda_1, \ldots, \lambda_5 = 0, 1, 2, 3$, and 4.

The matrix Λ is a pattern matrix with columns representing the loadings for the intercept and slope terms

$$\Lambda = \begin{pmatrix} 1 & 0 \\ 1 & 1 \\ 1 & 2 \\ \vdots & \vdots \\ 1 & T-1 \end{pmatrix}. \tag{15.13}$$

It should be noted that while this latent trajectory formulation has the general form of a factor model, the constraints placed on Λ give it a different and more specific purpose (the latent variables in this case serve as intercept and slope). Graphically, the basic latent trajectory model can be represented as in Fig. 15.3.

Nonlinear trajectories

Although there are several different ways we can modify Equation (15.11) to represent nonlinear trajectories, it is also possible to use it for certain situations where temporal changes are curvilinear or where time steps are unequal. First we will describe the latter, simple case, which involves estimating nonlinear factor loadings for the slope in the Λ matrix. In the example given above in Equation (15.13), the column of slope weightings is both linear and "fixed", with values $0, 1, 2, \ldots, T-1$ (note that specifying values for a parameter is often referred to as "fixing" the value of the parameter in SEM parlance). Rather than fixing the values for the slope loadings as shown in Equation (15.13), we can estimate some of the values, as represented in Equation (15.14), where the parameters for the third and subsequent time steps are estimated rather than given a fixed value:

$$\Lambda = \begin{pmatrix} 1 & 0 \\ 1 & 1 \\ 1 & \lambda_3 \\ \vdots & \vdots \\ 1 & \lambda_T \end{pmatrix}. \tag{15.14}$$

To establish the scale, we can, for example, fix the first loading (λ_1) at a value of 0 and the second loading (λ_2) at a value of 1 so that the first time interval has a unit length. Allowing the loadings for subsequent time periods to be freely estimated permits us to model the empirical data using a nonparametric approach similar to a spline. Rather than specifying the first two loadings, as just described, an alternative approach could be to fix the first and last loadings and allow the two intervening ones to be estimated.

A parametric approach to estimating nonlinear trajectories might be to use a polynomial series to represent slope parameters. To obtain a quadratic model, we expand Equation (15.8):

$$y_{it} = \alpha_i + \lambda_t \beta_{1i} + \lambda_t^2 \beta_{2i} + \varepsilon_{it}. \tag{15.15}$$

The introduced term λ_t^2 is the square of the time loading (for time period 2, it would be a value of 4) and β_{2i} represents the coefficient for the quadratic term for individual i. Thus, we are constructing our quadratic model through the squaring of the loadings instead of the squaring of beta (the slope term), allowing us to remain linear in our estimated parameters while modeling a nonlinear relationship. Using our convention of setting the first loading to 0 and second loading to 1 as in Equation (15.14), we can interpret α_i as the intercept, β_{1i} as the linear slope, and β_{2i} as a measurement of degree of curvature.

Certain additional nonlinear forms of the LTM have been proposed by du Toit and Cudeck (2001) and described in Bollen and Curran (2006, chapter 4, section 4). One nonlinear form of particular merit for modeling asymptotic change is the exponential,

$$y_{it} = \alpha_i + (1 - e^{-\upsilon \lambda_t})\beta_i + \varepsilon_{it}. \tag{15.16}$$

Exponential functions are convenient to use when trajectories tend toward an asymptotic value. In this case, β_i describes the expected total change in y_i while $e^{-\upsilon}$ represents the rate of deceleration of change over time. Typically, υ (upsilon) will be treated as a fixed effect across cases. The specification of the exponential model is similar to that shown in Equation (15.14) except $\lambda_2 = 1 - e^{-\upsilon(1)}$, $\lambda_3 = 1 - e^{-\upsilon(2)}$, and $\lambda_4 = 1 - e^{-\upsilon(3)}$. Additional information about nonlinear approaches to latent trajectory modeling can be found in Ram and Grimm (2007) and Grimm and Ram (2009).

The inclusion of covariates

One strength of the SEM approach is the capacity to include covariates in models in ways that can reflect the mechanistic processes that generate relationships in the data. In simple terms, we can think of the types of covariates that can be included as relating (directly or indirectly) to either the latent intercepts and slopes or to the observed responses themselves. Covariates that affect the slopes and intercepts of latent trajectory models are sometimes referred to as "time-invariant" while those that affect observed responses are commonly "time-varying" (having different values for each time period). In the ecological example that follows, we will include both types for illustrative purposes. Here we describe a few of the equational forms that can be used to represent covariates in SE models.

For covariates that affect the random slopes or intercepts of latent trajectory models, their impacts are not directly on the level 1 equation [Equation (15.8)], but instead, are in the level 2 equations:

$$\alpha_i = \mu_\alpha + \sum \gamma_\alpha \mathbf{x}_i + \zeta_{\alpha i} \qquad (15.17)$$

and

$$\beta_i = \mu_\beta + \sum \gamma_\beta \mathbf{x}_i + \zeta_{\beta i}. \qquad (15.18)$$

Here, γ_α and γ_β are vectors of coefficients for the covariates in vector $\mathbf{x}_i$ that affect either α_i or β_i.

We also may wish to include additional covariates that have specific time-varying effects. To do so requires expansion of our level 1 equation, which in vector form is

$$\mathbf{y}_{it} = \alpha_i + \lambda_t \beta_i + \gamma_t \mathbf{w}_{it} + \varepsilon_{it}, \qquad (15.19)$$

where $\mathbf{w}_{it}$ represents the time-specific covariates and γ_t represents their time-specific effects. To complete our presentation, we recognize that the covariates affecting slopes, intercepts, and observed responses can be allowed to be response variables in additional equations describing the overall causal network (as will be illustrated in Fig. 15.10). All of this can be represented mathematically using our general SEM framework, which is given by Equations (15.4)–(15.6).

The autoregressive latent trajectory model (ALT)

At the beginning of the chapter, we described the fundamental structure of autoregressive models [Equation (15.7), Fig. 15.2]. Such models are appropriate when we wish to represent the values of response variables at time t as being causally affected by their values at time $t - 1$. Curran and Bollen (2001) and Bollen and Curran (2004) have proposed a framework for merging both kinds of causal architectures (the ARM and the LTM) into a single model, which they call the autoregressive latent trajectory model (ALT).

When combining the latent trajectory structure with the autoregressive structure there is an incompatibility that arises that must be addressed. In the ARM model, the first measure in any chain of observations (e.g. y_1 in Fig. 15.2a as well as y_1 and w_1 in Fig. 15.2b) must be treated as predetermined. By allowing for these initial measures to be predetermined, they have determinant (fixed) values and we avoid the problem of "infinite regress" discussed by Bollen and Curran (2004). The problem of infinite regress comes into play when the measures being analyzed are part of a series of potential measures and our measure y_1 has a preceding value for which we have no estimate. If our y_1 is treated as endogenous in our model, there is a bias that can be propagated down the chain; therefore, autoregressive chains typically begin with the first observation period treated as exogenous. The incompatibility between ARM and LTM becomes apparent when we consider that the initial measures in an LTM are not considered exogenous because they serve as indicators for the estimate of the slope and intercept of the trajectory. To merge the two structures, it is usually appropriate to treat the first time measures as exogenous

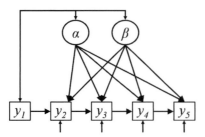

Figure 15.4 Autoregressive latent trajectory model (ALT). Trajectory described by y over time can be explained both by a common intercept α and slope β and also by autoregressive time-specific effects. In this version of the ALT model, the first observation of y is modeled as exogenous and simply correlated with the intercept and slope.

to the latent trajectory as shown in Fig. 15.4. Here, the relationships between the initial measure and the slope and intercept are treated as undirected covariances. The cost of such a model, which can be important in short time series, is that the first time interval (from y_1 to y_2) is not used in the estimation of the slope and intercept. In the exceptional case where the first measurement available is truly the first in the developmental series, we may consider modifying the structure of the ALT model so that the first observation is treated as endogenous to the slope and intercept (α and β). Further discussion of the ALT model can be found in Bollen and Curran (2006).

Issues of model identification

An issue of key importance in statistical modeling is parameter or model identification, which is required so that unique estimates are obtained. While problems with identification can arise for even simple models, complex models of the sort commonly found in SEM have a greater chance of including parameters that are not identified and unique estimates cannot be obtained. Here we provide some general background; a deeper discussion of the issues as they specifically relate to latent trajectory models can be found in Bollen and Curran (2006).

At its simplest level, the issue of identification can be understood from the perspective of the elementary principle of matrix algebra that to estimate the values of two unknown parameters requires two unique equations. So, for the relationship $y = ax_1 + bx_2$, where y, x_1, and x_2 are known, there exists an infinite number of combinations of values for a and b that will satisfy the relationship. However, if we also know that $y = 3ax_1 - bx_2$, we have two unique equations and in principle can derive unique values for a and b that will satisfy the pair of equations. When considering the full set of parameters included in a structural equation model, there is a subset whose values are readily identified from the data. This known set of parameters will typically include the means, variances, and covariances for the observed variables. If the unknown parameters (e.g. path coefficients) are unique functions of the known entities, they are identified. The unknown parameters for the simple LTM described by Equations (15.8)–(15.10) include the means for the latent intercept and latent slope, μ_α and μ_β, as well as the error variances $VAR(\varepsilon_{it})$, $VAR(\zeta_{\alpha i})$, and $VAR(\zeta_{\beta i})$, and the covariance between intercept and slope $COV(\zeta_{\alpha i}, \zeta_{\beta i})$. One simplifying assumption that is made is that the error variances for the individual cases are equal and thus $VAR(\varepsilon_{it}) = VAR(\varepsilon_t)$. There is no assumption made that the error variances are the same over time, however. In the case of the LTM, the loadings λ_t may or may not need to be estimated. In the linear model, all values for λ_t are specified, as shown in Equation (15.13). In models with nonlinear trajectories of fixed form (e.g. the exponential model), all elements of λ_t are again specified. However, in the nonparametric model shown in Equation (15.14), some loadings must be estimated. At a minimum, we should have at least two elements of λ_t specified.

As alluded to above, one condition for model identification that is generally necessary, although not always sufficient, is that there be at least as many known pieces of information as there are unknown parameters to estimate. This principle is helpful in allowing us to determine the number of temporal samples of data that are required for a model to be identified. For the linear model, at least three time points must be included in the dataset for identification. For more complex models, for example one with a polynomial slope structure, at least four time points must be included. Also, for the autoregressive latent trajectory model (Fig. 15.4), four time points will typically be required because the first sample will often be treated as exogenous to the rest of the series and not used to estimate the slope or intercept.

A final important point about identification is that it is possible for non-identification to occur because of empirical matters. For example, a model may be theoretically identified but the estimation process may be unable to obtain unique estimates for the parameters. This can easily happen in the situation where two variables in the data set are nearly perfectly correlated. In such a case, the two variables do not represent independent pieces of information, but instead, act as one. Other situations can produce non-identification problems and in practice they will be called to the investigator's attention by error messages from the software using for the analysis.

Estimation: classical and Bayesian approaches

In what we might call classical SEM (Jöreskog 1973, Bollen 1989), solution procedures have most commonly involved the use of maximum likelihood methods based on the analysis of covariances. A fundamental characteristic of the classical approach to estimating and evaluating the fit of SEMs involves a comparison of the observed variance–covariance matrix (for simplicity referred to simply as the covariance matrix) with the model-implied covariance matrix. This comparison is sometimes represented symbolically as

$$\Sigma = \Sigma(\Theta) \tag{15.20}$$

where Σ refers to the population covariance matrix of the measured variables, $\Sigma(\Theta)$ refers to the covariance matrix implied by the model as a result of the parameters of the model, and Θ refers to the full set of model parameters. The sample covariance matrix $\mathbf{S}$ is typically used as our best estimate of Σ, and $\hat{\Theta}$ refers to the estimated model parameters. There exists a similar relationship between the observed means μ and those implied by the model $\mu(\mathbf{\Theta})$.

Estimation in classical SEM involves the selection of values of $\hat{\mathbf{\Theta}}$ so as to make the estimated value of $\mathbf{\Sigma}(\mathbf{\Theta})$ as close to $\mathbf{S}$ as possible. Maximum likelihood methods are commonly used to arrive at the parameter estimates. A commonly used fitting function is described by the expression

$$F_{\mathbf{ML}} = \ln|\mathbf{\Sigma}(\mathbf{\Theta})| - \ln|\mathbf{S}| + tr[\mathbf{\Sigma}^{-1}(\mathbf{\Theta})\mathbf{S}] - \mathrm{p} - [\bar{\mathbf{y}} - \mu(\mathbf{\Theta})]'\mathbf{\Sigma}^{-1}(\mathbf{\Theta})[\bar{\mathbf{y}} - \mu(\mathbf{\Theta})],$$

$$(15.21)$$

where p is the number of observed variables in $\mathbf{S}$ and all other symbols are as previously defined. The fitting function has a minimum value of 0, which is obtained whenever the model is fully identified (number of unknown parameters equals number of known variances, covariances, or means). When the model is over-identified, and this will usually be when some pathways are omitted from the model, we can expect some discrepancy between $\mathbf{S}$ and $\mathbf{\Sigma}(\hat{\mathbf{\Theta}})$ and between μ and $\mu(\hat{\mathbf{\Theta}})$. Because the fitting function follows a chi-square distribution, the degree of discrepancy between observed and model-implied values can be summarized by a model chi-square and the magnitude of this chi-square can be evaluated for the probability that the observed data are consistent with values implied by the model. The chi-square discrepancy function, then, provides a basis for rejecting a model as inconsistent with the observed data. Such a rejection typically results in the search for a model that is consistent with the data and still theoretically justified. For more information about estimation, refer to Bollen (1989). More about the assessment of model fit and the search for properly specified models is presented below.

Adopting a Bayesian approach to estimating SEMs (Dunson *et al.* 2005, Lee 2007) permits a greater flexibility for modeling. First of all, a great variety of distributional forms for response variables is permitted, although many are permitted in classical modeling procedures as well. Second, equations can be nonlinear in their parameters and still be estimated using the Markov chain Monte Carlo (MCMC) methods often associated with Bayesian applications. Constraints on parameter identification still apply, but opportunities for estimating a greater variety of models open up with a Bayesian approach because the use of prior information can allow for identification of parameters in some cases. A more complete discussion of a Bayesian approach to SEM can be found in Lee (2007).

The treatment of missing data

Missing data are a commonly encountered situation in any analysis. In monitoring studies, missing data are frequently more of a problem than in single-sample studies. A variety of ways of dealing with missing data are used in practice, though not all are equal with regard to either efficiency or bias (Rubin 1976). Chapter 13 outlined methods for dealing with missing data when such "holes" would hinder an objective of producing index counts for a population each year. Here we give a brief presentation on the subject from the perspective of SEMs.

There are many different reasons why data might be missing in a data set. It is even possible that the value for an observation may not be missing initially, but instead, judged to be erroneous or an outlier. When encountering missing data, the reason the data are missing may influence the remedy employed (see also Chapter 13). To review, categories of missing data include:

- "Missing completely at random" (MCAR) refers to when the value of a single observation is missing with equal probability for any observation on any variable from any case in the data set.
- "Missing at random" (MAR) refers to a different situation. Here, we can consider a missing case as MAR even if there is a pattern to the missing data, as long as the mechanism causing missing data does not depend on the unobserved data. That is, MAR assumes that the pattern of missing data is not predicted by the unobserved values of the variables (and thus, that the missing data do not lead to a biased estimate of the predictors of missingness). Let's imagine, for example, that we have data from a sample of a species of grassland birds. Perhaps for some reason the data for a variable are more difficult to obtain reliably from the younger members of the species and this results in a pattern of missing values in the data set (we are more likely to have missing values in younger individuals). If we analyze the data as a single group and bird size or a correlated variable like bird age is used as a predictor in the model, then we classify those missing data as "missing at random" (MAR).
- The condition of "missing not completely at random" (MNAR) is when missing observations predict the pattern of missing values. This condition is one that typically leads to confounding of missing values with other factors and remedies are difficult for such a situation without additional information to use in correcting for the missing values (Schafer and Graham 2002). Remedies involving the MNAR case are beyond the scope of our presentation here.

A number of different approaches are sometimes used for dealing with data that are MCAR or MAR. One of the most common methods for dealing with missing values is casewise deletion, where entire records are removed when one value from those cases is missing. In some situations outside of the SEM context, this strategy would result in entire years of monitoring data being discarded from the analysis (Chapter 13). Apart from these situations, when the sample is large, the casewise deletion of single or a small percentage of cases has modest effects on statistical power. However, when a sizable number of cases are dropped through casewise deletion, this is a highly inefficient method as many data from the non-missing observations for those cases are wasted. It is possible to use pairwise deletion in the estimation of a corrected sample covariance matrix. Again, if only a small percentage of observations are missing, the consequences are not too severe. However, such an approach can be quite unsuitable for situations where there are substantial numbers of missing values, as the resulting matrix is not a coherent representation of the population. Sometimes missing data are addressed by inserting imputed values in place of the missing values (see Chapter 13). In the simplest situation, the mean for a variable is inserted in place of a missing value. Again, when imputed values are inserted for a small number of observations, the consequences are not severe. However, insertion of means has the effect of reducing the variance for a variable when the number of insertions is substantial. Imputation can involve the random selection of a value from the distribution of values for a variable when we wish to avoid inserting means into the dataset. Again, in moderation such an approach has small effects, but is inappropriate when the number of missing values for a variable is substantial. However, it is important to realize that methods such as listwise or pairwise deletion assume data are MCAR and if data are MAR, these deletion methods do not necessarily lead to consistent estimates of model parameters.

Full-information maximum likelihood approach to missing data

In 1996, Arbuckle presented a method for estimating structural equation models in the presence of missing data using all the nonmissing data and ignoring the missing values. This method, which is referred to as full-information maximum likelihood (FIML), is widely regarded as an excellent approach to analysis in the presence of missing data as long as the data can be assumed to be MAR or MCAR. [Note that early SEM references referred to the usual ML, maximum likelihood estimation, without missing data as FIML because it was a ML estimator that made use of all equations and all information in the system to estimate the parameters. Here we are using FIML in the sense proposed by Arbuckle (1996).] The FIML approach retains all the strengths of maximum likelihood while maintaining maximal efficiency of information and minimum bias. The reader is referred to Arbuckle (1996) for a further description of this method.

An overview of the structural equation modeling process

An ambition of the SEM enterprise is learning about causal processes. This involves more than simply the specification of models and estimation of parameters. SEM also involves a workflow process that is designed to advance scientific understanding and the strength of inference (Grace *et al.* 2010). Our presentation of models in the following example is designed to teach the reader, step by step, how to specify an SEM for temporal dynamics. In practice, the experienced SEM practitioner would start with a fully developed model (such as the one shown later in Fig. 15.11) and compare it to a set of related competing models (Box 15.2). Because our presentation in the next section is geared towards teaching technique, here we provide a brief description of the SEM model development and evaluation process.

SEM can be used in three different modes: exploratory, competing models (model comparison), and strictly confirmatory (Jöreskog 1977). Often when practitioners are first applying SEM or when one is approaching a new system or undeveloped topic, the application will be exploratory (we might also call it "model building"). The essential characteristic of an exploratory application is that a large number of models are examined and the final model can be very different from the ones first considered. The example presented in the next section is an illustration of an exploratory approach in that we demonstrate the building of a full model by starting with simple, incomplete models to which additional complexity is added.

> **Box 15.2** Common challenges: struggling to learn
>
> A major feature of SEM is the comparison between data covariances/correlations and those implied by the models used to evaluate that data. As a result of this feature, the experience of using SEM is one of discovery, where the analyst commonly finds that their models are inconsistent with the data because of some omitted relationship. This discovery process is both highly educational and demanding of patience at the same time. Simple SEM applications are not difficult to learn and apply. However, for more complex models, formal training in SEM or access to an expert in the methodology is commonly needed to navigate the process of discovering suitable models. Those analyzing monitoring data should obtain some training in SEM if they wish to use this methodology. SEM is capable of handling a great number of statistical complexities and difficulties because it is a highly flexible methodological framework rather than a specific analysis technique. Learning how to successfully apply SEM when faced with complex issues requires persistence. It is our plan to continue to provide materials that can help with the learning process.

In mature SEM practice, we generally strive for a competing models approach. In this approach, theoretical ideas are first translated into a conceptual model or "meta-model" (Grace *et al.* 2010) that defines the general ideas being addressed in the analysis. The second step involves the specification of SEMs based on the meta-model. This step requires both consideration of competing hypotheses about mechanisms and also consideration of data characteristics, especially non-Gaussian responses, nonlinear pathways, and hierarchical data structure. Then, specified models should be examined for mathematical suitability to ascertain whether all parameters are potentially identifiable (as discussed above). Once data have been used to estimate values for model parameters, two kinds of model evaluation take place.

The first and most primary form of model evaluation is the assessment of "absolute" fit. As described in the previous section on estimation, it is a characteristic of classical SEM that analyses are conducted on the matrix of covariances. This approach permits a comparison of model-implied to actual covariances. When models include fewer estimable parameters than the known pieces of information in the variance–covariance matrices, we have "model degrees of freedom". Having model degrees of freedom gives us the potential to detect model–data discrepancies. There are a number of different metrics that can be used to assess absolute model fit, though the most fundamental is the model chi-square. As stated previously, the chi-square discrepancy function provides a basis for rejecting a model as inconsistent with the observed data. Thus, it is possible (and common) for models to be found to be "unacceptable" because the data contain relationships not specified in the model. Such a rejection typically results in the search for a model that is consistent with the data and still theoretically justified. Only when we determine that a model has an adequate absolute fit do we trust the parameter estimates. For further discussion see Bollen (1989) or Grace (2006).

Model comparison and selection is, of course, also important. It is possible that we can have more than one model with acceptable fit. This is particularly the case when pathways are included that do not contribute to variance explanation. When comparing models that differ by one parameter (e.g. with and without a path included), competing models are considered to be "nested" and we can use single-degree of freedom tests. These can either be chi-square discrepancy tests as classically used in SEM or they can be conditional independence tests such as the d-sep test (Shipley 2009). It is also possible to use information-theoretic methods to compare models that contain different sets of variables. In this case, the concept of parsimony is given stronger weight in the evaluation. It is beyond our scope to go into the complex topic of model selection in detail here [more can be found in Bollen (1989) and Grace (2006)]. However, we provide a brief description of the model selection/evaluation process used in the example that follows.

In the example presentation below, we first rely on model chi-square to evaluate the overall adequacy of a model (absolute model fit). When a model possesses an associated chi-square that is small (relative to the number of degrees of freedom), this indicates that the cumulative discrepancy between observed covariances and the covariances implied by the model is small and model–data fit is good. In the classic case, we seek models for which the probability associated with a model chi-square given a certain number of degrees of freedom is > 0.05. Note that in null-hypothesis testing, we often declare a parameter to be significant when its associated P-value is *less than* 0.05. However, when assessing model fit, it is important for our confidence in the estimates that the model is not obviously mismatched with our data, so we seek models with low chi-square values (and P-values *greater than* 0.05).

One limitation to the use of the model chi-square is that as statistical power goes up along with the number of cases in a sample, our ability to detect small discrepancies increases. Generally this effect is non-problematic in situations where the sample size is less than, say, 100 or 200. In our example, the number of cases in the data set is 88, so we are comfortable relying on the model chi-square. In situations where there are a large number of cases in a sample, the SEM practitioner will often rely on sample-size-adjusted measures, such as the root mean square of approximation or proportionalized measures of discrepancy (e.g. comparative fit index).

In addition to examining overall model fit, the presentation below relies on single-degree-of-freedom chi-square tests to evaluate hypotheses about individual parameters. The single-degree-of-freedom chi-square difference of 3.84 is often used as a cut-off for declaring two models distinguishable, as the probability of obtaining a value larger than 3.84 from a chi-square distribution with 1 df is < 0.05. So, for example, if we relax a constraint in a model (perhaps we have been omitting a pathway, which effectively sets its value at 0) and now we examine a second model where we estimate that parameter, a drop in model chi-square greater than 3.84 indicates the second model is distinguishably better than the first.

It is worth mentioning that many different bases for model comparison have been offered, both from within SEM circles and from without. Generally, it is the philosophy of SEM to rely on multiple metrics when making a decision about the significance of a parameter (one might use parameter P-values to guide the conduct of chi-square tests, for example). The good news is that it has been shown recently (Murtaugh 2009) that there is a strong convergence between model comparison methods in some situations, which provides reassurance for the multi-metric approach. It is our view that "the decision problem" of statistics (is a parameter to be treated as significant or not?) ultimately requires both evidence and professional judgment, particularly when effect sizes are of interest. This view is particularly relevant to SEM practice where we are theory-oriented rather than null hypothesis-oriented.

To continue our description of the SEM workflow process, further evaluation of conclusions is recommended through the process of sequential learning. Model results are used to inform the theoretical ideas that are the basis for the next round of modeling. A key motivation for followup studies in causal modeling is to test hypothesized explanations for pathways through the evaluation of mediator variables (MacKinnon 2008). As an example, building on the findings from an initial SEM analysis of fire effects on shrublands (Grace and Keeley 2006), Keeley et al. (2008) used data from another set of fires to examine more complex SEMs that evaluated the roles of additional mechanisms influencing shrubland recovery. As this is meant to illustrate, SEM philosophy encourages subsequent studies to strengthen our basis for inferring causal interpretations.

An ecological example

Background

In the fall of 1993, major fires burned through southern California, USA, during a short period of time, representing a large-scale disturbance for the region's natural ecosystems. Keeley et al. (2005) established one 1000-m^2 plot in each of 90 sample sites in burned areas and began sampling in spring of the first postfire year, continuing for a total of 5 years. Objectives of the study included (i) to determine the degree of recovery of vegetation; (ii) to test certain theories about community temporal dynamics; (iii) to examine the role of year-to-year variation in rainfall in vegetation dynamics; and (iv) to determine the factors causing site-to-site differences in response. In addition to temporal dynamics, spatial variation in vegetation recovery and the environmental factors that influenced it were also of interest. At each site, the variables measured annually included herbaceous cover (as a percentage of ground surface) and plant species richness. Time-invariant spatial covariates that were measured included pre-fire age of the shrub stand (estimated from ring counts of stem samples), fire severity (based on skeletal remains of shrubs), and site abiotic characteristics.

One objective of the repeated sampling was to examine the dynamics of plant biodiversity so as to better understand the role of disturbance in maintaining diversity on the landscape. Previous experience suggested that immediately following fire there is a

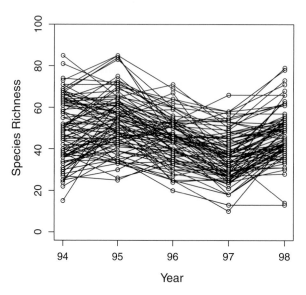

Figure 15.5 Observed values of herb species richness for the 88 plots in the example data set from southern California shrublands.

flush of plant germination resulting in peak diversity in the first year. Over time, diversity is expected to decline due to reduced germination and as competition intensifies. This "pulse-response" hypothesis was examined in an earlier analysis by Grace and Keeley (2006) that used latent trajectory modeling and SEM to study diversity responses to the fires. Here we examine the temporal dynamics of these data in greater detail to illustrate SEM methods.

The temporal dynamics of plant species richness in the 88 plots with complete data represents a collection of trajectories. We begin by considering the individual trajectories (Fig. 15.5) and some of their characteristics (Fig. 15.6). Richness showed a general decline over time, though with substantial fluctuations (Fig. 15.5a). A summary of the individual regression slopes (Fig. 15.6b) shows that the majority were negative. The raw data showed higher than average richness in 1995 and 1998 and it was observed that mean annual precipitation was higher than normal in those years (Fig. 15.6c). Once precipitation is considered, the decline over time is more conspicuous (Fig. 15.6d).

The SEM analyses presented in the following sections were conducted using the software Mplus (Muthén and Muthén 1998–2010). The data and software code for the final model are available as on online supplement to our chapter. There exist numerous software platforms for the analysis of SE models. Some of the more popular ones include Amos, LISREL, EQS, and Mplus. In addition, there are modules in the R software for SEM (e.g. "sem" and "lavaan"). All of the packages mentioned have the capabilities needed to analyze the models below, although their other capabilities and their ease of use differ.

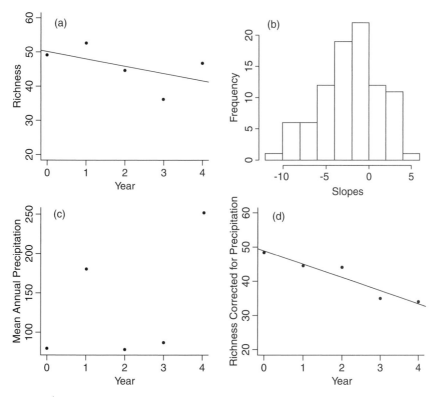

Figure 15.6 Some characteristics of the data being modeled in our extended example. Years 0–4 in the graphs correspond to 1994–1998: (a) mean richness over time, (b) histogram of individual slopes for the 88 trajectories, (c) mean annual precipitation values, and (d) plot of mean richness corrected for mean annual precipitation.

A linear latent trajectory model (LTM)

For didactic purposes, we begin by modeling the richness data using a simple, linear LTM [Equations (15.8)–(15.10) and (15.13)]. With this model we are fitting a linear relationship to the trajectories in Fig. 15.5 and seeking estimates of the means and variances for the slopes and intercepts of the individual cases (Fig. 15.7). Illustrated in the latter figure are some common characteristics of the LTM, such as the fact that the loadings connecting the intercept with the observed richness values from each time are all fixed at 1.0. The loadings connecting the slope variable with the observed richness values progress from 0 to 4 in increments of 1. As described earlier, this implementation is one where each individual trajectory is described by an intercept and a slope and the intercepts and slopes have a distribution of values across the various cases.

There are a total of five observed variables involved in this analysis; therefore, the number of known pieces of information includes their five means + their five variances + their 10 covariances, which equals a total of 20. Of the 25 parameters in the model (Table 15.1), 15 parameters have fixed values so only 10 parameters need to be

Table 15.1 Summary of parameters in simple linear LTM (Fig. 15.7).

Estimate type	Loadings[1]	Covariances	Variances	Means	Intercepts	Total
fixed value	15	0	0	0	0	15
estimated	0	1	7	2	0	10
total	15	1	7	2	0	25

[1] The 15 fixed loadings include the five loadings from the intercept to the five responses (all fixed to a value of 1), the five loadings from the slope to the five responses (fixed to values 0–4), and the five predicted intercepts for the responses r94–r98, which were all fixed to 0.

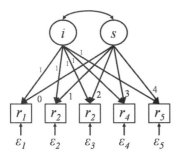

Figure 15.7 Linear latent trajectory model for species richness r_1–r_5 in the example. Summary of parameters from this model is given in Table 15.1 and results are presented in Table 15.2.

estimated (Table 15.2). This leaves $20 - 10 = 10$ model degrees of freedom for hypothesis testing.

Can a simple linear LTM provide an adequate representation of the data? Based on the large degree of discrepancy between observed and model-implied covariances and means (a model chi-square $= 162.4$ with 10 model degrees of freedom and *P*-value < 0.001), we conclude that the model is a very poor representation of the forces that shaped these data. Ignoring this important point for the time being, we can see from Table 15.2 that our estimate of the mean intercept for the 88 time trajectories is 50.94. The calculated standard deviation for the random intercepts equals 8.79 (the square root of the variance of the intercepts, 77.33). The estimated mean slope in this case is –2.66 and its variance is effectively 0. Assuming these values to be valid estimates, it is interesting that the variance of the slopes is so small, implying a great deal of uniformity in the average declines in richness over time in the different plots. Overall, the estimated intercept and slope derived from our LTM are generally consistent with the mean response for our trajectories (Fig. 15.6a). However, we have reason to suspect that our model does not capture the important factors controlling richness dynamics because of the poor model fit.

Do we have homogeneous variances over the time trajectory? While it is not mandatory in the SEMs, we wish to know whether we can represent our mean time trajectory more simply with a common error variance across the years. It is possible to estimate a single error variance for the trajectory and to test for whether there is a common error variance. To accomplish this we can estimate a single value for all ε_1–ε_5. When we

Table 15.2 Select results for the simple, linear latent trajectory model (Fig. 15.7). The codes "r94–98" refer to species richness per plot during the years 1994 through 1998. Number of cases (plots) in the data set = 88. The model chi-square was 162.4 with 10 model degrees of freedom and associated P-value < 0.001, all of which indicate very poor model-data fit.

Parameters[1]	Estimate	Std. error	Critical ratio	P-value
slope → r94	0 (fixed)			
slope → r95	1.0 (fixed)			
slope → r96	2.0 (fixed)			
slope → r97	3.0 (fixed)			
slope → r98	4.0 (fixed)			
mean of intercept, i	50.94	1.50	34.01	< 0.001
mean of slope, s	−2.66	0.58	−4.61	< 0.001
covariance between i and s	5.53	7.23	0.76	0.44
VAR(intercept)	77.33	26.95	2.87	0.004
VAR(slope)	−4.01[2]	2.65	−1.52	0.13
R^2 for r94	0.28			
R^2 for r95	0.51			
R^2 for r96	0.67			
R^2 for r97	0.44			
R^2 for r98	0.32			
Average R^2	0.44			

[1] Note that for the P-values associated with parameters, in contrast to those associated with the model chi-square, significant parameters have P-values LESS THAN 0.05.
[2] Note that the model-implied variance for the slope estimate was negative. A Wald test shows that the value of −4.014 was not significantly different from a value of zero (note associated P-value of 0.130). While a value of 0 for this variance estimate is technically acceptable, if our variance was significantly negative, that would be considered an inadmissible estimate that suggests model misspecification.

do this, we get an estimate of 90.60. However, our model chi-square increases from 162.4 to 183.9 (21.5 points) while releasing four degrees of freedom because four fewer parameters were estimated. This magnitude of chi-square increase indicates that a single error variance for the trajectory is an oversimplification for our data.

An LTM with time-varying covariates

What role does annual variation in precipitation play in year-to-year variations in richness? In addition to the common effects of trajectory intercept and slope, we have reason to believe that conditions that affect richness in any given year vary over time. One conspicuous time-varying feature that seems to correlate with the deviations of observed richness from a simple trajectory is annual precipitation. This covariate is time-varying (different values for each time) and thus our model corresponds to that described by Equation (15.19). However, our precipitation covariate in this case is made up of single study area-wide values, one for each year for all 88 cases, and is therefore a fixed rather than random effect. Figure 15.8 illustrates the form of the model in which we have allowed annual richness values to be a function of precipitation. Further, because

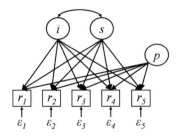

Figure 15.8 Latent trajectory model for species richness r_1–r_5 with covariate p, for precipitation, included. Results are summarized in Table 15.3.

our regional rainfall values are very approximate for our specific locations, we have used fixed loadings to indicate that 1995 was a wetter than average year and 1998 was much wetter than average (Fig. 15.6c), allowing "normal" years to have loadings of 0, the wet year a loading of 1, and the very wet year a loading of 2. Typically, we would allow the precipitation covariate to correlate with our latent intercept and slope. However, since precipitation has only one unique value for each year, we assigned it a very small fixed variance of 0.01 and fixed its covariance with other exogenous variables to zero to simplify the assumptions made in model. For this model, we obtained a chi-square of 43.9 with nine degrees of freedom and $P = 0.001$. We consider this model to have inadequate fit and in need of further modification before it can be used to draw interpretations.

One thing we noticed in looking at the results was a residual correlation between richness in 1994 and in 1995. This suggests the errors may be correlated over time, perhaps due to autocorrelation, which would be expected in these data. Including an error correlation in our model, we obtain a chi-square of 33.05 with eight degrees of freedom (a drop of 10.85). We retain the error correlation in our model for the time being, though the introduction of additional covariates could conceivably make it disappear. At this point, our model still does not fit the data very well and we continue examining our question about the role of precipitation in the next section where we allow precipitation to have a nonlinear effect.

An LTM with a nonlinear effect from the time-varying covariates

Examination of residuals from the previous model suggested that richness in 1996 was somewhat greater than our model predicted (see Fig. 15.6d for an approximate view). If theoretically justifiable, one way we could model this is to freely estimate the loading from precipitation for 1996, along the lines of what is represented in Equation (15.14). If we relax the constraint of a fixed loading from precipitation for that year, model fit is substantially improved, providing empirical (but not theoretical) justification for fitting a nonlinear relationship. A model allowing for the free estimation of a precipitation effect for 1996 explains the mean and covariance structure in the data much better (chi-square $= 13.401$ with seven degrees of freedom and a P-value $= 0.063$). Select results are presented in Table 15.3. The estimated value for the loading for 1996 using this

Table 15.3 Select results from nonlinear latent trajectory model (Fig. 15.8) with precipitation as time-varying covariate included. Model chi-square $= 13.401$ with seven degrees of freedom and a P-value $= 0.063$.

Parameters[1]	Estimate	Std. error	Critical ratio	P-value
slope → r94	0 (fixed)			
slope → r95	1.0 (fixed)			
slope → r96	2.0 (fixed)			
slope → r97	3.0 (fixed)			
slope → r98	4.0 (fixed)			
p → r94	0 (fixed)			
p → r95	1 (fixed)			
p → r96	0.550	0.113	4.851	< 0.001
p → r97	0 (fixed)			
p → r98	2 (fixed)			
mean of intercept, i	49.286	1.423	34.629	< 0.001
mean of slope, s	−4.352	0.344	−12.668	< 0.001
mean of precip, p	7.489	0.508	14.752	< 0.001
covariance between i and s	5.995	6.314	0.949	0.342
VAR(intercept)	76.30	27.34	2.790	0.005
VAR(slope)[1]	−2.353	1.988	−1.184	0.237
R^2 for r94	0.281			
R^2 for r95	0.577			
R^2 for r96	0.683			
R^2 for r97	0.752			
R^2 for r98	0.554			
Average R^2	0.569			

[1] Note again that the model-implied variance for the slope estimate was negative, but not significantly different from a value of zero. We conclude once again that the variance of the slope is approximately zero.

nonlinear approach is 0.55, which implies that conditions were favorable for richness in that year, but not as favorable as in 1995. It seems that this favorability for 1996 was not due to differences in total rainfall per se, so we speculate that perhaps spring temperatures, rainfall distribution, or some other particular set of circumstances favored richness in that year. We will revisit this issue later once we have illustrated more modeling possibilities.

The estimated intercept based on our model that includes precipitation is 49.29 species per plot, which is very close to the estimate from our simpler model (Table 15.2). However, our new estimate of slope is −4.35 (Table 15.3), indicating a steeper decline than estimated from our model that ignored precipitation (where the slope was −2.656). The reason the intercept is much more negative in our model including precipitation is because the elevated richness values in 1995 and 1998 are now being explained by increased precipitation in those years, which implies that if precipitation levels in those years had been comparable to the other years (1994, 1996, and 1997), we would have expected a loss of 4.35 species each year across time (for a total loss of 17.4 species

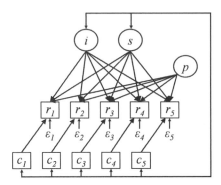

Figure 15.9 Latent trajectory model for species richness r_1-r_5 with covariate p, for precipitation, and covariates c_1-c_5, herb cover, included. Results are summarized in Table 15.4.

or 35% of the initial species during the 5-year period). In addition to the fact that our mean precipitation effect is significant, we can also see that our R^2 values for the individual richness responses each year have increased for the years 1995–1998. Average R^2 increased from 0.442 for our model without precipitation included (Table 15.2) to an average of 0.569 (Table 15.3). Taken together, these pieces of evidence support the interpretation that precipitation has a very powerful influence on richness in any given year. Once we control for annual variations in precipitation, we see there is a steady decline in richness following fire as predicted by theory.

Are annual variations in herb cover predictive of year-to-year variations in richness? It is useful for illustrative purposes to consider a more typical time-varying covariate that possesses individual values for each plot at each time of measurement. For this reason, we include in our model year-to-year variations in herb cover. There are actually several different ways we could include the herb-cover variables in the overall model of richness, two of which we will describe here (the choices described here are for purposes of illustrating SEM modeling methods and not a comprehensive evaluation of the possibilities). First, it would be possible to expand our latent trajectory structure to model a time trajectory for herb cover variations as we did with richness (using the architecture in Fig. 15.7, but with cover as the response variable). This would cause our total model to contain two interacting trajectories, along with the effect of precipitation. A further discussion of some of the possibilities for such models can be found in Bollen and Curran (2006, chapter 7).

Second, it would be possible to ignore the causal forces driving annual variations in herb cover and treat it as a simple time-varying covariate. This is the approach we have taken to address the question of whether cover variations help explain our richness trajectory, as well as to test whether there remains a monotonic decline in richness once we control for herb-cover temporal variations (Fig. 15.9). Initial results indicated the need to include a correlation between herb cover and richness for 1994. We again interpret this correlation as indicative of autocorrelation in the data, which is common for spatially arranged data such as these.

The inclusion of herb cover as a time-varying covariate of richness in the model raises the question of what kind of linkage between herb cover and richness we should

Table 15.4 Select results from nonlinear latent trajectory model (Fig. 15.9) with precipitation and herb cover as time-varying covariates. Model chi-square = 22.096 with 19 degrees of freedom and a *P*-value = 0.280. Note that a common parameter estimate describes the effect of herb cover on richness for 1995–1998.

Parameters	Estimate	Std. error	Critical ratio	*P*-value
slope → r94	0 (fixed)			
slope → r95	1.0 (fixed)			
slope → r96	2.0 (fixed)			
slope → r97	3.0 (fixed)			
slope → r98	4.0 (fixed)			
p → r94	0 (fixed)			
p → r95	1 (fixed)			
p → r96	0.596	0.115	5.183	< 0.001
p → r97	0 (fixed)			
p → r98	2 (fixed)			
herb cover 94 → r94	0.046	0.027	1.723	0.085
herb cover 95 → r95	0.035	0.012	2.978	0.003
herb cover 96 → r96	0.035	0.012	2.978	0.003
herb cover 97 → r97	0.035	0.012	2.978	0.003
herb cover 98 → r98	0.035	0.012	2.978	0.003
mean of intercept, *i*	45.832	1.888	24.282	< 0.001
mean of slope, *s*	−4.728	0.373	−12.69	< 0.001
mean of precip, *p*	7.343	0.523	14.04	< 0.001
covariance between *i* and *s*	7.029	5.823	0.207	0.227
VAR(intercept)	71.38	25.49	2.800	0.005
VAR(slope)	−2.374	1.851	−1.282	0.200
R^2 for r94	0.326			
R^2 for r95	0.610			
R^2 for r96	0.699			
R^2 for r97	0.749			
R^2 for r98	0.575			
Average R^2	0.592			

specify. We might theorize that there should be a consistent dependence of richness on cover over time, in which case we wish to estimate a single coefficient. What was found was that the dependence of richness on cover for 1995 through 1998 could indeed be represented by a single coefficient without a loss of information (i.e. including this constraint did not significantly increase model chi-square). However, the dependence of richness on cover in 1994 was of a different magnitude from the other years (Table 15.4). Collectively, the results from this analysis indicate that variation in herb cover is a significant time-varying covariate for our richness trajectory.

An LTM with time-varying and time-invariant covariates

How do spatial, time-invariant covariates play a role in our model? The earlier analysis of these data by Grace and Keeley (2006) illustrated that richness variations among sites

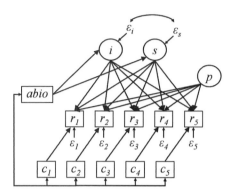

Figure 15.10 Latent trajectory model for species richness r_1–r_5 with time-varying covariates p, precipitation, and c_1–c_5, herb cover, as well as time-invariant covariate, *abio* (abiotic conditions). Results are summarized in Table 15.5.

in the first year following fire depend on spatial variations in environmental conditions. To illustrate further the integrative framework of SEM, we show how the effects of time-invariant covariates can be modeled. To illustrate the flexibility of the SEM methodology, we include a network of three time-invariant covariates: abiotic favorability (a moderating factor that potentially alters responses), age of stand that burned (a primary predictor of system response to fire), and fire severity (a hypothesized mediator of the stand-age impacts).

Regarding abiotic favorability, in the earlier study it was observed that sites (with one plot at each site) differed in their abiotic soil conditions and that some sites were more favorable for species richness compared to others. Thus, we might expect that variations in the intercepts for our individual trajectories can be related to the site-to-site variations in those soil conditions. Concomitant with differences in intercepts, rates of decline (slopes) might also vary among sites depending on abiotic favorability. Regarding stand age and fire severity, earlier work (Grace and Keeley 2006) demonstrated that in the first year after fire (1994) older stands experienced higher fire severities, reduced plant growth, and (indirectly) reduced richness.

The model specifying the effects of abiotic conditions, stand age, and fire severity is shown in Fig. 15.10. In this model we are able to control for variations in abiotic favorability while (i) testing the hypothesis that stand-age effects on richness are mediated by effects on fire severity and (ii) partitioning the direct and indirect (through herb cover) effects of fire severity on post-fire richness.

Results are summarized in Table 15.5. Model fit was good, with a chi-square = 44.88 with 38 degrees of freedom and a *P*-value = 0.205. However, in this model, we continue to allow the precipitation effect on richness to be modeled with free parameters. Specifically, we permit an empirically determined nonlinear response by richness in 1996 to precipitation. As mentioned earlier, we have no theoretical justification for such a nonparametric nonlinear effect, though up to this point in the chapter we have not had an alternative modeling option. In the next section, we illustrate the inclusion of an autoregressive effect in an LTM, which allows us a more meaningful way to explain the

Table 15.5 Select results from latent trajectory model with time-varying and time-invariant covariates (Fig. 15.10). Path from abio to slope was found to be non-significant and set to zero. Model chi-square = 44.88 with 38 degrees of freedom and a P-value = 0.205. Note that a common parameter estimate describes the effect of herb cover on richness for 1995–1998. Paths involving time-invariant covariates shown in bold.

Parameters[a]	Estimate	Std. error	Critical ratio	P-value
slope → r94	0 (fixed)			
slope → r95	1.0 (fixed)			
slope → r96	2.0 (fixed)			
slope → r97	3.0 (fixed)			
slope → r98	4.0 (fixed)			
p → r94	0 (fixed)			
p → r95	1 (fixed)			
p → r96	0.601	0.117	5.141	< 0.001
p → r97	0 (fixed)			
p → r98	2 (fixed)			
herb cover 94 → r94	0.155	0.026	5.970	< 0.001
herb cover 95 → r95	0.042	0.015	2.833	0.005
herb cover 96 → r96	0.042	0.015	2.833	0.005
herb cover 97 → r97	0.042	0.015	2.833	0.005
herb cover 98 → r98	0.042	0.015	2.833	0.005
fire -> r94	**−1.544**	**0.425**	**−3.631**	**< 0.001**
fire -> herb cover94	**−8.420**	**1.804**	**−4.666**	**< 0.001**
age -> fire	**0.061**	**0.013**	**4.841**	**< 0.001**
abio → intercept, i	**0.594**	**0.235**	**2.531**	**0.011**
intercept for intercept[a], i	16.222	11.79	1.376	0.169
mean of slope, s	−4.800	0.369	−13.02	< 0.001
mean of precip, p	7.308	0.526	13.90	< 0.001
mean for abio (fixed est)	48.93	–	–	–
covariance between i and s	3.218	5.657	0.569	0.569
VAR(intercept error)	83.33	25.33	3.290	0.001
VAR(slope)[b]	−2.328	1.828	−1.273	0.203
R^2 for r94	0.25			
R^2 for r95	0.67			
R^2 for r96	0.72			
R^2 for r97	0.73			
R^2 for r98	0.56			
Average R^2 for richness	0.59			
R^2 for herb cover94	0.20			
R^2 for fire	0.21			

[a] Since the LTM intercept latent variable is endogenous in this model, it has an estimated intercept based on its regression on abiotic conditions. To obtain an estimate of the intercept of the latent trajectory in this case, we calculate $16.222 + 48.93 \times 0.595 = 45.35$.
[b] Note that the model-implied variance for the slope estimate was negative. A Wald test showed that the value of −2.328 was not significantly different from a value of zero.

higher-than-expected richness in 1996. Therefore, we will reserve interpretive statements for the next, final model presented.

An autoregressive latent trajectory model (ALT)

As mentioned earlier, an alternative modeling approach is to explain temporal change using an autoregressive approach [Equation (15.7)]. With this approach, we model richness at each time as some function of richness the previous time as well as effects from covariates. To a degree, the latent trajectory and autoregressive models are two different ways of representing the same set of causal forces. In the latent trajectory model, we hypothesize a common factor controlling change over time. For example, if richness is stimulated by fire-induced release of species from the seed bank and the decline over time is a progressive loss of fire-dependent species, a latent trajectory captures this causal process. On the other hand, if richness in each year is dependent on the number of species the previous year because "diversity begets diversity" through seed production and plant recruitment in the following year, then there is a direct effect between richness in one year and richness in the next year. Of course, it is possible that both kinds of processes operate simultaneously, which means an autoregressive path could be included within a latent trajectory model. Here we address a question we ignored earlier about year-to-year effects of richness.

Does elevated richness in one year beget more richness in the next year, all other factors equal? We noticed, as mentioned earlier, that richness in 1996 was higher than expected based on a simple consideration of year-to-year variations in precipitation. Earlier we hypothesized that perhaps this was because weather conditions were favorable in that year in some way other than captured by the annual precipitation. To incorporate this effect, we allowed the loading of precipitation on richness in 1996 to be freely estimated.

An alternative mechanism to explain the elevated richness observed in 1996 is that the high levels of richness in 1995 were promoted by higher than normal precipitation that carried over to 1996. This year-to-year effect can be evaluated by modifying our latent trajectory model to (i) force the precipitation effect to be linear and then (ii) include an autoregressive relationship between richness in 1995 and 1996 (as shown in Fig. 15.11). To accomplish these two things, along with adding a path, we set the loading from precipitation to richness in 1996 to 0, making our model linearly related to the annual precipitation data. Estimation of this model yielded a chi-square of 48.68, with 38 df, and $P = 0.115$. We consider this model to have good fit and accept it as our final model in this demonstration of modeling options. Select results are presented in Table 15.6.

The autoregressive path from species richness in 1995 to richness in 1996 was found to be significant, with a raw path coefficient of 0.073 ($P < 0.001$). This result supports the conclusion that the elevated richness in 1996 can indeed be explained by a carryover effect from 1995, a wetter than average year. We estimate that there was about a 7% enhancement of richness in 1996 from richness in 1995 (coefficient for the path r95 → r96 was 0.073, which is measured in terms of species enhancement in 1996 per species

Table 15.6 Select results from autoregressive latent trajectory (ALT) model (Fig. 15.11). Path from abio to slope was found to be non-significant and set to zero. Model chi-square $= 48.68$ with 38 degrees of freedom and a P-value $= 0.115$. In this model, an effect of richness in 1995 on richness in 1996 was included and the effect of precipitation on richness in 1996 was fixed to a value of 0. The added autoregressive effect is shown in bold (other autoregressive parameters were not significant).

Parameters	Estimate	Std. error	Critical ratio	P-value
slope $\rightarrow$ r94	0 (fixed)			
slope $\rightarrow$ r95	1.0 (fixed)			
slope $\rightarrow$ r96	2.0 (fixed)			
slope $\rightarrow$ r97	3.0 (fixed)			
slope $\rightarrow$ r98	4.0 (fixed)			
p $\rightarrow$ r94	0 (fixed)			
p $\rightarrow$ r95	1 (fixed)			
p $\rightarrow$ r96	0 (fixed)			
p $\rightarrow$ r97	0 (fixed)			
p $\rightarrow$ r98	2 (fixed)			
herb cover 94 $\rightarrow$ r94	0.152	0.026	5.867	< 0.001
herb cover 95 $\rightarrow$ r95	0.037	0.015	2.519	0.012
herb cover 96 $\rightarrow$ r96	0.037	0.015	2.519	0.012
herb cover 97 $\rightarrow$ r97	0.037	0.015	2.519	0.012
herb cover 98 $\rightarrow$ r98	0.037	0.015	2.519	0.012
r95 $\rightarrow$ r96	**0.073**	**0.017**	**4.210**	**< 0.001**
fire -> r94	−1.625	0.424	−3.834	< 0.001
fire -> herb cover94	−8.434	1.803	−4.677	< 0.001
age -> fire	0.061	0.013	4.841	< 0.001
abio $\rightarrow$ intercept, i	0.607	0.228	2.662	0.008
intercept for intercept[a], i	16.338	11.428	1.430	0.153
mean of slope, s	−4.803	0.372	−12.92	< 0.001
mean of precip, p	7.200	0.532	13.54	< 0.001
mean for abio (fixed est)	48.93	–	–	–
covariance between i and s	6.465	5.789	1.117	0.264
VAR(intercept error)	68.34	25.21	2.711	0.007
VAR(slope)[b]	−3.181	1.886	−1.686	0.092
R^2 for r94	0.19			
R^2 for r95	0.63			
R^2 for r96	0.72			
R^2 for r97	0.72			
R^2 for r98	0.55			
Average R^2 for richness	0.56			
R^2 for herb cover94	0.20			
R^2 for fire	0.21			

[a] Since the LTM intercept latent variable is endogenous in this model, it has an estimated intercept based on its regression on abiotic conditions. To obtain an estimate of the intercept of the latent trajectory in this case, we calculate $16.338 + 48.93 \times 0.615 = 45.212$, where 48.93 is the mean for the abio variable.

[b] Note that the model-implied variance for the slope estimate was negative, but not significantly different from a value of zero.

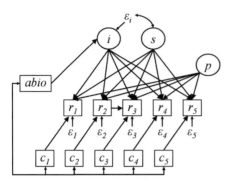

Figure 15.11 Autoregressive latent trajectory model for species richness r_1-r_5 with time-varying covariates p and c_1-c_5, as well as time-invariant covariate, *abio* (abiotic conditions) included. Only a single autoregressive effect $r_2 \rightarrow r_3$ was found to be significant. Results are summarized in Table 15.6.

in 1995). This year-to-year enhancement occurred at the same time that there was a general decline in richness following fire of 4.80 species per year (nearly a 50% total decline over the 5-year period).

In addition to the unstandardized results summarized in Table 15.6, we can also calculate standardized coefficients, which provide an additional way of looking at results. A complete set of results, including the standardized coefficients for indirect and total effects, can be found along with the data and Mplus code at the online supplement. Here we present a few summary findings for the effects of time-invariant covariates; specifically, we present the standardized direct, indirect, and total effects of stand age ("age") on spatial variation in species richness following fire "r_1" (richness in 1994).

First, the results confirm that age had no direct effect on richness in 1994 by comparing models with and without a direct path from age to richness. Inclusion of a path from age to r_1 in Fig. 15.11 leads to a decrease in model chi-square of only 0.46, well below the critical cutoff chi-square difference value of 3.84. This result supports the earlier interpretation (Grace and Keeley 2006) that the reason older stands that burned had lower richness was because of higher fuel loads and higher fire severities in those older stands. Second, we estimate that the standardized total effect of age on richness in 1994 = –0.153. This standardized total effect can be assembled from four standardized path coefficients: *age* → *fire* = 0.459, *fire* → c_1 = –0.443, c_1 → r_1 = 0.333, and *fire* → r_1 = –0.187. The indirect pathway *age* → *fire* → r_1 has a standardized effect of (0.459)(–0.187) = –0.085, while the other indirect pathway *age* → *fire* → c_1 → r_1 has a standardized effect of (0.459)(–0.443)(0.333) = –0.068. Together these add up to the total effect of age on richness of –0.153.

What is perhaps the most biologically interesting effect in this web of relationships is that fire has two different kinds of negative influence on richness, one through a suppression of plant cover following fire in high severity plots (*fire* → c_1 → r_1 = –0.148) and an additional reduction in richness that is independent of cover (*fire* → *r*– = –0.187). These two effects appear to be of roughly equal strength and combine to create a moderately strong overall effect (–0.335).

Future research and development

The SEM framework provides very substantial flexibility in model specification. The possibilities for modeling longitudinal data are extensive and we have only touched on a few of the more basic approaches. The interested reader is encouraged to examine Bollen and Curran (2006), which additionally considers (i) further approaches to modeling nonlinear trajectories, including oscillating dynamics; (ii) comparing trajectories among groups; (iii) mixture models in which group identity is initially unknown; (iv) models involving dichotomous and ordinal responses; (v) ways of dealing with measurement error; and (vi) various ways of dealing with missing data. The use of Bayesian methods in SEM potentially further increases our flexibility in specifying and solving models.

Along with the great variety of options for modeling within the SEM framework is an attendant complexity that can challenge those who wish to learn to use these methods. What we believe is most needed at this point to advance use of these methods is the development of documented examples and decision-support systems that can guide and aid model development and testing. Ecological data analysts encounter numerous complexities, such as spatial autocorrelation, nonlinearities, and hierarchical structures that can make the modeling of longitudinal data complex. This is coupled with the fact that those from the biometric tradition face a substantial learning curve to become proficient with SEM because it is not part of our usual training. Focus here needs to be not only on statistical issues but how to address the many interesting ecological questions that could be examined with these methods.

A somewhat separate area where further work is needed is with the deductive process of projecting the consequences of model results. Particularly deserving of further exploration is the interface between SEM and predictive probabilistic networks (Cowell *et al.* 2007). Both analysis traditions can be classified as forms of graphical modeling, but probabilistic networks (sometimes referred to as Bayesian networks) are specifically designed for forecasting and informing decisions in the face of uncertainty. Ultimately, society wants not only to know how systems work but also the likelihood of their exhibiting various states in the future. This extrapolative process requires us to apply the network structures and parameter estimates we discover with SEM to projections to new situations, something imminently possible, but infrequently considered.

Summary

There are many questions that can be asked of temporal data and the underlying causal forces that structure it. In the SEM tradition, we seek models that test the plausibility of underlying causal forces hypothesized in our models. We also seek to understand the interworkings of multiple processes that simultaneously influence system dynamics. A premium is placed on interpretability in SE models and the focus is rarely on simply arriving at a set of predictors. Because SEM is concerned with underlying causes, which can only rarely be measured directly, latent variables are often incorporated in such models. It is entirely possible to model temporal dynamics using SEM methods without latent variables. Autoregressive models, for example, allow for the examination of causal networks without latent variables. Latent variables in SE models permit a greater degree of abstraction and associated generality, however. In this chapter we emphasize the latent trajectory model (LTM) which can represent the characteristics of a set of repeated measures in a simple and elegant way. Such models seek to describe trajectories with a minimum of parameters, while allowing for complexities to be included as needed.

In the ecological example explored in this chapter, which deals with vegetation dynamics in a fire-prone landscape, we use LTMs to ask whether there is a decline in species richness over time (and how rapid this decline is) once the influences of time-varying covariates are controlled. The pulse-decline pattern studied in this example represents the influence of periodic fire on a fire-adapted community where occasional post-fire recruitment from the seedbank affects richness. The LTM allows us to not only evaluate the hypothesis of a post-fire decline and to examine factors that cause short-term deviations from that decline, but also to examine forces controlling the variations among individual plots within the population. The inclusion of autoregressive effects within the general trajectory model further permits us to determine that wet conditions in a year not only enhance diversity in that year but can have a carryover effect on the following year.

Overall, SEM provides a flexible framework for learning from monitoring data. It is not designed for purely descriptive analyses, but instead, works with models having some theoretical justification. A great variety of variations are possible and their use depends on the hypothesized mechanistic processes and the characteristics of the data. An immediate need is a greater illustration of the many possible uses of SEM with ecological data.

Acknowledgments

We are grateful to Robert Gitzen and Joshua Millspaugh for both guidance in constructing the chapter and for very helpful comments and suggestions on an earlier draft of the chapter. We also thank Michael J. Adams, Brian Mitchell, Diane Larson, Hardin Waddle, and two anonymous reviewers for comments on an earlier version of the manuscript. This work was supported in part by the USGS Inventory and Monitoring Program and the USGS Climate Change Program. The use of trade names is for descriptive purposes only and does not imply endorsement by the US Government.

Section IV

Advanced issues and applications

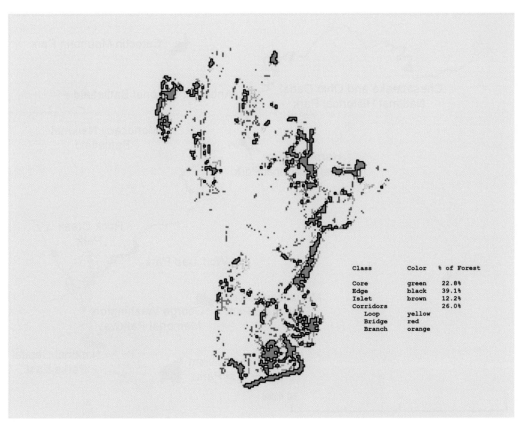

Figure 16.1 Representation of the connectivity of Antietam National Battlefield Park (Maryland, USA) landscape using morphological spatial pattern analysis. Landsat TM image analyzed using the GUIDOS (Graphical User Interface for the Description of image Objects and their Shapes) software version 1.2 (2008).

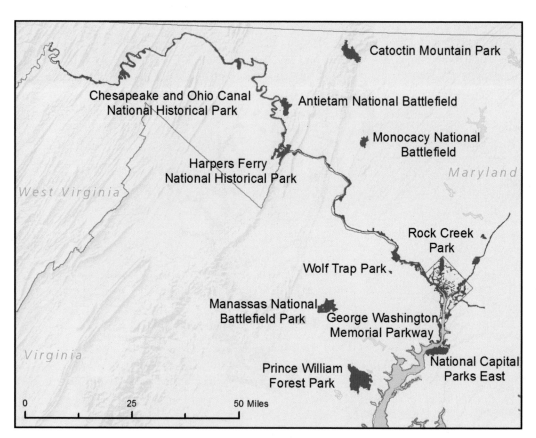

Figure 16.4 The US National Park Service National Capital Region I&M Network (NCRN) includes more than 75 000 acres distributed among 11 parks and is located in the urbanized landscape in and around Washington DC.

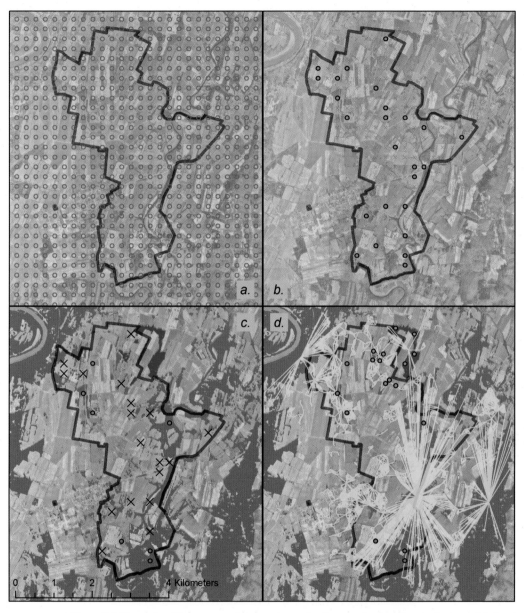

Figure 16.5 Example of the proposed approach to selecting monitoring sites in Antietam National Battlefield Park: (a) GIS was used to lay down a square grid of points, 250 m apart, over the entire NCRN. All points within park boundaries were treated as potential monitoring locations. (b) Of these potential locations, 757 were selected by the GRTS draw, 22 of which (red dots) fell within Antietam NBP. (c) These 22 sites were visited and the 7 that were in forest (red dots) were identified for monitoring. (d) Additional sample locations within potential bottleneck patches were identified by graph theory analysis. The graph representation of forest patches in and around Antietam NBP uses the D_{crit} value from Fig. 16.3 (i.e. 180 m) to define edges (yellow lines). Blue patches [shown in (c) and (d) for clarity] indicate priority locations for the 11 additional monitoring plots (green dots)

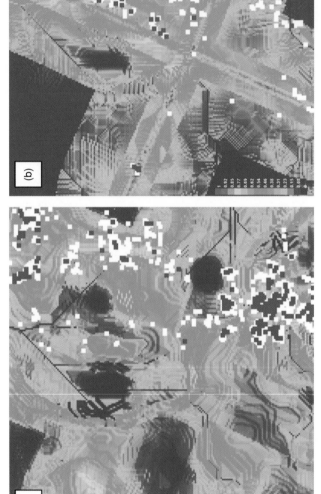

Figure 17.2 Map of the predicted probability of species occurrence overlaid with known species occurrence (squares) for (a) *Tamarix ramosissima* and (b) *Hedysarum scoparium* within the 1 km² study area. Purple squares are the subset of occurrences used in maximum entropy modeling. DEM-derived habitat variables were elevation, slope, distance to roads, solar insolation, and topographic convergence; see Appendix 17.1 for additional information.

16 GRTS and graphs

Monitoring natural resources in urban landscapes

Todd R. Lookingbill, John Paul Schmit, and Shawn L. Carter

Introduction

Environmental monitoring programs are an important tool for providing land managers with a scientific basis for management decisions. However, many ecological processes operate on spatial scales that transcend management boundaries (Schonewald-Cox 1988). For example, adjacent lands may influence protected-area resources via edge effects, source-sink dynamics, or invasion processes (Jones *et al.* 2009). Hydrologic alterations outside management units also may have profound effects on the integrity of resources being managed (Pringle 2000). The impacts of climate change are presenting challenges to resource management at local-to-global scales (Karl *et al.* 2009). This potential disparity between ecological and political boundaries presents an interesting dilemma for natural resource monitoring and is readily apparent in urban and agricultural environments, which tend to be dominated by external stressors (Collins *et al.* 2000). Despite their limited control over external land use, natural resource managers are concerned with processes such as development in the surrounding landscape, as these may lead to habitat loss and degradation that directly impair their resources. As a consequence, the management of the natural resources in and around parks and other areas requires a broad and dynamic understanding of the spatio-temporal patterns of environmental change. If monitoring is to be successful in providing data that inform management, information about regional and landscape context should play a critical role in designing monitoring strategies.

   Urban parks provide a useful example of the influence of external stressors on managed resources. These parks tend to be small in area and tend not to encompass complete ecological units (e.g. watersheds or ecosystems; Forsyth and Musacchio 2005), conditions that could pose significant challenges to natural resource monitoring (Shafer 1995). Despite these challenges, or perhaps because of them, the conservation value of protected areas in urban environments has been increasingly acknowledged (Niemela 1999, Miller and Hobbs 2002, Lookingbill *et al.* 2007), and the significance of urban parks as biological refuges will likely increase as urbanization results in continued land conversion of adjacent habitats.

Design and Analysis of Long-term Ecological Monitoring Studies, ed. R.A. Gitzen, J.J. Millspaugh, A.B. Cooper, and D.S. Licht. Published by Cambridge University Press. © Cambridge University Press 2012.

A more general challenge to long-term natural resource monitoring is to provide the different types of information needed to manage at short- and long-term temporal scales and multiple spatial scales (Levin 1992, Wiens *et al.* 2002). Monitoring to detect long-term trends is often accomplished by taking repeated measurements at permanent monitoring sites. However, data from these locations may provide little information on short-term urgent threats, which may not have been present when the monitoring program was initiated. Conversely, data collected to address urgent needs may not have the spatial or temporal coverage needed to determine broad-scale, long-term patterns. For example, data collected from remote sensing platforms are among the best options for monitoring landscape dynamics within parks and their surrounding ecosystems (Gross *et al.* 2009). Different imagery provides information at different temporal intervals and spatial grains and extents. Long-term records, such as those from moderate resolution Landsat imagery, can be leveraged to track broad-scale trends in development (Elmore and Guinn 2010) or phenology (McNeil *et al.* 2008). High-resolution imagery (e.g. Ikonos, Quickbird) may provide more valuable information for responding to more localized disturbance or other short-term monitoring needs (Lin *et al.* 2008). Approaches for integrating these diverse sources to provide a holistic assessment of natural resource condition are badly needed (see Gardner *et al.* 2008 for an example application of how this integration can be efficiently achieved).

In addition to scale considerations, the method of selecting monitoring locations is also an important determinant of the types of analyses that can be used on the data. For example, monitoring sites can be selected using either a probability-based or a model-based selection process (Edwards 1998; see Chapter 2). Randomized site selection based on probability sampling allows design-based inference that requires no assumptions about the population in order to produce valid estimates and a measure of the uncertainty of these estimates. Model-based selection occurs when locations are purposely (not randomly) chosen based on predictions that they have unique characteristics or have particular importance to the area being monitored. Importantly, model-based selection requires a rigorous theoretical or statistical model to take the place of, or supplement, randomization to guide the selection process and subsequent data analysis. Model-based selection has the advantage that it may meet survey objectives (e.g. achieving a specified precision) with less data collection (Urban 2000). However, the conclusions drawn from model-based selection are only valid if the underlying model itself is a valid representation of the system of interest.

In this chapter, we describe a hybrid approach to long-term monitoring that takes a regional perspective but does not ignore the specifics of local ecosystem dynamics (Box 16.1). The approach provides a probability-based sampling framework while allowing flexibility to include model-based samples that address more local, urgent management needs. This combined approach draws on the strengths of both design-based and model-based monitoring and addresses some of the limitations from which each suffers when applied individually. We illustrate the concepts with an example from our work monitoring forest dynamics within national parks of the National Capital Region in Maryland, Virginia, West Virginia, and Washington, DC, USA.

Box 16.1 Take-home messages for program managers

The status and trends of environmental resources are determined by local and regional influences that often transcend management boundaries. Therefore, long-term monitoring of landscapes may require information at multiple spatial scales and from multiple property owners to respond to varied and changing management objectives. Meeting such diverse needs requires a flexible, multi-layered approach. For example, the monitoring of landscape dynamics at multiple spatial scales can often benefit from the coordinated application of both direct field-based and remotely sensed observations. This chapter describes how a hybrid approach to monitoring can be used to consider specific landscape processes with reference to a larger, probability-based sample.

The first step of this hybrid approach selects permanent monitoring sites within a network of management units using probability sampling, as discussed throughout this volume. Data from these sites support unbiased estimation of regional status and trends for the overall population of interest, and detection of unanticipated patterns. The second step uses model output to locate additional monitoring sites in areas predicted to be of high importance for landscape-level processes. This chapter provides an illustrative example of the hybrid approach which incorporates use of a graph theory model applied to a regional land cover data set to identify forest patches of special management importance in maintaining habitat connectivity. Data from supplemental monitoring in these patches is integrated with monitoring data from network-wide monitoring to assess the condition of these high-priority patches.

The hybrid approach is especially useful when (i) information is needed at multiple resolutions and/or multiple extents (e.g. inside and outside the management unit boundaries); and/or (ii) information is needed to respond to multiple management challenges (e.g. some stressors are known, while others are not). Combining a model-based sample with probability sampling can give managers flexibility to address specific issues of high current importance while maintaining a surveillance program to flag unexpected environmental damage. Graph theory is especially useful for questions of spatial connectivity, but other model frameworks would also be appropriate based on the specific monitoring objectives. Location of the model-selected monitoring sites can change over time based on periodic reassessment of monitoring objectives, available information, and even land use change in and around the study area. In this sense, the dynamic, model-based and permanent-site, probability-based components of the monitoring are truly complementary. The designs are also complementary in that they facilitate partnerships between land managers with expertise and resources for long-term, repetitive measurement of their administrative units and researchers with complementary knowledge about the surrounding landscape context and interest in testing hypotheses about specific ecological variables.

A hybrid approach

Hybrid designs combining a fixed set of monitoring sites with additional, potentially "roving" sites whose locations are optimized to inform dynamic modeling of spatio-temporal processes and address shorter-term management priorities, have been advocated for the detection of change in natural resources and for understanding the underlying dynamics that produce change (Hooten *et al.* 2009a). We have observed that hybrid approaches can be useful when either of the following conditions are met (see also Brus and de Gruijter 1997): (i) information is needed at multiple resolutions and/or multiple extents (e.g. inside and outside the management unit boundaries); or (ii) information is needed to respond to multiple management challenges (e.g. some stressors are known, while others are not). We address each of these situations below.

Often, a significant drawback of a sampling strategy based on randomization alone is its inefficiency at capturing fine-scale spatial patterns over large spatial extents. Understanding these multi-scale patterns is important for protected areas that are interested in preserving spatially dependent processes, such as the movement of wildlife across the landscape. Data at regional scales may provide useful information on the constraints on these processes, but finer-scale data are required to understand mechanisms. One way of addressing this is through multi-stage sampling with design-based or model-assisted inference. For example, Nusser *et al.* (1998) provide an example of a two-stage sampling design in which land-cover measurements of primary units are used to improve estimates of variables measured at the secondary (sample-point) level and to detect changes that would not usually have been observable at the sample-point level. Alternatively, design- and model-based sampling can be combined. Models are useful tools for studying mechanisms, and model refinement (i.e. parameter optimization) is an additional benefit of model-based sampling. Thus, hybrid designs facilitate partnerships with a research community focused on testing hypothesis about specific ecological processes and variables (Jones *et al.* 2010).

Additionally, when the organism or process of interest is highly mobile or crosses an ecological/administrative boundary, different sampling strategies may be required at different locations on the landscape. Methods that efficiently identify the best sites for targeted sampling are especially important when those sites may lie outside of the direct administrative control of the monitoring agency. Adopting a monitoring approach that is sensitive to regional and landscape processes and stressors often requires coordinating efforts among multiple land owners. Hybrid approaches that use probability sampling for surveillance monitoring of regional trends, can use alternative methods to identify key neighbors for collaborative monitoring efforts.

A combination of sampling methods also can be used to respond to different monitoring challenges, such as the needs for assessing long-term changes in resource condition as well as addressing current management priorities and current hypotheses about the system (Chapters 3, 4, 22). Natural resource monitoring data can be used in either a retrospective or predictive manner (Yoccoz *et al.* 2001). The retrospective or post-hoc approach attempts to draw inferences from monitoring data after they have been collected, with no substantial effort to assess relationships a priori. A spatially balanced, probability sampling design lends itself to these types of evaluations, which can be useful for capturing unanticipated events such as the population decline of a particular species following the introduction of a novel pathogen. The predictive approach uses existing knowledge to guide data collection to address specific hypotheses. For example, model-based methods can apply information on habitat preferences and life history characteristics of a specific endangered species to identify habitat patches of special importance to populations of concern. Alternatively, model outcomes can be used to propose specific sites for potential management actions. These important patches may be missed entirely by the regional-level, random sampling. Predictive and retrospective uses of monitoring data both can provide managers with valuable information; whether it be through evaluating the impacts of past management actions or identifying the relative benefits of proposed actions. Whenever possible, however, predictive hypotheses should be developed because they allow for a more controlled examination of cause–effect relationships (Lookingbill *et al.* 2007).

Monitoring landscape dynamics

Our approach to forest vegetation sampling combines methods for generating a spatially balanced, probability sample of vegetation plots with model-based methods for identifying monitoring locations on the landscape that are particularly important to landscape processes and thus justify additional sampling effort. The randomized sampling provides an unbiased, coarse-scale assessment of regional trends. The model-based analysis, in contrast, targets forest patches that may have disproportionately large effects on ecosystem processes such as species dispersal. The model also identifies specific park neighbors whose properties potentially impact park resources and therefore helps to prioritize regional monitoring partnerships.

Probability designs are covered thoroughly in other chapters of this book (see Chapters 5 and 6), and we will not go into further detail here except to mention that monitoring of landscape dynamics at broad spatial scales can often benefit from the coordinated application of both direct, field-based and remotely sensed observations. We discuss above the importance of matching the spatial and temporal scale of imagery to the ecological pattern or process being assessed. An additional consideration in using remotely sensed data for monitoring is the choice of landscape metric. Literally hundreds of landscape pattern indices are available within the FRAGSTATS software package alone (McGarigal *et al.* 2002), and new metrics continue to be developed at a dizzying pace. The application of surface rather than patch-based metrics (McGarigal *et al.* 2009) and morphological spatial pattern analysis (Vogt *et al.* 2007) represent especially promising recent developments (Fig. 16.1). The Heinz Center (Heinz Center 2008) and US National Park Service (Gross *et al.* 2009), among others, have emphasized the value of landscape pattern indices in monitoring programs. A common pitfall is the selection of pattern indices for monitoring that are ecologically meaningful and independent. Cushman *et al.* (2008) provide guidance for metric selection based on seven fundamental properties of landscape configuration. Townsend *et al.* (2009) recommend a parsimonious

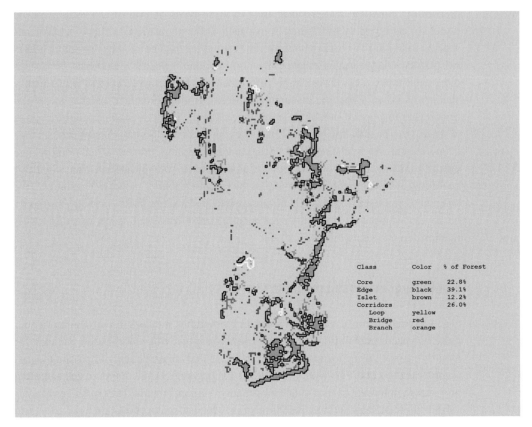

Class	Color	% of Forest
Core	green	22.8%
Edge	black	39.1%
Islet	brown	12.2%
Corridors		26.0%
Loop	yellow	
Bridge	red	
Branch	orange	

Figure 16.1 Representation of the connectivity of Antietam National Battlefield Park (Maryland, USA) landscape using morphological spatial pattern analysis. Landsat TM image analyzed using the GUIDOS (Graphical User Interface for the Description of image Objects and their Shapes) software version 1.2 (2008). See plate section for color version.

set of five metrics, including the graph theory-based metric described in our case study below, for monitoring landscapes confronted by fragmentation pressures.

A description of model-based approaches for hypothesis-driven monitoring requires additional attention, as these methods are traditionally less familiar to resource managers than probability-based designs (Gregoire 1998). With model-based frameworks, the model serves as a basis to make inference about a population parameter of interest. Urban (2000, 2002) provides some excellent examples of the application of habitat models, decision trees, and geostatistics to inform sample designs. Jobe and White (2009) provide another creative example using cost–distance modeling for human accessibility to assess vegetation monitoring plots accumulated over the last three decades in Great Smoky Mountains National Park.

Graph theory

Our case study describes the use of graph theory as a model for hypothesis-based, predictive modeling. Graph theory is an analytic technique for evaluating spatial properties

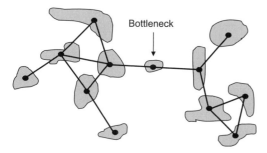

Figure 16.2 The connectivity of habitat patches in a landscape can be represented using graph theory. Pairs of patches are considered either connected or unconnected based on the distance between their edges. A graph can then be drawn that represents patches with points and connections between patches with lines. In this illustration, a bottleneck is shown that is highly important to linking two potential subcomponents of the graph.

of networks that has been applied for decades in fields such as transportation and communications (Harary 1969; Hayes 2000a, b). Recently, there have been an increasing number of applications of these and related connectivity methods such as circuit theory to assess the consequences of habitat modification and landscape change (e.g. Calabrese and Fagan 2004, Minor and Urban 2008, Rayfield *et al*. 2011). These approaches generally treat a landscape as a network of discrete habitat patches. The graph model considers pairs of these patches as either connected or unconnected based on some measure (Euclidean distance or other) of their spatial proximity (Fig. 16.2).

The graph-based model is appealing for large-scale monitoring because it provides a visually intuitive representation of landscape connectivity and provides a computationally efficient structure for analyzing data sets, e.g. by summarizing remotely sensed data collected from millions of pixels to a small subset of patch centroids for analysis and interpretation. A number of well-developed indices are available for quantifying landscape attributes based on properties of the landscape graph (see Pascual-Hortal and Saura 2006, Kindlmann and Burel 2008). One simple measure of the connectivity of a landscape is the proportion of total habitat area that is considered connected (A_{LC}; Ferrari *et al*. 2007). In addition to providing basic information about the overall landscape structure, the metric can be used to identify individual habitat patches of special importance by examining how selective patch removal changes the metric value (Urban and Keitt 2001). These critical bottlenecks for long-distance movement potential (Fig. 16.2) would be patches that if lost, damaged, or modified would greatly reduce the traversability of the landscape (e.g. result in significant decreases in A_{LC}).

A key challenge to the application of connectivity models to long-term monitoring is identifying the appropriate scale to parameterize the model. The construction of a landscape graph is organism-specific, and the same set of patches may yield different landscape graphs for species with shorter or longer dispersal capabilities. In instances where the dispersal characteristics (e.g. dispersal probability function, maximum dispersal distance) of an organism of concern are known, those attributes may be used to define patch connections (e.g. Goetz *et al*. 2009, Lookingbill *et al*. 2010). Otherwise, multiple dispersal distances can be systematically evaluated to determine the threshold

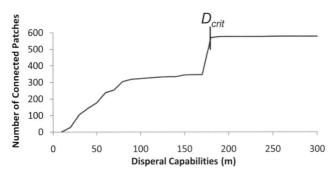

Figure 16.3 Connectivity (measured as number of connected patches) as a function of theoretical dispersal capabilities. A threshold of connectivity (D_{crit}) occurs at a dispersal capability of 180 m. Organisms capable of moving 180 m from one forest patch to another can move among nearly 100% of the patches in the landscape. This example is derived for the forests of Antietam National Battlefield Park.

distance at which the landscape may switch between being acceptably connected vs. disconnected (D_{crit}). Dispersal capabilities have been shown to be strongly nonlinear for most landscapes (Gardner *et al.* 1987), and D_{crit} values are often readily apparent from a curve of graph metrics (such as A_{LC} or the number of connected patches) versus dispersal distance (Fig. 16.3). One rule of thumb for assigning D_{crit} is the minimum distance a hypothetical organism would need to be able to disperse through non-habitat to be capable of moving among all habitat patches (i.e. for A_{LC} to equal 1.0). The value can be used to construct a graph by drawing lines between all patches separated by less than D_{crit}. The resulting graph represents the landscape as highly connected under current conditions, but highly sensitive to any loss or degradation of habitat. Graphs built using these D_{crit} threshold distances to identify patches of interest such as potential dispersal bottlenecks can be used for prioritizing site selection for monitoring purposes.

By identifying locally significant patches within a landscape context, the analysis provides a powerful tool for resource monitoring. Flexibility built into the sample design of long-term monitoring protocols allows the distribution of sample plots to be at least partially guided by specific natural resource concerns, such as preserving the overall landscape connectivity or the connectivity for a specific species. The contribution of a graph-theory analysis to the overall hybrid sampling strategy is thus to provide a complementary sample list frame for the assessment of targeted habitat changes through time.

The approach is tiered in that annual monitoring effort can be allocated first to the sampling of probability-based permanent plots. Remaining monitoring resources can then be directed towards the supplemental plots identified by the model analysis. Balancing total effort between these two components will depend on the overall objectives of the monitoring program and expected comparative value or importance of information produced by each component. For example, greater effort should be allocated to the model-based component when a single stressor is thought to be dominating the natural resource of concern or specific management actions are to be evaluated. When multiple or unknown stressors are thought to be dominating the system, in contrast, it would be appropriate to allocate greater sampling effort to the probability-based component of the sampling plan. In these instances, generating a sufficient model to test a hypothesis about one stressor is complicated by the variability introduced by the other stressors. The model-based component of the plan, though downplayed, is still important in attempting to disentangle the expected responses associated with the various stressors.

Additional considerations

Sample frames in relation to multi-scale monitoring

One of the first steps in any sample design is deciding on a target universe (target population) and related sample frame from which to sample to provide estimates for the target universe (Chapters 2, 5). In many cases these will correspond to simply the boundaries of the protected area or management unit being monitored. This will be commonly the case where the site is contiguous and has a more or less compact shape. In some cases, however, the issue is more complicated. Protected areas are frequently established as several discontinuous management units, such as a network of parks. The land between the units is not managed by the protected area, may be used for some other purpose (developed, agricultural, etc.), and may not be part of the target population. In this case, a sample could be selected from a single sample frame encompassing all management units (either including or excluding areas in between units), or independently from within each unit separately.

A similar situation occurs when a unit has an elongated shape, rather than a compact shape. This can occur when an area protects a linear natural feature, such as a river, a shoreline, or a barrier island. Conditions at one end of the site may differ drastically from conditions at the other end. When this occurs, it may be desirable to divide the area into sections with similar conditions, essentially leading to a separate sample frame for each section.

Sampling each management unit or section separately (i.e. with each of multiple smaller frames encompassing one management unit or section) may be preferred when the site or sites can be unambiguously divided up into smaller homogeneous units that are likely to be impacted by similar stressors and are the focus of management actions for the foreseeable future. This is essentially a stratified sampling approach if at least some of the same variables are measured in all units, with the goal in this case being to support independent estimates for each unit rather than to increase precision of estimates for the entire target population (Chapter 5). However, this independent sampling of each unit could decrease statistical power and complicate analysis when investigating issues at larger scales. Sampling multiple management units at once from one or more larger sample frames typically is preferred when it is difficult to divide the site into smaller units, or when stressors or management activities will cut across unit boundaries. For example, a monitoring program may be tasked with generating data for a network of a number of units in close proximity, rather than just one. In this case it may be desirable to identify conditions or trends that are common to all of the units rather than focus on issues unique to each one. Similarly, there may be a need to maintain maximum flexibility for future analyses combining subsets of data from each management unit for network-wide analyses. In these situations, a sample frame and sampling process that ignores the management unit "identity" is useful when generating a sample. On the other hand, it may be that few monitoring sites occur in any given area, and therefore there may be little power to examine status and trends on smaller scales. Regardless of what decisions are made about sampling frames, it is important to consider these issues ahead of time, and anticipate the need for analysis at a variety of scales, some of which may not yet be identified at the time the monitoring program is initiated.

Analytical considerations

As discussed elsewhere in this book (Chapter 2), specifying the approaches for analyzing the data is a critical early step in developing a quantitatively sound monitoring program. Selection of analysis methods typically follows closely from the determination of objectives. Model-based sampling designs provide specific hypotheses that can be evaluated, often with standard parametric statistics – e.g. comparing average metric values for different classes of locations on the landscape. Data analysis of probability designs, such as GRTS designs, are also specified elsewhere in the book (e.g. Chapters 6, 11, 14). More interesting, from the perspective of this chapter, is a discussion of the analytic framework for the hybrid approach.

There are two strategies for analyzing the data from the different monitoring components of the hybrid design. The first is to compare the data from the model-based approach to that of the random sampling. As an example, the fundamental benefit of a graph-theory analysis is that it will determine which habitat patches are likely to play a crucial role for some species, solely on the basis of the patch location in the landscape. The goal of the model-based portion of the monitoring could be to compare the habitat quality within these key patches to that of the protected area as a whole as determined from the probability-based sampling.

The second data-analysis strategy is to combine data from the two sampling frames to determine status and trends for the entire protected area. To do this, the data can be analyzed as an unequal probability stratified sample. For the analysis to be correct, it is important to keep in mind the location in the key patches could have been selected during either the first, probability sample or the second, model-based sample. Therefore the selection probability is the combined probability from each of these two components.

Forest monitoring of US National Capital Region parks

The US National Park Service's Inventory and Monitoring (NPS I&M) program was established to develop and implement a systematic and rigorous approach to monitoring natural resources in National Parks (Kaiser 2000, Fancy *et al.* 2009; see also Chapter 22).

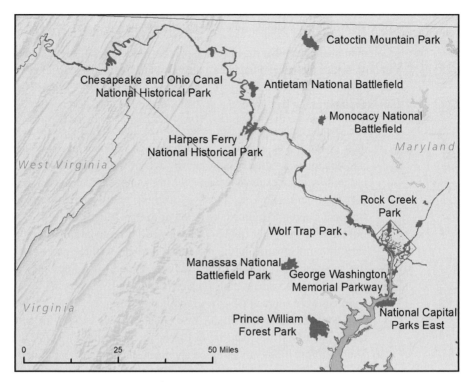

Figure 16.4 The US National Park Service National Capital Region I&M Network (NCRN) includes more than 75 000 acres distributed among 11 parks and is located in the urbanized landscape in and around Washington DC. See plate section for color version.

As part of the I&M program, parks of the eastern and midwestern US have collaborated to implement a consistent forest monitoring protocol based on the US Forest Service's Forest Inventory Analysis (FIA) and Forest Health Monitoring programs (Comiskey *et al.* 2009, Tierney *et al.* 2009). Key objectives of the program are to determine status and trends in (i) tree and shrub distribution and richness; (ii) tree and shrub basal area and density; (iii) volume of coarse woody debris (logs and large branches on the ground); (iv) presence and cover of exotic plant species; and (v) presence of certain forest pests and diseases.

The National Capital Region I&M Network (NCRN) has tested a strategy that combines the spatially balanced randomized sampling being conducted throughout the region with park-based modeling to define key locations for forest monitoring of landscape dynamics. We present an example of how a hybrid, two-step sampling design can be applied to forest monitoring of the 11 parks in the NCRN (Fig. 16.4). Our example also illustrates important decisions that must be addressed to implement the hybrid approach effectively (Box 16.2). Forests are the predominant natural vegetation cover for the parks in the NCRN, and most of the parks in the region have a specific mandate related to management of forests in their founding legislation. These include requirements to preserve natural forests, to preserve wildlife habitat, to protect watersheds, to provide recreation,

---

Box 16.2 Common challenges: hybridization issues

Common challenges that may accompany the application of the methods described in this chapter include: (i) balancing objectives for regional vs. local inference when designing the sampling strategy, (ii) identifying specific hypotheses to be examined, and (iii) integrating data from the probability and model-based components.

(i) For monitoring variables that require direct, field-based measurements, a common challenge in balancing local-scale and regional monitoring is whether samples for the region should be selected from a single sample frame ignoring unit boundaries or independently from smaller frames each encompassing a management unit. Our objectives working with small national parks in the National Capital Region Network led us to draw a network-wide GRTS sample from a large sampling frame. This allowed for regional inference but provided very little information at the individual park level (three parks had fewer than five samples each) and no information on resource condition in adjacent lands (establishing permanent plots outside park boundaries was not feasible). These concerns were addressed by the graph-theory sampling, which extended beyond park edges to watershed boundaries. Other possible extents for park sampling using remotely sensed data are described in Townsend *et al.* (2009).

(ii) The flexibility of the model-based approach can be a hidden pitfall if no leading hypothesis emerges to guide the sampling effort. Both the choice of model and parameterization of the model should be guided by the hypothesis defined at the outset of the sampling effort. The matching of appropriate model to hypothesis is a critical step in the process. The inferences drawn from the model-based sampling are highly dependent on the validity of the underlying model itself. An added benefit of the model-based sampling is that it will yield data that can be used to refine the model, and thus through continued iteration work toward design optimization (Hooten *et al.* 2009a).

(iii) Combining the data sets to provide integrated inferences at either the local scale or regional scales may be desirable, but should be undertaken with caution. For example, it would be inappropriate to simply lump the model-based sites with other sites as a single, statistical sample. Still, comparisons of data and estimates from each component may be of high value for examining and refining hypotheses of interest. Our case study provides an example of how the data from the model-based portion of the monitoring could be compared to data from the probability-based sampling to assess the habitat quality within key forest patches relative to that of the protected area as a whole.

---

and to protect scenic vistas. The most severe threats to park natural resources include high browsing pressure from white-tailed deer (*Odocoileus virginianus*), invasion by exotic plant species, loss of tree species such as eastern hemlock (*Tsuga canadensis*) and flowering dogwood (*Cornus florida*) due to pathogens, and regional changes in land use. Collectively, these threats have the potential to cause drastic changes in vegetation structure, species dominance and composition, and resources available to animal species. Effective monitoring should detect significant changes caused by known stressors and capture unanticipated trends in forest vegetation.

Site selection

Sample frame

The parks in the NCRN exemplify many of the challenges discussed above in deciding upon an appropriate sampling frame. In the Washington DC metropolitan area alone, the NPS is responsible for over 120 tracts of land of various sizes, approximately half of which are managed as natural areas. This large number makes it impractical to have a separate monitoring program for each tract. The forested tracts are managed by five different parks – George Washington Memorial Parkway, National Capital Parks East, Rock Creek Park, Wolf Trap Park, and part of the C&O Canal. The borders of these parks were established by legislation, and do not necessarily follow any natural boundary. All of these areas are impacted by urbanization and most have similar vegetation.

The C&O Canal is an example of a long, linear park. It stretches for over 290 km, from Washington, DC to Cumberland, Maryland, along the north bank of the Potomac River. Along its length, it borders dense urban areas, agricultural lands, and natural forests. The park has a common border with the George Washington Memorial Parkway and Rock Creek Park, cuts Harpers Ferry in half and passes less than 1 km from Antietam National Battlefield Park (NBP). In these areas, the lands managed by the C&O Canal often have more in common with the neighboring park than they do with the land in more distant parts of the canal.

For these and other similar reasons, it was decided that dividing the NCRN into park-based or other separate sample frames for forest monitoring would not be beneficial. Instead, the entire network of parks was included in a single sample frame and monitoring locations were chosen at random from this entire frame. When it is desirable to look at a specific park or other sub-area, the relevant data can be used for a more local analysis. However, a recognized drawback of our sample-frame decision is that the sample intensity in any particular area may be small.

Probability sampling

Once the appropriate sample frame was established, data for the first part of the sampling were collected following the NPS I&M forest monitoring protocol being uniformly implemented for eight I&M networks and three prototype parks in the eastern US (Comiskey *et al.* 2009). First, a list of potential sites was generated using a Geographic Information System (GIS) to establish a square grid of points, 250 m apart, over the entire region. All points that fell within park boundaries were treated as potential monitoring locations. Once the list of potential sites was determined, a Generalized Random Tessellation Stratified (GRTS; Stevens and Olsen 2004) design was used to randomly order the sites for monitoring (i.e. via the reverse hierarchical ordering approach described in Chapter 6). Sites were then visited in the random order generated by the

Table 16.1 The number of total grid points, sites visited, and sites selected for sampling per park in the US National Park Service National Capital Region Network using the GRTS design.

Park	Points in park	Points considered	Points monitored
Antietam National Battlefield Park	210	22	7
Catoctin Mountain Park	365	53	45
C&O Canal National Historical Park	1406	215	73
George Washington Memorial Parkway	332	68	33
Harpers Ferry National Historical Park	247	34	20
Manassas National Battlefield Park	284	45	15
Monocacy National Battlefield Park	106	18	3
National Capital Parks East	718	107	46
Prince William Forest Park	811	164	139
Rock Creek Park	195	30	18
Wolf Trap Park for the Performing Arts	9	1	1
National Capitol Region Network Total	4683	757	400

GRTS design process, and the first 400 sites on the list that were located in forest habitat and suitable for forest monitoring (e.g. safe for field crew, not overlapping with sensitive cultural resources, etc.) were selected (Table 16.1). These are being monitored with a 4-year serially alternating panel design (100 plots/year; see Chapter 7 for discussion of panel designs). Four hundred plots were chosen as a monitoring effort as this number is feasible given budgetary and staffing considerations. An initial power analysis indicated that 400 plots provides sufficient power to detect change in a wide variety of forest characteristics including tree density and basal area, density of coarse woody debris, and occupancy of exotic species (Schmit *et al.* 2009).

The flexibility of the GRTS approach was especially useful for working in NCRN parks, because it allowed sites to be excluded from monitoring based on the presence of vulnerable cultural/archeological resources or due to concerns with maintenance or visitor use. The GRTS design also allows the program to cope with unforeseen circumstances, such as changes in budget, or with potential monitoring sites which are inaccessible or unsuited for monitoring. These advantages make probability-based surveys such as GRTS a popular method for natural resource monitoring in urban landscapes such as then NCRN (e.g. Hope *et al.* 2003, Nowak 2008).

Combining probability and model-based sampling

For the entire region, the spatially balanced GRTS sampling provides a basis for statistical interpretation of broad-scale forest change. For any individual park, however, only a limited number of samples are collected (Table 16.1). The GRTS-selected sample sites are not necessarily located at locations within the landscape that are most sensitive to change or are of most interest to park managers; such information was not incorporated into the probability-sample design. However, the graph-based model analysis provides an efficient means for addressing these local management needs. The hybrid monitoring approach, therefore, allows for regional-level monitoring while also providing park-level flexibility to add samples that inform local management concerns. We illustrate this integration of park-based modeling information for Antietam NBP.

A consideration of spatial processes and landscape context in site selection is particularly important for the small, mixed land-use parks of the NCRN that can be heavily influenced by external stressors. Antietam NBP was established to preserve the site of the US Civil War battle of Antietam and is mandated to preserve the landscape as it was during the battle in 1862. Since the war, land use has changed considerably, requiring the park to undertake battlefield restoration activities to restore the historical vegetation (e.g. cutting regenerating forest to maintain open battlefields). The vegetation in the park is predominantly open fields, which surround a number of small woodlots (Fig. 16.5). Forest cover comprises 35% of the total area of the park, with significant forested areas occurring along Antietam Creek on the east side of the park, and the Potomac River, just to the west of the park. The land surrounding the park is a mixture of agricultural, forest, and urbanized areas.

As part of the region-wide forest monitoring in NPS units, 210 potential sampling sites were located within the park's 780-ha legislative boundary (Fig. 16.5a). Of these 210 potential sites, 22 points were sufficiently high on the GRTS ordered list of potential samples to be potentially included in the forest monitoring (Fig. 16.5b). Of these 22 points, 15 were eliminated, either because they fell on land which is not currently owned by the NPS or because they fell on NPS land which is maintained as an open field (Fig. 16.5c).

One of the objectives of forest monitoring in the park is to evaluate the condition of forest dispersal corridors for birds and small mammals. Additional sampling is required to understand the quality of corridors. In relatively small parks with high levels of fragmentation, like Antietam NBP, the most important corridors to promote park connectivity may not always be located within the park. Therefore, it is useful to also consider the quality of potential corridors that lie just outside the park.

Data for the model-based sampling were collected following the graph-theory methods outlined in Townsend et al. (2009). We first created a graph representation of the park using 10-m SPOT satellite imagery classified as forest/non-forest. The graph includes the 578 discrete forest patches contained within the park, along with 663 adjacent patches contained in small watersheds that feed into the park. By constructing graphs covering a range of potential dispersal capabilities, a critical dispersal threshold (D_{crit}) was identified (Fig. 16.3). This distance indicates the minimum distance an organism would need to be capable of traveling through non-forest to be able to move among all 578 patches in the park (i.e. dispersal capability of at least 180 m for the Antietam landscape). We used this D_{crit} value to construct a graph that represents park forests as fully connected (Fig. 16.5d). This approach assumes there is not a specific species of concern for the monitoring and applies the rule-of-thumb approach for creating the graph described above. If instead, a single species were the focus of the monitoring effort, then information on that species dispersal characteristics and other life history parameters could be used to build the graph (see Lookingbill et al. 2010 for an example of a single-species analysis). Separate graphs could also be created for multiple species with differing dispersal capabilities and the results overlaid to determine priority areas for monitoring.

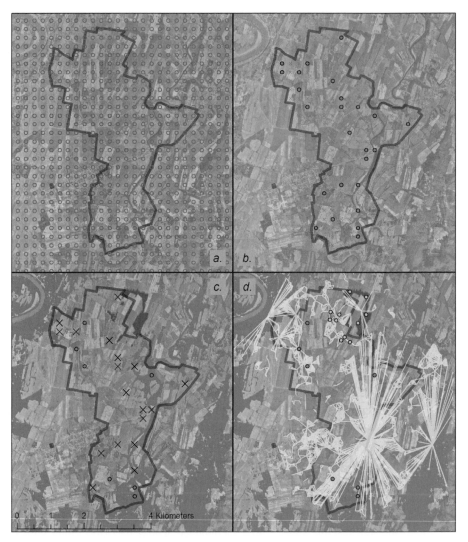

Figure 16.5 Example of the proposed approach to selecting monitoring sites in Antietam National Battlefield Park: (a) GIS was used to lay down a square grid of points, 250 m apart, over the entire NCRN. All points within park boundaries were treated as potential monitoring locations. (b) Of these potential locations, 757 were selected by the GRTS draw, 22 of which (red dots) fell within Antietam NBP. (c) These 22 sites were visited and the 7 that were in forest (red dots) were identified for monitoring. (d) Additional sample locations within potential bottleneck patches were identified by graph theory analysis. The graph representation of forest patches in and around Antietam NBP uses the D_{crit} value from Fig. 16.3 (i.e. 180 m) to define edges (yellow lines). Blue patches [shown in (c) and (d) for clarity] indicate priority locations for the 11 additional monitoring plots (green dots). See plate section for color version.

To identify priority areas in our Antietam example, we systematically removed each patch from our graph and then recalculated potential connectivity of the altered landscape. This iterative procedure identified specific patches whose loss would have the greatest effect on reducing the total area of connected habitat (A_{LC}) as shown in blue in Fig. 16.5d. When using our proposed method for developing a graph that represents the landscape as just barely connected (Fig. 16.3), there is generally an easily identifiable threshold change in A_{LC}. In these cases, removing a single patch can cause a change in connectivity of as much as 50%. For Antietam, these critical patches connect the Snavely Ford Woods along Antietam Creek in the eastern portion of the park to the riparian forest along the Potomac River northwest of the park. Those sites within the original sample grid frame (Fig. 16.5a) that fell within these forest patches but were not part of the original 400 sites selected were added to the proposed sampling effort. Also highlighted by the analysis was a relatively large patch of potential corridor forest just outside the park's current boundary. Because the park is continuously reassessing its holdings, this patch was added to the list of locations for monitoring. Sample points within the patch were located within the original GRTS framework, which was regional in scale and was not restricted to the current boundaries of NCRN parks.

Data collection

Forest vegetation plots were sampled at each site selected by the two-step design. The GRTS-based survey was conducted from 2006 to 2009 with the more focused, graph theory-based sampling conducted in 2008. For the GRTS-based component, data collected include identification of all woody plant species on the plot; measurement of trees (including saplings and seedlings), shrubs, and understory herbaceous plants; and quantification of coarse woody debris. These measurements were chosen as they provide information about the effects of deer browse, exotic invasive plants, and pathogens as well as about the quality of the habitat for wildlife. The graph theory-based component was concerned primarily with the quality of forest patches in terms of their invasive species composition and these plants were a focus of the data collection for that part of the sampling, as described below.

Testing management-relevant hypotheses with the hybrid design

The hypothesis to be tested by the model-based sampling was that the condition of important structural corridors in the parks differed from the overall quality of the parks' forests. Degraded corridors would be a cause for concern and potential management action, redirecting treatment to these critical resources. Alternatively, if superior quality corridors existed outside the park, it might be more efficient to focus conservation efforts on building strong partnerships with those neighboring landowners. We compared the invasive species communities found for the two different monitoring components as a means of testing our hypothesis. Eleven additional plots were sampled in patches identified by the graph-theory analysis (Minor *et al.* 2009). These were grouped in three clusters: two clusters comprised of a total of seven plots in critical patches within the park and one cluster of four plots within the potential corridor patch just outside the park's boundary. These data can be compared to the plots sampled in Antietam NBP as part of the GRTS design and the 400 plots sampled regionally.

As an example of our methodology, we compared the presence of invasive plant species on the seven forest plots from the GRTS sample with invasive species from the seven plots selected in Antietam NBP based on the graph-theory analysis. Our goal was to determine if there was a difference in the abundance of invasive plants between the two groups of plots. Within each plot, 12 quadrats (2 m long × 0.5 m wide) were surveyed for exotic invasive plants. Each plot was given an "invaded score", calculated by summing the number of invasive plants found on each quadrat of the plot. Thus, if a plot had two invasive plants, one found on eight quadrats and the other found on six quadrats, the invaded score for the plot would be 14. We determined the mean, variance, and confidence intervals of the invaded score for both the GRTS-based and the graph-based plots; analysis based on the GRTS design used the package "spsurvey" (Kincaid *et al.* 2009; see Chapters 6, 14) in R (R Development Core Team 2011). The plots from the GRTS-based survey had a significantly lower ($P < 0.05$) invaded score (17.1) than the seven critical corridor patches in the park (31.3). This is not surprising, as the plots selected based on the graph-theory model were, by design, located in smaller forest patches within the most fragmented part of the landscape. It was anticipated that these patches would be prone to invasion by edge-loving exotic plants, but the GRTS data provide a frame of reference for quantifying this degradation. If we take our exotic species metric as a reasonable measure of overall forest condition, we can conclude that these patches are unlikely to be serving a function as high quality forest corridors.

Interestingly, the four plots located in the patch outside the park boundary (Fig. 16.5d) had a mean invasive score (20.8) that did not differ significantly from the GRTS-based reference plots in the park. Given the greater forest cover outside the park (~42% in a 5-km buffer surrounding the park) than inside the park (~35% forest), the data fit with an expectation that larger, higher-quality forest corridors would exist outside park boundaries. It is worth noting that the lowest quality plot within this potential corridor (eight different exotic species with a total score of 26) was located closest to the park boundary. This observation leads to a secondary hypothesis to be tested by continued monitoring: invasive spread is occurring from the park into this potentially clean corridor. This concept was completely outside of our original conceptual model for the parks in this region, in which neighboring lands were viewed almost entirely as an external stressor and source of plant invasions. The next round of sampling could reallocate resources from the model-based portion of our sample design to focus on this new hypothesis.

Discussion

A review of the design of broad-scale monitoring programs found that most suffered from the lack of attention to the fundamental question: Why monitor? (Yoccoz *et al.* 2001). A clear and early statement of monitoring objectives is too often lacking (see also Chapters 2, 3, 18, 22). Without an explicit a priori objective statement, retrospective analyses of monitoring data permits only weak inference regarding the response of the system to management actions or proposed actions. Stronger inferences can be attained by comparing monitoring observations to existing model-based hypotheses. The graph-theory model proposed here provides a framework for this model-based sampling. The approach is particularly attractive for NCRN parks because it focuses monitoring around the issue of connectivity in fragmented landscapes, a topic of special concern for park management in these urbanizing settings. Future directions of this work include developing stronger linkages between the monitoring design and natural resource management activities within the parks (e.g. Lookingbill *et al.* 2008). The flexibility and hypothesis driven nature of the graph-theory approach provide a valuable tool for assessing effectiveness of management within an adaptive management framework.

The method of combining the design-based (based on the probabilistic GRTS design) and model-based (based on the graph-theory model) sampling addresses the inability of monitoring efforts to exhaustively survey large areas. In response to this challenge, a hybrid design produces: (i) a spatially balanced sample that is appropriate for regional trend detection and is not subject to biases produced by subjective selection of sites, and (ii) an efficient sample targeting sites that, based on the graph-theory model, are most critical to landscape-scale processes (in this case, species invasions). By providing information at the regional and landscape scales, data from a hybrid design can be valuable to management at multiple levels.

Future research and development

We have provided an example of how data from the two different sampling components can be compared. Further, the samples could be integrated regionally, for example, as an unequal probability stratified sample. One promising direction of future research would be to consider how the additional flexibility of the model design could be leveraged to provide a regional assessment that fluidly transcends spatial scale (e.g. through a correlogram or other similar spatial analysis).

The example application provided in this chapter also illustrates the selection of supplemental sites based on a hypothetical organism with movement capabilities equal to our rule-of-thumb D_{crit} distance that assumes full landscape connectance. Sites can also be identified as important in the context of managing for a particular wildlife species. For example, a number of small rodents and amphibians occupy the riparian forests of Antietam NBP, many of which have been shown to have dispersal abilities across non-habitat in the range of 180–500 m (Corry and Nassauer 2004). In cases where the focus is on a specific species, a reduced amount of data collection can take place at each of the plots, relating only to those aspects of the vegetation which are important to the species of interest. Additional refinements to the graph-theory model (e.g. multi-species, directional movement) or inclusion of altogether different model-based frameworks are other potential fruitful areas of research.

Summary

We have provided an overview of study design and analytic issues associated with long-term monitoring in mixed-use landscapes. The potential disparity between ecological and political boundaries in these landscapes poses a significant challenge for monitoring. An additional challenge is to provide information at relevant temporal and spatial scales to determine long-term trends and to address short-term urgent threats. We presented a hybrid method for confronting these challenges. The method combines spatially balanced, regional sampling with a model-based approach to address more local and immediate management needs. In the first step, locations for long-term monitoring are selected using a GRTS design. The randomized sampling provides an unbiased, coarse-scale assessment of regional trend. In the second step, a graph-theory model is applied to satellite imagery to identify additional monitoring locations to address more local concerns. The case study in the NPS National Capital Region illustrates how, in addition to providing information at different spatial scales, the sampling methods are complementary in their general approaches to monitoring. GRTS sampling provides a post hoc assessment of environmental changes that may or may not have been anticipated. The graph theory-based sampling provides an opportunity for a priori hypothesis testing of specific ecosystem processes – e.g. species dispersal and landscape connectivity. The graph-theory assessment additionally places park features into a broader landscape context by using maps of habitat within and around park units. The results are both hypothesis-based and provide new hypotheses for future monitoring via a flexible design.

Acknowledgments

We would like to thank Mark Lehman for his help in constructing figures. Samantha Tessel, Emily Minor, and Joseph Ferrari assisted with data collection for the case study in Antietam National Battlefield Park.

17 Incorporating predicted species distribution in adaptive and conventional sampling designs

David R. Smith, Yuancai Lei, Christopher A. Walter, and John A. Young

Introduction

Monitoring rare and clustered populations is challenging because of the large effort required to encounter occupied habitat and yield precise population estimates (McDonald 2004). Sampling designs are available to help reduce the effort required to encounter occupied habitat and increase precision, including stratified sampling, probability proportional to size (PPS) sampling, and various adaptive sampling designs (Thompson 2002). Use of these designs is motivated, in an intuitive sense, by each design's ability to allocate more sampling effort where target species are (or are likely to be) and less where they are not. This intuitive approach to allocation of effort can lead to increased precision when variability in the population tends to be higher in areas of high species density or abundance (Box 17.1). Conventional designs, such as stratified and PPS sampling, rely on prior information to allocate effort. For example, prior information could come from predicted species or habitat distributions (Guisan and Zimmermann 2000, Le Lay *et al.* 2010). Use of prior information is not a basic property of adaptive sampling designs, but these designs could use such information when available.

In this chapter, we demonstrate how prior information on species distribution can be incorporated into adaptive and conventional sampling designs. We start the chapter by introducing adaptive sampling, which remains a somewhat novel sampling design even though it was introduced by Thompson (1990) over 20 years ago. We then present a case study illustrating and evaluating the performance of conventional and adaptive sampling designs that either incorporate or ignore predicted species distributions. We examine how these designs compare in terms of efficiency, probability of sampling occupied habitat, and robustness to model inaccuracy. We end the chapter with recommendations and a discussion of future research and developments.

In contrast to many chapters in this volume, we focus on estimation of population parameters or population "state" rather than trend detection (Box 17.2). The purpose of monitoring in conservation contexts is to provide information for effective decision making. In many instances, decisions are based on population state rather than trend (Yoccoz *et al.* 2001, Nichols and Williams 2006, Lyons *et al.* 2008, Martin *et al.* 2009b).

Design and Analysis of Long-term Ecological Monitoring Studies, ed. R.A. Gitzen, J.J. Millspaugh, A.B. Cooper, and D.S. Licht. Published by Cambridge University Press. © Cambridge University Press 2012.

Box 17.1 Take-home messages for program managers

Monitoring rare and clustered species is challenging because it requires a large and expensive effort to sample rare individuals and their habitat. It makes intuitive sense to use sample designs that will allocate more effort where the rare species and habitat are and less where they are not. From a biological perspective there are rapidly diminishing benefits to sampling where target species are not found. From a statistical perspective, this intuitive sense does not negate the need for probability sampling of the entire area of interest if the program is seeking to draw inference from sample sites to this broader area. Moreover, obtaining information from both where a species is and where it is not is important for purposes such as accurately building predictive models. However, because variation and density often go hand-in-hand, there are statistical benefits when allocation of sampling effort can be increased in areas occupied by the rare species. In other words, allocation of effort to areas of higher density tends to be statistically efficient because it tends to increase sampling where variance is highest. Fortunately, there are numerous options for incorporating prior knowledge about species and habitat relationships into statistically valid designs that still produce objective estimates for the entire area of interest.

Either conventional or adaptive sampling designs can use predictions of species distribution to guide sample selection. Conventional designs, such as stratified random sampling with species predictions used to define strata, can help find more individuals and occupied habitat. However, if the underlying predictive model is inaccurate, then these designs can end up being no better and sometimes even worse than simple random sampling, which ignores prior information. Adaptive designs can also make use of species predictions to guide selection of the initial sample, but this sample may then be expanded, based on pre-specified conditions, to include additional units depending on what you find while sampling. For example, if field measurement of a selected sample site documents presence or some other threshold density of the species of interest, adjacent sample units may then be added to the sample and measured. The addition of adaptive sampling should make the sampling design less sensitive to inaccuracies of species predictions used to guide initial sample selection. Adaptive sampling also substantially increases the chance of encountering species and their occupied habitat.

Moreover, managers may need regular (e.g. annual) estimates of population state on a long-term basis to support reoccurring state-dependent decisions (Chapter 4) and for assessing temporal changes (Chapter 13).

Adaptive sampling

Adaptive sampling designs, which allow sampling effort to adapt to observations, are promising for clustered and rare populations because of potential increases in sampling

---

Box 17.2 Common challenges: addressing change

As stated simply by Gerber *et al.* (2005), "monitoring is the systematic acquisition of information over time", and the role of monitoring is to "gain information needed for management decisions". Although monitoring is at times equated with long-term trend estimation, there are many possible outcomes from monitoring that could feed into management decisions, including information on trends, population status to compare to decision or regulatory thresholds, and presence of rare species for endangered species consultation (Burgman 2005, Nichols and Williams 2006, Field *et al.* 2007, Lyons *et al.* 2008, Martin *et al.* 2009b).

Our evaluation of adaptive sampling designs is directed at providing accurate estimates of resource condition (e.g. density) where optimizing resource condition (e.g. maximizing or minimizing) is one of the conservation objectives. However, the performance of adaptive sampling for trend estimation is also of interest. Little has been published on the use of adaptive sampling to estimate trend. Magnussen *et al.* (2005) evaluated adaptive cluster sampling (ACS) for estimating simulated deforestation rates and concluded that ACS performed favorably relative to simple random sampling.

From an intuitive perspective, it seems likely that adaptive sampling would not estimate trends as well as designs that are specialized for trend estimation. Much work on fine tuning performance of panel designs for trend estimation focuses on temporal patterns and minimizing effects of spatial variation by incorporating periodic revisits to at least some sample sites (Urquhart and Kincaid 1999; Chapter 7). Trend or change can be estimated from periodic unbiased estimates of status (e.g. Chapters 11, 13, 14), and adaptive designs can be effective for estimating population status for rare species. Nevertheless, if trend estimation is the primary monitoring objective, then application of adaptive sampling should be first compared to alternative designs that are specialized for trend estimation.

---

efficiency, species detection, and probability of encountering occupied units (Thompson 2002, Smith *et al.* 2004, Philippi 2005, Turk and Borkowski 2005, Coggins *et al.* 2010). Adaptive sampling is most commonly associated with adaptive cluster sampling (Thompson 1990). However, a wide range of conventional sampling designs can be made adaptive by making effort dependent on sample observations (Thompson and Seber 1996). For example, adaptive sampling has been combined with single-stage, two-stage, two-phase, or stratified designs (Thompson 1990, 1991; Salehi and Seber 1997; Villella and Smith 2005; Conroy *et al.* 2008b). Recently, adaptive sequential sampling designs have been an active area of research and development (Christman 2004, Salehi and Smith 2005, Brown *et al.* 2008, Salehi and Brown 2010, Salehi *et al.* 2010).

Adaptive cluster sampling begins with an initial sample taken by some probability design (e.g. simple random sampling, stratified sampling; see Chapter 5 for an intro-duction to these conventional designs). If an initially selected sample unit is found to meet an a priori condition, then the neighborhood of that unit is adaptively added to the sample. In turn, if any neighboring unit meets the specified condition, then its neighborhood is sampled. This process is continued until no additional units meet the condition.

When adaptive sampling is applied to species assessment, the condition to adapt – the condition that triggers adding units to the sample – is typically based on the count of individuals or species within a sampling unit. For example, the condition could be to adapt if any individuals of a target species are found within a sample unit such as a quadrat. Alternatively, the condition could be based on an auxiliary variable, such as an easily measured habitat variable. The neighborhood definition can be quite flexible, but must adhere to a symmetry property so that if sampling unit i is in the neighborhood of sampling unit j then the reverse is true – sampling unit j is in the neighborhood of sampling unit i (Brown 2003). The symmetry property is necessary for the adaptive sampling estimators to be unbiased (Thompson 1990). A common neighborhood definition includes the four adjacent units and is cross-shaped. A variety of other neighborhood shapes can accommodate the symmetry property, such as a bookshelf pattern of linearly adjacent units (Woodby 1998, Bradbury 2000) or adaptive units that leap-frog adjacent units at regular intervals (Smith *et al.* 2004).

Because adaptive sampling designs are highly flexible, a number of design factors can affect performance (Turk and Borkowski 2005). Increased efficiency of adaptive sampling is highly dependent on the spatial distribution of the population of interest (Brown 2003). Adaptive sampling tends to be efficient when applied to a population with a spatial distribution comprised of compact networks that are geographically rare (Brown 2003). In contrast, increases in the total number of detections of the species and in the probability of encountering occupied units are robust results of adaptive sampling, which are seen with only a modest degree of clustering (Smith *et al.* 2003, Smith *et al.* 2011).

Brown (2003), Smith *et al.* (2004), and Turk and Borkowski (2005) review adaptive sampling and offer design recommendations. Magnussen *et al.* (2005) developed equations to predict standard error and expected sample size as functions of density and network size and shape for an application to estimation of deforestation rate. However, simple rules of thumb for application of adaptive sampling have been elusive because design factors interact to determine performance. Thus, recommendations are highly conditional on the specifics of the application. As a result, many have taken a simulation approach to evaluate performance of adaptive sampling for a population of interest prior to application, which is the approach we take in this study (Brown 2003, Morrison *et al.* 2008, Ojiambo and Scherm 2010, Smith *et al.* 2011).

Incorporating predicted species distribution into a sampling design

Predicted species distributions often are based on modeling the relationship between species occurrence and environmental variables hypothetically associated with the species' occurrence and habitat use (Guisan and Zimmermann 2000). Many statistical and mathematical modeling techniques are available for this purpose (Elith *et al.* 2006, Thuiller *et al.* 2009). Incorporating predicted species distributions into a probabilistic sampling approach is an example of the general category of model-assisted sampling (Särndal *et al.* 1992; see also Chapter 2). Model-assisted sample designs can be very efficient, i.e. have much higher precision for a given effort, but performance depends on model accuracy (Särndal *et al.* 1992). If the model is a poor predictor, then sample efficiency can be poor, and may be no better or even worse than simple random sampling. For example, predictions could be poor when the model is used to predict beyond the time and space of the original study, when data for model building are based on a biased sampling approach, or when species-specific models are applied inappropriately to other species. Thus, there is considerable value in identifying model-assisted designs that are robust to model inaccuracies.

Previous evaluations have incorporated predictions of species or habitat distribution into conventional stratified designs (Edwards *et al.* 2005, Guisan *et al.* 2006). Focus has been mostly on increasing species detection and has not considered estimation of population state. In this chapter, we incorporate predictions into a wider range of conventional and adaptive designs and broadly consider sampling design performance.

There are three basic approaches to incorporating predicted habitat in conventional and adaptive designs, all of which are demonstrated in the case study. First, predicted species distribution can be used as the condition to adapt in an adaptive cluster sampling design. This design was proposed by Ramsey and Sjamsoe'oed (1994) and was termed "GIS-adaptive sampling" by Morrison *et al.* (2008). In this design, a conventional sample of sites is selected, and then adaptive selection of additional sites is determined prior to field sampling. This assumes accurate model predictions but has the advantage that additional edge units do not need to be visited during field sampling.

Second, sampling units can be selected with probability proportional to predicted species distribution within a conventional or adaptive design (Roesch 1993, Thompson 2002). These unequal probability sample designs are referred to generically as PPS designs. When implemented in conventional sampling, all units in the sample are selected with probability proportional to predicted species distribution. When implemented in an adaptive design, only the initial sample is selected with probability proportional to predicted species distribution. Subsequent adaptive selection depends only on observations within the initially selected sample units.

Third, predicted species distribution can be used to determine strata for stratified sampling with either conventional or adaptive designs (Thompson 1991, Edwards *et al.* 2005). In adaptive stratified designs, adaptive selection within strata is determined by species observations after an initial selection of units. Strata boundaries can be ignored or not when adaptively sampling. We implemented both in the evaluation, but report results only for ignoring strata boundaries because results were similar. There was a slight increase in efficiency when adaptive sampling stopped at strata boundaries. However, stopping at strata boundaries would be difficult to implement in many field situations.

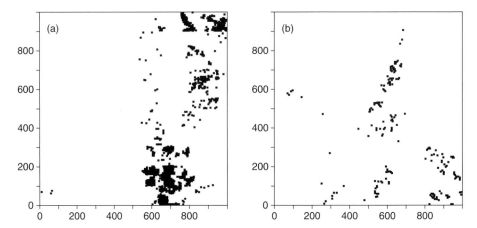

Figure 17.1 Distributions of (a) *Tamarix ramosissima* and (b) *Hedysarum scoparium* within 1-km² area in Dengkou County, western Inner Mongolia, in the Huanghe River valley.

Case study

Overview

We use populations of two plant species, *Hedysarum scoparium* and *Tamarix ramosissima*, from a 1-km² study area in the Inner Mongolia region of PR China to illustrate and evaluate the performance of various conventional and adaptive designs to estimate density and occupancy. *H. scoparium* is a sand-fixing sweetvetch that is useful for control of desertification (Hui-yong *et al.* 2008). *T. ramosissima* is native to China and invasive in North America (Sher *et al.* 2002, Si *et al.* 2005). Due to thorough stem counts of desert shrubs in the study area, we had complete information on the number and distribution of each species at the time of the study.

Because of the different spatial distributions and relationships to landscape variables, comparisons between *T. ramosissima* and *H. scoparium* will help to illustrate design robustness and the hazards of over-specifying designs in a multi-species monitoring program. The density of *T. ramosissima* was nearly an order of magnitude higher (stem density: 0.015 m⁻² vs. 0.002 m⁻²) and occupancy was 3 times higher than *H. scoparium* (6% vs. 1.7% of 10-m × 10-m grid cells occupied by at least one plant). *Tamarix ramosissima* was broadly distributed across the eastern half of the study area, and *H. scoparium* was most dense within the southeast quadrant and radiates along roads that intersect within the study area (Fig. 17.1). These species showed some degree of rarity and clustering, which is a condition when adaptive designs are efficient (Smith *et al.* 2004).

For each species, maximum entropy modeling was used to predict its distribution as a function of the environmental gradients in the study area (Fig. 17.2; Phillips *et al.* 2006). Predicted species distribution as measured by the predicted probability of species occurrence then was used as an auxiliary variable for sampling designs described below.

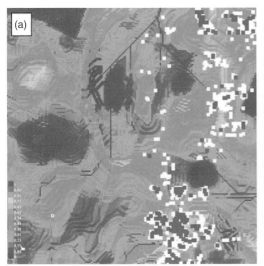

Figure 17.2 Map of the predicted probability of species occurrence overlaid with known species occurrence (squares) for (a) *Tamarix ramosissima* and (b) *Hedysarum scoparium* within the 1 km² study area. Purple squares are the subset of occurrences used in maximum entropy modeling. DEM-derived habitat variables were elevation, slope, distance to roads, solar insolation, and topographic convergence; see Appendix 17.1 for additional information. See plate section for color version.

See Appendix 17.1 for more information about the data set and species distribution models.

Using these populations with known parameters and the predicted distributions, we implemented simulations to demonstrate and compare performance of alternative designs incorporating adaptive sampling and/or model predictions. A simulation approach can be useful for evaluating sampling designs and the sampling distribution of estimators because it permits experimental comparison across populations and designs (Brown 2003, Morrison *et al.* 2008, Ojiambo and Scherm 2010, Smith *et al.* 2011; Chapters 9, 10, 19). It is often infeasible to analytically derive the sampling distribution for estimators across a range of populations and designs. Simulation comparisons across multiple populations and a broad range of designs can result in robust recommendations (Morrison *et al.* 2008).

Candidate sampling designs

For each population, sampling was simulated using the following specific designs.

(i) *Adaptive sampling ("Adaptive")*. This was the standard adaptive sampling design, typically called adaptive cluster sampling (Thompson 1990, 2002). In this adaptive design, an initial sample was selected by simple random sampling with subsequent adaptive selection depending only on species observations. The design factors were initial sample size, condition to adapt, and neighborhood. The condition to adapt was the stem count within a sampling unit; this triggered adaptive sampling in the neighborhood of the sampling unit. The neighborhood was cross-shaped and defined as all units that shared a side with the unit that met the condition to adapt; this neighborhood definition was used in all adaptive designs.

(ii) *GIS-based adaptive sampling ("GIS Adaptive")*. This was a simple modification of adaptive cluster sampling in which a GIS-derived auxiliary variable was the basis for the condition to adaptively sample (Ramsey and Sjamsoe'oed 1994, Morrison *et al.* 2008). The design factors were the auxiliary information used to trigger adaptive sampling, initial sample size, condition to adaptively sample, and neighborhood. In this application, the predicted probability of species occurrence was the auxiliary variable. The initial sample was selected by simple random sampling. The condition to adaptively sample was based entirely on the GIS-derived auxiliary variable (i.e. predicted species distribution) prior to field sampling. Thus, in practice additional edge units would not be visited in the field. With adaptive sampling based on such an auxiliary variable, the condition to adapt can take a number of forms depending on the expected relationship between the variable of interest (e.g. plant density) and the auxiliary variable in the ith unit (x_i). For example, condition (C) could be any of these three forms: $x_i < C, x_i < C$, or $C_L < x_i < C_U$. In this application the condition was $x_i > C$, where C was the threshold in predicted probability of species occurrence that maximized sensitivity plus specificity. Sensitivity is the probability that an occupied unit will be predicted to be occupied, and specificity is the probability that an unoccupied unit will be predicted to be unoccupied. The cross-shaped neighborhood was defined as all units that shared a side with the unit that met the condition to adapt.

(iii) *Conventional probability proportional to size sampling ("Conventional PPS")*. In this design, a fixed sample size was selected in proportion to an auxiliary variable (Thompson 2002). The design factors were the sample size and the auxiliary variable, predicted probability of species occurrence. Sample selection was without replacement.

(iv) *Adaptive probability proportional to size sampling ("Adaptive PPS")*. In this design, a fixed sample size was selected in proportion to an auxiliary variable followed by adaptive sampling triggered by species observations (Roesch 1993, Smith *et al.* 1995). The design factors were the initial sample size, auxiliary variable, condition to adapt, and neighborhood. The predicted probability of species occurrence was the auxiliary variable. The condition to adapt was the stem count within a sampling unit; this triggered adaptive sampling in the neighborhood of the sampling unit. The cross-shaped neighborhood was defined as all units that shared a side with the unit that met the condition to adapt.

(v) *Conventional stratified sampling ("Conventional Stratified")*. In this design, units were split into two strata according to an auxiliary variable (Edwards *et al.* 2005, Guisan *et al.* 2006). The design factors were the auxiliary variable used to create strata, the sample size, and sample allocation. The threshold in predicted occurrence that maximized sensitivity plus specificity was used as the auxiliary variable to set the boundaries for creating strata. Optimal allocation was used to allocate the fixed sample size among strata (Thompson 2002).

(vi) *Adaptive stratified sampling ("Adaptive Stratified").* In this design, strata were created as in the Conventional Stratified design, and units were adaptively sampled if a condition based on observations was met while sampling (Thompson 1991). The design factors were the auxiliary variable to create strata, initial sample size, allocation of the initial sample, condition to adapt, and neighborhood. The condition to adapt was the stem count within a sampling unit; this triggered adaptive sampling in the neighborhood of the sampling unit. The neighborhood was cross-shaped and defined as all units that shared a side with the unit that met the condition to adapt. Strata boundaries were ignored while adaptively sampling. Optimal allocation was used to allocate the initial sample size among strata (Thompson 2002).

The predicted distribution of each species, as measured by the predicted probability of its occurrence from the maximum entropy model, was used as an auxiliary variable in GIS adaptive, adaptive PPS, and stratified sampling designs for that species. As described below, we evaluated the robustness of each design to model inaccuracies by also simulating each design model-assisted design but "erroneously" incorporating predictions from the model developed for the other species. We also note that because distance to roads was an important contributor to the predictive model for *H. scoparium* (Appendix 17.1), application of the *H. scoparium* model to designs for *T. ramosissima* mimicked road-biased sampling for the latter species.

For all designs, the area of each unit in the sample frame was 100 m^2. Sample size (fixed sample size for conventional and expected final sample size for adaptive designs) covered a range from proportions > 0 to < 0.05 of the total number of units in the population. Additional details on the sampling designs and estimators can be found in Thompson and Seber (1996), Thompson (2002), and Morrison *et al.* (2008). In all adaptive designs, a modified Horvitz–Thompson estimator was used to estimate density (Thompson and Seber 1996).

Simulation and attributes for comparison of sampling designs

We used the software program SAMPLE (http://profile.usgs.gov/drsmith) that was designed for simulating and comparing sampling designs, particularly adaptive designs (Morrison *et al.* 2008, Smith *et al.* 2011). Each design was simulated 1000 times.

We considered three objectives when evaluating sampling design. The first objective was maximizing the accuracy of estimates. Accuracy incorporates bias and precision (Chapters 2, 3), but all of the designs and estimators considered here are practically unbiased. Thus, we used efficiency as a performance measure for accuracy. Efficiency is the ratio of variance from a simple random sampling design to variance from the candidate design with final sample size equal between the two designs. Final sample size is fixed for conventional designs, but is random in adaptive designs. Thus, for adaptive designs the expected sample size was the average of final sample sizes over the

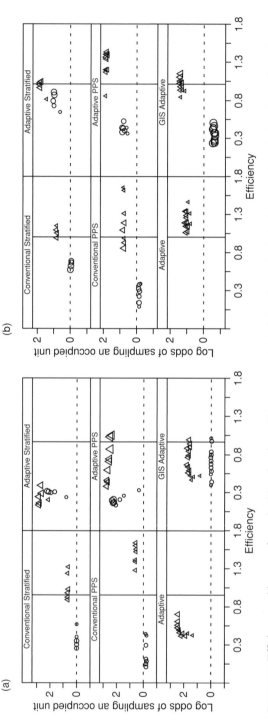

Figure 17.3 Efficiency and log-transformed odds ratio of sampling an occupied unit for six candidate designs based on simulated sampling of (a) *Tamarix ramosissima* and (b) *Hedysarum scoparium* within the 1-km² study area. Predicted species distribution was incorporated into all designs, except for the null Adaptive design. Odds ratio is odds of sampling an occupied unit for a candidate design relative to simple random sampling. Triangles indicate that the (correct) species' predictive model was incorporated, and circles indicate that the other (incorrect) species' predictive model was incorporated. The symbol size varies with final sample size. Reference lines shown for efficiency = 1 (vertical line) and log odds ratio = 0 (horizontal line).

1000 simulations. We also reported the coefficient of variation (CV: standard error of the density estimate/density estimate) as another measure of efficiency and precision.

The second objective was maximizing observations of individuals within a sampling unit and of occupied units. For this objective, we used relative risk as a performance measure. Relative risk is the ratio of the proportion of occupied units in the final sample (p_1) relative to that of the population (p_2). Thus, relative risk of sampling an occupied unit = p_1/p_2. We also report log(odds ratio) as another measure of the success in encountering occupied units, where odds ratio = relative risk $\times$ $((1 - p_2)/1 - p_1))$ (Agresti 1990).

The third objective was maximizing robustness to model inaccuracy. All candidate designs, except for the null Adaptive design, were model-assisted and thus sensitive to accuracy of model predictions. Inaccurate models can affect both efficiency and relative risk. Thus, we used a multivariate distance as a performance measure for the third objective. The multivariate distance was the Euclidian distance between the worst possible performance (i.e. efficiency and relative risk equal to zero) and performance when the other species' (i.e. incorrect) predictive model was incorporated in the design.

With multiple objectives, a design might perform well on one objective, but poorly on another. So, we needed a method to analyze the trade-offs and score each candidate design across all the objectives. We used the simple multi-attribute rating technique (SMART) to identify the optimal candidate sampling design for the study populations (Goodwin and Wright 2004). The performance measures within each objective (efficiency, relative risk, and the multivariate distance metric) were normalized and weighted. The weights represent relative preference for each objective. We considered weighting schemes with equal weighting and schemes with each individual objective receiving twice the weight of other objectives. The sum of the weighted, normalized scores was the overall score for each candidate design. The higher the overall score, the more completely the design met the objectives.

Performance of sampling designs

All candidate designs increased the probability of sampling occupied units compared to simple random sampling when the correct species model was applied (Fig. 17.3, Table 17.1). The highest probabilities of sampling occupied units were obtained with adaptive designs (i.e. Adaptive, Adaptive PPS, and Adaptive Stratified). When the incorrect species model was applied, performance declined except for the Adaptive design, which did not incorporate model predictions. Adaptive PPS and Adaptive Stratified designs maintained an elevated probability of encountering occupied units even when the model was incorrect. For the other designs, when the model was incorrect the probability of sampling occupied units dropped to or below the probability for simple random sampling (Fig. 17.3).

The PPS designs tended to be the most efficient designs (Fig. 17.3, Table 17.1). Efficiency was high for the Conventional Stratified design, but dropped considerably when the incorrect species model was applied. When the model was incorrect, efficiency dropped for most designs, but the Adaptive Stratified design showed only a moderate drop. The Adaptive design was highly efficient when applied to *H. scoparium*, which

Table 17.1 Average performance of candidate designs applied to simulated sampling of *Tamarix ramosissima* and *Hedysarum scoparium* in the case study area. Relative risk and efficiency are shown for when the correct model was used to predict species distribution. Efficiency is the ratio of variance from simple random sampling over variance from a candidate design for the equivalent final sample size. Relative risk is the probability of encountering an occupied unit from a candidate design relative to simple random sampling. Robustness is measured by the Euclidian distance between the worst possible performance (i.e. efficiency and relative risk equal to zero) and performance when the other species' predictive model was incorporated in the design. The goal is to maximize each objective.

| Species | Objectives | Candidate design | | | | | |
		Adaptive	GIS Adaptive	Conventional PPS	Adaptive PPS	Conventional Stratified	Adaptive Stratified
T. ramosissima	Relative risk	6.44	4.10	1.73	7.44	1.80	8.18
	Efficiency	0.55	0.77	1.56	0.69	1.11	0.27
	Robustness	6.46	1.20	0.81	5.42	1.02	4.64
H. scoparium	Relative risk	2.84	4.12	2.21	6.10	2.20	5.64
	Efficiency	1.18	1.00	1.15	1.22	1.03	0.96
	Robustness	3.08	0.65	0.87	2.12	1.12	2.49

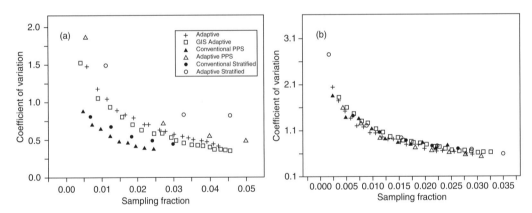

Figure 17.4 Coefficient of variation (CV) for six candidate designs based on simulated sampling of (a) *Tamarix ramosissima* and (b) *Hedysarum scoparium* within the 1 km² study area. Results are shown for the case where the correct model for predicted species distribution was incorporated in sample selection. CV was higher for model-based designs when the incorrect model was incorporated; this increase was proportionate to the change in efficiency shown in Fig. 17.3.

was rarer than *T. ramosissima*. Adaptive PPS and Adaptive Stratified were robust to model inaccuracy (Table 17.1). Conventional Stratified and GIS Adaptive were the most sensitive to model inaccuracy.

Precision (CV) varied among designs when applied to *T. ramosissima*, but varied little among designs when applied to *H. scoparium* (Fig. 17.4). When applied to *T. ramosissima*, Conventional PPS and Conventional Stratified resulted in the lowest CVs, and Adaptive PPS and Adaptive Stratified resulted in the highest CVs across a range of sampling fractions.

Table 17.2 Overall multiple objective scores for each candidate design applied to simulated sampling of *Tamarix ramosissima* and *Hedysarum scoparium* within the case study area. The performance measures for the objectives were efficiency, relative risk of sampling an occupied unit, and robustness (see text, Table 17.1). The performance measures were normalized and weighted within objectives, and then summed within a candidate design to arrive at the overall score. The weighting schemes reflected equal weighting among objectives or an individual objective receiving twice the weight of the other two objectives. For example, in the efficiency dominant weighting scheme, efficiency received twice the weight as relative risk and robustness. The top score within each row is highlighted in bold.

| Species | Weighting scheme | Candidate designs | | | | | |
		Adaptive	GIS Adaptive	Conventional PPS	Adaptive PPS	Conventional Stratified	Adaptive Stratified
T. ramosissima	Equal	0.65	0.27	0.33	**0.68**	0.23	0.56
	Efficiency dominant	0.54	0.30	0.50	**0.59**	0.34	0.42
	Relative risk dominant	0.67	0.30	0.25	**0.73**	0.18	0.67
	Robustness dominant	**0.74**	0.22	0.25	0.71	0.18	0.59
H. scoparium	Equal	0.67	0.22	0.27	**0.87**	0.15	0.55
	Efficiency dominant	0.71	0.20	0.39	**0.90**	0.18	0.41
	Relative risk dominant	0.54	0.28	0.21	**0.90**	0.11	0.63
	Robustness dominant	0.75	0.16	0.23	**0.80**	0.16	0.60

The multi-objective trade-off analysis indicated that the Adaptive PPS design performed best overall for both species and regardless of which objective was weighted heaviest (Table 17.2). The Adaptive and Adaptive Stratified designs also had high overall scores. The Adaptive design tended to have the second highest score, and thus was preferred over the Adaptive Stratified design. The other designs (i.e. Conventional Stratified, GIS Adaptive, and Conventional PPS) consistently received low overall scores.

Discussion

In this chapter, we demonstrated and evaluated both conventional and adaptive sampling designs that can be used with model-based predictions of species distribution. The conventional designs that we considered, Conventional Stratified and Conventional PPS, were outperformed in terms of efficiency, probability of sampling occupied units, and robustness to model inaccuracy. Although the Conventional Stratified design increased the probability of encountering occupied units, the design's performance was highly sensitive to model accuracy.

The conventional designs were improved by including adaptive sampling. Adaptive PPS and Adaptive Stratified designs were more robust to model inaccuracy than their conventional counterparts. In these designs, the initial sample selection takes into account a priori model predictions, and then the addition of adaptive sampling allows the final sample to reflect "on-the-ground" species observations. Ramsey and Sjamsoe'oed (1994) found that adaptive sampling resulted in a sample that better represented habitat types than simple random sampling. This would also be the case if model predictions inaccurately represented habitat types, but subsequent adaptive selection made adjustments during the sampling. However, this is not the case with the GIS Adaptive design, in which the condition to adapt is based only on model predictions rather than field observations. Like the conventional designs, the GIS Adaptive design was sensitive to model inaccuracies. The Adaptive design was not model-based and therefore was unaffected by model inaccuracy.

A general recommended sampling-design strategy that emerged from the evaluation was to begin by finding the best conventional design, and then include adaptive sampling into that conventional design. Conventional versions of PPS and Stratified performed well when applied with the correct model, and both were improved by including adaptive sampling. Consistent improvements were shown for increasing the probability of sampling occupied units and robustness to model inaccuracy. Encountering occupied units is often a major objective in rare species monitoring (Ramsey and Sjamsoe'oed 1994, Edwards *et al.* 2005, Guisan *et al.* 2006). Improvement in efficiency was inconsistent and depended on the species.

Although we have focused on robustness to model inaccuracies during one "round" of predictions, it is important to consider the potential for iterative learning in model-assisted or model-based sampling. Samples can be used to improve and update predictive models (Ramsey and Sjamsoe'oed 1994, Reese *et al.* 2005, Guisan *et al.* 2006), which in turn can be used to improve future sampling and monitoring of rare species (Edwards *et al.* 2005, Hirzel and Guisan 2002). Whether the focus is on species occurrence or other natural resource characteristics, incorporating model-based predictions into sampling designs and monitoring programs can lead to improved inference when management decisions and actions are used to test underlying models and theories (Särndal *et al.* 1992, Lyons *et al.* 2008, Lindenmayer and Likens 2009; Chapter 4).

Future research and developments

There has been some research on which sampling designs are useful for building and updating species distribution or habitat models (Ramsey and Sjamsoe'oed 1994, Hirzel and Guisan 2002, Vaughan and Ormerod 2003, Reese *et al.* 2005). However, modeling techniques have propagated rapidly in recent years (Thuiller *et al.* 2009), and the current range of candidate designs is potentially large. A systematic approach to evaluating sampling designs for model building and updating would be timely and would contribute to the growing application of adaptive resource management (Lyons *et al.* 2008; Chapter 4) and adaptive monitoring (Lindenmayer and Likens 2009).

Detectability was not an issue in sampling desert shrubs. However, detectability is an issue when sampling many ecological populations and can affect design performance (Thompson and Seber 1994, Ramsey and Sjamsoe'oed 1994; see also Chapters 4, 18, 19). Imperfect detectability has been shown to reduce the benefits of adaptive sampling (Smith *et al.* 2010). Numerous approaches exist for estimating and accounting for imperfect detectability in general (Chapter 18); continued development of strategies to incorporate these in adaptive designs is warranted.

Software for comparing performance of sampling designs or computing estimates given survey observations can be enormously helpful. In addition to the software used in this study, which is available at http://profile.usgs.gov/drsmith, an R package for adaptive cluster sampling is being developed by Arthur Dryver (National Institute of Development Administration, Bangkok, Thailand) and currently available at http://learnviaweb.com/moodle.

Summary

We demonstrated and evaluated approaches for incorporating predicted species distributions into conventional and adaptive sampling designs. We considered three overarching objectives for using these designs: to increase efficiency, increase probability of sampling occupied units, and increase robustness to model inaccuracy. Two populations of desert shrubs in Inner Mongolia were used as test populations in a series of simulations. Our evaluations indicated the following.

- Among candidate designs for populations under consideration, Adaptive PPS best met the performance objectives. The Adaptive Stratified and null Adaptive also performed well.
- Conventional Stratified, Conventional PPS, and GIS Adaptive were sensitive to model inaccuracy. When predictions came from an incorrect species model, these designs were inefficient or probability of sampling occupied units was not better and sometimes worse than simple random sampling.
- A recommended strategy is to begin by finding the best conventional design, and then include adaptive sampling into that conventional design. The addition of adaptive sampling consistently increased the probability of encountering occupied units and increased robustness to model inaccuracy.
- Simulation of sampling to evaluate designs is useful because design performance can be sensitive to specific population characteristics. A population can be generated that approximates the characteristics of the population of interest (see also Chapter 10). The cost to conduct a simulation-based evaluation of candidate designs for a population of interest is trivial compared to the cost of long-term monitoring.

Acknowledgments

The authors express their appreciation to the Ministry of Science and Technology and National Natural Sciences Foundation of China for fiscal support in the field work (Project Research Grants 2005DIB5JI42 and 31170588). Doug Nichols provided programming assistance. We thank Frank van Manen, Jennifer Brown, and Bill Thompson for helpful comments on an early version of the chapter.

Appendix 17.1. Additional information about the case study

The study populations for the example in this chapter came from an area in Dengkou County, PR China along the Huanghe River valley within the arid regions of western Inner Mongolia. During 2006, the Chinese Academy of Forestry counted all desert shrubs, including the two species that are the focus of this example, within a 1-km^2 area centered on latitude 40.26° and longitude 106.94° (Fig. 17.1). Population density of stems was 0.015 m^{-2} with variance of 0.0138 for *T. ramosissima* and 0.002 m^{-2} with variance of 0.0005 for *H. scoparium*. Population occupancy (% of 10 m × 10 m grid cells occupied by at least one plant) was 6% for *T. ramosissima* and 1.7% for *H. scoparium*.

We selected an Advanced Spaceborne Thermal Emission and Reflection Radiometer (ASTER) Digital Elevation Model (DEM) from NASA Jet Propulsion Laboratory that was scanned on 28 December 2006 and resampled to a 10 m × 10 m cell to match the scale of the species and habitat variables (Vaughan and Ormerod 2003). The topography-based environmental gradients, which we derived from the DEM, included elevation, slope, distance from roads, topographic convergence index (TCI), and solar insolation. TCI is an index of relative soil moisture and was calculated as the upslope contributing area in relation to slope expressed as percent rise (Beven and Kirkby 1979). Solar insolation uses solar azimuth and elevation during the summer solstice to compute solar radiation energy received on a given surface area.

Maximum entropy modeling was used to predict species distribution as a function of the environmental gradients (Fig. 17.2; Phillips *et al.* 2006). Modeling was implemented in the MaxEnt software program (http://www.cs.princeton.edu/~schapire/maxent/). Random occupied and unoccupied locations were selected within the study area to generate presence/absence data of *T. ramosissima* (117 present, 98 absent) and *H. scoparium* (102 present, 101 absent). The presence/absence subset and environmental variables were the inputs to build the species-specific predictive models. Model fit was evaluated using the area under a receiver operating characteristic curve (Phillips *et al.* 2006). The MaxEnt model for *T. ramosissima* resulted in an area under the receiver operating characteristic curve (AUC) of 0.84 for the test data. Elevation was the most important variable, and distance to road and solar insolation were moderately important. The MaxEnt model for *H. scoparium* resulted in an AUC of 0.85 for the test data. Distance to road and elevation were the most important variables in the model. Other environmental variables contributed relatively little to the final model. Predicted species distribution as measured by the predicted probability of species occurrence was used as an auxiliary variable in GIS adaptive, PPS, and stratified sampling designs demonstrated in this chapter.

18 Study design and analysis options for demographic and species occurrence dynamics

Darryl I. MacKenzie

Introduction

There are a wide range of approaches available for investigating the dynamics of the demographics and occurrence of ecological populations. So many that it would take an entire book, or more, to cover the important issues and options in sufficient detail. In this single chapter it is clearly impossible for me to go into detail on specific approaches, so I instead focus more on outlining some of the options available for addressing different types of questions and on general considerations, particularly with respect to program design. Inherently, because of those whom I have been fortunate enough to work with and learn from to this point in my career, most of the methods I discuss assume that detection of the items of interest (whether it be individual animals or plants, or of a species as a whole) will be imperfect, i.e. it will not be observed with certainty whenever you venture into the field to find it. However, many of the issues I will discuss are still relevant even with perfect detection. Recommended readings for further details on the topics I cover are Williams *et al*. (2002), Amstrup *et al*. (2005) and MacKenzie *et al*. (2006).

Before launching into the main thrust of this chapter, I am going to make a few comments (some might even say a rant) about the importance of thinking hard about *Why*, *What*, and *How* during the conception stage of *any* monitoring program. Although these are fairly basic questions that have been discussed early in this volume, their fundamental importance cannot be emphasized enough.

Why monitor?

It has been my experience that ecological monitoring programs, or studies in general, tend to fail because insufficient forethought has been devoted to these questions (Box 18.1). *Why* are we attempting to collect this information from the field in the first place? Are we mainly attempting to learn about how a certain aspect of the system is functioning, so our interest is primarily scientific, or is there more of a conservation/management context where decisions need to be made about whether certain management actions are to be undertaken. *What* information needs to be collected from

Design and Analysis of Long-term Ecological Monitoring Studies, ed. R.A. Gitzen, J.J. Millspaugh, A.B. Cooper, and D.S. Licht. Published by Cambridge University Press. © Cambridge University Press 2012.

Box 18.1 Take-home messages for program managers

There are a wide range of options available for monitoring programs that have some focus on the dynamics of demographic and occurrence processes. Before giving a lot of thought to the more specific aspects of the monitoring, it is vitally important to have clear understanding of why the monitoring is being implemented and how that information is going to be used to inform decision makers about progress towards the desired objectives. I believe that designing monitoring programs around the notion of detecting trends of a certain magnitude within some arbitrary timeframe is not a particularly useful approach from a management perspective as: (i) timeframes of at least 10 years are likely to be required; (ii) it does not provide information on why a trend might be occurring, with different causes potentially requiring different management actions; and (iii) it simply delays the very important discussions about how we really want the system to be performing.

Survey designs for monitoring demographic and species occurrence dynamics usually must address issues involved with incomplete and variable detectability. These issues often apply to monitoring of flora as well as fauna. Unaccounted for, imperfect detection can lead to biased estimates of dynamic processes and misleading conclusions about system performance. Basic sampling requirements should also be adhered to, in particular, that all items within the population of interest (however that is to be defined) should have some chance of being selected for monitoring (see Chapters 2, 5, 6). Clearly there are trade-offs in terms of how much effort should be devoted at various scales (for a fixed cost, increasing the number of monitoring sites requires decreasing the effort allocated to measuring each site). While there is some guidance within the literature for some situations, program designers may have to explore various options on a case-by-case basis. It must be realized, however, that quantity does not trump quality and that the "Garbage In – Garbage Out" principle will always apply.

---

the system in order to satisfy our reasons for looking at the system? Is interest directed towards individuals and their place within the system, or is there more of a community-level focus? Does information need to be collected at small or large spatial scales? *How* is the required information going to be collected and analyzed? Which of the myriad of available techniques are feasible to consider given the desired output and logistical constraints? Without due consideration to these questions, monitoring programs produce data that are either inappropriate for addressing the question of interest, or that no one has a clear idea of how to analyze, or both (see also Chapter 2).

There is clearly a sequential, and possibly even cyclical, process for considering them (i.e. you may have to revisit *Why* if after addressing the *How* it is found the program would not be feasible to implement). Ultimately, *Why* is the most important question to resolve before any consideration is given to field work. Failure to adequately articulate the reasons for implementing a study or monitoring program will likely lead to protocols that are inefficient at best, and a complete waste of valuable resources at worst.

"Forty-two" might be the answer, but what was the question in the first place (Adams 1979)?

Data collection is not an inherently useful pursuit, and only becomes useful once given a context or intended purpose (e.g. Nichols and Williams 2006). For me, monitoring for the sake of monitoring, i.e. with no stated intent of how that information is going to be useful, falls into the *So what* category. Claims that such data sets might one day be useful, e.g. to retrospectively examine possible causes of the decline, have some substance, yet for every data set that proves useful, how many do not? Arguments that such programs may become useful for some unspecified purpose at some uncertain future time point carry little weight relative to the resources they absorb and lost opportunity of implementing a program with more immediate information returns.

Many monitoring programs are implemented within a conservation or management context with an intent, either implicitly or explicitly stated, that actions will be taken if an undesirable outcome is observed. Such programs are much more useful as they are tied to the concept of providing information to some larger process, but oftentimes the act of the monitoring has undue focus over the desired conservation or management outcomes. Monitoring is simply a tool for informing us about the current state of the system such that we can decide what, if any, actions are presently required, and learn about the consequences of previous actions, in our progression towards those desired outcomes. Monitoring is not management. For example, cars contain many monitoring systems (speedometer, oil pressure indicator, rev counter, fuel gauge, voltmeter, etc.), but the actual act of monitoring those variables does not make the car magically perform any better. Those monitoring systems simply provide feedback to the driver, or onboard computer, such that the necessary actions can be taken so the car performs in the desired manner, e.g. slow down, add more fuel to the tank, or change gear. It is the mechanics of the car and the actions of the driver that make the car run, not the monitoring. It should be no different in ecological situations; the monitoring program only provides feedback about how the ecosystem is performing such that managers can make decisions about the type of actions that should be taken. Monitoring does not directly influence how the system is performing.

With this in mind, monitoring programs should not have goals, per se. It is the managers that should have the goals with the monitoring program being designed to provide the required information to assess whether that goal is being met. Many monitoring programs are designed around the notion of detecting a statistically significant trend in some metric over a certain timeframe. I have three prime reasons why I believe that designing a program around the notion of trend detection is less than ideal. First, longer time series (e.g. a minimum of 10 years) are likely to be required in order for any statistical test to have a reasonable level of power to detect a trend (e.g. Chapters 7, 10). I find the notion that a monitoring program is going to be implemented now, and may, or may not, be able to give you an answer in 10 years' time disagreeable, especially if management actions are likely to be required prior to that. By the time sufficient data have been collected in order to declare an observed trend statistically significant, the window of opportunity for positive action may have already passed. Second, while the question of a trend is a natural one to ask, often the more important question is why a trend might be occurring. For example, if a population is in decline is it because of increased emigration, lower survival rates, or depressed recruitment rates? Each of these possibilities may necessitate quite different management actions, hence the monitoring program would need to be designed to provide information about each of these different processes. Designing solely from a trend detection perspective is likely to be insufficient. Finally, focusing on the detection of a "significant" trend avoids the obvious question of what should be the next step. If no action is to be taken, then I would argue that the last 10 years of monitoring has been a waste of resources that could have been more usefully utilized elsewhere. If the intent is to take some action to reverse, or mitigate, the trend, then by how much? At what point are you going to decide that the job is completed and no further action is required? I would suggest that discussing what might be an acceptable or ideal level right at the beginning of the program, and then managing the population to maintain that level, are going to be much better uses of resources in the long term than having those discussions in 10, 15, or 20 years' time once a problem has been "confirmed". In the interim, what additional costs (both monetary and otherwise) have been incurred that could have been avoided? These types of issues and considerations are ultimately much more relevant for managing ecosystems than the detection of trends.

The what and how of it

Once it has been determined why a monitoring program is to be established, attention can be devoted to consider what to monitor and how it should be done. Determining what to monitor should follow quite naturally from the stated objective for the program, but there are three broad categories of state variables (metrics that quantify the current state of the system) that might be considered (e.g. Yoccoz et al. 2001): (i) individual-level, e.g. abundance; (ii) species-level, e.g. distribution; and (iii) community-level, e.g. species richness. I shall concentrate primarily on the first two options; however, it should be noted that some of the analytic tools that I will cover can be applied to community-level situations as well (e.g. Burnham and Overton 1979, Bunge and Fitzpatrick 1993, Boulinier et al. 1998, Nichols et al. 1998, Cam et al. 2000, MacKenzie et al. 2006, Dorazio et al. 2006). The defined objective should not only determine which state variables are of interest, but also for which population processes reliable information is required.

Individual-level state variables include metrics such as abundance, population size, or density of individuals within the defined region of interest. The underlying biological processes that make these variables change through time include survival/mortality (both "natural" and otherwise), birth, recruitment, and movement. The net effect of these processes is the population growth rate.

Species-level state variables move focus from "How many individuals on the landscape?" to "Where are they on the landscape?", e.g. species range or distribution and proportion of area/patches occupied. Generally, such metrics involve surveying a number of points within the landscape and determining the presence or absence (or perhaps other categories of population status) of the species at those points. The associated processes tend to have slightly different names depending upon the exact application, but basically describe how the status of the population at a specific location (e.g. presence or absence) changes from one time period to another. Examples are species-level probabilities of local extinction, persistence (the complement of extinction), and colonization. In combination, these components determine range expansions and contractions.

Community-level metrics such as species richness essentially quantify the number of species present within an area. One variation is to assign a value to each species to recognize inherent differences in the worth of each species to the current application (e.g. Yoccoz *et al.* 2001). Processes that result in changes in the community composition (in terms of which species are present, not the relative abundance of species within the community) include local extinction and colonization of specific species.

An important monitoring consideration for all three state variables is detectability. Typically our field methods will not provide us with a complete census of all the individuals or species present during a survey, either due to insufficient effort being available to adequately search the entire area of interest, or because the species is genuinely difficult to locate due to its size, coloration, or behavior. What we observe will tend to be an unknown fraction of the total items of interest, hence raw counts are an underestimate of the true state variable (see also Chapters 4, 11, 19). In order to make reliable inferences about the state variable of interest, we need to be able to correct, or calibrate, the count by accounting for the imperfection of the field methods. Furthermore, detectability is likely to change over time, hence inferences about state variables based on just count data may be misleading, particularly if detectability is changing in a systematic way (e.g. declining due to increasing vehicular activity along roadside survey routes). Covariates may be used to partially account for variation in the counts caused by detectability issues, but such covariates must be independent of the state variable otherwise it is impossible to determine what source of variation in the counts is being explained by the covariates. For example, observers are likely to differ in their ability to detect individuals, hence including observer ID as a covariate on a series of counts may account for some of that variation in detectability. However, both detectability and the number of individuals within an area may vary by habitat type. Therefore, including habitat type as a covariate on the counts may explain variation caused by detection, number of individuals, or a combination of both, resulting in uncertain inferences.

Detectability is also an issue when interest is directed at the underlying biological processes, regardless of which state variable is of interest. For example, at the individual level, survival rates will tend to be underestimated while the bias in population growth and movements may be in either direction (Williams *et al.* 2002). At the species level, estimates of colonization probabilities may also be positively or negatively biased, with extinction probabilities and turnover rates being overestimated (Moilenan 2002, MacKenzie 2005a, MacKenzie *et al.* 2006).

A second important methodological consideration is how to sample space. Typically it is not going to be possible to census the entire region; therefore, a procedure has to be developed to identify locations from which data are going to be collected and allows generalization of the results to unsampled locations. Essentially a probabilistic sampling scheme should be used, of which there are many varieties (e.g. Thompson 2002; Chapters 2, 5, 6), with the realization that even within the same region, for different objectives, different types of sampling schemes may be preferable. I will not go into the details of the available options (but see above for suggested starting references), but a key element of all is that each location to which the conclusions are going to apply must have had a non-zero probability of being selected for surveying. For example, if data collection is restricted to easy to access locations, there is no statistical justification to generalize the results to hard to access locations. The only justification is an assumption that "We think they're about the same", which is an untestable assumption because there will be no data available to provide evidence to the contrary. If inferences are going to be made to less-accessible locations then there needs to be some sampling effort allocated to those areas, or alternatively, be honest about the fact that inferences are restricted to the accessible areas (see also Chapter 5).

An associated issue is the trade-off between number of spatial locations, or monitoring stations, and level of survey effort within locations. Generally, what might be the most appropriate partitioning of effort will depend upon the degree of variation in the quantity of interest at the respective levels and, importantly, the overall objective(s) (see Chapter 8). It must be realized that quantity does not always trump quality (Box 18.2). Having a large number of monitoring stations providing data that contain a high degree of uncertainty about what conditions are actually like at specific locations is likely to be much less efficient than having fewer stations providing more accurate data. This issue is highly relevant when considering whether a monitoring program should be designed to incorporate methods that account for detection at the likely expense of reducing the number of monitoring stations. However, I am aware, likely through my own ignorance, of relatively few situations where detailed studies have been done that can provide some clear general guidance on such trade-offs. One exception within an individual-level context is Pollock *et al.* (2002), who discuss the use of double-sampling results to consider the relative effort that should be devoted at different scales.

Within a species occurrence, or species distribution, context a number of authors have explicitly shown that there is an ideal level of effort that should be conducted within a location to determine the presence of the target species and counter the effects of imperfect detection (MacKenzie and Royle 2005, Field *et al.* 2005, Guillera-Arroita *et al.* 2010). Using less effort at a specific location and attempting to reallocate it to another location (to increase spatial replication) actually turns out to be counterproductive and result in occurrence estimates that are less precise. It is my interpretation that the main benefit of conducting the additional fieldwork at each location is not that it provides you with the data to incorporate detection into the analytic methods (although this certainly is important), but that it increases the likelihood of confirming the species is present at the location if it was actually there. That is, higher-quality data have been collected that more accurately reflect the true current state of each location compared to a design with less effort per location, where the possibility of a false absence will be greater. My intuition suggests that similar results will likely hold in other contexts as well, although designers of monitoring programs are advised to carefully explore such trade-offs through analytic and simulation studies.

Box 18.2 Common challenges: quality vs. quantity

It is my experience that people either try to do *too much* or *too little* within a monitoring program.

- *Too much* in the sense that they have a wish list of the questions they expect the monitoring data to be able to address, with a consequence that the data answer none of them adequately. It would be preferable to instead focus attention upon a smaller number of questions and address them properly using sound methodologies. Similarly, *too much* in the sense that they attempt to collect as many different pieces of information from as many different places as possible. Collecting *more* data is not inherently useful, it's about collecting the *right* data. Consuming resources by collecting data that you do not have a clear idea of how to use is likely to be wasteful. Collecting a lot of low-quality data is likely to be less efficient than collecting less high-quality data.
- *Too little* in the sense that sample sizes will often be too small, in some regard, for the type and range of questions of interest. This will often stem from program designers having unrealistic expectations of the data that are to be collected and the analyses that will be applied to them.

The success or failure of a monitoring program that has some focus on the dynamics of demographic or occurrence processes, as with many monitoring programs, relies on the attention to detail provided during the design stage. Issues such as: why is the program being implemented in the first place; what information is expected to result from the process; and how are the data to be collected and analyzed, need to be very carefully considered. In the context of this chapter, this is particularly important for occupancy-based programs where there is a great deal of flexibility in terms of the different types of data that could be collected, what should be used as our basic sampling unit, over what duration should sampling be conducted, and so on. Failure to adequately address these issues in the design phase will produce results that may be difficult to interpret in a biologically meaningful way. Guidance on these issues comes from the overall objective.

Estimating individual-level demographic processes

As noted above, the dynamic processes that operate on the demographics of a population at an individual level could be described in a number of different ways. In this section I first discuss population growth rate, which is the overall combination of finer-scale processes, but can be estimated from successive estimates of the state variable rather than having to collect data at the resolution of the underlying processes. I give a brief overview of several approaches not based on mark–recapture before focusing more extensively on mark–recapture methods and how they can be used to estimate the more specific demographic processes, pointing out some of the prime design considerations.

Population growth rate and non-capture–recapture methods

Population growth rate is the coarsest individual-level demographic parameter, which can be estimated as the ratio of the population size at two times, i.e.

$$\hat{\lambda}_t = \frac{\hat{N}_{t+1}}{\hat{N}_t} \tag{18.1}$$

hence any field technique that enables abundance or population size to be estimated could potentially be used to estimate population growth rate. More generally, population growth rate can be assessed through an analysis of a series of abundance estimates (e.g. estimate a trend in population size through time). In some instances this may involve only direct counts, but methods such as distance sampling or capture–recapture will sometimes be required. There are also techniques such as double observer and time to first detection methods that I regard as essentially capture–recapture analytic methods (because of the underlying statistical model), except the data are collected without any physical capture.

Counts with perfect detection

Typically it will not be logistically possible to obtain a perfect count of all individuals within an area of interest. However, in some circumstances it may be possible to count the number of individuals at sample locations within that area where the sampling locations are sufficiently small to search exhaustively. For example, to estimate the number of breeding pairs within a seabird breeding colony, rather than attempting to survey the entire colony, a grid could be overlaid across the colony and cells randomly selected and searched intensively for breeding pairs. The mean and variance of the cell-specific counts can hence be used to obtain an estimate and associated standard error of the colony total (e.g. Williams *et al.* 2002). Note the important consideration here is that cell-specific counts are perfect. There must be absolutely no potential for systematic undercounting of the number of individuals within a cell due to factors such as being obscured from the view of the observers or temporary absence from the cell (e.g. due to foraging). Otherwise the count will represent an unknown fraction of the total number of individuals that truly reside within the cell.

Distance sampling

One form of imperfect detection is that individuals located further away from the observer will be harder to detect than individuals located closer to the observer. Distance sampling is a set of techniques that were developed to account for this form of imperfect detectability. Initially developed for situations where an observer would move along a transect and would record perpendicular distances from the transect line out to a defined maximum distance, it has since been extended to situations where observers are stationary and recording distances from a fixed point (e.g. Buckland *et al.* 1993, 2001, 2004). Where only a single observer is being used, it is assumed that detection along the line or at the point is perfect, but if two or more observers are searching simultaneously then this assumption can be relaxed. A detection function is then used to describe how the detectability of the individuals changes (typically monotonically decreasing) as distance increases. Once the parameters of the detection function are estimated, it is possible to estimate the number of individuals that were within the survey area that were not detected.

A key assumption of distance sampling is that the number of individuals at different distances from the point or transect line is uniformly, or evenly, distributed which is required to enable estimation of the number of undetected individuals. Practically, this can be achieved by the random placement and orientation of transects, hence individuals are guaranteed to be located at random distances. Without the random placement of transects then it must be assumed that individuals are randomly located spatially, which is a questionable assumption, particularly if transects are placed along topographical features such as tracks or ridgelines.

Distance sampling estimates the number of individuals that were available to be detected during the surveying; hence, individuals that had no chance of being detected in the survey, but were within the survey area, are not included in the estimate of abundance. For example, there may have been individuals obscured from view from the transect line throughout the survey, such as animals that may not have emerged from their burrows during the survey.

Repeated counts with imperfect detection

Royle (2004) developed an analytical method to account for imperfect detection in situations where surveys are conducted at multiple stations within an area of interest, and repeated counts of individuals are performed at each station within a relatively short time-frame to ensure the true number of individuals at the station is unchanged. Not all individuals will be observed in each count, but the maximum count will indicate the minimum abundance of the species at a station. By assuming that the true number of individuals per station can be well described by a discrete valued statistical distribution (e.g. Poisson or negative binomial), Royle (2004) showed how the repeated count data could be used to estimate the total abundance across all stations, or possibly across the entire region of interest.

There are, however, a number of key assumptions that must be met.

(i) Counts are of unique individuals, i.e. a count of five indicates five different individuals, not one individual observed five times (or something in between).
(ii) Individuals are counted independently of one another. That is, if one individual is observed the chances of observing a second individual is unchanged. This might be violated if members of breeding pairs or family groups tend to be observed simultaneously.
(iii) The number of individuals available to be counted at a station over the duration of the repeated counts is unchanging.
(iv) The same individual cannot be counted at multiple stations, and if extrapolating beyond the monitoring stations to elsewhere within the area of interest (e.g. to get

an estimate of overall total abundance), individuals must be restricted to a single station.
(v) All individuals at each station have the same probability of being included in a count.

The violation of any of these above assumptions (plus the assumption of how individuals are spatially distributed) will likely lead to biased estimates of abundance, or at least have repercussions on how the resulting estimate of "abundance" should be interpreted. From the perspective of estimating growth rates, if the nature and consequences of those assumption violations are consistent across time, reasonable inferences may still be possible.

Capture–recapture-based approaches

The basic premise of capture–recapture methods is that individuals from the population are randomly sampled on multiple occasions within a survey period (e.g. on each of several sequential trap nights during an annual monitoring visit) and on each occasion it is possible to determine which individuals have been captured previously. Individuals are identified through marks, either natural or artificial, that may be applied on each survey occasion. Physical capture of individuals is not always required, particularly when using natural markings. Unique marking of individuals is not always necessary depending upon how the data are to be analyzed; hence, in some cases batch marking may be appropriate (i.e. all individuals captured or observed at a particular survey occasion receive the same type of mark).

Capture–recapture approaches are traditionally divided into closed- and open-population situations. In a closed-population situation it is assumed that for the duration of the sampling the population within the area of interest is closed to changes, i.e. there are no births, deaths, immigration, or emigration of individuals. Closed-population techniques can be used to estimate the abundance or density of individuals at a certain point in time. When there are changes in the population during the sampling period, the population is said to be open. Open-population capture–recapture approaches are therefore used to estimate the demographic processes that cause the change in the population. While different sets of methods were originally developed for closed- and open-population situations, more modern approaches combine the two when data have been collected at two different temporal scales (e.g. Pollock 1982, Kendall et al. 1995). However, the distinction is still conceptually important, particularly when planning a program and deciding what type of information is required given the objectives.

Mark–recapture assumptions

There are a number of fundamental design assumptions that are common to most situations where capture–recapture methods are to be applied. Careful consideration of them is important as an inadequate design may lead to unreliable conclusions.

(i) All individuals within the population of interest have a non-zero probability of being captured at each sampling session

It is of prime importance that the population of interest be clearly defined and suitable field protocols employed to ensure this assumption is met. Resulting estimates and inferences do not extend to individuals beyond the effective area of the sampling. Similarly, inferences do not extend to individuals within the sampling area that have no chance of being captured due to permanent unavailability during the sampling (e.g. individuals that had not emerged from underground refuges), inappropriately sized capture devices (e.g. large individuals that physically cannot fit within a trap), or behavioral responses (e.g. individuals that will not enter an enclosed, metal space).

Problems may also occur if there is undue focus on recapturing previously marked individuals when unmarked members of the population are also of interest and they may have systematically different demographic rates. For example, suppose we wish to estimate the probability of adult survival over a 5-year period. We consider two possible study designs: (a) randomly select a number of adult individuals from the population in year 1 and subsequently maximize recapture rates by only attempting to recaptured those marked individuals; or (b) randomly select a number of adult individuals from the population each year (i.e. attempt to capture both marked and unmarked individuals), marking them on first capture and recording subsequent recaptures. If the defined population of interest is "adult individuals in year 1" then the first proposed design may be appropriate. However, as a cohort, those animals marked in year 1 will be getting progressively older and hence may not be representative of the overall population in subsequent years, presuming ongoing recruitment. Therefore, if the defined population of interest is "adult individuals in each year", the first proposed design may yield biased results, and the second design would be preferable.

A different situation where this basic consideration applies is when the sampling or trapping effort is devoted to only a portion of the area within which an individual may move. If animals move and mix within the overall area relatively freely (i.e. randomly), then all individuals have some a priori chance of being within the smaller study area at the time when the sampling will occur. However, if the animals are highly territorial or display a high degree of site fidelity to certain areas, but there is some degree of movement or dispersal, marked individuals that permanently move outside of the study area now have no chance of being recaptured in the future. As a result, mortality and permanent emigration become confounded, and survival rates will be underestimated. Furthermore, technically the population of interest is now restricted to those individuals within the portion of the area where sampling is occurring.

(ii) Individuals are captured independently of one another

Captures of individuals are not independent when the probability of capturing one individual becomes more, or less, likely if another individual within the population is captured. This may arise with breeding pairs, for example. If the female of the pair is captured in a sampling session, then in some circumstances the male may almost certainly be captured as well (or conversely, may be unlikely to be captured because he is foraging elsewhere). Similarly for animals that travel in family groups; if one individual is captured then all (or most) family members may be captured, otherwise none of them are. The lack of independence may result in biased estimates in some circumstances. In other cases the estimate may be accurate, but the reported standard error is inappropriately small as the effective sample size is smaller than the nominal sample size, and hence reported standard errors need to be inflated (e.g. Burnham and Anderson 2002).

(iii) Marked individuals are correctly identified each sampling session
Tag loss is one specific cause of misidentification, as a previously marked individual will subsequently appear to be unmarked. It will cause bias in demographic parameter estimates (e.g. survival will be underestimated because tag loss means the marked individual is removed from the marked portion of the population, much like mortality), particularly if individuals that have lost their marks become indistinguishable from individuals that have never been marked. In some situations, individuals that have lost marks may be recognizable through scarring which may be used as a form of batch mark and augmented into analyses (Williams *et al.* 2002). Using multiple marking methods (e.g. double tagging) can also be useful as it allows individuals to remain identifiable provided at least one mark is still in place, and also enables mark loss rates to be estimated and accounted for during analysis.

Incorrect identification of marks may also occur through misreading of tag numbers or failing to identify telling features when using natural markings. This may result in either a recapture being recorded for the wrong marked individual, or the same individual being incorrectly recorded as multiple unique individuals in the data. Clearly, either situation is undesirable. Aside from using sound verification procedures in the field and when transcribing records, attention should be devoted to practical aspects such as considering font size and type on tags – is there potential confusion among certain letter and number combinations (e.g. a lower case "l" being misread as the number "1"), or could a worn or partially obscured "8" be misread as a "3"? Some natural markings, e.g. scar patterns, may change or evolve through time, possibly resulting in a previously sighted individual being misreported as a new individual. Misidentification can also occur through genotyping errors when using DNA to identify individuals (e.g. Lukacs and Burnham 2005). While new methods are continually being developed to account for misidentification of individuals (Wright *et al.* 2009, Yoshizaki *et al.* 2009), a basic rule of thumb is that during analysis, if there is uncertainty about the identification of a recaptured marked individual, results are likely to be more robust by ignoring that uncertain recapture provided there is a possibility of correctly identifying the animal in the future. The underlying logic is that capture–recapture methods explicitly account for not detecting an individual even when it is still in the population (which may be due to not being able to accurately identify it), but incorrectly assigning the recapture to an individual that has already left the population, or incorrectly assigning a new identity to a previously identified individual, is more problematic.

(iv) There is no unmodeled heterogeneity in capture probabilities
Heterogeneous capture probabilities among individuals could be caused by attributes of the individuals themselves (e.g. age, gender, size, curiosity) or through aspects of

the field protocols (e.g. an animal that has one capture device within its home range will have a lower capture probability than an animal with four devices within its home range). Unaccounted for, heterogeneity in capture probabilities can create bias in esti- mated demographic parameters. It is relatively well known that heterogeneity will lead to underestimates of abundance (e.g. Otis *et al.* 1978) and can also result in under- estimates of survival (Carothers 1973). If a sufficient number of individuals have been captured, they may be stratified into known groups that are thought to have similar capture probabilities, or characteristics of the individuals used as covariates for the capture probabilities to account for some of the heterogeneity. In the context of abun- dance estimation, there are other approaches that are either non-parametric (Burnham and Overton 1979, Chao and Lee 1992) or assume a distributional form for the capture probabilities within the population (e.g. Pledger 2000). It has been shown, however, that while a range of different approaches to account for heterogeneity may provide a similar fit to the data, they may yield quite different estimates of population size (Link 2003). Hence, it is important to take all practical steps to limit individual variation in capture probabilities, especially those induced through study design. It is worth noting that the bias associated with heterogeneity is less when the average overall capture probability (i.e. the probability of being captured at least once during the study period) is higher, which suggests it is advantageous to use field methods that are as effective as possi- ble. Increasing the number of sampling occasions will also increase the overall capture probability of individuals and therefore also reduce any bias.

Closed-population methods

A series of abundance or density estimates obtained through closed-population capture– recapture methods can be used to obtain estimates of population growth rate through time. Where it is possible to recapture the same individuals across the longer timescale (e.g. annually), then more advanced methods may be used to estimate the demographic processes more explicitly. If few individuals are likely to be present in multiple time periods, such that the population is relatively distinct each time period, using a series of closed-population estimates may be the only viable alternative.

The Lincoln–Petersen estimator was one of the first capture–recapture techniques, developed for the simplest mark–recapture situation with two samples from the pop- ulation. All individuals from the first survey are marked and released back into the population. After a suitable time period to allow the marked individuals to fully mix with the unmarked individuals, a second sample is drawn and the ratio of marked to unmarked individuals is used to determine the population size. The method has also been extended to account for multiple recapture events with no additional marking (Bartmann *et al.* 1987, White and Garrott 1990), which may be applicable when "recaptures" con- stitute resightings of previously marked individuals rather than physical captures. The advantage of this approach is to improve precision of the estimated abundance over the two-sample Lincoln–Petersen method.

A wide range of methods have been developed to analyze data from closed-population capture–recapture studies where marks are applied to previously unmarked individuals when first captured. The primary focus, until relatively recently, has been on estimation of abundance or population size within a defined area of interest. The seminal monograph of

Otis *et al.* (1978) provided an excellent review of the state of knowledge in the late 1970s, and while there has been much progress in some areas, a lot of the basic fundamental information provided by Otis *et al.* (1978) is still relevant. In the 30-odd years since there has been advancements in a number of areas including incorporating covariate information from individuals (e.g. Huggins 1991), accounting for heterogeneous capture probabilities (Pledger 2000), and dealing with individual misidentification (Lukacs and Burnham 2005, Wright *et al.* 2009, Yoshizaki *et al.* 2009).

Attention has more recently turned to estimation of the density of individuals given that the edge of an area of interest is often ill-defined, which creates uncertainty about the exact extent of the population being sampled (see also Chapter 19). For example, with a trapping grid placed in a contiguous habitat, what is the effective extent of the area being trapped? Do animals immediately beyond that edge of the grid have no chance of being captured, or are animals that are 10, 50, or 100 m away from the edge of the grid likely to be caught at some stage? This "edge effect" has long been recognized – since at least Otis *et al.* (1978) – and ad hoc approaches for dealing with it have been suggested. More rigorous approaches have now been developed that explicitly utilize the information available from the spatial locations of captures and recaptures of individuals (e.g. Efford 2004, Borchers and Efford 2008, Royle and Young 2008). These spatially explicit capture–recapture approaches can be applied to any spatial arrangement of trapping devices.

Violations of the closure assumption may be due to either in situ processes such as births and deaths of individuals, or movement in and out of the area where individuals are at risk of capture. Practical steps to avoid violating the closure assumption will often depend upon the species under consideration, but primarily revolve around the basic principle of keeping the sampling session relatively short compared to the timescales at which demographic processes operate. For some species, it may be possible to target times of year when the population is relatively static, e.g. avoiding periods when animals may disperse or migrate across an area. Some violations of the closure assumption do not create a systematic bias per se, but redefine the exact interpretation of the resulting estimates (Kendall 1999). For the purpose of estimating population growth rates, such reinterpretations may not have a serious consequence in situations where the effect of violating the closure assumption is consistent through time. However, in other situations the consequences may be more dire and lead to inappropriate conclusions about population growth rates.

Variation in capture probabilities
There are three broad categories of effects on capture probabilities that should be considered during the study design phase: temporal, behavioral response to capture, and individual. Temporal variation in capture probabilities may be caused by factors such as changes in environmental and weather conditions, or, for some species, changes in activity levels, during the sampling period at a site (e.g. over the several trap nights making up that year's survey of the site). Some individuals may exhibit a behavioral response to being captured. For example, if an individual receives a food reward for entering a trap and allowing itself to be captured, it may be more likely to be recaptured

subsequently (trap happy), while if the trapping experience was generally negative, an individual may be less likely to allow itself to be captured in the future (trap shy). As discussed above, individual heterogeneity in capture probabilities can be very problematic for capture–recapture studies. A key assumption for abundance and density estimation is that information on capture probabilities collected from marked individuals applies to unmarked individuals. If there are systematic differences between marked and unmarked individuals (i.e. the unmarked individuals are less likely to be caught), biased estimates will result.

While it is possible to account for some combinations of these effects during the analysis phase (e.g. see Chapter 19), it is highly recommended that suitable field protocols be used to minimize them to the extent possible. Some forms of temporal variation may be controlled by only searching or trapping under certain conditions, although some forms will be impossible to control. In some circumstances behavioral effects may be accounted for by using techniques such as pre-baiting to acclimatize, or pre-condition, the population of interest prior to the collection of the data that will be used for the abundance estimation (Otis *et al.* 1978). An important realization is that when the probability of first capture of an individual is different from the probability of recapture, the recaptures contain little information that is relevant to the unmarked individuals in the population. Therefore, recaptures provide no additional information for abundance or density estimation, unless the effects of other sources of variation are similar on both first capture and recapture probabilities. Ideas on ways to limit the effect of heterogeneous capture probabilities have already been noted above.

Estimation of demographic processes using open-population methods

A wide range of demographic processes can be estimated using open-population capture–recapture approaches, e.g. population growth rate, survival or mortality, recruitment, and movement. Which processes are of prime interest will largely dictate how the study is to be designed, and which particular set of analytical methods would be appropriate. Due to the vast breadth of the topic, I will not go into the specific detail of any open-population methods here except to illustrate aspects of study design considerations, as such details are readily available elsewhere (e.g. Williams *et al.* 2002, Amstrup *et al.* 2005).

There are few options besides open-population capture–recapture techniques for studying and estimating demographic processes when individuals are detected imperfectly. Perhaps the only common alternative is radio (or satellite) telemetry, although I would argue telemetry-based approaches are essentially capture–recapture methods, just with very high recapture probabilities. Therefore, some of the suggestions below will apply equally to telemetry studies. Methods such as age- or harvest-ratios and life tables can be used to estimate demographic parameters; however, if capture or detection rates differ for the portions of the population of interest (e.g. are different for different age classes), biased estimates will result (Williams *et al.* 2002). Other assumptions may also be required in some instances, such as a stationary and stable age distribution, that may be unlikely to hold in practice.

The basic idea of open-population capture–recapture approaches is that the population of interest is sampled at systematic points in time, where the elapsed time between sampling sessions is sufficient to allow the demographic processes of interest to occur. Obviously, what might be a "sufficient" time period between sampling sessions will depend upon the questions and species of interest. Individuals are marked in one sampling session, with attempts to recapture them in later sessions. When, and where, individuals are recaptured provides the required information to estimate the demographic rates. Marking of previously unmarked individuals may continue for the duration of the study.

Survival

Survival is often one of the key demographic parameters of interest in population monitoring and assessment. Estimation of survival within a capture-recapture framework was formalized in the Cormack–Jolly–Seber (CJS) model (Cormack 1964, Jolly 1965, Seber 1965). At systematic points in time the population is sampled once each session. Marks are applied to unmarked individuals that have been captured for the first time and records kept of captures of previously marked individuals. Following capture, the processed animals are released back into the population, although occasionally there may be losses on capture (e.g. animals that might be critically injured during capture). In the CJS model, the information associated with the processes resulting in the first capture of an individual is not used. Only the data following the first release of an individual are used to estimate survival. The methods of Jolly (1965) and Seber (1965) are sometimes separately referred to as the Jolly–Seber model as, unlike the approach of Cormack (1964), the information of first capture is also used to estimate population size for the second sampling session onwards.

Band recovery models (e.g. Brownie *et al.* 1985) are another common method for estimating survival, with the distinction from the CJS model being that in a band recovery model each individual is only ever recaptured (at most) once. Band recovery models are typically applied to harvested populations with the single recapture being a fatal event for the banded animal.

There have been many extensions of these basic frameworks, oftentimes through incorporation of additional sources of information, e.g. recaptures with dead recoveries (Burnham 1993); recaptures with dead recoveries and live resightings (Barker 1997, 1999); age structures (Pollock 1981, Stokes 1984); geographic movement (Arnason 1973, Brownie *et al.* 1993, Schwarz *et al.* 1993); and multiple capture attempts at each systematic point in time (Pollock 1982, Kendall *et al.* 1995, Kendall and Bjorkland 2001), just to name a few. Many of these extensions enable estimation of other demographic parameters, but oftentimes have stemmed from the quest to obtain improved estimates of survival.

Given that many capture–recapture studies operate on areas that do not encompass the entire region where a marked animal may reside, the estimated survival parameter is often termed local survival. That is, it is the probability that an animal survives the intervening time period and stays within the region where it is at risk of capture. If an animal permanently (relative to the duration of data collection) emigrates or disperses to an area outside of the region where capture attempts are made, the permanent movement is indistinguishable from mortality. If ancillary dead recovery and/or live resighting information have been collected from throughout the entire region where marked individuals may move, true survival may be estimated (Burnham 1993; Barker 1997, 1999). Another option to account for movement of individuals to immediately beyond the primary study area is the use of a "buffer zone", within which attempts are made to only recapture previously marked individuals with no marking of individuals captured for the first time (Kendall *et al.* 2009). The buffer zone would need to be wide enough to recapture those individuals that have moved away from the main study area; hence, such an approach may only be suitable for situations where individuals permanently move only relatively short distances. If a random subset of individuals are fitted with a tracking device, the associated movement information may be used to determine the fraction of marked individuals that remain within the study area, provided that the timescale of the telemetry data is comparable to that desired for the survival estimate, and with sufficient spatial coverage. Similarly, the extension of spatially explicit capture–recapture models to open populations may enable the movement information gathered from the spatial locations of the recaptures to be used to account for the potential of individuals moving permanently off the study area, provided that the study area is extensive enough to identify situations where individuals permanently relocate within the study area. Finally, when there may be a limited number of other regions outside the primary study area to where an individual may move (e.g. other breeding colonies), it may be advantageous to consider also conducting capture–recapture studies at those places as well. By doing so, multi-state (or multi-strata) capture–recapture models could be used (Brownie *et al.* 1993, Schwarz *et al.* 1993), with each area being used as a different stratum.

Reproduction and recruitment to breeding population
Appropriateness of field and analytic methods for reproduction estimation will often be relatively species specific. As such, my comments on monitoring reproduction are primarily of a general nature. Despite the variation in available methods, there are a number of common considerations.

The first question is, what stages of the reproductive and recruitment process are of interest? In a management context, part of that consideration can be rephrased as: for which stages are there going to be potential management actions that could be implemented, and what is their likely impact compared to actions that could be implemented on other life stages? If it is unlikely that there are going to be any useful actions for certain reproductive stages, the potential benefits of the information gained should be weighed against the additional effort required to gain such information. For example, in a bird species, is it worth the additional field effort required to separate clutch size, hatching success, and fledging success if there are no potential actions that could be applied to alter one of those specific components? Clearly there will often be scientific interest in the separate components, but could the field effort that would be required be better spent on other aspects of the monitoring or on other projects? If the answer is "yes" then an ultimately more efficient use of resources may be to instead focus monitoring on a more general combination of the processes; for instance, as an extreme example, ignoring many of the early stages of the reproductive process and instead designing the monitoring to simply estimate the number of new recruits to the breeding population.

Once the stages of interest have been decided upon, attention should turn to the more practical aspects associated with monitoring those stages. The types of questions that will need to be considered will depend upon the ecology of the species of interest, but include issues such as detectability, mobility of the individuals, and whether the outcome of interest is likely to be fully observed for all selected individuals before the end of the field season (i.e. potential censoring of the observations). Detectability may often be an issue. For example, underestimation of the number of nests within an area or failure to locate young after birth to determine their initial survival could lead to biased estimates of reproduction or recruitment. Variation in detectability can be particularly problematic when detectability and the likely outcome of the reproductive stage of interest vary together. For example, nests that are easier to find by field workers may also be those that are more likely to be predated, resulting in overestimation of predation rates. Where detectability is an issue, techniques based upon capture–recapture or distance sampling might be appropriate (e.g. Nichols *et al.* 1986). If individuals are mobile during the reproductive stage of interest, this mobility may cause a form of imperfect detection as animals may be difficult to relocate. Other types of movement such as individuals permanently dispersing outside the monitoring area prior to the stage of interest being reached may cause problems, particularly if the reason for dispersing is a failed breeding attempt which is unidentified by the monitoring program. When the field work might end prior to the stage of interest being completed by all individuals, the outcome for those individuals is censored, e.g. the field season ends before it can be determined whether all chicks have successfully fledged. When such censoring occurs, the analytic methods employed need to account for it (e.g. Heisey and Nordheim 1995). The basic message is to fully consider the vagaries of both the species ecology and behavior, and the likely field methods that are to be employed, identifying potential shortcomings relevant to the objective of the monitoring program. Then implement design-based solutions to overcome them and use analytic methods that are appropriate for the task.

In most circumstances, ultimately it is the number of recruits to the adult or breeding population that should be of interest, which is the outcome of numerous stages in the reproductive and life cycle. As noted earlier, part of the initial considerations should be to determine the level of detail that is necessary for the situation at hand. While there may be long-standing expectations from natural history and scientific studies to use certain protocols and collect information on specific life stages, it may not always be strictly necessary, or efficient, to do so in a monitoring program. This would particularly be the case in situations where the benefits gained, in relation to the monitoring objectives, for the effort required are small.

It is a relatively common practice in natural history and scientific studies, across many different species, to mark young individuals and subsequently collect information on juvenile dispersal and movement, survival, age of first reproduction, etc. It also creates cohorts of known-age individuals. There are obvious benefits in gaining a better understanding of how populations function; however, in order to gain a clear understanding, the demographic parameters must be estimated accurately hence the data requirements must be carefully considered. For example, to separate out juvenile survival from permanent dispersal from a natal area would require data to be collected at a broad scale to determine dispersal of marked survivors.

Recruitment to the adult or breeding population can also be determined by capture–recapture being performed on only the adult population, either by itself using a reverse-time analysis (Pollock *et al.* 1974, Williams *et al.* 2002) or in combination with esti-mation of survival and other demographic parameters (e.g. Jolly 1965, Seber 1965, Pradel 1996, Crosbie and Manly 1985, Schwarz and Arnason 1996). A basic require-ment, however, is that adults are captured at random from the population, with marked and unmarked individuals having the same capture probability. Essentially, imagining a situation with perfect detection, an individual would be first captured at the first sam-pling occasion following its recruitment to the adult population; hence, it would be relatively straightforward to determine the number of new recruits at each time period. Given that detection is not perfect, it is not possible to determine the exact time of recruitment to the population, but it can be accounted for probabilistically, just as the exact time of death usually is unknown in the estimation of survival (e.g. as in the CJS model).

Finally, where adult individuals may transition between a breeding and non-breeding state, the rates of transition may also be of interest. These may be quantified using a multi-state capture–recapture approach (e.g. Nichols and Kendall 1995), which also allows differing survival probabilities for breeders and non-breeders when it is possible to capture individuals in both states. In situations where it is only possible to capture individuals in one of those states, methods that account for an "unobservable" state could be used for estimation (Kendall and Nichols 2002). However, to use such methods, it is necessary to make some potentially restrictive assumptions, e.g. survival is equal for individuals in the observable and unobservable states.

Growth rates
Any open-population capture–recapture method that can be used to estimate time-specific estimates of population size can be used to either estimate directly, or derive through secondary calculations, population growth rates [Equation (18.1)]. These methods include the Jolly–Seber (Jolly 1965, Seber 1965), Crosbie–Manly–Arnason–Schwarz (Crosbie and Manly 1985, Schwarz and Arnason 1996), Pollock's robust design (Pollock 1982), and Pradel (Pradel 1996) models. An important design consideration with these models is that there need to be attempts to capture previously unmarked individuals at random from the population of interest. For example, if interest is in the growth rate of the adult breeding population, efforts should be made to capture (and mark) currently unmarked adults, not just focus on previously marked adult individuals. This is particularly important if there might be immigration into the area from other natal areas.

When interest is in the direct estimation of growth rates it may not be necessary to separate out the effects of permanent emigration from survival as the emigration will be offset by the effect of permanent immigration into the area (Nichols and Hines 2002). That is, while estimated local survival of individuals within an area will be lower than true survival rate due to some individuals permanently emigrating away from the area, the estimated recruitment rate will be higher than the in situ recruitment rate as it will also include immigrants that are permanent emigrants from outside of the area of interest.

Species occurrence and occurrence dynamics

Rather than focusing on individual-level metrics, species-level metrics often may be more appropriate or feasible to consider. Instead of attempting to estimate the number or density of individuals of a species within an area, and how that changes through time, determining the presence or absence of the species at locations within that area may be a suitable alternative. Questions associated with species range or distributions fall quite naturally into such a category, as do questions concerning metapopulations, habitat modeling, and resource selection (e.g. MacKenzie 2005a, MacKenzie *et al.* 2006). Here, I am going to generically refer to these types of questions as ones of species occurrence, or occupancy.

With attention shifted to inference at the species level, the need to capture, or even just observe, individuals is greatly reduced. Therefore, there can be a great deal of flexibility in the types of field information that could be used to register as a detection of the species at the location of interest, e.g. scat, tracks, calls, vocalizations, etc., in addition to physical capture or observation of individuals of that species. There is also a great deal of flexibility for what may practically constitute a "location" at which the presence of the species is being determined, which I return to below. This flexibility is both a blessing and a curse. It means that the general approach has great practicality as it may be applied in a multitude of different ways to different situations. On the flip side, it also means that one could potentially misuse the approach, or use data that nominally appear reasonable but are actually inappropriate for addressing the biological questions of interest. The key to ensuring there is not a mismatch between the field data collected and desired questions of interest is good study design.

A species occurrence perspective is not a new idea, though it has long been recognized that there is often the potential for false absences, which can lead to bias in estimates of current occurrence or the underlying processes of change in occurrence (Moilenan 2002, MacKenzie *et al.* 2006). Analytic methods for estimating species occurrence that explicitly allow for imperfect detection are relatively new. While there was some initial development of ideas in the late 1980s (Geissler and Fuller 1987, Azuma *et al.* 1990), more generalized approaches were independently developed in the early 2000s (MacKenzie *et al.* 2002, Tyre *et al.* 2003, Wintle *et al.* 2004, Stauffer *et al.* 2004) with subsequent extensions (e.g. MacKenzie *et al.* 2003; Royle and Nichols 2003; Royle 2004, 2006; Nichols *et al.* 2007; MacKenzie *et al.* 2009). These methods tend to be very similar to those used for the analysis of capture–recapture data. Important work has also been conducted associated with study design issues (MacKenzie and Royle 2005, Field *et al.* 2005, MacKenzie *et al.* 2006, Bailey *et al.* 2007, Guillera-Arroita *et al.* 2010). Recently, attention has been devoted to understanding the relative sensitivity of species occurrence to the underlying dynamic processes (Martin *et al.* 2009a), similar to examining the effect of individual-level demographic processes on asymptotic population growth rates.

Accounting for imperfect detection is achieved by conducting multiple surveys at each sampled location, where each survey is a single attempt at detecting the target species if it is present at the location over a specified time period. Essentially, the repeated surveys are a method of quantifying the effort expended to detect the species at each location. There are a number of practical ways in which the multiple surveys may be conducted (e.g. temporally, spatially, or with different detection methods), with the prime consideration being how the nature of the repeat surveys affect the biological interpretation of "occurrence" as it is being measured in the field. Some attention has also been recently devoted to addressing situations where detections of the species are recorded in continuous time or space (Guillera-Arroita *et al.* 2011; see below). Importantly, at each time point where occurrence, or changes in occurrence, are to be assessed, these repeated surveys need to be conducted unless potentially restrictive assumptions are to be made during analysis.

The basic dynamic processes of species occurrence are local extinction and colonization. These processes determine the probability of the species being present at a location given it was present or absent from the location at the previous time point. When coupled with an initial level of occurrence, many other processes can be described using these basic building blocks, such as distribution or range changes, turnover rates, and rates of change in occurrence.

Below I outline a number of general considerations that I have found to be useful when assisting people with the design of monitoring programs and other studies. They focus on what I regard as some of the key aspects that need to be carefully, and appropriately, addressed in order for a monitoring program to yield reliable data. But first, a brief review of some available analytic approaches.

Analytic methods

Occupancy as a state variable

As with inferences at the individual-level (e.g. population growth rate), a series of estimates of species occurrence at different points in time could be used to coarsely describe demographic processes. Of the methods that have been developed for assessing occurrence at a single point in time, while accounting for potential false absences, the approach detailed by MacKenzie *et al.* (2002; see also MacKenzie *et al.* 2006) is arguably the most flexible, as it accounts for unequal sampling effort and incorporation of covariates, or predictor variables, on both occurrence and detection probabilities (the methods described by other authors could be regarded as special cases of the MacKenzie *et al.* 2002 approach). Essentially, it is a form of generalized logistic regression where simultaneous logistic regression analyses are conducted on both occurrence and detection, such that the regression coefficients for the occurrence component of the model are automatically corrected for the effects of imperfect detection. Once an analysis has been completed, the regression coefficients can be used to predict the probability of occurrence (or detection) at unsampled locations within the area of interest based upon available covariate values, just as with standard logistic regression. In a very simple situation where no covariates are included in the analysis, the approach of MacKenzie *et al.* (2002) reduces to a zero-inflated binomial model. While the method of analysis proposed by MacKenzie *et al.* (2002) was cast within a maximum likelihood framework, placing it within a Bayesian framework is also possible (e.g. MacKenzie *et al.* 2006, Royle and Dorazio 2008).

There have been a number of notable extensions to the methods of MacKenzie *et al.* (2002) including those that could be used to account for additional heterogeneity in detection probabilities (Royle and Nichols 2003, Royle 2006), lack of independence and/or violation of assumptions for the detection process (Hines *et al.* 2010), co-occurrence of two or more species (MacKenzie *et al.* 2004), and multiple types of "occurrence", e.g. categories of relative abundance (Royle and Link 2005) or species absence/species present without reproduction/species present with reproduction (Nichols *et al.* 2007). This final extension is particularly interesting as it potentially enables broad-scale questions to be addressed at a finer scale than presence/absence without progressing to the higher resolution level that would require information from marked individuals. Guillera-Arroita *et al.* (2011) have recently extended the methods to situations where, rather than detections being recorded at discrete repeat surveys of a location, detections may be recorded in continuous time, such as with camera traps. The advantage of the extension is that it avoids the necessity of arbitrarily discretizing the data for use with other approaches, which if done inappropriately can result in biased estimates. Methods for accounting for spatial correlation in the occurrence of the species on the landscape have also been developed (Sargeant *et al.* 2005, Magoun *et al.* 2007, Royle and Dorazio 2008), although doing so may not always be strictly necessary (more below).

Occurrence dynamics

Through time, the presence of the target species at units across the landscape will change. Where the area of interest is sampled at systematic points in time to determine the status of the units at each occasion, it is possible to describe the processes of those changes. Here the processes will be referred to as the probabilities of local extinction and colonization. Admitting the potential for false absences, then there will be ambiguity as to whether an extinction or colonization event has actually occurred. MacKenzie *et al.* (2003) extended the single time point method of MacKenzie *et al.* (2002), enabling the dynamic rate parameters of occurrence to be estimated. Key to the method is that at each systematic point in time, repeat surveys are conducted to enable separation of detection from the underlying biological processes of interest. Similar methodology was also used by Barbraud *et al.* (2003). Similar to the case of the MacKenzie *et al.* (2002) model, covariates or predictor variables can be introduced to relate the dynamic vital rates to characteristics of each location to explain variation in those rates across the landscape (e.g. due to elevation, annual rainfall at each location, etc.) or temporally (e.g. generalized weather patterns or implementation of certain management actions). MacKenzie *et al.* (2006) describe how the basic model of MacKenzie *et al.* (2003) could be reparameterized to address alternative metrics of interest including overall occurrence at each time point, rates of change in occurrence, or turnover rates, and

Royle and Kéry (2007) provided another parameterization using Bayesian methods of inference. MacKenzie *et al.* (2009) have recently extended the methods of MacKenzie *et al.* (2003) to situations where at each point in time the occurrence of the species could be classified into more than two discrete states.

Large-scale monitoring programs for avian influenza and the chytrid fungus motivated an extension by McClintock *et al.* (2010) to investigate occurrence dynamics at multiple spatial scales. For example, surveys may be conducted for the presence of avian influenza at a water body within a watershed, which is one of many watersheds within the region of interest. In some instances, monitoring may be desired of the occurrence dynamics at both the watershed and the waterbody scale. The approach outlined in McClintock *et al.* (2010) enables the dynamic processes at both scales to be estimated, while also accounting for imperfect detection.

A final extension allows the simultaneous estimation of both species occurrence and habitat dynamics (MacKenzie *et al.* 2011). It applies to a situation where, at each location of interest, habitat can be considered in terms of discrete types, or categories, which may potentially change through time. The presence or absence of the species at a location could consequently change due to the change in habitat type. Conversely, the rate of habitat change may also vary given the presence or absence of the species, either because the species modifies the habitat directly (so when it is present at a location the habitat is more likely to change), or the species is smart enough to preferentially select places that have different habitat dynamics even though the places appear similar to us in our assessment of habitat type (e.g. selecting ponds that are less likely to dry). In the face of such a situation, solely focusing upon species-occurrence dynamics may give an incomplete view of the system. For example, if there is an overall decline in occupancy through time, is that due to a general decline in occurrence while the available habitat is remaining relatively stable? Or within each habitat type the level of occurrence is stable, but the system is gradually becoming predominated by habitat types that have a lower level of occurrence? Having the ability to more clearly understand why any change is occurring will be beneficial to managers as the most appropriate management action relative to the desired objective could be different for each of the two possibilities. While previous methods can incorporate habitat information in the form of covariates, the integrated approach is likely to be more mechanistic and enables simultaneous prediction about the future state of both species occurrence and habitat. Such a tool is likely to be very valuable in the face of a changing world.

General design considerations

MacKenzie *et al.* (2006) devote an entire chapter to study design issues for estimating occurrence at a single time period, with additional suggestions related to estimating dynamic rates in another chapter. It provides comprehensive coverage of many of the issues that one must work through when designing a study or monitoring program based upon occurrence as a state variable. Here, I summarize many of these issues, but readers are directed there for further reading. In some areas, my thoughts on study design have evolved since 2006; hence, where any of the below suggestions appear at conflict with those in MacKenzie *et al.* (2006), the ones here should take precedence. Other notable texts that discuss design issues include MacKenzie and Royle (2005), Field *et al.* (2005), Bailey *et al.* 2007 and Guillera-Arroita *et al.* (2010).

An important factor to consider early during study design is that occurrence is not abundance. Some arrive at considering occurrence or presence/absence as a state variable because the requirements to reliably estimate abundance or density at the same spatial scale are infeasible given the resources available. While clearly the two concepts are inextricably linked, they are not equivalent and there will typically be subtle differences in the types of questions that can be addressed with each state variable: a change in one does not necessarily result in a change in the other. Therefore, moving to a coarser scale requires a change in mind set and expectations as to what information can be reliably collected from the field, which may result in overall objectives also having to be reassessed and rephrased.

Equally important is that it must be clearly articulated what the desired biological interpretation of "presence" or occurrence represents. There are often implicit spatial and temporal assumptions that have been made in the past when using species presence/absence data ignoring issues of detection. When dots are placed on a map to indicate known instances where the species is present, how big an area is the dot supposed to represent? How far away from the dot should it be regarded that species occurrence is unknown? Do we assume that the dot indicates that the species will always be present at that location, or should we assume it only indicates presence at a certain time period? Consideration of such questions leads quite naturally to what should be regarded as a sampling unit and a sampling season.

Defining the sample unit

The sample unit is the basic landscape unit for which we are attempting to determine the presence or absence of the species. At the scale of the sampling unit, occurrence is simply a binary outcome: yes or no. A sampling unit may be naturally or arbitrarily defined, e.g. ponds and habitat patches or grid cells. Where a sampling unit is "naturally" defined, some consideration should be given to whether that unit is likely to be relatively static in time, and if not consider the potential consequences of it changing. For example, what may be considered a single vernal pool in the current year could turn into multiple pools following a dry year, or amalgamate with other nearby pools after a wet year. Therefore, how should a change in occurrence be regarded when the defined sampling units themselves are dynamic?

When a grid-based sampling approach is being considered, the spatial scale of the defined cells determines the resolution with which occurrence is to be measured, with larger cells resulting in a courser metric or grainier image. With larger cells there may also be the possibility of the species occurring within only part of the cell, yet if detected anywhere within the cell the species will be regarded as "present" in the entire cell. If that is undesirable, that would suggest that a cell should be defined at a smaller spatial scale. When dealing with a territorial species, the suspected home range size of an individual could (but does not have to) be used to define the size of a cell, although it must be realized that as individuals have no knowledge of the arbitrary grid lines, the number of occupied cells is unlikely to match the number of individuals in the area (i.e. the territory of a single individual may overlap the boundary of multiple grid cells, thus occupying all of them).

The collection of sampling units that have a non-zero probability of being selected for sampling defines the statistical population of interest. Potential sampling units that have no chance of being selected (e.g. because they are considered inaccessible) are outside of the scope of inference, hence there is no statistical basis for generalizing the final results to these places (see also Chapters 2, 5). An important change in mind set when moving from an abundance- to occurrence-based state variable is with occurrence, the scope of the sampling is no longer of individuals on the landscape, but of the landscape itself. That is, the primary sampling consideration within the general area of interest is not to find individuals of the target species, but to select spatial sampling units from within that area. Finding evidence of the target species if it is within a unit is a secondary sampling consideration. The presence or absence of the species is just a characteristic of a sampling unit, much like the habitat type or elevation. As such, there is information in absences and, generally speaking, selecting only units where the species is believed to be present is undesirable, unless only that subset of units are of specific interest.

Related to this, if the intent of the monitoring program is to identify range changes, then the statistical population of interest cannot be defined simply as the area currently occupied by the species. Sampling should be conducted in areas that are considered to be outside the current species range (i.e. where the species is currently absent) as the species may occur there at some point in the future. Similarly, if the species has a particular habitat requirement in order to be present at a location (e.g. requires a certain vegetation type), and that habitat is dynamic, then the long-term sampling frame needs to include areas that are currently non-habitat as they may became habitat, and subsequently become occupied at a later time point.

While spatial correlation (e.g. where the probability of a unit being occupied might depend upon number of occupied neighboring units) is likely to be a biological reality in many situations, often it may not be necessary to get overly concerned about it. If the monitoring data are being used to obtain an estimate of the overall probability of occupancy within the population of interest (i.e. a single numerical value), and simple random sampling is used, unbiased estimates will be obtained with completely appro-priate standard errors using standard methods that do not account for spatial correlation. The reason is that by using random sampling, spatial correlation on the landscape has no influence on the observation process. Figuratively, with random sampling the landscape is being broken into the individual sampling units which are then placed into a (really big) hat and mixed up, then units are drawn out independently. Regardless as to whether there is really spatial correlation or not on the landscape, sampling theory guarantees us an appropriate answer (including a completely appropriate standard error). Similar results hold if interest is in the dynamic vital rates; the overall estimate will be correct even if there is an element of spatial correlation in the underlying processes, provided random sampling is used. I have recently confirmed this via simulation, although I have yet to publish the results. However, if non-random sampling is used, then it would seem likely that a systematic bias may be induced.

When interest is primarily directed towards producing accurate maps, or inference is required for specific units, then consideration of spatial correlation will be more necessary. Some forms of spatial correlation may be well explained by covariates. This may be either because of a true underlying relationship but it is actually the covariate that is spatially correlated (e.g. a species is clustered on the landscape because preferred habitat types are clustered), or correlation may arise from a more mechanistic process (e.g. distance from a presumed center of the species range). Inclusion of suitable covariates may therefore (partially) account for any spatial correlation, or there have been a number of methods that have been developed for explicitly modeling spatial correlation while investigating occupancy-type problems and accounting for imperfect detection, all of which relay on Bayesian methods of analysis (e.g. Sargeant *et al.* 2005, Magoun *et al.* 2007, Royle and Dorazio 2008).

Defining the sampling season and repeat surveys

One of the most important assumptions with occupancy estimation is that during the period of repeated survey, sampling units are closed to changes in occurrence. That is, species are either always present or always absent at a unit for the duration of the repeat surveys (e.g. during the repeated surveys conducted within a portion of each year as part of a monitoring program). This period is referred to by MacKenzie *et al.* (2006) as a sampling season. The closure assumption can be relaxed to allow temporary absences provided they are at random with respect to the temporal frequency of the repeat surveys. In this case, occurrence is measuring "use"; that is, the species is present at some point during the season, but may not always be there for the entire season (MacKenzie 2005b).

While recent extensions of the basic models have been developed to account for situations where the closure assumption is violated (e.g. Hines *et al.* 2010), it is advisable to make all practical efforts to meet the closure assumption as this keeps the modeling simpler and final inferences will be more robust. It is important to recall that closure is at the species level, hence there can be movement of individuals, but provided at least one individual is always present at a unit the closure assumption is being met. It must also be realized that if the required evidence to detect the species at a unit is sign, scat, or something similar, then the closure assumption relates to the evidence being searched for rather than actual individuals.

Repeat surveys do not have to constitute discrete visits to the sampling units, but could be achieved through multiple surveys conducted during a single visit by one or multiple observers. In a large sample unit, the repeat surveys could also be surveys of multiple subplots within the unit, although selection of the subplots should be at random and may have to be with replacement if the number of subplots within a unit is limited (Kendall and White 2009). I am not aware of any formal investigations evaluating whether it is best to randomly subsample every year, or only in the first year, as part of a long-term monitoring program. When considering different options for practical ways to conduct the repeat surveys, it is important to keep in mind that how the surveys will be conducted relate to the assumption of the modeling, and to the desired biological interpretation of occurrence.

It is also important to keep in mind that the time between sampling seasons determines the interpretation of the dynamic parameters. The data are silent with respect to any processes that may occur at a finer timescale. For example, if the species is detected at a unit in consecutive sampling seasons, that would be interpreted as the species persisting at that unit. Regardless of whether it is a biological reality or not, it cannot be determined from data at that scale whether the species has actually persisted at the unit for the full duration of the interval between sampling seasons, or has gone locally extinct with a subsequent recolonization. If it is suspected that there are finer temporal processes, and those are of interest, then this can only be captured through more frequent sampling.

Future research and development

Analytic techniques for investigating demographic processes are constantly evolving. New analytic techniques are frequently being developed either to provide new answers to old questions, or an answer to new questions. This development is an organic process that is driven by a number of factors including the appropriateness of existing methods, available resources, technological developments, and, most importantly, the questions being asked. In this field, the analytical developments and the types of questions being asked are often in a symbiotic relationship, with each driving and advancing the other forward through time. As such, it is difficult for me to envision in which direction future advancements will be made.

If anything, I believe that most analytical developments will be centered around technological advancements. For example, telemetry and satellite tags are constantly becoming smaller, smarter, and cheaper, with the ability to supply large quantities of data over various timescales. The analytical challenge will be extending current methods to efficiently deal with these large data sets, taking into aspects such as finer-scale movements, and developing software packages or routines that make such approaches available to mere mortals. Incorporation of landscape-level information with individual-level demographic processes is another area where I believe further developments will be made, particularly as computer speed continues to improve enabling faster processing of the data sets.

In terms of future research, there are two main areas where I see work is required. The first is methods to assess how well a model fits the data. Traditionally this is an area that lags behind the development of new and improved analytical methods, but verification that the model is a reasonable representation of reality is still vitally important in order to have faith in the reliability of final conclusions. For many analytical approaches for estimating demographic process, such assessment methods are still lacking. The second relates to data collection and general study design. Analytical extensions are sometimes motivated by trying to extract information from data sources that are less than ideal for use with current methods. In such situations, resulting inferences can be very model-dependent and may heavily rely on a series of assumptions. Which begs the question, just because we can, does that mean we should? Could the program have been designed differently to provide a design-based solution allowing a simpler modeling approach to be used? What are the trade-offs in terms of robustness of inferences and cost effectiveness? My personal philosophy is that if I identify a design-based and a model-based solution to a problem, then, where practical, I will always recommend the design-based solution first, as I will still have modeling as a back up if required. Whereas, if I opted for a modeling approach right from the start, I will be reliant on getting the modeling correct, with only further modeling as a back up. Particularly with respect to occurrence-based monitoring programs, research is also required around the relative efficiency of different sampling methods (e.g. completely random sampling, unequal probability sampling, etc.) in order to provide practical, real-world advice to program designer. However, I would suggest that such research should not only focus on questions of efficiency, but also robustness. What happens if any assumptions made while developing the sampling scheme are invalid? Are there any trade-offs in terms of efficiency versus robustness of the data?

Summary

Why is, in my view, the most important part of any monitoring program. Not having a clear understanding of why the data are going to be collected in the first place is likely to lead to inappropriate monitoring programs that may ultimately be a major waste of resources. Monitoring should be there to provide feedback on some larger endeavor, with that information being used to inform some decision-making process. If the objective of the endeavor changes through time, then the monitoring may also have to change, and this evolution should not be avoided.

Monitoring programs that require information on demographic processes will frequently require some form of a capture–recapture approach to be implemented to account for imperfect detection, particularly if operating over longer timescales. Approaches that do not account for detection issues may provide misleading results (Williams *et al.* 2002). Specific approaches will have specific sampling requirements, but some of the key common elements are: (i) all individuals within the population of interest have a non-zero probability of being captured at each sampling session; (ii) individuals are captured independently of one another; (iii) marked individuals are correctly identified each sampling session; and (iv) there is no unmodeled heterogeneity in capture probabilities. Here I have focused on some of the key demographic processes (population growth rates, survival, and reproduction/recruitment) and discussed some of the analytical techniques appropriate to them.

Rather than being focused on data being collected at the individual-level, the focus of monitoring programs may instead be shifting to a species-level. At the species-level, the question has changed from "How many individuals are there?" to "Where are they?" Techniques that account for imperfect detection in this situation are much newer, with the vast majority being developed within the last decade (as of 2011). As such, their scope is currently more limited, but growing rapidly. Switching to more of a species presence/absence measure is both a blessing and a curse. There is much more flexibility in the type of data that could be collected to detect the species at a location, but the added flexibility can also introduce uncertainties in terms of how exactly parameter estimates should be interpreted. The cure is to very carefully work through the various issues when designing a program, with the key being to have a very clear understanding of what biological quantity is of interest.

In this chapter I have provided a bit of guidance on the different options available and highlighted some of the key considerations from a design perspective. I have only covered the tip of the iceberg on this topic. Readers should regard this as the first word on these approaches, not the last, and are strongly encouraged to undertake (and pay attention to) further research.

19 Dealing with incomplete and variable detectability in multi-year, multi-site monitoring of ecological populations

Sarah J. Converse and J. Andrew Royle

Introduction

An ecological monitoring program should be viewed as a component of a larger framework designed to advance science and/or management, rather than as a stand-alone activity. Monitoring targets (the ecological variables of interest; e.g. abundance or occurrence of a species) should be set based on the needs of that framework (Nichols and Williams 2006; e.g. Chapters 2–4). Once such monitoring targets are set, the subsequent step in monitoring design involves consideration of the field and analytical methods that will be used to measure monitoring targets with adequate accuracy and precision. Long-term monitoring programs will involve replication of measurements over time, and possibly over space; that is, one location or each of multiple locations will be monitored multiple times, producing a collection of site visits (replicates). Clearly this replication is important for addressing spatial and temporal variability in the ecological resources of interest (Chapters 7–10), but it is worth considering how this replication can further be exploited to increase the effectiveness of monitoring.

In particular, defensible monitoring of the majority of animal, and to a lesser degree plant, populations and communities will generally require investigators to account for imperfect detection (Chapters 4, 18). Raw indices of population state variables, such as abundance or occupancy (*sensu* MacKenzie *et al.* 2002), are rarely defensible when detection probabilities are < 1, because in those cases detection may vary over time and space in unpredictable ways. Myriad authors have discussed the risks inherent in making inference from monitoring data while failing to correct for differences in detection, resulting in indices that have an unknown relationship to the parameters of interest (e.g. Nichols 1992, Anderson 2001, MacKenzie *et al.* 2002, Williams *et al.* 2002, Anderson 2003, White 2005, Kéry and Schmidt 2008). While others have argued that indices may be preferable in some cases due to the challenges associated with estimating detection probabilities (e.g. McKelvey and Pearson 2001, Johnson 2008), we do not attempt to resolve this debate here. Rather, we are more apt to agree with MacKenzie and Kendall (2002) that the burden of proof ought to be on the assertion that detection probabilities *are*

Design and Analysis of Long-term Ecological Monitoring Studies, ed. R.A. Gitzen, J.J. Millspaugh, A.B. Cooper, and D.S. Licht. Published by Cambridge University Press. © Cambridge University Press 2012.

constant. Furthermore, given the wide variety of field methods available for estimating detection probabilities and the inability for an investigator to know, a priori, if detection probabilities will be constant over time and space, we believe that development of monitoring programs ought to include field and analytical methods to account for the imperfect detection of organisms.

When accounting for imperfect detection in monitoring programs, an investigator is faced with the need to estimate and account for a potentially large number of detection parameters that are not of primary interest. Uncertainty in estimated detection probabilities will necessarily increase uncertainty in estimates of monitoring targets. Therefore, it is important to employ efficient statistical methods for estimating detection probabilities. This will ensure that the results of monitoring (on which a fair amount of money is typically spent) will be as informative as possible.

In spatially and/or temporally replicated monitoring studies, investigators can increase the precision of estimates by sharing information across replicates when estimating detection (Box 19.1; e.g. Bowden *et al*. 2003, White 2005, Converse *et al*. 2006) instead of analyzing each individual time and/or site-specific data set in isolation. Also, sparse data in individual replicates may not support independent estimation of detection for each replicate, encouraging some to fall back on indices (e.g. McKelvey and Pearson 2001). Sharing information across replicates involves combining data in a single analysis, and then modeling detection probabilities with parsimonious models that account for variation in these parameters with either random effects or reduced numbers of fixed effects. This approach is particularly useful when applied to studies of uncommon species. Uncommon species are those for which often, paradoxically, there is both the greatest ecological interest (i.e. endangered species) as well as the least information due to small sample sizes.

Here, we demonstrate the use of hierarchical models for combining information among spatio-temporal replicates, in order to estimate, using a single synthetic analysis, both detection probabilities and monitoring targets. In this context, a hierarchical model is one that recognizes data as arising from two conceptually distinct processes: one is the ecological process that is of primary interest, such as the ecological process that determines the abundance of animals at a particular space and time, and the other is the observation process that samples a particular population at that space and time.

We focus on estimation and modeling of abundance parameters using capture–mark–recapture (CMR) to understand ecological processes such as influence of management on abundance and trend in abundance over time. We note that use of hierarchical models for estimation and modeling of other sorts of demographic parameters (such as occupancy, survival, movements, etc.; see Chapter 18) or using other sorts of data collection techniques for estimating abundance (e.g. mark–resight, distance sampling, multiple-observer, and repeated-count approaches) is conceptually very similar.

Hierarchical models are useful for two specific purposes. First, there is often direct interest in modeling variation in spatially or temporally indexed population parameters, such as abundance among populations. Hierarchical models can be used to impose hypothesized structure directly on model parameters and therefore estimate and model abundance in one synthetic analysis, as opposed to a post-hoc analysis of estimates

---

Box 19.1 Take-home messages for program managers

Most ecological surveys that count or assess the presence/absence of animals (and in some cases plants) will detect much lower than 100% of the individuals present or species occurrences in sample units. When designing monitoring programs, particularly for animal populations, it is critical that investigators design field methods to allow use of analytical methods that account for imperfect and variable detection probabilities. This is essential for producing defensible results and to avoid producing biased and misleading indices of population parameters such as abundance or site occupancy – faulty results which in turn may lead to faulty or inefficient management responses (see also Chapters 4, 18). However, even when appropriate field methods are used in multi-year, multi-site monitoring, it may not be feasible (if data are sparse) or statistically efficient to analyze data from each site and time separately. This problem can be addressed using analytical methods that allow investigators to share information across years and/or sites when estimating detection probabilities, while simultaneously estimating and modeling the parameters of true ecological interest. The development of hierarchical statistical models facilitates this sharing of information, resulting in more precise estimates of the parameters of ecological interest. This practice can be applied when estimating a wide range of demographic parameters, including state parameters (occupancy, abundance) and rate parameters (survival, fecundity, state transitions, etc.), and is particularly useful when applied to studies of uncommon species, or species that are highly variable in their density across the sampled area. In the case of abundance modeling, the hierarchical model includes components describing the observation process (which incorporates detection probability in, for example, a capture–mark–recapture model), the ecological process producing spatio-temporal patterns in underlying abundance (including, perhaps, a trend parameter), and the statistical distributions of the parameters in the observation and process submodels. This analytical approach can produce more statistically efficient estimates of detection probabilities and improvements (increased precision and decreased bias) in the estimates of demographic parameters.

---

obtained from some initial estimation procedure (e.g. Converse *et al.* 2006). Second, hierarchical models provide a mechanism for simplifying the observation portion of the model used to estimate detection probabilities. In particular, with hierarchical models, detection probabilities can be modeled using mixed models, containing random effects, so that it is possible to specify a probability distribution for how detection probability varies across replicates. In such cases, hierarchical models yield a compromise between overly complex models (e.g. site- and time-specific detection parameters p from an "independent" analysis of each replicate), which might be poorly informed by sparse replicate-specific data, and overly simple models (e.g. constant p from a "combined" analysis), which may omit important variation in detection probabilities over replicates. Hierarchical models thus provide a mechanism for imposing relatively weak stochastic constraints among parameters (e.g. Link and Sauer 2002).

We begin with a case study as a demonstration of concept, follow that with results of simulations to further illustrate the benefits of combined analyses, and then move on to a more general discussion. We illustrate the benefits, in terms of bias and precision of abundance estimates and abundance model parameters, of analyses of multiple CMR data sets in which data are shared across replicates to estimate detection probabilities, versus analyses where data are not shared.

Example: small mammal responses to forest thinning

Converse *et al.* (2006) collected CMR data on small mammal species to understand the ecological impacts of forest fuel reduction treatments in US southwestern ponderosa pine (*Pinus ponderosa*) forests as part of a larger nationwide study (McIver *et al.* 2008). Here we reconsider the data collected at the Southwest Plateau Study Area, near Flagstaff, Arizona. We were interested in obtaining abundance estimates and estimating the effect of the forest thinning treatment on abundance. We present three different hierarchical models for analysis of these data: one model where no information is shared across replicates to estimate detection probabilities, and two models that involve sharing data across replicates to estimate detection probabilities in a more parsimonious fashion. In each of the models, we also obtain abundance estimates and an estimate of the effect of the thinning treatment on abundance. We compare these three models based on the precision of the abundance and treatment effect estimates. Our goal in presenting this case study is to demonstrate the value of hierarchical models in developing statistically efficient estimates of parameters of interest from monitoring data that must be corrected for imperfect detection.

Design and field methods

The Southwest Plateau Study Area was composed of three study blocks, each composed of four experimental units. Grids of live traps were used to catch small mammals in each of these 12 experimental units in each year from 2000 to 2003, producing 48 total spatio-temporal replicates (four yearly measurements on each of 12 experimental units). Before data were collected in 2003, thinning treatments were applied to two experimental units at each block.

The taxa with the highest overall number of individuals captured was *Peromyscus* spp. (primarily deer mice, *P. maniculatus*, with a small number of brush mice, *P. boylii*), with 489 total individuals (range = 0–45 per experimental unit per year). We focus on this *Peromyscus* spp. data set – a relatively large data set – while noting that our methods are equally applicable, and in fact are even more critical to use for sharing information across replicates, when analyzing the sparser data sets which arise from monitoring uncommon species, or when analyzing data sets from the high number of replicates which may result from long-term, especially spatially replicated, monitoring.

Trapping occurred in July and August each year during five-day sessions. Two different trapping intensities were used (traps spaced every 50 m on grids for low intensity vs.

traps spaced every 25 m for high intensity), with the higher intensity applied based on the results of a pilot study conducted in the second year of data collection (Converse et al. 2004). Sherman live-traps (H.B. Sherman Traps, Tallahassee, Florida, USA; use of trade or product names does not imply endorsement by the US Government) were baited with grain, and were checked both morning and afternoon to yield 10 trapping occasions per replicate. In 2000, six of the experimental units were trapped for only nine occasions, and so detection probabilities were set to 0 in the analysis for the last occasion. Captured individuals received two uniquely numbered ear tags.

Specification of the modeling approach

We present analyses of this dataset based on a hierarchical model of abundance (see Royle 2004, Dorazio et al. 2005, Royle and Dorazio 2006, Royle and Dorazio 2008). We use a hierarchical model developed for spatially and temporally stratified sampling of populations (Royle et al. in prep). We assumed a model in which the size of the population in replicate j, N_j, has a Poisson distribution with mean λ_j. Population-level covariates are modeled on the log-mean scale, as in conventional Poisson regression. Royle et al. (in prep.) showed how this model describing variation in abundance among populations can be expressed as an individual-level CMR model (i.e. based on individual encounter histories). In particular, by introducing an individual covariate for "population membership" g_i, a specific multinomial distribution for g_i preserves the Poisson prior distribution on N_j. That is, Royle et al. (in prep) established that the choice of distribution for the individual covariate is directly linked to the choice of prior distribution for local population size N_j. For the Poisson model, the individual covariate is assumed to have a multinomial distribution with state-space $g_i \in (1, 2, \ldots, J)$ where J is equal to the total number of replicates, and with cell probabilities π_j, defined by:

$$\pi_j = \frac{\lambda_j}{\sum_J J\lambda_j} \qquad (19.1)$$

The total number of individuals across the 48 replicates was treated as an unknown parameter to estimate. For this parameter, we assume a uniform prior distribution on $[0, M]$ for some suitably large integer M.

The model can be analyzed using Markov chain Monte Carlo (MCMC) methods with data augmentation (Royle et al. 2007), which allows for Bayesian analysis of hierarchical models and is conveniently analyzed in WinBUGS (Gilks et al. 1996). To analyze the model by data augmentation, the total number of individuals captured, n, is augmented by some arbitrarily large number of "pseudo-individuals" (resulting in M total individuals in the dataset). The task is then to partition these individuals into sampling zeros (individuals that exist in one of the replicate populations but were not captured) and structural zeros (individuals that are not members of any replicate population). The model is fitted by recognizing that the augmented data can be represented as a zero-inflated version of the known-N model. Accommodating the zero-inflation is done by introducing a binary latent variable, z_i, which is Bernoulli-distributed with parameter Ψ. In this context, the parameter $\Psi = \Pr(z_i = 1)$ is the probability that an individual in the augmented data set belongs to the super-population of size $\sum_J N_j = N$. The use of data augmentation for abundance estimation is described in detail elsewhere (Royle and Dorazio 2008) and in the context of CMR models from replicated studies in Royle *et al.* (in prep.).

Outputs from the model then include g_i and z_i. The replicate membership, $g_i \in (1, 2, \ldots, J)$, is known for the captured individuals, but is unknown for the uncaptured individuals and is estimated in the analysis, while the z_i values are indicators for whether an individual in the augmented dataset (of length M) exists in the super-population N. Including only those i for which $z_i = 1$, the frequency distribution of the individual covariate g_i provides estimates of population size in each site in each year, N_j.

Necessary input data include a matrix (Y) of dimension $i \times k$ (where k is capture occasions per replicate) containing $i = M$ individual capture histories (1 on occasions when an individual is captured, 0 otherwise, and 0 for all occasions for the pseudo-individuals) and a vector of length i (g) that describes replicate membership and is "NA" for the pseudo-individuals.

We present three analyses of the data set, which we refer to as INDEPENDENT, REDUCED, and RANDOM. These analyses differ in the modeling of parameters associated with detection (as described below), but they have a common generalized linear model of abundance. In particular, we modeled the log of λ_j [from Equation (19.1); the means of the Poisson distributions controlling replicate-specific abundance] as a function of block (three different study blocks), year (four different years), and treatment (unthinned versus thinned). In this way, we obtain estimates of abundance for each site visit as well as an estimate of the treatment effects on abundance. For each of the analyses, we calculated the Coefficient of Variation (CV) of the treatment effect estimate and the abundance estimates (e.g. $\text{CV}_N = \frac{\hat{N}}{SE}$).

In the first analysis (INDEPENDENT), we estimated detection probabilities for each replicate independently; that is, we did not share information across yearly visits to each site or across sites to estimate detection probabilities. This can be seen as analogous to an analysis of 48 independent CMR data sets for estimation of abundance in each data set (although we note that in the hierarchical model, the data sets are linked together through the model of λ_j as we have described). We estimated three different detection effects for each of the 48 replicates a replicate-specific intercept, a replicate-specific behavior effect, and a replicate-specific time effect, i.e. we built detection models for each replicate analogous to model M_{tb} from Otis *et al.* (1978). The behavior effect modeled a different detection probability for an individual after the first time that individual was captured. The time effect was a modified version of the full time effect allowing for a different detection probability at each trapping occasion; here, the time effect differentiated only between trapping occasions that occurred in the morning and those that occurred in the afternoon (because *Peromyscus* spp. are nocturnal, they are more likely to be present in traps at morning trap checks). The INDEPENDENT analysis, then, included $48 \times 3 = 144$ parameters associated with the estimation of detection probabilities.

In the second analysis method (REDUCED), we shared information across replicates to estimate detection probabilities. Specifically, we included the 48 replicate-specific intercepts as in the INDEPENDENT analysis, but we only included a single common time effect and a single common behavior effect. Thus, the detection model in the REDUCED analysis included $48 + 2 = 50$ parameters. This model assumed that detection could vary across sites and years, but that the relative changes in detection across occasions within a session and as a result of previous capture were constant across sites and years.

In the third analysis (RANDOM), we shared information across replicates to estimate detection probabilities through the use of random effects. We modeled each of the replicate-specific detection intercepts as drawn from a normal distribution with mean μ_{int} and standard deviation σ_{int}; each of the replicate-specific behavior effects came from $N(\mu_b, \sigma_b)$; and each of the replicate-specific time effects came from $N(\mu_k, \sigma_k)$.

We note that, conceptually, the difference between independent and combined analyses of monitoring data is analogous to the difference between an analysis of replicated CMR data for abundance estimation with the closed population models offered in Program CAPTURE (Otis *et al.* 1978), which requires each replicate to be analyzed independently, versus an analysis that Program MARK (White and Burnham 1999) makes possible, where data may be combined across replicates for estimation. We refer investigators preferring the likelihood-based approach and the user-friendliness of Program MARK to White (2005) and Converse *et al.* (2006), who illustrate combined analysis approaches using Program MARK. However, investigators are frequently interested not just in obtaining abundance estimates, but also understanding the processes influencing abundance, such as treatment effects or time trends. In these cases, the investigator must develop models of abundance outside of Program MARK and must use methods that appropriately propagate estimation error through the abundance modeling process. For example, Converse *et al.* (2006) illustrate weighted least-squares regression as an approach to modeling abundance estimates obtained in Program MARK.

We specified prior distributions for block, year, and treatment effects, as well as fixed effects in the detection models (INDEPENDENT and REDUCED), as Unif(−5, 5). For the RANDOM analysis, we specified priors of Unif(−5, 5) for μ_{int}, μ_b, and μ_k, and Unif(0, 100) for σ_{int}, σ_b, and σ_k. The use of prior information in Bayesian analyses is discussed extensively elsewhere (e.g. Link and Barker 2010); in our case, we used priors that reflect the absence of prior information.

We implemented each of the analyses in WinBUGS (Gilks *et al.* 1996) using the R2WinBUGS package (Sturtz *et al.* 2005) in the R programming environment (R Development Core Team 2011). For each of three independent Markov chains, we ran 8000 samples and discarded the first 3000 samples, for a total of 15 000 samples from which we made inference. These sampling parameters were adequate to obtain good model convergence, based on visual inspection of the chains and the Brooks–Gelman–Rubin statistic, $\hat{R} < 1.2$, as recommended by Gelman *et al.* (2004). We used the posterior modes as point estimates of parameters of interest.

Example study results

The average CV for the abundance estimates was smallest for the REDUCED analysis (CV = 0.23), followed by the RANDOM analysis (CV = 0.25), and was largest for the INDEPENDENT analysis (CV = 0.31). As noted previously, a major benefit of hierarchical models is the ability to simultaneously estimate and model abundance; in this case, we estimated the effect of forest thinning on abundance of *Peromyscus* spp. This was a strongly positive effect, ranging from 0.75 to 0.84 (on the log scale) across the three analyses, indicating higher abundance on thinned than on unthinned units (Table 19.1; also the general conclusion of Converse *et al.* 2006). The CV on the treatment effect estimate was 0.23 for the REDUCED, 0.22 for the RANDOM, and 0.25 for the INDEPENDENT analyses. Abundance point estimates and SE for each replicate are presented in Table 19.1. While we note that the differences in CV for the treatment effect are not large, we suggest that the use of REDUCED- or RANDOM-type models will be increasingly valuable as the sample size per replicate declines (i.e. in monitoring uncommon species) or as the number of replicates increases (i.e. in long-term monitoring data sets; see the following simulation assessment).

We also note the large differences in detection effects across replicates, as well as the large behavior effect, which lend further credence to our argument about the importance of accounting for detection probabilities in monitoring programs. Using results from the independent analysis, which are least constrained, we found that the lowest daily detection probability – on one replicate during the afternoon trapping sessions and for an animal's first capture – was 0.016 (SD = 0.012), while the highest – on a different replicate, during the morning occasions and for a previously captured individual – was 0.347 (SD = 0.213).

Simulation assessment for a monitoring scenario

Although the results of the data analysis presented above demonstrate that the precision of abundance estimates and estimates of the effect of treatment on abundance (as measured by the CV) can be increased when information is shared to estimate detection probabilities, one might reasonably be concerned that the increase in precision would be paid for by an increase in bias resulting from specification of an overly simple model. Therefore, we conducted a simulation study to evaluate comparative bias and precision under these three analyses. To demonstrate the applicability of this approach for long-term monitoring, we focused on a trend-estimation scenario.

Simulation methods

We simulated data in the R programming environment and analyzed the simulated data using the three general approaches described above. We envisioned a monitoring program where replicates represented consecutive years of data collected on a single

Table 19.1 Example study results: abundance estimates (SE, CV) for mice (*Peromyscus* spp.) captured at the Southwest Plateau Study Area, Arizona, based on three different estimation approaches. The INDEPENDENT analysis estimates detection probabilities separately for each of the 48 unit/year replicates, the RANDOM analysis estimates detection probabilities for each replicate using Bayesian random effects, and the REDUCED analysis utilizes a reduced number of fixed effects compared to the INDEPENDENT analysis and shares information across replicates to estimate detection. M_{t+1} = number of individual mice captured.

Unit/year	M_{t+1}	INDEPENDENT	REDUCED	RANDOM
U1/Y1	5	5 (0.0, 0.01)	5 (0.7, 0.14)	5 (1.0, 0.19)
U2/Y1	6	6 (1.6, 0.26)	6 (1.0, 0.16)	6 (1.3, 0.22)
U3/Y1	3	4 (2.0, 0.50)	4 (1.6, 0.41)	5 (1.9, 0.37)
U4/Y1	10	10 (0.5, 0.05)	10 (1.1, 0.11)	10 (1.2, 0.12)
U5/Y1	9	9 (2.5, 0.28)	14 (2.9, 0.21)	13 (3.2, 0.25)
U6/Y1	4	11 (4.0, 0.36)	10 (3.1, 0.31)	12 (3.8, 0.32)
U7/Y1	4	11 (4.1, 0.37)	10 (3.0, 0.30)	12 (3.8, 0.31)
U8/Y1	13	14 (2.3, 0.16)	18 (2.9, 0.16)	17 (2.9, 0.17)
U9/Y1	9	10 (3.1, 0.31)	17 (3.3, 0.20)	16 (3.7, 0.23)
U10/Y1	16	16 (1.7, 0.11)	20 (2.7, 0.14)	19 (2.7, 0.14)
U11/Y1	12	14 (2.7, 0.20)	15 (2.9, 0.20)	17 (3.0, 0.18)
U12/Y1	8	9 (3.5, 0.38)	15 (3.4, 0.22)	16 (3.8, 0.24)
U1/Y2	2	2 (1.3, 0.67)	2 (1.2, 0.58)	3 (1.4, 0.47)
U2/Y2	5	5 (0.7, 0.14)	5 (0.6, 0.13)	5 (1.0, 0.21)
U3/Y2	2	3 (1.7, 0.57)	2 (1.3, 0.65)	3 (1.6, 0.52)
U4/Y2	6	6 (0.5, 0.08)	6 (1.0, 0.16)	6 (1.1, 0.18)
U5/Y2	3	3 (3.4, 1.12)	6 (2.4, 0.40)	7 (2.9, 0.41)
U6/Y2	14	14 (0.8, 0.06)	14 (1.1, 0.08)	14 (1.4, 0.10)
U7/Y2	3	3 (3.0, 1.00)	6 (2.4, 0.41)	7 (2.9, 0.41)
U8/Y2	6	7 (2.7, 0.38)	9 (2.4, 0.26)	10 (2.8, 0.28)
U9/Y2	15	16 (1.5, 0.09)	16 (1.4, 0.09)	16 (1.8, 0.11)
U10/Y2	5	7 (3.0, 0.43)	9 (2.7, 0.30)	10 (3.2, 0.32)
U11/Y2	6	6 (2.7, 0.45)	11 (2.7, 0.24)	11 (3.1, 0.28)
U12/Y2	4	4 (3.5, 0.87)	8 (2.7, 0.34)	9 (3.2, 0.35)
U1/Y3	2	7 (3.3, 0.47)	5 (2.1, 0.42)	6 (2.6, 0.44)
U2/Y3	3	7 (3.1, 0.44)	5 (2.0, 0.41)	6 (2.5, 0.41)
U3/Y3	5	6 (2.6, 0.43)	6 (1.9, 0.32)	7 (2.1, 0.30)
U4/Y3	0	8 (3.3. 0.41)	0 (2.4, NA)[1]	5 (2.8, 0.56)
U5/Y3	20	24 (3.4, 0.14)	24 (2.9, 0.12)	25 (3.2, 0.13)
U6/Y3	19	21 (3.2, 0.15)	22 (2.5, 0.11)	23 (2.9, 0.13)
U7/Y3	9	21 (5.0, 0.24)	18 (3.7, 0.20)	21 (4.4, 0.21)
U8/Y3	21	27 (3.7, 0.14)	25 (2.7, 0.11)	26 (3.1, 0.12)
U9/Y3	14	22 (4.5, 0.21)	24 (4.1, 0.17)	23 (4.2, 0.18)
U10/Y3	18	21 (4.0, 0.19)	27 (4.0, 0.15)	27 (4.2, 0.16)
U11/Y3	13	24 (4.8, 0.20)	24 (4.1, 0.17)	24 (4.6, 0.19)
U12/Y3	12	21 (5.0, 0.24)	22 (4.1, 0.19)	23 (4.6, 0.20)
U1/Y4	7	7 (4.4, 0.63)	8 (2.0, 0.25)	9 (2.8, 0.31)
U2/Y4	11	12 (3.1, 0.26)	12 (1.9, 0.16)	13 (2.5, 0.19)
U3/Y4	8	8 (1.1, 0.14)	8 (0.5, 0.07)	8 (0.9, 0.11)
U4/Y4	3	6 (2.7, 0.45)	4 (1.7, 0.42)	5 (2.0, 0.41)
U5/Y4	3	16 (5.4, 0.34)	9 (3.2, 0.35)	11 (4.2, 0.38)

Table 19.1 (*cont.*)

Unit/year	M_{t+1}	INDEPENDENT	REDUCED	RANDOM
U6/Y4	17	20 (3.6, 0.18)	20 (2.4, 0.12)	20 (2.8, 0.14)
U7/Y4	38	44 (4.3, 0.10)	42 (2.7, 0.06)	44 (3.6, 0.08)
U8/Y4	1	37 (7.9, 0.21)	17 (4.7, 0.28)	28 (7.2, 0.26)
U9/Y4	16	16 (2.5, 0.15)	16 (1.4, 0.09)	17 (2.1, 0.12)
U10/Y4	26	37 (6.1, 0.16)	31 (3.3, 0.11)	35 (5.3, 0.15)
U11/Y4	45	50 (3.6, 0.07)	49 (2.9, 0.06)	52 (3.7, 0.07)
U12/Y4	8	17 (4.9, 0.29)	14 (3.5, 0.25)	15 (4.0, 0.27)
Average CV		0.31	0.23	0.25
Treatment effect estimate (SE)		0.79 (0.20)	0.75 (0.17)	0.84 (0.19)
(95% CI)		(0.49, 1.18)	(0.44, 1.11)	(0.46, 1.20)
CV		0.25	0.23	0.22

[1] Excluded from calculation of average CV.

population. We included two different simulated scenarios, with five replicate years in the first scenario and 10 replicate years in the second scenario, to explore the impact of the number of replicates on results. For both scenarios, we generated λ values for each of 10 years (analogous to the replicates j in the case study, $j \in 1{:}10$) using a log-linear trend model:

$$\lambda_j = e^{\beta_0 + \beta_1^* j} \tag{19.2}$$

where $\beta_0 = 4$ and $\beta_1 = 0.2$. We then generated abundance using random draws from Poisson distributions with parameters λ_j. For the first simulation scenario, we discarded years 6–10, while in the second scenario, we retained all 10 years. Therefore, the 10-year scenario is a continuation of the 5-year scenario. The mean true abundance from our simulated data sets was $\bar{N}_1 : \bar{N}_5 = (22, 27, 33, 41, 50)$ for the 5-year scenario and $\bar{N}_1 : \bar{N}_{10} = (22, 27, 33, 41, 49, 60, 74, 88, 109, 134)$ for the 10-year scenario (small differences between the mean true abundance for the 5-year vs. the 10-year scenario are due to Poisson variation).

We generated unique detection probabilities ($p_{j,k}$) for each of the annual replicates (j), at each of 10 encounter occasions (k) within the replicate from:

$$\log\mathrm{it}(p_{j,k}) = \alpha_0 + \alpha_1^* j + \alpha_2^* k. \tag{19.3}$$

For each simulated data set, $\alpha_0 = -2$, while α_1 and α_2 were drawn from Unif(0.02, 0.04). We calculated recapture probabilities as $c_{j,k} = p_{j,k} + c.eff_{j,k}$, where $c.eff_{j,k}$ was also generated using Equation (19.3) except $\alpha_0 = -3$, while α_1 and α_2 were drawn from Unif(0.01, 0.02). This resulted in a "trap happy" effect (i.e. $c > p$; after first capture, animals became more trappable) though the magnitude of the effect varied over years and capture occasions. Therefore, the true detection model was one where detection varied, interactively, by year, occasion, and whether or not an individual had been captured previously.

We analyzed the generated data as described for the *Peromyscus* spp. data set. The model for the INDEPENDENT analysis estimated a separate first capture and recapture effect for each of the years (i.e. a "behavior" model for each year, with 10 parameters for the 5-year scenario and 20 for the 10-year scenario). In the model for the REDUCED analysis, we estimated separate first capture effects for each replicate but only one common recapture effect (6 parameters for 5-year scenario, 11 parameters for 10-year scenario). In the RANDOM model, we fit random effects for both capture and for recapture effects, each from normal distributions. Priors were set at Unif(-5, 5) for all parameters except the standard deviations of the random effects, which were Unif(0,100).

We also estimated and modeled the abundance parameters, as well as the trend in the abundance parameters, under each of the three analyses for both the 5-year and 10-year scenarios. That is, we specified a model on λ_j where mean abundance for replicate j is a product of year and a trend effect, β, the true value of which was 0.2.

We present simulation results for each of several metrics. For each abundance estimate, under each simulation, we recorded the percent relative bias ($\text{PRB}_N = \frac{|\hat{N} - N|}{N}$), the Coefficient of Variation, and the relative root mean square error ($\text{RMSE}_N = \frac{\sqrt{(\hat{N} - N)^2 + SE^2}}{N}$). We averaged these metrics over each set of simulated monitoring years (either 5 or 10). We report similar metrics for the estimate of the time trend in abundance from Equation (19.2): PRB_β, CV_β, and RMSE_β.

We discarded results of any simulation runs where convergence was poor, i.e. $\hat{R} > 1.5$, which is slightly more liberal than $\hat{R} > 1.2$ as recommended by Gelman *et al.* (2004), but allowed us to include a substantially greater number of our simulations while still completing the simulations in a reasonable amount of time. We discarded 5 of 100 simulations from the 5-year scenario and 22 of 100 from the 10-year scenario. The large $\hat{R}$ values resulted from poor mixing of the MCMC chains, and only occurred in the RANDOM analysis (although we discarded those simulations for all three analyses).

Simulation results

Perhaps counter-intuitively, we found that PRB_N and PRB_β increased with increasing number of parameters in the model (i.e. bias was always largest in the INDEPENDENT analysis) for both the 5- and 10-year scenarios (Table 19.2). The expectation of greater bias in models with fewer parameters is predicated on asymptotic arguments, while in this data set, we found that the sample sizes involved were inadequate to support highly parameterized models. This led to unreliable estimates consistent with non-identifiable or weakly identifiable parameters due to small sample sizes. Also, PRB_N and PRB_β were greater in the REDUCED than in the RANDOM analysis (except PRB_β in the 5-year scenario, a result that may simply be due to Monte Carlo error), but the difference between the REDUCED and RANDOM analyses in terms of bias was substantially smaller than the difference between the INDEPENDENT and REDUCED analyses for both the 5- and 10-year scenarios.

For both variance (as measured by CV_N and CV_β) and total error (RMSE_N and RMSE_β), the INDEPENDENT analysis always performed worst, followed by the

Table 19.2 Diagnostic metrics for models estimating abundance, N, and trend in abundance, β. Metrics include percent relative bias (PRB), coefficient of variation (CV), and relative root mean square error (RRMSE) recorded for closed capture data simulated 100 times each over 5 years and 10 years for one site, with 10 capture occasions per year.

Parameter	Metric	INDEPENDENT	REDUCED	RANDOM
			5 years	
N	PRB	0.1745	0.1232	0.1164
	CV	0.5611	0.2308	0.2069
	RRMSE	0.5955	0.2771	0.2482
β	PRB	0.3379	0.2441	0.2526
	CV	0.5909	0.3803	0.3563
	RRMSE	0.7125	0.4463	0.4461
			10 years	
N	PRB	0.1409	0.0921	0.0818
	CV	0.2203	0.1154	0.0951
	RRMSE	0.2823	0.1563	0.1316
β	PRB	0.0829	0.0729	0.0684
	CV	0.1141	0.0840	0.0810
	RRMSE	0.1424	0.1196	0.1155

REDUCED and then the RANDOM analyses, for both the 5- and 10-year scenarios (Table 19.2). Again, however, the difference was much greater between the INDEPENDENT and the REDUCED than the REDUCED and RANDOM analyses.

As the number of years increased (i.e. for the 10-year versus the 5-year scenario), the RANDOM analysis performed increasingly well by all measures compared to the REDUCED in estimating abundance. The pattern was less clear in estimation of the trend parameter. The greater number of replicates improved the relative performance of the RANDOM analysis over the REDUCED in terms of variance (i.e. CV_β), but the opposite pattern held for bias; in fact, the one place where the REDUCED analysis performed best was for PRB_β in the 5-year scenario, but again, we speculate that this may be due to Monte Carlo error. However, although the relative performance of the models changed from 5 to 10 years, all three models had much lower bias and higher precision after 10 years, as expected.

Discussion

We describe three approaches to the analysis of data arising from designs that allow for estimation of detection probabilities, and our results indicate that combining data across replicates, using hierarchical models, can result in greater efficiency in estimating detection probabilities as well as, more importantly, the parameters of primary ecological interest – in our case, abundance, treatment effects on abundance, and time trends in abundance. Unfortunately, examples of long-term monitoring programs that utilize designs allowing for estimation of imperfect detection are relatively few, although there are several notable examples. These include the Monitoring Avian Productivity and Survivorship Program (MAPS; Saracco *et al.* 2009), the North American Waterfowl Breeding Population and Habitat Survey (Smith 1995), the Swiss Breeding Bird Survey (Kéry and Schmid 2006), and, more recently, tiger monitoring efforts in India (Karanth *et al.* 2006).

Our results suggest that the INDEPENDENT-type analysis we present will not produce the best results, at least at the sample sizes we explored. Monitoring uncommon species will often result in even sparser data sets than those we present, which would suggest an even greater value in the use of integrated analyses with reduced fixed effects (the REDUCED analysis) or random effects (the RANDOM analysis). There are additional advantages of the REDUCED and RANDOM analysis approaches when estimating demographic parameters. For the REDUCED approach, given that multiple models can be considered, an investigator can allow the data to sort out the level of model complexity that is best supported (i.e. the optimal number and type of parameters to be fitted) based on model selection metrics (Burnham and Anderson 2002, Link and Barker 2006), although we note the difficulties involved in hierarchical model selection below. For the RANDOM approach, results should improve as the number of replicates (sites and/or years) in a data set increases. Finally, another benefit of RANDOM-type models is that they can easily accommodate variation in the observation process over replicates. For instance, if one fits behavior effects to all of the replicates in a data set using a random effect, and for some replicates this behavior effect was negative, while for others it was positive, and for others still the effect was negligible, this would be accommodated by the random effect distribution, resulting in a relatively large variance associated with the random effect.

With abundance or density estimation (Box 19.2), one question an investigator must confront is how to model the abundance estimates themselves, as presumably there is an interest in deciphering patterns in the abundance estimates. Unlike the analyses we demonstrate, Program MARK, which is a very common platform for obtaining abundance estimates from CMR data, does not allow abundance estimates to be modeled directly as part of the analysis (although synthetic estimation and modeling of other types of parameters, such as survival, state transitions, and occupancy, is possible in Program MARK). In this case, modeling of abundance estimates must be a separate step after abundance estimation. If an investigator obtains abundance estimates using REDUCED- or RANDOM-type models, then use of the estimates as "data" in general linear models clearly violates basic assumptions of those models – that the data are independent and identically distributed – because combining data across replicates will induce non-zero sampling covariance between the estimates. We suggest that even when the abundance estimates are obtained using the INDEPENDENT approach, taking those results into a general linear model framework, without dealing explicitly with sampling variances, is not ideal, as the general linear model assumption of equal variances may well be violated, while these variances are accounted for explicitly in the integrated approach we

---

Box 19.2 Common challenges: density estimation

In monitoring animal populations, density (organisms/unit area) frequently is the target variable of primary interest, so often investigators are interested in converting abundance estimates to density estimates. This has been a frequent stumbling block of capture–mark–recapture (CMR) studies focused on abundance estimation. Density estimation is challenging because, for vagile organisms sampled passively (i.e. using live traps, camera traps, or similar methods), the area to which the abundance estimate applies is not known. For example, animals whose home ranges are partially outside a trapping grid may be captured on the grid.

Several approaches have been taken to deal with this problem, including both CMR-based approaches where an independent estimate of trapping area must be obtained, as well as distance-sampling-based approaches involving trapping webs (see Parmenter *et al.* 2003 for a review). Converse *et al.* (2006) use the former approach, and specifically estimate mean maximum distance moved (MMDM; Wilson and Anderson 1985) in an analysis that also combined data over space and time. However, the MMDM approach is ad hoc.

Recently developed spatially explicit CMR models represent an improvement on historical methods of adjusting abundance estimates obtained by ordinary (i.e. non-spatial) CMR models. Spatially explicit CMR models were posed by Efford (2004) and recent developments in statistical inference have been made using both classical likelihood (Borchers and Efford 2008) and Bayesian (Royle and Young 2008) methods.

Investigators should, when estimating abundance using passive sampling technologies, determine how they will convert these abundance estimates to density estimates that represent a more meaningful measure of population status. Spatially explicit CMR models can be implemented in WinBUGS (e.g. Royle and Young 2008, Gardner *et al.* 2010). While these methods have not yet been demonstrated in combination with modeling spatially replicated populations as demonstrated here, the integration of these components is certainly possible.

---

describe. Converse *et al.* (2006) present a multi-step approach to modeling abundance estimates obtained from Program MARK.

Our approach was to estimate and model abundance simultaneously using a Bayesian hierarchical model, where the propagation of the variance (within and among populations) is implicit in the procedure. Use of a Bayesian hierarchical model also allows the investigator the flexibility to consider both random and fixed effects in the abundance estimation phase. Royle *et al.* (in prep.) provide example code for hierarchical modeling of spatially and temporally stratified abundance estimates which integrates the RANDOM abundance estimation approach described here and demonstrated in Appendix 19.1.

When designing long-term monitoring studies, investigators should begin by considering the field methodologies best suited to their system and to the parameters they wish to estimate (Chapter 2). There is a wide variety of advice available on the design of monitoring studies accounting for imperfect detection and a growing number of available field methods that can be employed (e.g. Thompson *et al.* 1998, Williams *et al.* 2002, Thompson 2004; Chapter 18). Our analyses here focused on CMR data, but the use of hierarchical models with combined data is by no means restricted to CMR data sets.

The most powerful tools available for designing surveys are pilot studies and simulation studies (Chapters 8, 10). With a 1–2-year pilot study, an investigator can get a sense of detection probabilities, as well as the variation in detection probabilities across time and space. This information can be used to simulate data and the data can then be analyzed with the three model types shown here. The investigator can then evaluate whether the estimates produced from the best-performing analysis are of the needed precision. If not, detection probabilities or the number of sampling occasions can be modified, simulating a more intensive design; the investigator can use such an iterative approach to determine the level of sampling intensity required to obtain ecological parameter estimates of the required precision. Again, we emphasize the importance of placing monitoring firmly within a framework of management decision-making or scientific learning (Nichols and Williams 2006); the needs of this framework should determine the type and precision of information required of a monitoring program.

Recommendations for future research and development

When adopting hierarchical models for modeling abundance, one of the challenges is developing general and flexible models for the abundance parameters, N_j, for a collection of replicates $j = 1, \ldots, J$. We provide a hierarchical model which allows a Poisson assumption on the abundance parameters. Development of more general assumptions for the abundance parameters would be valuable. For example, the model we describe does not allow for extra-Poisson variation (overdispersion) in abundance (e.g. Chapter 20), which, if present, would result in underestimation of variance. A related extension would be development of models to allow for spatial dependence in abundance across populations or replicates. More general assumptions regarding the distribution of the abundance parameters would provide greater flexibility to analysts.

A broader issue in need of more development is Bayesian model selection (Kéry 2010). Program MARK (White and Burnham 1999), in particular, has offered demographic analysts ready access to model selection metrics including AIC_c and $QAIC_c$ (Akaike 1973, Hurvich and Tsai 1989, Burnham and Anderson 2002). A Bayesian analog, the deviance information criterion (DIC) exists and is calculated by WinBUGS (Spiegelhalter *et al.* 2002). However, DIC as computed by WinBUGS is not reliable in hierarchical models (Millar 2009, Kéry 2010). Alternatives exist, including reversible-jump MCMC (King *et al.* 2009) as well as model expansion using binary indicator variables of whether predictor variables are in the model (Link and Barker 2006, Royle and Dorazio 2008,

Smith *et al.* 2011). However, several complications exist, including the sensitivity of results, in Bayesian model selection, to the specification of priors on parameters (Link and Barker 2006). It is apparent that model selection (in either likelihood or Bayesian analyses) is not resolved (Link and Barker 2006) and we anticipate that model selection methods, particularly in Bayesian analysis, will continue to evolve in the near future.

Finally, an important technical challenge to the widespread adoption of Bayesian hierarchical models is the lack of fast, stable, and user-friendly software. Currently, WinBUGS (Gilks *et al.* 1996) and other implementations of the BUGS language are the most widespread options, but we advocate for development of additional packages that will facilitate use of these models by a larger community of analysts who may not have sustained access to a collaborator with expertise in Bayesian hierarchical modeling.

Summary

When studies of animal populations include replication over space and time, there is great potential to share information to increase the efficiency of statistical estimation of demographic parameters. We found that estimation approaches that involved sharing information for the estimation of detection probabilities, and hence abundance, either through fixed effect covariates or through random effects, resulted in less biased, more precise abundance estimates than when information was not shared across space and time. We further demonstrated the utility of Bayesian hierarchical modeling for not just estimating abundance, but for modeling abundance (e.g. for experimental comparisons or trend analyses) in a single synthetic analysis. Bayesian hierarchical modeling offers tremendous flexibility to investigators because of its ability to integrate fixed as well as random effects, and, with the availability of WinBUGS (Gilks *et al.* 1996), Bayesian approaches are accessible to a wider array of investigators. We emphasize that, although our analyses focused on CMR data and estimation of abundance, the general principles we present – the use of hierarchical models with combined data for estimation of detection probabilities – are relevant to a wide variety of field approaches and demographic parameters.

Appendix 19.1. Example WinBUGS model code for estimating replicate-specific abundance with the random analysis described in text

```
model{
#PRIORS
  mu.rep ~ dunif(0,10)
  tau.rep <- 1/(sigma.rep*sigma.rep)
  sigma.rep ~ dunif(0,100)
  mu.cap ~ dunif(0,10)
```

```
tau.cap <- 1/(sigma.cap*sigma.cap)
sigma.cap ~ dunif(0,100)

psi ~ dunif(0,1)
for (j in 1:nreps){
  lam[j] ~ dgamma(1,1)
}

#LIKELIHOOD
  for(j in 1:nreps){
    gprobs[j]<- lam[j]/sum(lam[1:nsites])
  }
  for(i in 1:M){
    g[i] ~ dcat(gprobs[])
    z[i] ~ dbern(psi)
    for(j in 1:nreps){
      for(k in 1:lastcap[i]){
        Y[i,k] ~ dbern(mu[i,j,k])
        mu[i,j,k] <- z[i]*p[i,j,k]
        logit(p[i,j,k]) <- rep.p[g[i]] +
cap.p[g[i]]*reencounter[i,k]
      }
    }
  }
  for(j in 1:nreps){
    rep.p[j] ~ dnorm(mu.rep,tau.rep) #random effect,
replicate-specific
 #detection intercept
    cap.p[j] ~ dnorm(mu.cap,tau.cap) #random effect,
replicate-specific
 #capture effect
  }
}
```

20 Optimal spatio-temporal monitoring designs for characterizing population trends

Mevin B. Hooten, Beth E. Ross, and Christopher K. Wikle

Introduction

Spatio-temporal modeling

Spatio-temporal statistical models are being used increasingly across a wide variety of scientific disciplines to describe and predict spatially explicit processes that evolve over time. Correspondingly, in recent years there has been a significant amount of research on new statistical methodology for such models. Although descriptive models that approach the problem from the second-order (covariance) perspective are important, and innovative work is being done in this regard, many real-world processes are dynamic, and it can be more efficient in some cases to characterize the associated spatio-temporal dependence by the use of dynamical models. The chief challenge with the specification of such dynamical models has been related to the curse of dimensionality. Even in fairly simple linear, first-order Markovian, Gaussian error settings, statistical models are often over parameterized. Hierarchical models have proven invaluable in their ability to deal to some extent with this issue by allowing dependency among groups of parameters (Cressie *et al.* 2009). In addition, this framework has allowed for the specification of science-based (i.e. based on knowledge and hypotheses about the ecological system) parameterizations (and associated prior distributions) in which classes of deterministic dynamical models [e.g. partial differential equations (PDEs), integro-difference equations (IDEs), matrix models, and agent-based models] are used to guide specific parameterizations (Wikle and Hooten 2010).

Therefore, two of the main questions to ask in early stages of building a statistical model for a spatio-temporal ecological process are:

- Does enough a priori scientific information exist about the process to specify an intelligent mechanistic model that can mimic it?
- If the answer to the above question is "yes", then do sufficient data exist (or can they be collected) to fit such a model?

If the answer to either of these questions is "no", then perhaps a more naïve statistical model is warranted – one that is still spatially and temporally explicit, but that is

Design and Analysis of Long-term Ecological Monitoring Studies, ed. R.A. Gitzen, J.J. Millspaugh, A.B. Cooper, and D.S. Licht. Published by Cambridge University Press. © Cambridge University Press 2012.

sufficiently parsimonious to enable statistical learning. In that situation, because the actual model structure is limited, it is critical to consider potential lurking sources of latent autocorrelation, both spatially and temporally.

Optimal design

In many ecological studies, cost can be a limiting factor, and as available resources and budgets change over time, an adaptive sampling design can be of value (Hooten *et al.* 2009a; Box 20.1). Optimal adaptive sampling has been shown to increase sampling efficiency in biological populations (Brown *et al.* 2008, Salehi and Brown 2010; Chapter 17), although this may be related to factors such as quadrat size, initial sample size, and the method of selection (Smith *et al.* 1995, Giudice *et al.* 2010). Monitoring strategies based on adaptive sampling have been utilized across a wide variety of specific applications; for example, in capture–mark–recapture studies (Conroy *et al.* 2008b), occupancy studies (Field *et al.* 2005, MacKenzie and Royle 2005), and combinations of these (Witczuk *et al.* 2008). Adaptive sampling strategies have also been utilized in waterfowl surveys (Smith *et al.* 1995, Pearse *et al.* 2009), and many of these studies highlight the importance of considering the desired inference prior to collecting the data and examining ways to improve precision or decrease costs (e.g. by adding a second observer in aerial surveys to double the transect strip width; Pearse *et al.* 2009).

The basic concept of optimal monitoring involves the following steps.

  (i) Identify a critical modeled quantity of interest about which improved inference is desired (e.g. predictions, forecasts, or parameter estimates).

 (ii) Specify a design criterion that quantifies the form of improvement desired (e.g. precision).

(iii) Pose a model that explicitly considers the data that you plan to collect optimally in future monitoring efforts.

(iv) Fit the model, given available data, and optimize the design for new data collection with respect to the design criterion.

Diggle and Lophaven (2006) describe such two-part adaptive monitoring schemes as using a prospective sample (i.e. current data) to identify an optimal retrospective sample and therefore to optimize future sampling. Some studies have taken this basic theme and extended it to the situation where the process of interest is dynamic and thus the optimal design itself is dynamic (e.g. Wikle and Royle 2005, Fuentes *et al.* 2007, Hooten *et al.* 2009a).

Background for example application: scaup breeding pair trends

Scaup are the most abundant and widespread diving duck in North America, and an important game species in many areas (Austin *et al.* 1998). Since 1978, the continental population of scaup (lesser scaup, *Aythya affinis*, and greater scaup, *A. marila*, collectively) has been declining at an average rate of 105 000 scaup per year, and numbers are

---

Box 20.1 Take-home messages for program managers

Because nearly all ecological data are collected over time and space, contemporary methods of analysis need to consider these aspects explicitly. Hierarchical models provide a convenient framework for specifying both the observational process by which monitoring data are collected and complicated underlying ecological spatio-temporal process behavior (Cressie *et al.* 2009) such as spatial variability in trends and spatial/temporal correlation among monitoring units within and across time. Spatio-temporal behavior can be modeled as a latent "nuisance" source of error or as an integral part of the underlying dynamic process describing changes in the resource of interest (Wikle and Hooten 2010). In either case, statistical structure can be exploited to fuse the traditionally distinct acts of monitoring and modeling for improved inference and conservation of resources. Although these benefits come at the cost of increased analytical complexity, the necessary analytical building blocks are increasingly becoming accessible (e.g. Royle and Dorazio 2008).

The basic premise behind optimal monitoring is to use the existing data (or other sources of a priori information about the process under study) to help you put monitoring effort in places that will aid inference the most. Adaptive design refers to the process of altering a study in light of new information to increase precision or decrease costs, whereas dynamic design (e.g. Wikle and Royle 2005, Hooten *et al.* 2009a) is also adaptive but assumes an underlying dynamic model for the ecological resource and relies on the model to predict future states of the system and therefore guide monitoring decisions. In either case, the ultimate type of inference and information desired from the monitoring effort should be considered explicitly early during survey design (Chapter 2) and throughout the data-collection phase to help ensure that the data are collected in a manner that decreases uncertainty (Box 20.2). This allows the manager to do more with less.

---

currently well below the goal set in the North American Waterfowl Management Plan (Afton and Anderson 2001; USFWS 2009a, b). Additional modeling of the population has indicated further decline since 1997, sparking concern amongst hunters, management agencies, and conservation groups. Several hypotheses have been proposed regarding the cause of the decline [e.g. The Spring Condition Hypothesis, which suggests that there has been a decline in the condition of females when they arrive on breeding grounds (Anteau and Afton 2004), and hypotheses related to increasing levels of heavy metals (Anteau *et al.* 2007)], but contradictory evidence has been found, indicating a lack of certainty concerning these potential causes (DeVink *et al.* 2008a, b). It appears that the decline is more pronounced in the boreal forest breeding areas (Afton and Anderson 2001), emphasizing the importance of understanding the population dynamics across a broad spatial scale. There is strong rationale for the use of an explicit spatio-temporal model to better understand the scaup breeding population (Austin *et al.* 2006).

For continental-level studies on ducks, most researchers use the Waterfowl Breeding Population and Habitat Survey (Canadian Wildlife Service and US Fish and Wildlife

---

Box 20.2 Common challenges: necessary decisions

The biggest and most important decisions to make when building a spatio-temporal hierarchical model for optimal adaptive monitoring are as follows.

- What type of questions do you want to answer after the data are collected? For example, are you interested in learning about population size or change? Is your focus going to be centered on estimating survival or measurement characteristics such as detection bias?
- Do you know enough about the system under study to construct a scientific physical process model? For example, is a partial differential equation a good motivating mechanistic model for the system you are studying, or do you not have enough existing knowledge about it to tell? If not, perhaps you just want to account for spatial and temporal variation without the interactions required to mimic a more complicated process model?
- What type of design criterion will best help you answer your initial question? For example, are you most interested in reducing uncertainty about specific parameters in the model you have specified (i.e. trend coefficients), or do you wish to focus primarily on forecasting with minimal uncertainty?
- What flexibility do you have in modifying the monitoring design? Suppose that, for consistency reasons, you must maintain a former static monitoring design, but you are also given resources for additional monitoring. In this case, how many extra monitoring locations do you need, and are there any restrictions to where you put them?

These are all questions that do not have easy answers and will be specific to each individual project. Experts on the study system can help you address many of them, and simulation can be used, to some extent, for providing a general idea of how the decisions may play out in altering monitoring strategies and inference. Also, depending on the novelty in the modeling and monitoring methods, there is a rapidly growing body of literature that can provide guidance.

Service 1987). This survey has been conducted every May through June since 1955 using aerial transects covering northern portions of North America (Smith 1995). Pilots and observers record each duck species observed, and whether or not the ducks are paired (with a mate), single drakes, or in mixed-sex groups. Surveys are flown at approximately 120 miles per hour and an altitude of 90–100 feet.

To illustrate the utility of spatio-temporal modeling in the context of long-term monitoring as well as the process of optimizing a monitoring design, we make use of these data to (i) estimate spatially explicit trends in scaup breeding pairs and identify significant increasing and decreasing regions; and (ii) obtain an adaptive monitoring design for future observation periods that allows for optimal trend characterization in budget-limited scenarios. The second objective is addressed as an example to illustrate the savings that can be achieved when using an approach that explicitly couples monitoring and modeling.

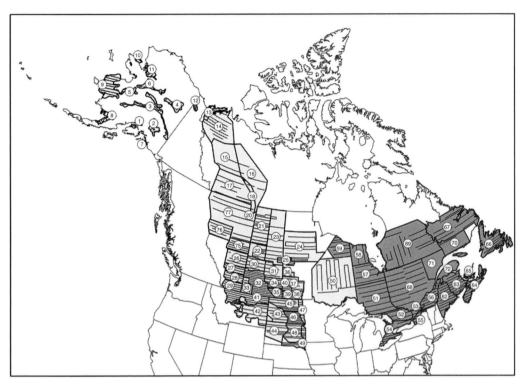

Figure 20.1 Aerial survey units (strata) in the Waterfowl Breeding Population and Habitat Survey (USFWS 2009a).

Developing and fitting the spatio-temporal model

For the scaup example, following the general approach of Gardner *et al.* (2007), we specify a model for breeding pair trend analysis using a hierarchical specification involving three modeling stages: data, process, and parameter. By setting up the model in this manner, we can use an algorithm to approximate the intractable posterior distribution of the process and parameters given the data (Berliner 1996). In our case, the latent process determines the spatio-temporal patterns in mean breeding pairs, while the model parameters are trend coefficients and variance components. We show that these latter quantities can then be used to construct optimal strategies for conserving monitoring resources (e.g. funding) before new data have been collected. Afton and Anderson (2001) and Gardner *et al.* (2007) constructed similar models for analyzing scaup trends, and although the latter study was focused on total scaup from years 1975 to 2005, many of the general model components they describe can be found in our specification. In constructing our model, where the focus is on long-term scaup breeding pairs, we allow our latent process to have much coarser spatial units (i.e. strata; Fig. 20.1) and specify a temporal regression that allows the strata to be correlated. Gardner *et al.* (2007) also described some of the difficulties in Bayesian analyses when using such large data sets and spatial domains. In light of this, we use a computational approach based on integrated nested

Laplace approximations (INLA; Rue *et al.* 2009), rather than the more common Markov chain Monte Carlo (MCMC) methods.

Data model

In order to accommodate potentially overdispersed count data, we specify a negative binomial distribution for the breeding pair observations: $y_{i,j,t} \sim \text{NegBinom}(\mu_{j,t}, \phi)$ for segments $i = 1, \ldots, n_j$ in stratum $j = 1, \ldots, m$ during observation period $t = 1, \ldots T$ (i.e. years 1957–2009). In this specification, $E(y_{i,j,t}) = \mu_{j,t}$. Two alternatives to this data model are the Poisson and quasi-Poisson distributions (e.g. Ver Hoef and Boveng 2007). Although the Poisson model is more parsimonious and conserves degrees of freedom for the estimation of other model parameters, the quasi-Poisson and negative binomial are more general and can better model overdispersed count data. Based on an exploratory analysis and model comparison for this study, we employ the negative binomial instead of the Poisson data model as other studies have done (e.g. Gardner *et al.* 2007). We found that the overdispersion parameter (ϕ) allows for a better fit to the data as indicated by a lower deviance information criterion (i.e. DIC) in our preliminary model fits.

Process model

The traditional negative binomial generalized linear model employs a log link function to connect the count means ($\mu_{j,t}$) to a set of covariates (Chapter 11). In our case then, introducing new variables, we specify a log-linear process model:

$$z_{j,t} = \log(\mu_{j,t}) = \beta_{0,j} + \beta_{1,j}t + \varepsilon_{j,t} + \eta_{j,t}, \tag{20.1}$$

where $\boldsymbol{\varepsilon}_t = (\varepsilon_{1,t}, \ldots, \varepsilon_{m,t})' \sim \text{N}(\mathbf{0}, \boldsymbol{\Sigma}_\varepsilon)$ for all $t = 1, \ldots T$ represents spatially correlated errors and $\boldsymbol{\eta}_j = (\eta_{j,1}, \ldots, \eta_{j,T})' \sim \text{N}(\mathbf{0}, \boldsymbol{\Sigma}_\eta)$ for all $j = 1, \ldots m$ represents temporally correlated errors. This linear model, operating in the log-space of the mean breeding pairs, provides for fairly simple inference on the general trend of breeding pairs by stratum and allows for nonlinear growth on the natural support of the counts themselves. An alternative process specification could use a discrete dynamic model that would allow for the estimation of growth and carrying capacity parameters, $z_{j,t} = f(z_{j,t-1}, r_j, k_j) + \varepsilon_{j,t}$, where the function f, for example, could be in the form of Ricker or Beverton–Holt density-dependent growth (Turchin 2003). For examples of these latter forms of model specifications, see Boomer and Johnson (2007), Hooten *et al.* (2007), Hooten *et al.* (2009b), and Wikle and Hooten (2010).

Ideally, the correlated fields, $\boldsymbol{\varepsilon}_t$ and $\boldsymbol{\eta}_j$, in the given model specification, will capture unexplained spatial and temporal dependence in the process not due to the general breeding pair trends. In general, accommodating these sources of uncertainty (i.e. spatial and temporal correlated random effects) in the model is critical because they allow for valid inference on the remaining parameters (i.e. $\boldsymbol{\beta}$).

The strata in this example are irregularly shaped, sized, and spaced spatial regions (Fig. 20.1), thus we need to be cautious in specifying an appropriate form of spatial dependence in the model (i.e. $\boldsymbol{\Sigma}_\varepsilon$). A commonly used form of dependence for areal processes is referred to as a conditional autoregressive structure (CAR; Brook 1964, Besag 1974). CAR models allow the spatial process (i.e. ε) to depend on itself based on the neighborhood structure of the spatial regions or strata. Assuming a CAR spatial structure, one can then write: $\varepsilon_{j,t}|\{\varepsilon_{k,t}, \forall k \neq j\} \sim N(\sum_k w_{j,k}\varepsilon_{k,t}, \sigma^2)$ where the $w_{j,k}$ describe the neighborhood connectivity of stratum j and stratum k. The simplest specification would be to let $w_{k,j} = 1$ if stratum k is a neighbor of stratum j, and zero otherwise. The model is then referred to as an Intrinsic-CAR (ICAR) model if we write the distribution of the entire spatial field as $\boldsymbol{\varepsilon}_t \sim N(\mathbf{0}, \boldsymbol{\Sigma}_\varepsilon)$, where $\boldsymbol{\Sigma}_\varepsilon \equiv \sigma_\varepsilon^2(\mathbf{D} - \mathbf{W})^{-1}$, $\mathbf{W}$ is a matrix containing all of the $w_{j,k}$, and the matrix $\mathbf{D}$ is diagonal with the row sums of $\mathbf{W}$ as diagonal elements (Banerjee *et al.* 2004).

In a Bayesian framework, the ICAR model serves as a prior for the spatial fields $\boldsymbol{\varepsilon}_t$ and is not technically proper in that it will not integrate to 1. We note that, under some relatively mild conditions, the posterior distribution for ε will indeed integrate to 1 as required (Rue and Held 2005).

We also need to specify a model that will accommodate potential temporal dependence in the $\boldsymbol{\eta}_j$. Here, we assume a time-series version of the CAR model described above where we let $\eta_{j,t} \sim N(\alpha\eta_{j,t-1}, \sigma_\eta^2)$. This model induces a correlation structure on the vectors $\boldsymbol{\eta}_j$ such that $\boldsymbol{\eta}_j \sim N(\mathbf{0}, \boldsymbol{\Sigma}_\eta)$. Now, due to our careful specification, the entire latent process model for $z_{j,t}$ [(Equation (20.1)] is a Gaussian Markov random field; this can be advantageous when fitting the model to observed data.

Parameter model

The hierarchical model specified above contains $2m + 4$ parameters that need prior distributions to complete the model formulation. The overdispersion parameter is specified as $\phi = \log(n)$, where n is the original negative binomial size parameter, and then modeled as $\phi \sim N(0, 100)$. We use exchangeable conjugate Gaussian priors for the m sets of regression parameters: $\beta_{0,j}, \beta_{1,j} \sim N(0, 1000)$, for $j = 1, \ldots, m$. We used a conjugate inverse gamma prior for the variance component: $\sigma_\varepsilon^{-2} \sim \text{Gamma}(1, 1/20\,000)$. Finally, we have the parameters in the temporal autocorrelation model. We reparameterize these such that $\theta_1 = (1 - \alpha^2)/\sigma_\eta^2$ and $\theta_2 = (1 + \alpha)/(1 - \alpha)$ and then assign the following priors: $\theta_1 \sim \text{Gamma}(1, 1/20\,000)$ and $\theta_2 \sim N(0, 5)$. An alternative set of priors could be specified for the precision and autoregressive parameters directly, although the reparameterization described above is preferred for the form of implementation we describe next (Rue and Martino 2009).

Implementation

At this point, we are able to make inference on the underlying process and parameters by finding their posterior distribution conditioned on the observed data:

$$
\begin{aligned}
[\{\boldsymbol{\mu}_t\}, \phi, \{\boldsymbol{\beta}_t\}, \sigma_\varepsilon^2, \sigma_\eta^2, \alpha|\{y_{i,j,t}\}] \propto \prod_i \prod_j \prod_t [y_{i,j,t}|\mu_{j,t}, \phi] \prod_t [\boldsymbol{\mu}_t|\boldsymbol{\beta}, \sigma_\varepsilon^2, \sigma_\eta^2, \alpha] \\
\times [\phi][\boldsymbol{\beta}][\alpha][\sigma_\varepsilon^2][\sigma_\eta^2].
\end{aligned}
\tag{20.2}
$$

Note, in the above equation, we use a conventional Bayesian notation where the square brackets " [·] " refer to a probability distribution. Fitting the above model is relatively trivial and can be accomplished using an MCMC algorithm to sample from each of the full-conditional distributions of the process and parameters. Software packages such as BUGS (Lunn *et al.* 2009) or JAGS (Plummer 2003) could be employed to do this and Gardner *et al.* (2007) provide computer code to implement a similar MCMC algorithm.

A preliminary simulation study indicated that MCMC methods can be quite slow for these large data sets with correlated spatial effects, as suggested by Gardner *et al.* (2007). Thus, we employ a new method for fitting Bayesian Markov random field models called INLA, an acronym standing for "integrated nested Laplace approximations" (Rue *et al.* 2009). The INLA algorithm is not an iterative stochastic method like MCMC, but rather a sequence of techniques to approximate the marginal posterior distributions for well-posed latent Gaussian models (like the one we have specified in the preceding sections) with a closed-form expression. It has been shown to be very accurate (Rue *et al.* 2009), and although not as flexible as MCMC, it is generally much faster for a specific class of models. A version of INLA has been implemented in R (R Development Core Team 2011) and is highly accessible.

The INLA algorithm proceeds by approximating the marginal posterior distributions for each unknown Gaussian variable of interest given the data. This approximation is accomplished by integrating the "hyperparameters" (i.e. all non-Gaussian unknown model parameters) out of the joint posterior distribution. In practice, the required integration is intractable but since the integrand can be factored into a full-conditional distribution of the Gaussian parameters and a marginal posterior that can be well-approximated by either a Gaussian or Laplace distribution, it is possible to carry out the calculation numerically. Depending on the desired inference, a potential disadvantage of the INLA approach to fitting hierarchical Bayesian models is that we are restricted to marginal and linear combinations of marginal posterior distributions. However, many studies only require this form of inference and thus INLA could be widely used in spatio-temporal analysis of ecological data.

Model fit and parameter estimation

We used the "INLA" R package (Rue and Martino 2009) to fit the model described above and obtain parameter estimates. We provide a basic outline for the INLA syntax required to fit the model described herein. Although this package is very easy to use as demonstrated below, we caution that, as with automatic MCMC algorithms, care needs to be taken to ensure the correct model is being specified and fit. In this case, after loading the INLA package and having the data already organized, one can simply issue the following command:

```
inla(y ~ X + f(j,model="besag",graph.file=W,replicate=t)
+f(t,model="ar1",replicate=j), model="nbinomial", E=E)
```

where **y** represents the summed counts by stratum for all years, **X** is a "design" matrix consisting of the entire set of submatrices $\{\mathbf{I}, t \cdot \mathbf{I}\}$ for all $t = 1, \ldots, T$ (where $\mathbf{I} =$ the identity matrix), and the 2 **f** model components correspond to the spatial and temporal

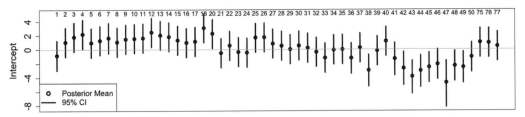

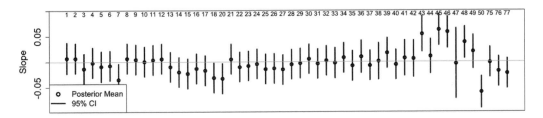

Figure 20.2 Marginal posterior means and 95% credible intervals for each of the strata intercept and slope parameters for the scaup trend analysis example.

autocorrelated errors, respectively. The graph file **W** specifies the spatial connectivity among strata, "nbinomial" specifies the data model to be used, and **E** corresponds to a known scaling factor in $\mu_{j,t}$; in this case, the scaling factor adjusts for the differing numbers of segments per stratum. The "replicate" option in the latent error models (**f**) specifies over which index the hyperparameters should be shared.

In fitting the model to scaup data, we focused on the 52 western-most strata (Fig. 20.1), containing a total of 103 266 scaup counts in these strata over a period of 53 years (1957–2009). The fitting algorithm took approximately 27 minutes on a 2×2.93 GHz 6-Core Intel Xeon workstation. Based on preliminary comparisons with an MCMC algorithm, the INLA algorithm appears to be an order of magnitude faster than MCMC for this model and data set.

In constructing the proximity matrix **W**, we first created a Euclidean distance matrix between all strata centroids and then denoted all pairs of strata as neighbors if they were within a threshold distance of 7.75 decimal degrees of each other. This number represented the minimum distance at which all strata had at least one neighbor. Because some of the northern strata are more isolated than the southern strata, we felt that this connectivity should be expressed in the proximity matrix.

In terms of model results, the posterior distributions for the model coefficients are of primary interest (Fig. 20.2). Focusing on the marginal posterior credible intervals for the slope parameters in the strata trends (lower plot in Fig. 20.2), these results indicate that we do not have enough evidence to detect a log-linear trend in many of the strata (i.e. the credible intervals overlap zero); however, some (e.g. strata 46 and 50) do

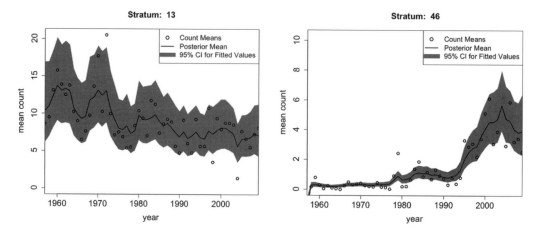

Figure 20.3 Posterior means and 95% credible intervals for scaup breeding pairs in two example strata (13: McKenzie River Delta and 46: Drift Prairie and Missouri Coteau).

indicate both positive and negative trends. Even though it is not immediately obvious, one of the advantages of being able to accommodate latent forms of dependence in our hierarchical model is that we reduce the chance of making Type I errors in our inference, as discussed in Chapter 11. That is, a failure to appropriately account for latent forms of dependence can lead to a reduction in posterior variance and the incorrect detection of non-significant trends. In fact, in comparison with a preliminary study that used a simpler, uncorrelated, latent model (not included here), we see a decided increase in credible interval width pertaining to the marginal posterior distributions for the trend coefficients when considering the spatial and temporal autocorrelation.

As developed below, the optimal design depends most on the estimated variance component σ_ε^2. The INLA algorithms actually model the precision (i.e. $\frac{1}{\sigma_\varepsilon^2}$) instead of the variance, thus, the marginal posterior mean and standard deviation for the precision parameter are 190.4 and 66.1, respectively. The posterior mean and standard deviation for the temporal model precision were 26.6 and 3.3, while the autoregressive parameter α had a posterior mean and variance of 0.93 and 0.02, respectively. Finally, the overdispersion parameter in the negative binomial likelihood, if parameterized in terms of the "size" (n), had a posterior mean and standard deviation of 7.4 and 0.38.

In addition to parameter estimates, it is helpful to connect the posterior parameter distributions back to the process under study. In this case, we are modeling scaup breeding pairs, thus to illustrate two potential scenarios in abundance trends, we highlight model predictions for stratum 13 and stratum 46 (Fig. 20.3), which are decreasing and increasing in abundance, respectively.

Defining and selecting an optimal monitoring design

Specifying the design criterion and implementation

Using the scaup example and the spatio-temporal model developed above, we next illustrate the process of optimizing a monitoring design (see also Box 20.2). In terms of choosing an optimal monitoring strategy for a future observation period $(T + 1)$ when available resources may be a limiting factor, in the scaup example we would need to select a subset (S_{T+1} with dimension m_{T+1}) of the total strata (S with dimension m, Fig. 20.1) that maximizes the information we receive from the data while minimizing the uncertainty concerning our inference on the breeding pair trends. Thus, we want to construct a sampling design for observation period $T + 1$ that allows us to minimize the posterior variance of the coefficients β while monitoring only a subset of total strata.

Consider, then, a subset of strata $S_{T+1} \subseteq S$ where $m_{T+1} \leq m$, being the resource limited monitoring scenario during the observation period $T + 1$. This implies that we can only obtain additional information, during observation period $T + 1$, about the process at the subset of strata [$\mu(S_{T+1})$, or $z(S_{T+1})$ in log space]. We then specify the design criterion as a measure of total posterior parameter variance given the new design S_{T+1}:

$$q(S_{T+1}) = \text{tr}(\text{var}(\beta|z(S_{T+1}), \cdot)). \tag{20.3}$$

That is, the design criterion is the sum, across all retained strata in the new design, of the posterior variances of each stratum's coefficients in the log-linear process (trend) model specified by Equation (20.1). The full conditional random parameters $(\beta|z(S_{T+1}), \cdot)$ have a distribution whose variance can be found analytically. To see this, we write the full process model for all μ_t, where $\mu \equiv (\mu_1, \ldots, \mu_T)'$ and $z \equiv \log(\mu)$. Now, we augment the process vector z with the process in a future monitoring period, i.e. $z_{T+1} \equiv z(S_{T+1})$, such that $z_+ = (z', z'_{T+1})'$ and write the full process specification as a linear model: $z_+ = X_+\beta + \varepsilon_+ + \eta_+$. In this formulation, $X_+ = (X'_1, X'_2, \ldots, X'_T, X'_{T+1})'$, where each X_t is $m \times 2m$ and contains an identity matrix I in the first m columns and the remainder of columns equals $t \cdot I$. For time $T + 1$, the matrix X_{T+1} is only $m_{T+1} \times 2m$ dimensional since we are assuming only a subset of the process components will be able to influence the posterior distribution of β which contains the global intercept and trend coefficients. The final terms, ε_+ and η_+, represent the spatially and temporally correlated error for all locations and times. Using this augmented model specification, the full conditional distribution for β is proportional to the product of the augmented process model and the prior:

$$[\beta|z(S_{T+1}), \cdot] \propto [z_+|\beta, \sigma^2][\beta]$$

$$\propto \exp\left(-\frac{1}{2}(z_+ - X_+\beta)'\Sigma_+^{-1}(z_+ - X_+\beta)\right) \times \exp\left(-\frac{1}{2}\beta'\Sigma_\beta^{-1}\beta\right)$$

$$= N(A^{-1}b, A^{-1})$$

where the covariance matrix $\Sigma_+ = (I \otimes \Sigma_\varepsilon + \Sigma_\eta \otimes I)$ and $A^{-1} = (X'_+\Sigma_+^{-1}X_+ + \Sigma_\beta^{-1})^{-1}$. Thus, the design criterion $q(S_{T+1})$ is not a function of the future process $z(S_{T+1})$, but rather, only a function of the observation periods t, the design S_{T+1} (which is embedded in the augmented matrix X_+ and the spatio-temporal covariance matrix Σ_+), and the prior covariance matrix Σ_β.

If an MCMC algorithm is chosen to fit the model, then the implementation of the optimal design for period $T + 1$ within the algorithm proceeds by first sampling from each of the parameter full-conditional distributions, and then finding the design S_{T+1} that minimizes the criterion $q(S_{T+1})$ on each MCMC iteration. In this scenario, the optimal design itself is stochastic with a posterior distribution indicating which subsampling scheme in observation period $T+1$ provides the greatest probability for a reduction of posterior parameter variance. In terms of settling on a fixed optimal design, one would then select an a priori dimension (i.e. m_{T+1}) for the optimal subsample of strata, and then identify the set of strata that maximize the posterior design distribution.

The implementation process outlined above requires that model fitting and finding the optimal retrospective design occur in a single analysis. The process will be inefficient if the space of potential designs is large (because of the nested maximization loop within the MCMC algorithm), and moreover, could not be implemented within the INLA approach directly since the full-conditionals are never directly sampled from. Therefore, we suggest using an alternative quasi-criterion that relies on the posterior mean covariance of the process error, $E(\Sigma_+|\mathbf{y})$:

$$q(S_{T+1}) = \mathrm{tr}\left(\left(\mathbf{X}'_+ E(\Sigma_+|\mathbf{y})^{-1}\mathbf{X}_+ + \Sigma_\beta^{-1}\right)^{-1}\right). \tag{20.4}$$

With this new criterion, we can now fit the model (as described earlier) and find the optimal retrospective sampling design separately. The use of the INLA software and the decoupling of these two steps in the analysis make for a fast and easily modifiable approach for finding model-based optimal designs, and the results that follow provide an illustrative example of how these adaptive sampling approaches could be used to conserve monitoring resources.

An important point to consider when seeking optimal designs is that there are an infinite number of different types of optimality. That is, because we optimize with respect to the design criterion, its form controls the type of design we will end up with. The form we have chosen [Equation (20.4)] is referred to as "A-optimality", and minimizes the average variance of the coefficients. A-optimal designs are commonly used in the literature (i.e. Wikle and Royle 2005, Hooten *et al.* 2009a), but many other options exist. For example, two other common forms involving the covariance of the coefficients are "E-optimality" and "D-optimality", minimizing the maximum (via the spectral radius or maximum eigenvalue) and generalized variance (via the determinant), respectively (Gentle 2007). Yet another approach to finding optimal designs involves optimizing over the entropy of the distribution containing the design (e.g. Le and Zidek 2006). Each of these different choices for design criteria has advantages and disadvantages, and moreover, will most certainly influence the outcome; although, it has been argued that any type of well thought out design criterion will result in more efficient data collection and conservation of resources.

Optimal monitoring design for scaup

Based on the results of the model fit for the current scaup data set, we searched the space of designs twice, using 10 000 random designs each time, for the monitoring year 2010 over a range of strata subsample sizes (m_{T+1}) from 5 to 51, with 5 being the most restrictive case in terms of limited resources for future sampling and 51 being the best case scenario. We used the case where all 52 strata were hypothetically monitored in year 2010 as a baseline to compare the design criterion and proportion of budget across all possible subsample sizes.

In situations where there is large variation in the optimal design search results, a further refinement of the optimality is recommended. Various exchange algorithms (e.g. Nychka and Saltzman 1998) can be employed in these cases to test a sequence of swaps between neighboring strata for further increases in quasi-optimality. In this study, the resulting optimal designs from each set of searches were similar enough to combine results and forego further measures to increase quasi-optimality. That is, in the few cases where the optimal design differed between our two searches, we retained the one that yielded a smaller $q(\boldsymbol{S}_{T+1})$.

The first obvious result to report is the design criterion $q(\boldsymbol{S}_{T+1})$ and how it changes with increasing subsample size m_{T+1}. Figure 20.4a illustrates the gain in mean coefficient variance, over that of a random (i.e. non-optimal) subsampling scheme, with increasing future monitoring sample size. In this case, as the potential future subsample becomes more and more restricted, we see an exponential reduction in parameter uncertainty (i.e. an increase in the difference of criteria between random subsampling and optimal subsampling). That is, at smaller future subsample sizes, the benefits of optimal monitoring increase dramatically.

It is unlikely that a resource-limited situation would arise in such a way that the actual number of strata to be sampled is the limiting quantity. The more likely scenario is a limiting quantity based on available budget. As such, it is first important to be able to view the effect on budget by calculating the actual dollars saved when using the optimal design versus a random subsample of strata. Figure 20.4b expresses how these savings fluctuate depending on the number of strata retained in the optimal design. In this study, dollar amounts are calculated based on an average cost of $246 per segment monitored. Notice that, in some cases, even though the number of strata are being reduced, there is actually an increase in cost for the optimal design as compared with a random design. This is due to the fact that some of the strata contain many more segments than others, and thus the optimal designs for those scenarios contain more total segments.

Under a restricted budget, we can see from Figure 20.4c that, although the budget generally declines with a decreasing number of strata retained in the optimal design, there are distinct optimal design sizes where the budget is reduced more than would be expected in a purely linear relationship between budget and sample size. For example, from Figure 20.4b it is clear that the savings are extraordinary when retaining 22 strata (approx. $80 000) in the monitoring design as opposed to retaining 21 strata, where it only saves $20 000 to monitor optimally. Although both designs save money, in this specific case, with the scaup breeding pair trends, it actually saves more to go with a larger sample size of 22 over the optimal design for the smaller sample size of 21, given that they both yield nearly the same reduction of uncertainty (Figures 20.4a, c).

We can also conclude, from Figure 20.4, that optimal monitoring schemes, in this breeding pair study, are most effective when the number of strata and/or budget is reduced to 50% and lower. Although optimal designs are able to decrease uncertainty at every level of budget restriction, the difference between random subsampling and optimal subsampling increases substantially at the smaller subsample sizes.

(a) Gain in Coefficient Precision

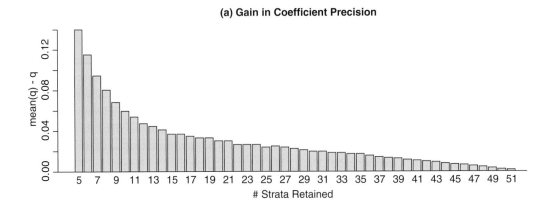

(b) Optimal Design Savings

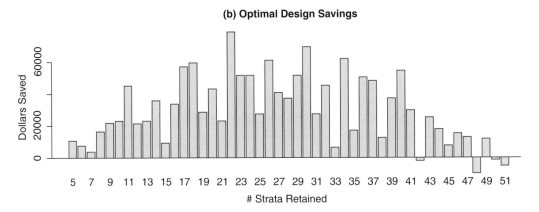

(c) Decrease in Budget with # of Monitored Strata

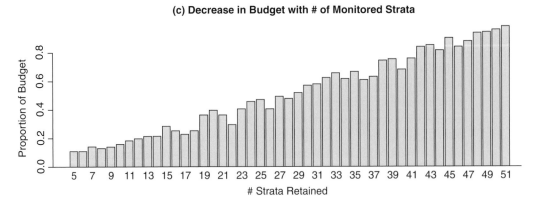

Figure 20.4 Optimal monitoring results for the scaup monitoring example. (a) The mean optimality criterion under random subsampling minus the criterion under optimal subsampling. A positive difference in this setting indicates a gain in parameter precision under optimal sampling. (b) Dollars saved by number of strata retained in each optimal design. (c) The cost of optimal monitoring as a proportion of maximum budget for a given number of retained strata.

Table 20.1 Optimal monitoring with respect to budget-restriction scenarios (expressed as percentage of the full budget needed to monitor all 52 possible strata) for the scaup example. Results shown are the optimal subsample size m_{T+1} for the monitoring design that reduces parameter uncertainty the most at each level of budget restriction, along with the actual cost of the optimal sample.

Available budget (%)	Optimal m_{T+1} (% of 52 total strata)	Actual cost (%)
50	28 (54%)	\$306 316 (48%)
60	31 (60%)	\$372 607 (58%)
70	40 (77%)	\$435 940 (68%)
80	41 (79%)	\$487 444 (76%)
90	47 (90%)	\$562 114 (88%)

Given the output from this optimal design study, it is a relatively simple task to find the optimal design given a specified budget restriction. The procedure to do this requires only two steps. First, for each optimal subsample size, identify the optimal designs that cost less than the available budget. Second, with these new designs, keep the one that provides the greatest reduction in mean coefficient variance (i.e. the smallest design criterion q).

Based on our scaup breeding pair analysis, we used this procedure to find the best optimal designs for each 10% increment of budget proportion from 50% to 90% (Table 20.1). These findings indicate that the number of strata to optimally monitor are proportionally at least as large as the available percentage budget and that the actual cost of the optimal monitoring is below the available budget, even sometimes reducing the budget by as much as an additional 4%, as in the case of an 80% available budget.

Future research and development

Overall, there is substantial evidence supporting the use of a model-based optimal monitoring design in the assessment of scaup breeding pairs. Although uncertainty about the trend parameters increases in general with a decreased monitoring effort, we can see that there is a distinct gain in coefficient precision (i.e. reduction in uncertainty) when using an optimal design as opposed to a random subsampling design (Fig. 20.4). This approach to optimal design, in general, has enjoyed much success in other fields, but, to our knowledge, this is the first time it has been utilized to conserve resources in the study of waterfowl breeding pair trends.

More generally, we have presented an approach for constructing adaptive optimal sampling designs for a spatio-temporal ecological process using a relatively simple trend model within a Bayesian framework. The scaup breeding pair data set that we used as an example is fairly large (at over 100 000 records) and we found significant improvement in computational efficiency using the integrated nested Laplace approximation approach described by Rue *et al.* (2009). For hierarchical models that contain a latent Gaussian Markov random field process (like the one presented herein), the INLA methods can be utilized and have been shown to be quite accurate. In our specific example, we focused on the situation where resources for monitoring may be limited in future observation periods and show that significant savings can be obtained by exploiting the natural spatial dependence in the process under study.

The methods and software we have used in this study are very accessible and could easily be modified for use with other species, design objectives, or limiting resource constraints. A few examples of potential modifications and extensions to the approach we present here include the following.

- Use of a likelihood that explicitly accounts for imperfect detectability: although our negative binomial allows for overdispersed counts, it could be generalized to explicitly accommodate an observation bias, as in the hierarchical model outlined in Chapter 19.
- Different forms for spatial and temporal dependence (e.g. geostatistical spatial model or higher-order time-series model): while the use of a continuous spatial model for latent autocorrelation would prohibit us from using INLA to fit the model, it also makes MCMC methods more feasible for these problems.
- More scientifically motivated dynamic process model (e.g. Wikle and Hooten 2010): specific physically based process models, if appropriate, can provide better predictions and forecasts as well as intuitive scientific learning about the process under study.
- Different design objective (e.g. prediction error variance): if our focus was on forecasting, we may want to use a design criterion that represents prediction uncertainty.
- Consideration of more complicated forms of incurred cost: in our scaup example, we assumed a constant per-segment rate; however, the actual aerial survey is much more complicated than that, and it may be beneficial to consider factors like actual flight paths and re-fueling stops.
- Planning or evaluation process of long-term survey development: we described a framework for acute optimality in terms of desired short-term inference; however, the model and/or design criterion could be modified to consider objectives with a greater temporal extent. An example might be a combined objective including inference on long-term population trends in addition to acute management decisions such as setting regional harvest limits.

Summary

We presented a Bayesian hierarchical modeling framework for optimally monitoring population trends while accounting for overdispersion in count data and spatio-temporal dependence. We provide an example pertaining to scaup, using breeding pair counts from the North American aerial waterfowl survey. Specifically, the dynamic process is modeled as a Markov random field, where scaup relative abundance in each stratum in the survey increases or decreases with respect to a general trend, and is subject to additional temporal and spatial dependence due to similarities between strata that remain unexplained by the trend alone. The understanding of these trends and residual correlation in scaup abundance allows us to exploit the dependence in the process and construct an optimal adaptive monitoring strategy for future observation periods. The strategy we present is meant as an example only, and is focused on the scenario where the available monitoring budget is reduced during future monitoring efforts. Overall, the methods we present could be applied to nearly any spatio-temporal ecological or environmental process where inference on general trends is of primary interest. Although it requires upfront effort to develop a model-based optimal monitoring program, the long-term benefits of potential cost reduction and increased learning can definitely make this a worthwhile endeavor.

Acknowledgments

The authors thank David Koons, Beth Gardner, Andy Royle, Scott Boomer, Sara Martino, and Håvard Rue for providing data and helpful suggestions pertaining to this project. The authors also acknowledge the support of Delta Waterfowl and C. Wikle acknowledges the support of NSF grant OCE-0814934. Any use of trade names is for descriptive purposes only and does not imply endorsement by the US Government.

21   Use of citizen-science monitoring for pattern discovery and biological inference

Wesley M. Hochachka, Daniel Fink, and Benjamin Zuckerberg

Introduction

Citizen science, broadly speaking, is the involvement of amateur (i.e. unpaid, but not necessarily unskilled) participants in the process of scientific studies, typically with data collection or data processing (Cooper *et al*. 2007, Silvertown 2009). The benefits of citizen-science monitoring programs for public engagement and education are often emphasized (Chapter 23), but here we focus on programs sharing the same primary motivation as those discussed in the rest of this volume: obtaining data and information needed for managing and understanding ecological systems. Often, "amateur" monitoring may be the only or the most cost-effective approach for obtaining such information. For example, an important role for citizen science in monitoring of animals or plants follows from the fact that identification of species is currently difficult or impossible to automate, although technological inroads are being made (e.g. Brower 2006, Sarpola *et al*. 2008). Thus, if the goal is to monitor over large geographical areas such as most or all of a species' range, data will often need to be collected by a large number of people, and volunteer-dependent monitoring will be the only logistically feasible approach (with very few exceptions, e.g. Smith 1995).

While creating opportunities for monitoring, volunteer-based approaches also can introduce challenges for the design of studies and analyses of the resultant data (Box 21.1). Challenges are created by three common characteristics of data that come from citizen-science projects. First, the wide geographic and temporal scope of studies mean that attention must be paid to maintaining consistency in data-collection protocols and checking for potential spatial and temporal sources of bias. Second, data often must be collected using protocols that are potentially not as demanding of the data collectors as are protocols for professionally collected data. Third, data are often collected for surveillance monitoring and the general nature of such data may allow their re-use for multiple purposes, necessitating additional steps in analyses to investigate the possibility that biases exist for which the study design does not adequately control. Although we do not wish to imply that volunteer-collected data will always have these characteristics, we believe that these traits are present frequently enough in citizen-science monitoring to

Design and Analysis of Long-term Ecological Monitoring Studies, ed. R.A. Gitzen, J.J. Millspaugh, A.B. Cooper, and D.S. Licht. Published by Cambridge University Press. © Cambridge University Press 2012.

Box 21.1 Take-home messages for program managers

Volunteer data collection ("citizen science") can be well suited for monitoring, particularly over very large areas or long time periods, for which it may be infeasible to rely on paid personnel to collect data. However, the potential availability of volunteers is only one consideration in deciding whether to use a citizen-science approach. More importantly, the objectives of monitoring need to be clearly understood in order to determine whether citizen-science monitoring is appropriate. These objectives will determine the types of data that need to be collected (Chapter 2), and the appropriateness of volunteer data collection. Considerations include whether suitable data can be collected without need for highly specialized skills, and whether the time and effort requirements can be kept modest enough so as not to overwhelm the majority of potential volunteers.

Citizen-science approaches often may lead to large-scale data sets most useful for discovery of new patterns of interest. Therefore, methods of analysis intended for exploratory analyses, many not widely used by ecologists, may be more appropriate than parametric statistical analyses. However, when analyses are used to explore and discover patterns, these need to be viewed as hypotheses to be validated formally with subsequent research (either analyses of existing and independent data, or through collection of additional data).

warrant their place in motivating the topics discussed in this chapter. In contrast, if data from a citizen-science project can be collected with a tightly controlled survey design and are well-suited to answering a focused question using a well thought out parametric statistical model or with a design-based analysis, then there is nothing to distinguish such a project from any other monitoring initiative discussed in this volume. Aside from issues discussed under *Data Collection and Management*, below, we will not discuss these "targeted" projects.

We start this chapter by considering when volunteer-based monitoring is appropriate, and the logistics of setting up and running a citizen-science survey. This discussion is based on our own experience and the experience of colleagues at the Cornell Lab of Ornithology (Ithaca, New York, USA) and elsewhere in developing and running various citizen-science projects over the last two decades. We follow with a discussion of issues involved in the collection and management of citizen-science data. Next, because citizen-science data are such a rich source of information for pattern discovery and hypothesis generation, we discuss the use of these data with analytical methods geared towards exploratory analysis. The flexible data-collection protocols often used for citizen-science projects demand careful attention to not just analysis (construction of summary models from the data) but also validation of models that are constructed. As a result, we conclude by describing approaches to model validation in the face of partly uncontrolled spatial and temporal variation in density of sampling of the population of interest – i.e. variation in the representativeness of sampling.

Developing citizen-science monitoring

The development of any monitoring program faces numerous questions that should be addressed through careful, integrated planning (Chapter 2). In this section, our objective is not to repeat important recommendations made in other chapters of this volume; rather, we highlight a few considerations of particular importance in citizen-science programs.

When is citizen science appropriate?

The use of volunteer-gathered data for monitoring is not always appropriate, and objectives for monitoring need to be articulated clearly before the suitability of a citizen-science project can be assessed. While many factors could influence the utility of citizen-science monitoring, we see three principal questions that need to be answered:

Do we need to collect data over a large geographical area or long time period? One of the main benefits of using volunteer assistance for monitoring is monetary: the use of volunteers allows for a larger number of observers in the field for a given cost of project maintenance than if all field personnel were paid employees. Given the invaluable role that monitoring has played in assessing long-term environmental changes, such as changes in species distribution and abundance (e.g. Robbins *et al.* 1986, Thomas and Lennon 1999), and the challenges of obtaining long-term funding to have this work conducted by paid staff, the use of volunteers to collect data is very appealing, if not necessary. However, the larger number of observers will necessarily add variance to the observation process, with multiple factors affecting the measurement errors of each observer (e.g. Sauer *et al.* 1994, Simons *et al.* 2007; Chapter 11). Thus, if a monitoring project is focused on a relatively small geographical area, then the use of a smaller number of paid observers has advantages. Conversely, the larger the geographical area to be covered, the more compelling is the case for volunteer-collected data. It is not surprising that country- or continent-wide monitoring programs (e.g. Robbins *et al.* 1986, Gibbons *et al.* 1993, Freeman *et al.* 2007) rely on volunteer observers – generally there is no other option! For monitoring on smaller scales, the appropriateness of using volunteers to collect monitoring data will depend on multiple trade-offs that will need to be evaluated in each situation. For example, there may be a trade-off between quality and quantity: does the increased sample size achievable with volunteers increase the chances of detecting biologically relevant patterns, or is this advantage outweighed by the increased variability among data collectors coupled with larger numbers of data collectors (e.g. Munson *et al.* 2010)?

Is the geographical area heavily or lightly populated? A pool of volunteers has to be available in order to make citizen science tractable, and generally the size of the pool in any area will be directly related to human population density (Fig. 21.1). Hence, if monitoring is required in areas of low human population density, such as much of the Great Plains of North America, there may not be enough volunteers to make volunteer monitoring as effective or efficient as paying employees to collect monitoring data.

Figure 21.1 Density of data varies with human population density when observers are allowed to select locations. Plotted are locations from which observations were provided by volunteers in 2009 from a single citizen-science monitoring project, eBird, a project in which participants select locations at which data are collected. The area mapped is southern Canada to northern Mexico. Note the uneven distribution of data, with major urban areas (e.g. northeastern US seaboard, Chicago, Salt Lake City) clearly visible due to the higher density of observation locations from these areas.

Can methodological difficulties be addressed with limited training and equipment? For example, are the taxa of interest readily detected and identifiable, and the required information about these taxa feasible to collect by volunteers? Any observer can incorrectly identify an organism, and such "false positive" records in monitoring data are a source of error that may be difficult or impossible to correct (e.g. Fitzpatrick *et al.* 2009). Even experts are challenged by difficulties in identifying some taxa (e.g. aquatic insect larvae; see Sarpola *et al.* 2008) and such taxa may not be amenable to monitoring through citizen-science projects. Similarly, some physical characteristics of organisms or environments can be difficult to measure accurately. We suspect that there is less error associated with documenting the occurrence of a species than there is in documenting other characteristics, such as age or sex. In still other cases, it may be impossible for a large number of volunteers to collect data for reasons of permitting if not training; for example, the collection of tissue samples. However, in this last case, on a relatively small scale the collaboration of volunteers and paid staff in the field may make the collection of these types of data possible (e.g. Gómez *et al.* 2008).

Formal training can be used to increase the skills of volunteers, and should be given serious consideration as a method for reducing identification and measurement errors (e.g. Koss *et al.* 2009). However, technological aids also can sometimes be used to reduce errors. One example from our own experience is the use of standardized audio playbacks to attract birds and bring them closer to observers (e.g. Hames *et al.* 2001), thus increasing detection probabilities as well as reducing identification errors. Additionally, separating collection of the observation from identification or measurement may also be possible if voucher specimens, photographs, sound recordings, or other information relevant for species identification can be collected. Collection of raw information (e.g. recordings, specimens) for subsequent processing cannot eliminate all need for training, because volunteers will still need to be trained to follow appropriate sample-collection techniques. In summary, some training will always be needed, but if it is not feasible to provide sufficient training required to maintain high data accuracy then volunteer data collection may not be useful.

More generally, it should be recognized that participants in citizen-science projects are volunteers, and that there are limits to what volunteers will enjoy or be willing to do. Our own experience at the Cornell Lab of Ornithology is that a trade-off exists between the work required to collect data for a citizen-science project and the number of people who will participate in a project. We have seen this where projects requiring minimal and relatively unconstrained and opportunistic participant effort (Project FeederWatch, eBird) can attract thousands of participants from across North America, whereas projects with more rigorously constrained protocols have received data from under 500 participants annually (e.g. Hames *et al.* 2001). In the latter cases, the time commitment (multiple visits to a site each year) and amount and types of effort requested of volunteers (detailed descriptions of habitat, in addition to surveys for birds) deterred potential participants. We suspect that citizen-science monitoring will generally face difficulties in successfully implementing carefully structured spatial and temporal designs when data collection relies on the participation of a large number of volunteers. Although exceptions exist where data collection was well-proscribed in advance (e.g. Robbins *et al.* 1986, Kéry *et al.* 2005) or data could be coerced into an appropriate design (e.g. Kéry *et al.* 2010a, b), analyses of data from citizen-science monitoring programs will often require particular attention to irregular and opportunistic sampling, both spatially and temporally.

Building from previous citizen-science monitoring

The three questions listed above will need to be answered before deciding to start a new citizen-science monitoring project. Additionally, pre-existing citizen-science data, where available, can play a role in designing new monitoring programs (volunteer-based or otherwise). Possible uses would be to identify major sources of bias and variability (e.g. time of day or year, observer variability), or generate ecological hypotheses to be examined. Exploratory analyses of even relatively low-quality data, particularly when large volumes are available (Munson *et al.* 2010), may provide information useful for these purposes (Hochachka *et al.* 2007). Bayesian approaches (e.g. Royle and Dorazio

2008, Kéry 2010) offer another possible avenue for using citizen-science data to inform subsequent monitoring: using knowledge gained from citizen-science data to make informative priors in analyses of data from other sources. If done appropriately, the result would be more accurate models emerging from the analyses.

Existing citizen-science projects may also be useful as the logistical foundation on which to build entirely new monitoring programs. In our own research, we were able to take advantage of a long-term survey, Project FeederWatch (e.g. Lepage and Francis 2002) and its established base of observers to begin quickly collecting data to describe the geographic spread of an emerging disease across North America (Dhondt *et al.* 2005). This second monitoring project resulted in experimental studies designed to understand ecology and evolution of host–disease interactions at a mechanistic level (e.g. Hawley *et al.* 2010). A citizen-science monitoring project provides an initial core group of people willing to collect data, who then can be called upon to help address a series of additional and more focused questions.

Data collection and management

All monitoring programs need to manage potential sources of error during the measurement, data management, and analysis processes (Chapters 2, 11). In a volunteer-based monitoring program, our experience suggests that data collection and management need to be designed with three issues in mind: (i) ensuring that information on observer effort is collected, (ii) motivating and retaining volunteers, and (iii) dealing with error detection and correction. In regards to the latter two issues, our own experience is that effective coordination requires at least one and probably two or more full-time employees for any project of a continental scope. This is the minimum staffing level needed to continuously monitor volunteers' work on projects and provide feedback to participants as quickly as possible. Such contact is needed to validate incoming data in a timely fashion and quickly answer participants' questions regarding the protocol, as well as to explicitly remind volunteers that their continued work is valuable and appreciated.

Recording observer effort

As noted above, to maintain volunteer motivation and interest, the survey protocol usually cannot be overly restrictive. One facet of this flexibility may be in the effort required for collecting data. When observers are able to vary the amount of effort that they expend at each location and time, metrics of observer effort need to be recorded for use during analyses to help account for potential variation in detection probabilities or other sources of error (Box 21.2). For example, time spent making observations, distance traveled, and number of observers are potentially important types of effort information.

Volunteer motivation

Motivation should be maintained by reinforcing participants' awareness that the data that they provide are valuable. We see this motivation as particularly important for monitoring of species' presence or numbers where some participants may have few organisms to monitor. These data are valuable and approaches that we have taken to reinforce this

---

Box 21.2 Common challenges: modeling issues

In our chapter, we have noted that the loosely structured protocols for collection of data from many citizen-science projects can result in uneven coverage across the geographical region of interest and over the course of data collection. Our own experience is that unequal data "density" can result in models that are not accurate representations of reality, as patterns in regions of higher data density can overwhelm information from regions of lower data density. To guard against erroneous inferences, we recommend both forcing appropriate spatio-temporal structure on models and validation of models using methods that give equal weight to patterns in all spatial and temporal regions. When predictive accuracy is the primary motivation of the analysis and when there are very high sample sizes typical of large-scale citizen-science data sets, we advocate cross-validation (using an independent subset of the data specifically withheld for validation) for evaluating model performance.

For many citizen-science programs, another possible aspect of loosely structured data-collection protocols is observer effort and detection probabilities that vary among sampling events. Where protocols cannot be structured to allow formal estimation of detection probabilities, factors affecting detectability may be included in analytical models to control for this source of variation. Therefore, data on all factors likely to affect the detection process (e.g. time of day, time spent, and distances traveled by observers) need to be gathered as part of the data-collection process. However, this does not solve the problem if both detection probabilities and the underlying ecological variable or process of interest are simultaneously affected by a given factor (see Chapter 18). Biological intuition of the analyst, supplemental analyses (e.g. Chapter 11), and formal validation of models using independent data may increase confidence, to some extent, that such potentially confounding effects were absent or unimportant. However, the better solution is to structure protocols from the start, if at all possible, to allow estimation of detection probabilities (Chapters 4, 11, 18, 19).

Specifically regarding nonparametric analyses, a constraint on inferences is that confidence intervals around effects are not directly produced. Bootstrapping is currently the only obvious way to estimate confidence intervals from machine-learning analyses. However, this process is computationally expensive and as a result would greatly extend the time taken to build models. We know of no good solution to this issue.

---

notion to volunteers include: explicitly communicating that detecting nothing is just as important as detecting something; and making sure that participants always enter something for each observation period, such as "I saw no birds during this observation period".

Error trapping and correction

Observers, whether professional or volunteer, make errors both during field observation and data entry, and formal mechanisms are needed for detecting and correcting such errors. Once data have been collected, dealing with potential errors generally occurs in two phases: at the time of data entry, and immediately prior to analysis of data.

Many steps can be taken to reduce some types of data-entry errors. Use of computer programs or websites for data entry greatly facilitate error reduction, relative to data submission on paper forms that are subsequently scanned or keypunched by paid staff. We feel that online data entry is superior to using programs that reside on volunteers' own computers because data are immediately available for program staff to peruse, allowing project staff to quickly identify potential problems and communicate with volunteers to resolve issues such as misunderstanding of protocols. Among the design features that have evolved at the Cornell Lab of Ornithology with its online data entry systems are:

(i) Log-in systems requiring a user-name–password combination, that automatically attach participant information (e.g. location, observer identify) to new records. In our experience with paper data forms, participant errors in entering identification information on multiple forms are inevitable, and time-consuming to detect and correct.

(ii) Forms designed to minimize the possibility of accidentally entering data in the wrong data fields.

(iii) Data-entry systems that flag potential errors in a way that recognizes that these observations may be either true errors or unusual but legitimate observations. As an example, automatic error messages can ask participants to "provide required data" instead of to "correct problems", and to "confirm unusual observation" instead of to "fix errors". We recommend that databases be designed so that no information provided by participants is discarded. In our own database systems we have fields that flag unusual records (that are confirmed by participants) for easy detection. These flags are initially used by project staff to identify participants with whom staff can communicate in order to confirm records (e.g. participants can send photos to confirm the identities of unusual species), and our databases include flag fields to indicate whether staff have verified unusual records. Secondarily, these flags need to be used by data analysts to make their own decisions about which data to include in analyses.

Exploration in the analysis and interpretation of citizen-science data

The purpose of data analysis and subsequent interpretation can vary along a gradient from hypothesis testing or parameter estimation, to pattern discovery and exploration. A theme running through this book is that parameter estimation and hypothesis testing are best done using a highly structured survey design linked to carefully specified analyses to address well thought-out questions. However, this framework assumes that the program has fairly tight control over the survey design, and that sufficient prior knowledge of a system is available in order to structure the collection of data and analyses appropriately. This is true regardless of whether the data are collected by professionals or volunteers. For the remainder of this chapter, we will instead consider a different framework focused on discovery of patterns through exploratory analysis of data. We feel that there are three reasons why citizen-science data, in particular, will lend themselves to use in exploration.

- *Data re-purposing.* Large-scale volunteer-collected monitoring data are more likely to be useable for multiple and initially unanticipated purposes than data from smaller-scale studies. The range of possible uses is large. For example, using the National Audubon Society's Christmas Bird Count, researchers have examined topics as varied as the effects of temperature and climate change on wintering bird distributions (e.g. Root 1988, La Sorte *et al.* 2009) and dynamics of emerging disease (Hochachka and Dhondt 2000). Similarly, occurrence data collected by volunteers have been used to test hypotheses regarding the ecological effects of habitat loss and fragmentation (e.g. Trzcinski *et al.* 1999, Vallecillo *et al.* 2009) and climate change (e.g. Thomas and Lennon 1999, Zuckerberg *et al.* 2009). With such initially unanticipated uses, some level of exploration may be required to identify the most appropriate model(s) to fit to the data.
- *Quantity–quality trade-offs.* As discussed above, volunteer monitoring often involves a trade-off leading to use of a less rigorous data-collection protocol, but one that yields a larger set of data with which to work (Munson *et al.* 2010). However, with less control imposed by the design of the data-collection protocol, unanticipated biases may need to be discovered as part of the analysis, necessitating exploratory approaches.
- *Lack of prior knowledge.* Empirically obtaining the insights needed to create well-designed data-collection protocols (e.g. determining sample sizes needed) and formulate parametric models is an explicitly exploratory activity. If the ultimate goal is to design an effective large-scale monitoring scheme, we suggest that the most probable existing sources of data for such explorations will be citizen-science projects. Alternatively, if new preliminary data need to be collected over a large area in order to gain needed insights, then the most cost-effective way of collecting these data potentially is with the assistance of volunteers. In either case, exploration of citizen-science data would be required.

Large-scale citizen-science data – sources of bias and dependence

In ecological surveys, we can view the survey design and analyses as needing to accomplish two general tasks to produce accurate inferences. First, there is the need to ensure that the data are representative of the population(s) on which inferences are desired (Chapters 2, 5, 6). Second, the design and analysis need to isolate the effect of interest – the population parameter about which inferences are required – from variation due to other biologically real influences as well as from sampling variability and other sources of error. Our suspicion is that the more exploratory an analysis is and the lower the control over the survey design, the greater the burden that is placed on the analysis phase for completing these tasks. Below, we discuss what we believe are the major considerations in exploratory analyses of citizen-science data.

Defining the temporal and spatial scope of interest

In any monitoring effort, the data sample needs to reflect the population upon which inferences are to be made, spatially and temporally. Whenever data are re-purposed to examine questions for which the survey design was not tailored, or when there is only loose control over when and where data were collected, then there is potential mismatch between the focus of the analysis and the spatial and temporal frames from which data were collected. This is true regardless of whether volunteers or professionals collected the data. Proper use of the data will require assessment of whether the entire data set should be used or only a subset selected to ensure that the sampling frame matches the population of interest relevant to the specific analysis. For example, Kéry *et al.* (2010a) note that in a topographically complex country such as Switzerland, elevation and its correlates will explain much of the variation in distribution of species. In such a case, the analyst would need to define clearly whether their objective is to model distributions across the entire country, or only within the elevation band in which the species is found. Relevant to this example, Lobo *et al.* (2008) suggest that relatively high accuracy often seen for predicting distributions of rare species may be an artifact of modeling the biologically trivial problem of predicting something that is mostly not there.

Incorporating covariate information

Given the complexity of ecological systems, any variable being monitored is affected by many aspects of the environment, although it is possible that typically only a small number will have substantial influence (e.g. Burnham and Anderson 2002: 21). For some well-understood systems, scientists have already identified the most important ecological and environmental predictors as well as the most important predictors of variation in the observation–detection process, or other nuisance effects. With prior information or hypotheses, appropriate survey designs and parametric statistical models can be applied to account for nuisance sources of bias and variation. As noted earlier, in these circumstances data from citizen-science monitoring will not differ in their collection and analysis from any other monitoring data. Although effects of these nuisance variables may be averaged out across the available data in some cases (e.g. Bas *et al.* 2008), we see no reason for assuming that this will always be the case. A variety of regression techniques have been developed to harness covariate information for prediction and inference. The decision about the analysis method to use will depend on how much is known about the effects of the covariates (see Table 21.1).

In less well-understood systems, a situation that we expect to be common in using (or re-using) citizen-science data, the most important environmental covariates of the system may not be known or be only partially understood. In these situations, nonparametric and semi-parametric regression techniques aid in the discovery of new relationships, often more efficiently than through the use of parametric statistical models. This means that the analyst does not need to make assumptions as to the most important predictors and their functional relationship with the response variable.

If some important influences are known in advance, but others are only suspected, then a "semi-parametric" analysis method can be used (Table 21.1). In this analytical

Table 21.1 General classes of methods for analysis that are likely applicable to the analysis of monitoring data collected by volunteers. The methods are organized in the table from those most suitable for parameter estimation and hypothesis testing (top) to those most suitable for exploration and pattern discovery (bottom). The references cited are not primary references for the description of methods, but are sample references chosen for their relevance for the analysis of ecological data.

Method	Remarks	References
Fitting a fully parametric statistical model	Typical method for analysis of monitoring data, e.g. generalized linear mixed models, occupancy models Both the parameters and forms of relationships fixed in advance Model parameters typically estimated using maximum likelihood or Bayesian methods Best approach when ecological system known well enough to accurately specify model structure in advance	Williams *et al.* 2002 MacKenzie *et al.* 2006 Kéry 2010
Multi-model inference	Combines information from a small set of fully parametric models to produce parameter estimates unconditional *within the set of models* Parameter estimation and predictions can be calculated across several competing models Typical approach is to use Akaike's Information Criterion (AIC) to assess relative support of data for models fit by likelihood; for empirical Bayesian analyses equivalent model selection criteria exist (DIC scores, Bayesian Information Criterion, Bayes Factors), or reversible-jump MCMC may be possible Useful when enough is known about the system such that a small set of candidate models representing alternative explicit hypotheses can be formulated and compared Appropriateness of using model-selection criteria with mixed/hierarchical models is unclear	Burnham and Anderson 2002 King and Brooks 2002 Spiegelhalter *et al.* 2002 Gurka 2006 Link and Barker 2006 Plummer 2008
Semi-parametric models	Some aspects of model structure are not specified in advance, but are "discovered" as part of the analysis Widely known example is generalized additive models (GAMs) or generalized additive mixed models (GAMMs) in which all predictors are specified, but forms of continuous relationships can be left unspecified and fit with splines (useful for discovering forms of relationships between predictors and response when these are unknown). Cox proportional hazard models are another form of semi-parametric model: predictors of relative differences in survival rate are specified, but the actual shape of a survival function is determined in the analysis Useful when some aspects of a system are well known and can be specified accurately, but other aspects are less clearly known	Gienapp *et al.* 2005 Wood 2006
Nonparametric (machine-learning) models	Both the most important predictors and the forms of relationships between predictors and response are discovered in the analysis Many methods exist, e.g. gradient boosting, Random Forests Useful for discovering potentially important predictors and forms of relationships between predictors and response Most appropriate when analyses are explicitly intended to be exploratory (i.e. for hypothesis generation rather than hypothesis testing) and little is known with certainty about a system	Hochachka *et al.* 2007 Elith *et al.* 2008 Marmion *et al.* 2009

framework any known important predictor variable is treated as in a fully parametric statistical model, but where uncertainty exists the suspect predictors are treated in various ways that allow the important predictors to be identified automatically in the course of the analysis.

When essentially nothing is known for certain about a system, or when the primary goal of analyses is only to make accurate predictions, nonparametric regression can be used. These techniques originate from machine learning and have only recently received attention from ecologists (Table 21.1). These methods are capable of efficiently handling very large data sets in terms of the number of observations and number of covariates. Predictor variables describing a very large set of potentially important environmental influences can be screened in analyses that identify both important predictors and forms of relationships. The major limitations of machine-learning methods are: (i) that extraction of information on important predictors requires additional computation and even then no simple equivalents of regression parameters exist (as with GAMs); (ii) no easily computable confidence limits are produced by the analysis and resampling methods like bootstrapping are required to generate information on confidence; and (iii) use of large numbers of predictor variables will inevitably lead to the inclusion of multiple correlated predictors in models; therefore, isolating important underlying environmental influences may be difficult if not impossible.

Managing biases inherent in the sampling process

Monitoring surveys face many potential sources of bias (Chapter 2, 11, and others). An appropriate survey design can (and should) be used to deal with many of these; for example, ensuring that representative data are collected along all of the environmental gradients of interest and using measurement methods that are consistent across all sampling events. However, other biases can only be dealt with using a combination of appropriate sampling design and analysis.

In monitoring studies, repeated sampling at the same location means that the resulting observations are not independent (Chapter 11). In addition, when monitoring data are analyzed using models which assume that locations are independent, then spatial correlation among nearby locations is a concern (Chapter 20). Such non-independence at best will result in overly optimistic estimates of model accuracy, such as artificially low estimates of standard errors associated with parameter estimates and confidence limits bounding predictions that are too narrow (McCulloch *et al.* 2008). At worst, when this non-independence is ignored and combined with a highly uneven number of measurements across the study area and between sites and times, parameter estimates and predictions may be subject to significant bias. This would occur because data in the regions, sites, or time periods with greatest proportion of the data will also have the greatest proportion of influence on the model that is constructed during analysis.

Tests for spatial (Dormann *et al.* 2007, Beale *et al.* 2010) and temporal (Cryer and Chan 2010) autocorrelation are well documented in the statistical literature, and software are readily available to perform diagnostic tests in major software packages [e.g. in R (R Development Core Team 2011) the "ncf" package (Bjornstad 2009) and *acf*() function]. Where non-independence exists, an appropriate analytical approach is needed.

One analytical solution was already noted in the *Incorporating Covariate Information* section: if the environmental features underlying non-independence (e.g. similar habitat in nearby sites) can be identified and their values measured, predictors describing these features can be included in analysis and autocorrelations ("induced spatial dependence"; Fortin and Dale 2005) eliminated. When this is not possible, parametric statistical models need to account for these autocorrelations using other methods such as mixed effects/hierarchical models (e.g. Bolker 2008, Fortin and Dale 2005; Chapters 7, 11, 20). Note that when samples measured nearby in space are also collected in close temporal proximity, temporal autocorrelations can be mistakenly identified as spatial autocorrelations (Hochachka *et al.* 2009), obfuscating the appropriate approach to eliminating these autocorrelations.

Another major, and well described, bias in monitoring data is imperfect detection of the organisms and species present during a sampling event, which is an issue even for large plants (Chen *et al.* 2009) or for monitoring characteristics of organisms as well as their presence or abundance (Cooch *et al.* 2012). Acknowledging and formally dealing with imperfect detection can be important for any trait being monitored; for example, monitoring of size or disease status can be confounded by size- and health-specific differences in detection probability. Other chapters in this volume focus on study designs and statistical models used to estimate and account for imperfect detectability (Chapters 18, 19), so we will keep our discussion of this topic brief. In general, for a citizen-science monitoring project, we believe that the easiest of these methods to implement typically will be occupancy modeling (MacKenzie *et al.* 2006), which can use relatively coarse information (species detected or not detected over a series of repeat surveys at each site during each monitoring occasion) to estimate the probability of detecting a species given it was present at a site (see Chapter 18). Even data from monitoring not initially intended for occupancy modeling sometimes can be structured into the appropriate form (Kéry *et al* 2010b, Zuckerberg *et al.* 2011).

As indicated by other chapters in this volume, citizen-science monitoring is not unique in requiring that data be collected and analyzed so that biases can be minimized. However, to the extent that a complex or highly constrained sampling design is required to remove biases during the sampling phase, recruiting a large number of participants for data collection may not be possible. In such cases, analysts of citizen-science projects may have to fall back on less than ideal data-collection protocols and analysis methods. For example, instead of directly estimating detection probabilities, correlations of detection such as duration of observation period or density of vegetation would be entered into analyses in an effort to statistically control for biases, assuming the factors correlated with detection probability are not also related to the underlying variable of interest (Chapter 18). Perhaps, the most important weakness of these simplified statistical controls is that an absolute estimate of quantities such as prevalence of a species cannot be calculated, only estimates conditional on specific values of the factors that affect detection rates (Royle and Dorazio 2008). The limitations of bias control during data analysis need to be considered when planning new citizen-science projects and acknowledged when using data from existing projects.

Validation of models and resultant inferences

After models are fit, assessing the validity of these models is essential for producing defensible inferences and conclusions. For analyses that have exploratory components and that produce complex models, as we are suggesting is often the case when using data from citizen-science programs, model validation requires special care and appreciation of the limitations of validation methods. No single technique of model validation will provide all of the relevant information needed to judge the accuracy of complex models. Multi-model inference only identifies the best single model or combination of models from a set of pre-specified models, all of which could potentially be poor abstractions of reality (Burnham and Anderson 2002). Hypothesis tests and confidence limits effectively only tell an analyst that a model is better than chance. For citizen-science data sets that are very large, relatively narrow confidence limits and small P-values are to be expected even for models that are not very good descriptions of reality; i.e. there is high power to detect potentially trivial or meaningless effects. More generally, conventional statistical evaluations of a model's predictive ability are typically done on the same data set used to create the model, whereas confirmation of patterns in independent data sets (i.e. replication) will provide stronger confirmation. For example, Fisher envisioned independent confirmation as an important aspect of using P-values to assess validity of findings (see Robinson and Wainer 2002).

In this section, we discuss a method for model validation, relatively unfamiliar to many ecologists, which can be a useful complement to more commonly used validation procedures, particularly when large data sets such as those from many citizen-science projects are available for analysis: cross-validation.

Cross-validation

Cross-validation is a long-established technique (e.g. Stone 1974) for determining the robustness and accuracy of statistical models. While it is only recently that cross-validation has started becoming a routinely used approach in ecology (e.g. Boyce *et al.* 2002), it is a standard part of evaluation of machine-learning analyses in computer science. In cross-validation, one part of a data set, often termed the "testing set", is put aside while the other ("training") set is used for the analysis. Then, the model built from the training data is used to predict the response values from the testing set. The disparity between the predicted and observed response values for the testing test is used to quantify the model's predictive accuracy, as measured by one or more of various available metrics (e.g. Fielding and Bell 1997, Mouton *et al.* 2010).

Multiple variations on this general theme exist. A single split of data and single evaluation is the most basic. Another possibility is the creation of multiple data splits and repetition of the analysis for each split in order to produce a measure of model performance that averages across the random noise within individual single data splits. A variation on this idea is k-fold cross-validation in which the data are divided into multiple groups, and in each of k iterations of the process the subset of data from one of these groups forms the testing set while the data from the other $k - 1$ groups are used in the analysis for model construction. Chapter 5 of Harrell (2001) provides a more detailed discussion of cross-validation and related methods for assessing model validity.

Using cross-validation for model assessment has both strengths and weaknesses depending on the objectives of the analysis. If the purpose of an analysis is parameter estimation (i.e. quantifying the influence of some factor(s) on the population attribute being monitored), and especially if the analyst has a strong prior knowledge allowing a realistic parametric model to be specified, then cross-validation is arguably not important. In such cases, using a cross-validation approach that sets aside one part of the data set results in a lower sample size for model building and therefore produces greater uncertainty in parameter estimates. In contrast, if an explicit goal of analysis and model construction is to interpolate or extrapolate outside of the data sample, for example to predict a species' global distribution, and particularly if the analyst does not have a strongly justified prior belief of appropriate model structure, then an appropriate cross-validation is important. In this latter situation, assessment of model appropriateness by cross-validation is a direct test of whether the model serves its purpose of making accurate predictions outside of the sampling locations and time periods. If analyses are done using machine-learning methods, then cross-validation may be the only computationally feasible approach to model assessment (see Elith *et al.* 2008 for an ecological example of cross-validation with machine-learning analyses).

Selecting testing data for cross-validation

Conventionally, selecting data for model building and testing is done by randomly partitioning and reserving a user-defined proportion of the available data (e.g. 30%) for testing. However, in the context of ecological monitoring data, we believe that a purely random split of data is almost certainly inappropriate, because of spatial and temporal structuring in the data. That is, observations taken from locations near to each other or at the same location at close time intervals will be more similar than would occur by chance alone, and may even be measurements of the same individuals. Thus, evaluations based on a purely random division of data into training and testing subsets will overestimate the accuracy of predictions (Araújo *et al.* 2005; Fig. 21.2).

The process of separating data into training and testing subsets needs to create independence between these two subsets. Spatial independence is sometimes fostered by creating a checkerboard (e.g. Munson *et al.* 2010), using "black" squares for model building and "white" for validation, for example. Similarly, creating separate blocks of time for model building and validation can be used to remove temporal overlap in the data used to create and validate models.

A further complication arises if data are not uniformly distributed (Fig. 21.1), because regions of high data density will have a greater impact on the overall index of model accuracy. Subsampling a larger testing data subset to ensure uniform representation of validation through space and time (Fink *et al.* 2010) can be used in these instances.

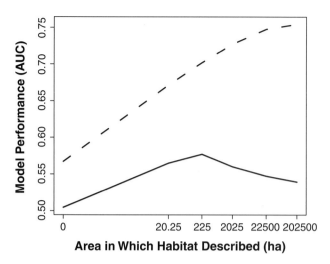

Figure 21.2 Model validity should be assessed using a set of data that is independent of the data used to generate the model. Graphed are measures of model performance [Area Under the Receiving Operating Characteristic Curve (AUC) values; larger values indicate better model fit] for models predicting the distribution of horned larks (*Eremophila alpestris*) in shortgrass prairie habitat in the western USA. Models were created in which habitat composition and configuration were described in areas of different sizes around each location. The upper, dashed, line shows changes in predictive performance of models using a validation data set that was composed of a random 20% of the data; data from the same and nearby locations were found in both the data set used to generate the species distribution models and in the data used to validate the models. The lower, solid line shows changes in predictive performance when independent data (validation data were from sites no closer than 45 km away from any data point used to generate the model) were used for model validation. Independent validation data led not only to lower estimates of model performance, but a qualitatively different association between model performance and the area over which habitat was described.

Future research and development

Although we believe that machine-learning approaches can be useful for exploratory analyses of large-scale citizen-science data sets, we are not aware of any established conventions for using machine-learning analysis techniques for these types of spatially and temporally correlated data. Development of machine-learning techniques for analysis of ecological monitoring data is currently an area in which active research is needed. We believe that four areas in particular need to be explored systematically: (i) adapting existing machine-learning methods (i.e. turning nonparametric techniques into semi-parametric techniques) for spatially and temporally autocorrelated data without reducing the chances that patterns of interest will emerge as part of analyses; (ii) developing appropriate techniques for evaluating, through some form of cross-validation, the performance of a machine-learning analysis of data produced by unequal sampling through space or time, as we believe will often occur for citizen-science data; (iii) identifying computationally efficient methods for estimation of confidence intervals; and (iv) systematically evaluating the performance of machine-learning methods for modeling abundance of organisms. Our own work, and research in machine learning in general, has concentrated on the problem of binary classification (occurrence), so that we do not have strong intuition about how well machine-learning models of abundance would typically perform when developed using existing analysis techniques.

We also see the need to identify the most efficient ways to integrate citizen-science monitoring into targeted and controlled study designs and analyses. We believe that there is greater scope to use citizen-science surveys to collect targeted monitoring data than is currently practiced (Wiersma 2010). More experience is needed in determining how best to design highly structured citizen-science data-collection protocols, and in setting up and managing citizen-science projects that collect targeted monitoring data. For example, we need to learn how to best motivate volunteers to make the sorts of repeated observations that could be used for explicit estimation of detection probabilities. We are intrigued by the possibility of more formal integration of surveillance monitoring data from citizen-science surveys with data from other, more targeted monitoring efforts, either in analyses involving direct combination of data from multiple sources (data fusion; e.g. Freeman *et al.* 2007, Link and Sauer 2007), or with citizen-science data used to generate informative priors for Bayesian analyses of other data.

Summary

When a targeted monitoring objective exists and a citizen-science protocol has been designed to collect data using a carefully structured survey design, standard statistical inference methods can be used, as described in the rest of this volume. We have instead discussed issues surrounding use of citizen-science data that were collected with protocols that were not specifically designed to answer the questions at hand, and that may have had only loose control over the survey design. Such uses have been varied and common (e.g. Root 1988, Hochachka and Dhondt 2000, Zuckerberg *et al.* 2011), and are likely to continue to be so. In our experience with such unanticipated uses, the greatest challenge to making robust inferences from citizen-science data is in defining the structure (predictor variables and forms of relationships) of the statistical model(s) to be fit. As a result, some manner of exploratory analysis will often be required in developing and validating statistical models. We have briefly described how diverse methods for data analysis (Table 21.1) can be viewed as existing along a gradient of increasing emphasis on exploration, and discussed the process of validation of models that emerge from exploratory analyses. We believe that cross-validation is under-used as a tool for validation of such models, or more generally when the intent of analyses is to produce models for predicting into new areas or time periods.

We have emphasized the utility of citizen-science data as a tool for exploratory analysis when prior knowledge of a system is not sufficient to construct carefully structured data collection and analysis protocols. However, a carefully structured data collection and analysis protocol will produce the most conclusive insights into ecological systems and prior knowledge is key to a well-designed study (e.g. Burnham and Anderson 2002). In this context, data from citizen-science monitoring may be analyzed to create the prior knowledge needed to design new data-collection protocols or to develop hypotheses that can be examined through subsequent analyses of additional and independent existing data.

Acknowledgments

We have had the pleasure of working with colleagues in the computer science departments of Cornell University, Northeastern University, and the University of Oregon who have shaped our understanding of machine learning: R. Caruana, A. Munson, B. Shaby, D. Sorokina, M. Riedewald, and T. Dietterich. Our discussion of citizen-science projects is based on work with many colleagues at the Cornell Lab of Ornithology, with the following people directly providing information appearing to this chapter: D. Bonter, C. Cooper, J. Dickinson, S. Hames, J. Lowe, and K. Rosenberg. Motivating much of our discussion was work related to the Avian Knowledge Network, an initiative guided by S. Kelling and supported by the information sciences unit of the Lab of Ornithology. Funding for this work has been provided by the National Science Foundation (grants ITR-0427914, DBI-0542868, DUE-0734857, IIS-074826, and IIS-0832782), The Leon Levy Foundation, and The Wolf Creek Foundation. Comments from the editors of this book, W. Venables, and one anonymous reviewer greatly improved this chapter.

Section V

Conclusion

22 Institutionalizing an effective long-term monitoring program in the US National Park Service

Steven G. Fancy and Robert E. Bennetts

Introduction

The need for long-term ecological monitoring

Managers of protected areas have increasingly recognized the value and need for credible, scientific information as a basis for making management decisions and working with partners and the public to conserve natural resources. The management of natural resources has become increasingly complex, both technically and politically. Managers need reliable data and information on the status and trends in the condition of key resources that they manage as a basis for conservation planning, determining whether current management practices are having the desired effect, and informing stakeholders and the general public of changes in the condition of natural resources that may be caused by stressors operating at regional or global scales.

Long-term ecological monitoring provides information needed to understand and identify change in natural systems characterized by complexity, variability, and surprises. This information can be used to assess whether observed changes are within natural levels of variability or may be the result of unwanted human influences. Data collected in a consistent way over long periods are fundamental to conservation and management because they provide the context for interpreting observed changes, and may provide the basis for initiating new management practices or changing existing practices (Carpenter 1998, Lovett *et al.* 2007). For example, reliable and consistently collected data from long-term studies are currently in high demand for developing quantitative models to inform conservation and action plans for addressing the ecological consequences of rapid climate change.

The legacy of long-term monitoring programs

Despite the importance of reliable, relevant long-term monitoring data, the track record for initiating and sustaining effective long-term monitoring has been poor (Mulder and Palmer 1999, Reid 2001, Noon 2003, Nichols and Williams 2006, Lindenmayer and

Design and Analysis of Long-term Ecological Monitoring Studies, ed. R.A. Gitzen, J.J. Millspaugh, A.B. Cooper, and D.S. Licht. Published by Cambridge University Press. © Cambridge University Press 2012.

Likens 2009). Monitoring programs are often hurriedly planned and implemented in response to a short-term funding opportunity or political directive, are often insufficiently funded, and historically have been one of the first programs to be cut in times of budget reductions. Large-scale monitoring programs designed to provide inferences at a regional or national scale often do not provide sufficient information relevant to the highest priorities of land managers at the local level. For example, the number of sampling sites in a management unit may be too small to provide reliable local-scale information. Therefore, on-the-ground managers are often not enthusiastic about such large-scale efforts.

The legacy of science and monitoring in the US National Park Service

The nearly 400 units managed by the US National Park Service (NPS) include many of the nation's most treasured natural landscapes and historical and cultural sites. The NPS mission, as defined by the 1916 Organic Act, is "To conserve the scenery and the natural and historic objects and the wild life therein and to provide for the enjoyment of the same in such manner and by such means as will leave them unimpaired for the enjoyment of future generations." NPS management policies state that "The Service will also strive to ensure that park resources and values are passed on to future generations in a condition that is as good as, or better than, the conditions that exist today", and that "Decision makers and planners will use the best available scientific and technical information and scholarly analysis to identify appropriate management actions for protection and use of park resources" (NPS 2006). As natural laboratories and long-term monitoring sites, NPS units can serve as reference sites and places where effects of regional and global changes may be detected without many of the smaller-scale confounding influences found on other public and private lands.

An overarching natural resource *management* objective for all of the parks is to pass the resources on to "future generations in a condition that is as good as, or better than, the conditions that exist today". To address this management objective, park managers and planners clearly need reliable scientific information about the condition and trends of the natural resources for their park. However, until recently almost all of the NPS workforce and budget was focused on traditional scenery and tourism management (Sellars 1997), with relatively little attention and funding given to science as a basis for resource management. The Natural Resource Challenge, initiated in 2000 (NPS 1999), resulted in increased funding and a commitment by the NPS leadership to strengthen resource preservation and restoration through strong, science-based programs. Although funding was allocated to an ecological monitoring program, development of this program faced significant challenges due to the relatively low funding level and the agency's decentralized organizational structure. Furthermore, the diversity of ecological systems included in the NPS (e.g. coral reefs, deserts, arctic tundra, prairie grasslands, caves, rivers, and tropical rainforests) made it difficult to design a monitoring program that would be relevant and effective to parks throughout the system.

Despite these challenges, the NPS has successfully designed a long-term ecological monitoring program that continues to gain support and acceptance as a key component of natural resource stewardship. In this chapter we share some of the key lessons learned in designing, implementing, and institutionalizing a long-term monitoring program in a natural resource management agency. At one level, we emphasize the importance of these lessons for effective implementation of a scientifically credible program. We have effectively incorporated many of the statistical recommendations made throughout this volume into the design of the program; NPS has worked with statisticians and subject-matter experts to develop and disseminate more than 100 peer-reviewed monitoring protocols applying a variety of statistical tools. However, we also provide a broader context. We emphasize that although sound statistical design and analytical approaches are an essential underpinning of the scientific credibility and reliability of any monitoring program, they are just one component of developing and institutionalizing an effective monitoring program.

Critical elements for institutionalizing a monitoring program: relevance, reliability, and commitment

The term "institutionalize" implies broad acceptance of the program as an integral part of an agency's operation. Mere longevity does not constitute institutionalization. To be truly institutionalized, a monitoring program needs to be sufficiently and formally integrated into the key operations (e.g. decision-making and planning), such that it helps the agency achieve its mission and goals. Only a handful of large-scale ecological monitoring programs have achieved relative longevity and even fewer have become institutionalized. Probably the best examples come from regulatory contexts where monitoring for compliance has a legal mandate as part of an organization's day-to-day operations (e.g. water quality monitoring by state governments to address the Clean Water Act in the USA). If a monitoring program is not legally mandated, it can achieve institutionalization only if it has long-term support from all levels of an organization.

In practice, the level of integration within an organization's operations and the roles played by a monitoring program in these operations can be highly variable. All too often, the primary means of integrating scientific information into management is limited to opportunistic gleaning of information if a manager happens to be aware of potentially relevant science. If this is the primary use of information produced by monitoring, the program clearly has not been institutionalized effectively. Rather, the information gathered via the monitoring program needs to be an integral part of the planning and operations of the organization. Such integration requires a systematic and routine process for communication of relevant science and a process that enables effective consideration of that science and the potential for its incorporation into management decisions.

There are many aspects of a monitoring program that contribute to its success. There are, however, a few broad themes without which a program has virtually no chance of becoming an integral part of an organization: *relevancy*, *reliability*, and *commitment* (Box 22.1). The primary focus of this chapter will be to discuss those three themes. As part of this discussion, we will consider some of the trade-offs we made and how the NPS monitoring program sought, and is still seeking, to become an integral part of the planning and operations within NPS. We also discuss how quantitative topics of other chapters in this volume relate to these three themes.

Box 22.1 Take-home messages for program managers

Based on our experience in designing a long-term monitoring program for the US National Park Service, there are three broad themes determining whether a monitoring program has a chance of becoming a widely accepted, integral part of an organization: *relevancy*, *reliability*, and *commitment*.

Relevancy

The establishment of clearly defined goals and objectives and the selection of indicators are the most critical components of making the monitoring program relevant. To ensure relevancy, the program must have a carefully structured process that allows both natural resource managers and scientists to have input into developing these objectives and selecting indicators. Understanding the information needs of an organization is an essential first step. Such a process also begins establishing what should be a long-standing partnership among scientists and managers.

Reliability

The reliability of scientific information used in management, planning, and policy decision-making partly determines the credibility of these decisions and support from stakeholders, local communities, and the general public. Key elements that contribute to the reliability of a long-term ecological monitoring program are the development of the following:

(i) clear, specific, and measurable monitoring objectives;
(ii) conceptual models that describe important components of the ecosystem and the interactions among them, and that help to justify and interpret the ecological measurements;
(iii) well-documented, peer-reviewed protocols that describe how data are to be collected, managed, analyzed, and reported;
(iv) survey designs to ensure that data collected are representative of the target populations and sufficient to allow defensible conclusions to be derived about the resources of interest; and
(v) procedures that ensure that data are properly managed, routinely analyzed, and are readily available to key audiences in a usable and timely manner.

Commitment

A prerequisite for a successful long-term monitoring program is an agency's commitment and solid funding base to sustain the program. Institutional commitment is best achieved through clear demonstration of a program's value in supporting credible decisions and sound management of natural resources, and by showing that the benefits derived from the program are worth the expense. Developing a partnership between scientists and managers from the outset will help ensure that the monitoring program addresses questions that are relevant to management and policy issues, and that the monitoring can effectively answer those questions in a scientifically defensible manner. This step alone will go a long way toward gaining support at the ground level – support that is essential if the program is to become an integral part of the organization's operations.

Relevance of the monitoring program

The importance of clearly defining relevant goals and objectives

There is virtually universal consensus that setting realistic, clear, specific, and measurable monitoring objectives is a critical first step in developing a monitoring program (e.g. Spellerberg 1991, Elzinga *et al.* 1998; Chapters 2, 3, 18, 21, and others in this volume). Olsen *et al.* (1999) noted that "Most of the thought that goes into a monitoring program should occur at this preliminary planning stage. The objectives guide, if not completely determine, the scope of inference of the study and the data collected, both of which are crucial for attaining the stated objectives." They stated that a "clear and concise statement of monitoring objectives is essential to realize the necessary compromises, select appropriate locations for inclusion in the study, take relevant and meaningful measurements at these locations, and perform analyses that will provide a basis for the conclusions necessary for meeting the stated objectives." Chapter 2 further discusses the importance of well-defined objectives in guiding all other quantitative decisions required in the development of a monitoring program.

Recognizing these critical roles, the NPS monitoring program put considerable emphasis on the task of formulating monitoring objectives that meet the test of being realistic, specific, and measurable (Fancy *et al.* 2009). We found that there was confusion initially about the differences between goals, management objectives, monitoring objectives, and sampling objectives. Therefore, we developed guidance for writing monitoring objectives that was effective in improving the quality and consistency of objectives in monitoring plans and sampling protocols (NPS 2008a).

Management-oriented monitoring is most efficiently accomplished when clearly defined management objectives exist and are accompanied by clearly defined monitoring objectives. Relevant to the agency's overarching management objective, the NPS long-term ecological monitoring program was established to address whether natural resources in parks are being passed on to future generations in a condition that is as good as or better than the conditions that exist today. Most parks also have more specific and usually shorter-term management objectives for specific restoration projects and other management actions, but any monitoring associated with those specific management actions requires other funding sources (see *Future Research and Development* section below for further discussion of overall condition vs. effectiveness monitoring).

Top down or bottom up? Trade-offs between relevancy, efficiency, and inference

There are numerous advantages for designing an agency-wide "top-down" monitoring program with an interchangeable set of indicators and sampling protocols applied to all units so that data can be "rolled up" to address questions at different geographic and/or organizational scales. The development of sampling protocols, databases, analysis routines, and reporting structures is more efficient if a consistent set of indicators and protocols is used at all sites, and there are scientific advantages in being able to provide context and better interpret the monitoring results from a site by comparing them with data from other sites.

We initially evaluated, but rejected, the strategy of selecting a core set of indicators that every park would measure in a similar way. The "information-rich" indicators that best characterized park ecosystems differed greatly among ecological systems, and very few relevant measures were common across parks. Moreover, partnership opportunities (and the appropriate ecological indicators and sampling methodologies associated with them) available to parks differed greatly throughout the National Park System (Fancy *et al.* 2009). The development of partnerships with other government agencies and non-governmental organizations is important for developing political relevance, which in turn contributes to agency commitment. A top-down, "one size fits all" approach to monitoring design would not have been effective or supported in the NPS because of the tremendous variability among parks in ecological context and in park sizes and management capabilities, and because individual parks have very different resource issues, information needs, and partnership opportunities. Thus, in order to gain acceptance at the park level, some sacrifice of national consistency was necessary. We balanced park-specific relevance with organizational efficiency by grouping the more than 270 parks with significant natural resources into 32 Inventory and Monitoring (I&M) networks (see the map at http://science.nature.nps.gov/im/networks.cfm) that each share a professional staff and funding to conduct long-term monitoring (Fancy *et al.* 2009). Each network worked with partners and subject-matter experts to prioritize and select park-specific or network-wide indicators based on their most critical information needs and local partnership opportunities, as described in the next section. Thus, the issue of considering a top-down versus bottom-up program extends beyond gaining the acceptance of the individual parks, but also is an essential consideration for developing an efficient and cost-effective program that takes advantage of partnership and cost-leveraging opportunities.

Selecting indicators

The selection of indicators is a critical step in ensuring the relevancy and usefulness of monitoring data. Consequently, this topic has been widely addressed in the literature and has an entire journal (*Ecological Indicators*) devoted to the topic. Ecological indicators serve three primary purposes (National Research Council 2000): (i) to quantify information in such a way as to illustrate its significance, (ii) to simplify information about otherwise complex phenomena (Hammond *et al.* 1995), and (iii) to serve as a cost-effective alternative to monitoring a larger suite of species and processes (Landres 1992). In order to achieve these purposes and to maximize the relevance of the monitoring results to meeting the stated objectives, it is essential to use a careful, structured process that allows both managers and scientists to have input into the indicator selection process (Jackson *et al.* 2000).

The process for selecting indicators for each of the 32 I&M networks began with a scoping process to identify park issues, monitoring questions, and data needs. The scoping process identified focal resources (including ecological processes) important to each park; agents of change or stressors that are known or suspected to cause changes in the focal resources over time; and key properties and processes of ecosystem condition (e.g. weather, soil nutrients). Conceptual models were then developed to help organize and communicate the information compiled during scoping, and to identify known or hypothesized cause–effect relationships between some of the stressors and response variables. The scoping and conceptual modeling efforts resulted in a long list of potential indicators, which were prioritized using a set of criteria for management and ecological significance, and a scoring system agreed upon by each network (see Fancy *et al.* 2009 for details). The final step in the process incorporated other criteria such as efficient use of personnel, cost and logistical feasibility, partnership opportunities with other programs, and a large dose of common sense to select the initial set of indicators for each network's monitoring program (Table 22.1). We obtained best results when networks treated prioritization and selection of indicators as two separate steps in the process (Fancy *et al.* 2009).

Reliability of the monitoring program

Management, policy, and planning decisions based on reliable scientific data and information generate credibility and support from stakeholders, local communities, and the general public. Reliability entails all program elements that enable the user to develop an expectation of credible and available information. The key elements for the reliability of a long-term ecological monitoring program that we will emphasize here include the development of the following:

  (i) clear, specific, and measurable monitoring objectives;
 (ii) conceptual models that describe the important components of the ecosystem and the interactions among them, and that help to justify, interpret, and communicate the ecological measurements;
(iii) well-documented, peer-reviewed protocols that describe how data are to be collected, managed, analyzed, and reported;
 (iv) survey designs to ensure that data collected are representative of the target populations and sufficient to allow defensible conclusions to be derived about the resources of interest; and
  (v) procedures that ensure that data are properly managed, routinely analyzed, and are readily available to key audiences in a usable and timely manner.

We discussed the development of objectives above; therefore, we focus now on each of the last four key elements. We also consider an additional challenge relevant to all of these elements – the need for adaptability over time.

Table 22.1 Summary of the most common indicators of natural resource condition and examples of specific measures that are being monitored by the US National Park Service long-term ecological monitoring program.

Indicator category	Example measures	Number of parks
Weather and climate	Temperature, precipitation, wind speed, ice on/off dates	246
Water chemistry	Dissolved oxygen, pH, temperature, conductivity	211
Land cover and use	Area in each land cover and use type; patch size and pattern	203
Invasive/exotic plants	Early detection, presence/absence, area	200
Birds	Species composition, distribution, abundance	189
Surface water dynamics	Discharge/flow rates, gauge/stage height, lake elevation, spring/seep volume, sea level rise	158
Ozone	Ozone concentration, damage to sensitive vegetation	140
Wet and dry deposition	Wet deposition chemistry, sulfur dioxide concentrations	114
Visibility and particulate matter	Visibility and fine particles [IMPROVE (Interagency Monitoring of Protected Visual Environments) network]	113
Fire and fuel dynamics	Fire frequency, average fire size, average burn severity, total area affected by fire	105
Vegetation complexes	Plant community diversity, relative species/guild abundance, structure/age class, incidence of disease	101
Mammals	Species composition, distribution, abundance	93
Forest/woodland communities	Community diversity; coverage and abundance; condition and vigor classes; regeneration	93
Soil function and dynamics	Soil nutrients; cover and composition of biological soil crust communities; soil aggregate stability	91
Stream/river channel characteristics	Channel width, depth, and gradient; sinuosity; channel cross-section; pool frequency and depth; particle size	89
Aquatic macroinvertebrates	Species composition and abundance	86
Threatened and endangered species and communities	Abundance; distribution; sex and age ratios	85
Air contaminants	Concentrations of semi-volatile organic compounds, PCBs, DDT, mercury	71
Groundwater dynamics	Flow rate, depth to ground water, withdrawal rates, recharge rates, volume in aquifer	69
Amphibians and reptiles	Species distribution, species abundance, population age/size structure, species diversity, percentage area occupied	54
Grassland/herb communities	Composition, structure, abundance, changes in treeline	51
Fishes	Community composition, abundance, distribution, age classes, occupancy, invasive species	50
Insect pests	Extent of insect-related mortality, distribution and extent of standing dead/stressed/diseased trees, early detection	50
Riparian communities	Species composition and percent cover; distribution and density of selected plants; canopy height	45
Nutrient dynamics	Nitrate, ammonia, dissolved organic nitrogen, nitrite, orthophosphate, total potassium	45
Primary production	Normalized differential vegetation index (NDVI), change in length of growing season, carbon fixation	41
Wetland communities	Species composition and percent cover; distribution and density of selected plants; canopy height; areal extent	40
Microorganisms	Fecal coliforms, E. coli, cyanobacteria	30
Water toxics	Organic and inorganic toxics; heavy metals	30
Invasive/exotic animals	Invasive species present, distribution, vegetation types invaded, early detection at invasion points	29
Coastal/oceanographic features and processes	Rate of shoreline change; sea surface elevations; area and degree of subsidence through relative elevation data	29

Conceptual models

Conceptual models are an essential tool for framing the right questions and aiding in the selection of relevant indicators for any ecological monitoring program (Barber 1994, National Research Council 1995, Noon *et al*. 1999; see also Chapter 2). Conceptual models help us to understand the key components and processes deemed important in an ecosystem (Manley *et al*. 2000, Gross 2003). These models help us identify assumptions about how components and processes are related, and to identify gaps in our knowledge. Essentially, they are a representation of working hypotheses about system form and function (Huggett 1993, Manley *et al*. 2000). A well-constructed conceptual model will provide a scientific foundation for the monitoring program and aid in the selection of relevant indicators (Gross 2003). More narrowly, by helping identify links among attributes and important factors affecting attributes, conceptual models can help increase the ecological insights gained from monitoring and can indicate supplementary covariates that could be incorporated into eventual statistical analyses to reduce unexplained variability and increase statistical power and precision. Figure 22.1 is an example of a conceptual model that serves several of these purposes.

Monitoring protocol development and peer review

A monitoring protocol is a detailed study plan that describes how data are to be collected, managed, analyzed, and reported. It is a key component of quality assurance for natural resource monitoring programs (Oakley *et al*. 2003). Protocols that are well-documented, peer-reviewed, and tested can demonstrate that any detected trends are actually occurring in nature, and are not simply a result of measurements being taken by different people or in slightly different ways. Protocol development is often an expensive, time-consuming process involving a research component. In order to promote consistency and data comparability and to reduce costs, a program should adopt or modify existing protocols developed by other programs and agencies whenever monitoring objectives are similar.

The survey design as a central component of scientific credibility

A carefully planned survey design seeks to ensure that the data collected are representative of the target populations and sufficient to allow defensible conclusions to be derived about the resources of interest (USEPA 2002; Chapters 2, 5–7). Sample sizes for any monitoring program will usually be limited by shortages of funding and personnel, and it is critical to be able to make inferences to larger areas from data collected at a relatively small subset of potential sampling locations.

During the design of the NPS ecological monitoring program, the I&M networks were required to develop survey designs with the goals of (i) making unbiased and defensible inferences from sample observations to the intended target populations, and (ii) encouraging the co-location of sampling sites and events among indicators to improve efficiency and depth of ecological understanding. In the early planning and design phases, we promoted the following four basic principles for developing an overall survey design for each of the I&M networks that has been effective in the development of credible, practical approaches for selecting sampling locations and revisit designs (Fancy *et al*. 2009).

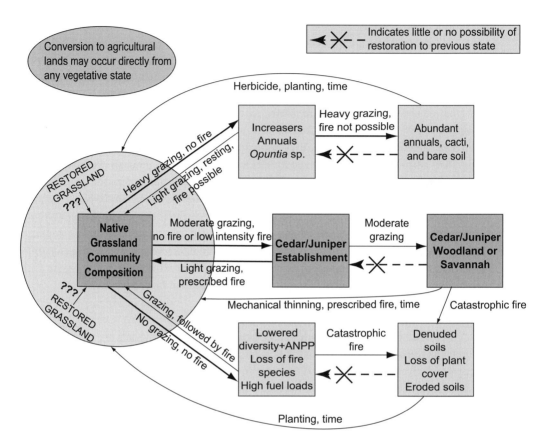

Figure 22.1 Example of conceptual model developed by the National Park Service Southern Plains Inventory and Monitoring Network (NPS 2008b) for NPS units in the Southern Great Plains region of the USA. The model depicts three potential pathways for community composition changes that result from interactions of fire and grazing, as well as a fourth pathway that results in the conversion of any grassland community to agricultural lands.

- Wherever possible, a probability design should always be used for selecting sample locations. Probability designs, where each unit in the target population has a known, non-zero probability of being included in the sample, and a random component is included in the selection of sampling sites, allow for unbiased statistical inference from sampled sites to unsampled elements of the resource of interest (Hansen *et al.* 1983, McDonald 2003; Chapters 2, 5). Probability designs provide more reliable and defensible parameter estimates than convenience or judgment samples (Olsen *et al.* 1999, Schreuder *et al.* 2004), and provide measures of the precision of population estimates (Stevens and Olsen 2003). The most common probability design that has been used by almost all of the NPS I&M networks for a wide range of indicators in both aquatic and terrestrial systems is the spatially balanced Generalized Random Tessellation Stratified Design (GRTS; Stevens and Olsen 2003, 2004; Chapters 6, 10, 16).

- Judgment samples that use "representative" sites selected by experts are not recommended because they may produce biased, unreliable information (Olsen *et al.* 1999; Chapters 2, 5) and can often be easily discredited by critics.
- Stratification of the park using vegetation maps or other biological data or models is problematic because features used to establish stratum boundaries at the start of monitoring will change over time. A vegetation map is a model based on remote sensing and field data, and map boundaries will change as classification models improve or as additional ground-truthing data becomes available. Using these units to define strata will limit (and greatly complicate) long-term uses of the data by restricting future park managers' abilities to include new information into the sampling framework (e.g. estimating current status for dynamic domains or subpopulations of interest such as a vegetation cover type for which the permanently specified strata boundaries are no longer meaningful). It is legitimate, and better, to delineate areas of special interest such as riparian or alpine areas based on physical characteristics such as terrain, and use these judiciously to define either strata or areas to sample with higher probability (see also Chapters 2, 3, 5).
- Permanent sampling sites, revisited over time, are recommended for monitoring, because the objective is to detect changes over time. In most situations, revisiting the same sites increases the precision of change estimates, and therefore increases power to detect temporal changes (Chapters 7–8).

Another key aspect of the survey design involves consideration of whether sufficient sample sizes can be obtained to enable distinguishing the signal of interest from the noise (background variation). Although statistical power in the ecological sciences is seldom what we would like it to be, we at least need to ensure the cost and effort we expend on data collection have a reasonable probability of producing interpretable results (Chapter 8). If we do not have confidence in our results, then we are not providing managers with the information they need to make informed decisions.

Data management, analysis, and reporting

Data and information are the primary products of an ecological monitoring program (see also Chapter 2). Efforts to provide organized, well-documented data and information to key audiences will largely determine the monitoring program's efficacy and image among critics, peers, and advocates. Monitoring information is "wasted if it is not analyzed correctly, archived well, reported timely or communicated appropriately" (Gibbs *et al.* 1999). Information is created from data as a result of processing, manipulating, synthesizing, or organizing data in a way that provides interpretation or meaning.

Therefore, a critical component of the reliability of the monitoring program is to ensure that data and information are managed and analyzed so that they can be easily found and obtained, are subjected to full quality control before release, and are accompanied by complete metadata. The program needs to provide data and information in formats that are useful to end users. Finally, sensitive data (e.g. locations of some legally protected species) need to be identified and protected from unauthorized access and distribution.

The appropriate analysis of monitoring data, which is a focus of many chapters in this volume, is directly linked to the monitoring objectives, the survey design, the intended audiences, and the management uses of the data. As emphasized in Chapter 2, analysis methods need to be considered when the objectives are identified and as the survey design is developed, rather than after data are collected. Each monitoring protocol should contain detailed information on analytical tools and approaches for data analysis and interpretation, including the rationale for a particular approach, advantages and limitations of each procedure, and Standard Operating Procedures for each prescribed analysis (Oakley *et al.* 2003). Four general levels of data analysis that typically occur with a monitoring program are: (i) descriptive and summary statistical analysis; (ii) determination of resource status; (iii) determination of trends in condition over time for a monitored resource; and (iv) synthesis of status and trend information across multiple resources over time to depict larger-scale aspects of ecosystem structure and function.

Monitoring results must be reported and communicated using a variety of products and approaches in order to effectively convey results to key audiences. Park managers sometimes complain that scientists "tend to know (and communicate) too much" (Lewis 2007: 39), and managers usually prefer the Cliff's Notes® version of scientific reports in the form of resource briefs or other short summary documents. The short summary documents, however, must be backed up by detailed technical reports and protocol documents so that the scientific credibility and reliability of the results can be established. Therefore, to deliver monitoring results to key audiences, each of the 32 NPS I&M networks produces a suite of products including 1- or 2-page resource briefs, simple data summary reports, more detailed technical reports, journal articles, and trend analysis and synthesis reports. Internet and intranet websites are the primary outlet for delivering monitoring results to park managers, planners, the scientific community, and the public (see http://science.nature.nps.gov/im).

The need for adaptability

One of the trade-offs that will inevitably be encountered is whether to change the monitoring program, and if so, how and when. The peer-review process helps us correct initial flaws or weaknesses in our approach, but other potential changes will need to be considered. Over time, new technologies for sampling emerge, analytical approaches are advancing, and even the questions being asked can change or evolve as we gain knowledge or have shifts in priorities. A successful monitoring program needs to be able to adapt to such changes without losing the long-term integrity of the data (Lindenmayer and Likens 2009). This decision requires careful consideration of the potential trade-offs between long-term data integrity and the potential need or benefits of change (Chapter 3). Analogous to Albert Einstein's well-known portrayal of model complexity, monitoring programs should accommodate as much change as is needed, but no more.

Commitment to the monitoring program

One of the most important prerequisites for a successful long-term monitoring program is an agency's commitment and solid funding base to sustain the program (Strayer *et al.* 1986). Institutional commitment is best achieved through clear demonstration of a program's value as a means of supporting sound science-based decisions and protecting resources, and the ability to demonstrate that the benefits derived from the program are worth the expense. In the NPS monitoring program, achieving commitment and support from the individual parks largely boils down to whether or not the program provides relevant and timely information that helps managers make decisions and work with other agencies and the public for the long-term protection of the park's resources and values.

Developing a partnership between scientists and managers from the outset

Any monitoring program intended to support management decisions must first understand the information needs of the managers (Chapters 2–4 and others). It is not enough for scientists to plan the effort based on what they believe a manager should know, rather than taking the time to work with managers to determine their priority information needs. Similarly, managers often will not know how to frame questions in such a way that these questions can effectively be addressed through long-term research or monitoring (Lindenmayer and Likens 2009). Developing a partnership between scientists and managers from the outset will help ensure that the questions being asked are relevant to the management and policy issues, and that the monitoring is designed to answer those questions effectively and in a scientifically defensible manner. This step alone will go a long way toward gaining support at the ground level – support without which the program has little chance of becoming an integral part of the organization's operations.

The NPS monitoring program attempted to establish a partnership between scientists and managers from the outset. An important step for achieving this was to involve both groups in the scoping process that led to the selection of appropriate indicators, as described earlier. This partnership has continued through the use of advisory and oversight committees comprised of park superintendants and park resource managers. Through these committees, park managers and other park staff work with I&M network scientists to provide advice and feedback on the relevancy and effectiveness of the I&M network's efforts. Another important step has been to hire regional and network program managers with a solid scientific background but who understand the importance of establishing a strong long-term partnership between scientists and managers.

With good reason, many managers may be wary of new programs that promise much, but are slow to deliver results that make their job any easier. One of the key elements for maintaining support beyond the initial acceptance is to keep expectations realistic. It is important to be clear from the outset what can be delivered, but also what cannot be delivered (Chapter 3). A common expression in the consulting arena reminds us that it is far better to under-promise and over-deliver than the reverse.

Future research and development

Based on initial program reviews of most of the 32 I&M networks, and a review by the National Academy of Public Administration (NAPA 2010), the NPS monitoring program seems to be meeting its goals of providing credible, scientific data and information to park managers and other key audiences. All organizational levels of the agency view the monitoring program as successful. Although it may be several years (especially for slowly changing resources) before parks reap the direct benefits of long-term monitoring, we are already seeing the indirect benefits of having more scientists available who are familiar with a particular park's resources and issues (e.g. involvement in park planning, helping to identify research needs, etc.). The strategic decisions made and the guidance provided in the early years of planning and designing are paying dividends, but in hindsight there are also some things we would have done differently, as well as opportunities for improvement.

Integration of the NPS monitoring effort into planning and decision processes

Although the NPS monitoring program has gained considerable acceptance throughout our organization, we still have some work ahead of us regarding full integration of the program into the planning and decision processes (NAPA 2010). The true measure of whether a monitoring program is successful is if it routinely produces reliable information that is perceived as being essential toward achieving the agency's mission and goals. Being an integral component of the planning at all organizational levels from routine park management to agency policy decisions is the best way to achieve such success. Park managers have provided many examples of how the scientific data and understanding generated by the monitoring program have already been incorporated into park operations and planning documents. However, throughout the agency there are still many challenges and opportunities for integrating monitoring data and information into the planning and decision-making operations of the organization. The NPS leadership recently decided to begin developing a State of the Park Report for each park that will summarize status and trend information for important park resources and services. This decision is an important step towards institutionalizing scientific information into routine park operations.

Effectiveness monitoring versus overall condition monitoring

The NPS monitoring program, by design, provides information on the overall condition of each park and the long-term effectiveness of management regimes based on changes in the status and trend of selected park resources. In contrast, many monitoring programs have an emphasis on monitoring the effectiveness of specific management actions, often with short-term "experiments" in an adaptive management framework (Mulder and Palmer 1999). Because of the NPS mission and its "leave unimpaired for future generations" mandate, specific management actions that are conducive to such shorter-term effectiveness monitoring are more limited in NPS units compared to those of other agencies that more actively manage natural resources to meet specific objectives. Even in the NPS, though, both types of monitoring are needed. Some of the contexts within the National Park Service that are quite amenable to an adaptive management

Box 22.2 Common challenges: we've detected a problem

Long-term condition monitoring is designed to assess current status and trends for important resources and ecological processes, or human-caused stressors to the ecosystem. When a problem is detected, additional research studies or monitoring generally will be needed to determine likely causes of the problem. However, condition monitoring often can lead to relatively rapid and effective detection and mitigation of a problem, even when supplemental data collection is needed to assess causes of the problem. Moreover, although some benefits of a long-term monitoring program may not be realized until it has been operational for a decade or more, usually the program will almost immediately begin providing information highly relevant to managers. The following example demonstrates how water quality monitoring at Zion National Park (Utah, USA) relatively quickly confirmed the occurrence of an ongoing problem and led to management actions to try to reduce the problem.

The Northern Colorado Plateau Inventory and Monitoring network began monitoring the water quality of the North Fork Virgin River upstream from Zion National Park in 2006. Each summer for three years, the monitoring detected elevated *Escherichia coli* levels (see figure below) at a site where the river flows through irrigated pastures grazed by cattle (Van Grinsven *et al*. 2010). The results were reported to park staff. In 2009, hikers entering the Virgin River upstream of the park were issued warnings about these elevated levels. Monitoring was increased to isolate

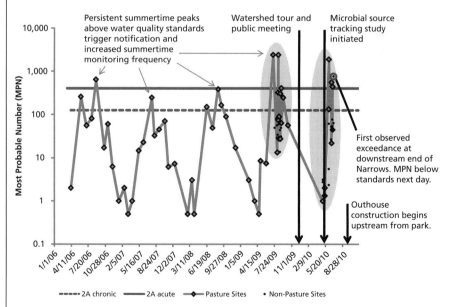

Estimates of *Escherichia coli* population levels (Most Probable Number) for monitoring sites on the North Fork Virgin River, Utah, USA. Horizontal lines indicate water quality standards (exceedence thresholds) for the river.

the source of contamination and help determine if the problem was persistent or intermittent. The increased monitoring included a higher frequency of data collection at the original pasture site and the addition of three non-pasture sites: two sites 1–2 miles upstream, plus a heavily used visitor site at the mouth of the Zion Narrows, 16 miles downstream. Results from 2009 confirmed that the contamination source was near the pasture site, and that elevated *E. coli* levels were persistent during the summer season.

The Utah Division of Water Quality (UDWQ) subsequently hosted a public meeting with staff from NPS, the Bureau of Land Management (BLM), and the county Conservation District. The outcomes of the meeting included additional intensive monitoring and research study to determine if the source of the bacteria was human, bovine, or wildlife. Definitively identifying the source of contamination is complicated by several factors. Cattle graze the pastures adjacent to the stream in summer, but hundreds of visitors also use the stream, and there are second homes and wildlife in the watershed. After the meeting, the UDWQ provided the BLM with funds to build an outhouse at the trailhead upstream from the park, to mitigate potential bacterial inputs from park visitors. Continuing cooperation with the UDWQ, BLM, and Conservation District will help determine the source of bacteria and eventually help arrive at solutions that will involve federal and state agencies, as well as local land owners.

approach, particularly when accompanied by effectiveness monitoring, are fire management, invasive plant management, and visitor management (how many, where, when, etc.).

However, funding that was available for the NPS long-term monitoring program was enough to fund only one professional-level position per park, on average, plus some operating funds. We chose to focus on monitoring of long-term changes in the overall condition of park resources because it is integral to our agency's mission, and we assumed that funding from specific programs and projects would support short-term effectiveness monitoring as necessary. In retrospect, such funding has rarely materialized and this continues to be a real and unmet agency need.

If additional funding were available, the addition of a research complement to long-term monitoring would greatly facilitate our ability to use the information being generated by our monitoring efforts. Our current programmatic direction allows us to examine changes in the status and trends of resources in our parks, but without additional study does not effectively allow us to examine the causes of any changes we observe (Box 22.2). Incorporation of key covariates in our monitoring often leads to correlative results that can further lead to the development of hypotheses about such causes and effects, but to be truly effective, a complementary effort specifically designed for this purpose is needed. As important changes in status and trend of resources are observed, a complementary research effort focused on understanding the causes would greatly enhance our ability to incorporate information about such changes into sound management decisions.

Summary

Managers of natural resources need reliable scientific data and information on the status and trends in the condition of key resources as a basis for conservation planning, determining whether current management practices are having the desired effect, and informing stakeholders and the general public of changes in the condition of park resources that may be caused by stressors operating at regional or global scales. Data collected in a consistent way over long periods are fundamental to conservation and management because they provide the context for interpreting observed changes, and may provide the basis for initiating new management practices or changing existing practices. Despite this need, the track record for institutionalizing long-term programs that provide such information is poor. In this chapter, we have provided examples and "key lessons learned" from our experience in planning and designing a long-term ecological monitoring program for more than 270 parks in the US National Park Service. We emphasize three critical elements for a long-term monitoring program in a resource management agency: relevance, reliability, and commitment. These elements are essential to institutionalizing a monitoring program such that it becomes an integral component in the planning and operations of the agency. The quantitative recommendations presented in this volume are an essential underpinning of the scientific credibility of any monitoring program, but they are just one component of designing, implementing, and institutionalizing a long-term ecological monitoring program.

23 Choosing among long-term ecological monitoring programs and knowing when to stop

Hugh P. Possingham, Richard A. Fuller, and Liana N. Joseph

Introduction

Long-term ecological monitoring is generally considered an essential tool for the effective management of biodiversity (Strayer *et al.* 1986, Lindenmayer and Likens 2010a). There is a large body of scientific literature describing why long-term ecological monitoring is important and describing sophisticated and useful approaches for conducting it. In this chapter, however, we ask two slightly different and somewhat controversial questions. First, how do we identify which long-term ecological monitoring programs are more worthwhile than others? There are limited resources to fund monitoring and choices must be made. Second, how do we know when we should stop monitoring, if ever? So few examples of successful long-term monitoring exist that the idea of stopping monitoring sounds like heresy, but if the benefits are diminishing and the effort could be better used elsewhere, why not? To answer these questions we need to be able to classify and quantify the benefits of long-term monitoring and hence we pull together many of the ideas on this issue from the literature and other chapters in this book.

Monitoring has well-known benefits and is likely to play an increasingly important role as we try to determine how large-scale anthropogenic changes, in the context of a rapidly changing climate, alter the way in which we should manage ecosystems (Balmford *et al.* 2005, Field *et al.* 2007). Effective monitoring is a critical link in the cycle of adaptive management that aims to iteratively improve conservation actions over time (Holling 1978, Walters 1986) and is important for influencing government policy and investment in environmental programs (GAO 2006, OECD 2006).

However, because resources are limited, we have to make choices among long-term monitoring projects when we invest these resources. Despite increasing interest and investment in long-term ecological monitoring, we will never have enough money to do everything. For example the long-term impact of grazing on micro-fungal dynamics in the Australian rangelands is unlikely ever to be known at a broad scale, so some difficult choices about the best locations, intensities, and approaches to monitoring have to be made. Globally, only a small proportion of the species threatened with extinction are monitored (Baillie *et al.* 2004) and, in many cases, the monitoring is insufficient to

Design and Analysis of Long-term Ecological Monitoring Studies, ed. R.A. Gitzen, J.J. Millspaugh, A.B. Cooper, and D.S. Licht. Published by Cambridge University Press. © Cambridge University Press 2012.

detect real changes in population parameters and responses to management intervention (Taylor and Gerrodette 1993, Field *et al.* 2004). Consequently, we need to choose among the many aspects of an ecosystem that we could monitor, and the techniques we use to monitor them. Essentially, we must decide how to allocate resources among alternative possible long-term ecological monitoring programs to maximize conservation outcomes. Furthermore, we need to trade off investment in monitoring with investment in management. We need to be receptive to the idea that, in many cases, despite the potential benefits of monitoring, the best choice could be not to invest in monitoring at all (McDonald-Madden *et al.* 2010).

A second consequence of limited resources is that we might wish to consider whether some monitoring programs should stop. Because there are so few long-term ecological monitoring programs in Australia and most other countries, we generally assume that the ones we have are precious, indeed invaluable – we would never choose to stop them. However, if we slavishly continue every ecological monitoring program that has started, we will end up not being able to support new programs. This is analogous to a situation in an ancient city, such as Rome, where so many historic sites and buildings are protected that the city itself is in danger of stagnation.

To investigate both of these issues we need at least two pieces of information. First, we need to know the current and expected cost of a monitoring program in terms of time and money. This is a messy but intellectually trivial task and we will assume that most organizations can accurately estimate these costs. Second, we need to determine the benefits of a long-term ecological monitoring program. This is an extremely difficult task, because the benefits are uncertain and hard to quantify, in both when they appear and how big they are. To add to this complexity, there are at least five kinds of practical benefit to long-term ecological monitoring to quantify, aside from the benefit of knowledge for knowledge's sake. Here we attempt to define those benefits and discuss how they might be quantified and hence combined with costs to answer our two primary questions (Box 23.1).

The practical benefits of long-term ecological monitoring

To decide when we should be starting and stopping long-term ecological monitoring programs we need to know what benefits we expect to gain, and how these benefits change over time. In short, what is monitoring for? We identify five separate conservation, scientific, and social benefits of long-term monitoring (Nichols and Williams 2006, Salzer and Salafsky 2006, McDonald-Madden *et al.* 2010):

(i) monitoring the state of a system to enable us to choose the best management actions;
(ii) learning how an ecological system functions and responds to management, as part of the adaptive management cycle;
(iii) communicating to the public and policy makers about long-term ecological changes;
(iv) engaging the public in ecological issues thereby leveraging effort and support; and
(v) uncovering unexpected events serendipitously.

Box 23.1 Take-home messages for program managers

Most chapters in this volume have focused on decisions (e.g. how should data be analyzed?) affecting the relevance, efficiency, and quality of information produced by an individual monitoring program. However, government agencies and other organizations that develop and support monitoring programs must also address two larger-scale questions: which new long-term ecological monitoring programs should we invest in, and when should we stop an existing monitoring program? To make these broad-scale choices rationally, we must measure and compare the long-term costs and benefits of potential long-term ecological monitoring programs. Measuring the costs of monitoring involves identifying and budgeting for the people and infrastructure required to implement monitoring, something that is done frequently and competently by many practitioners. Predicting benefits is more difficult. Benefits of long-term ecological monitoring can include:

(i) helping us to decide which management action to implement where the choice of action depends on the state of the ecosystem. For example, the level of a quota or limit to fishing effort will depend on the size and trend of the population being harvested;

(ii) learning about the causes of species declines or understanding how species respond to natural environmental cycles (e.g. weather) or ecological factors (e.g. population density, predators, or infectious disease) and thus informing future management;

(iii) informing policy makers or the public about threats to biodiversity to catalyze change;

(iv) engaging the public in the act of monitoring and thus leveraging further effort and support; and

(v) serendipitously uncovering novel and unpredictable threats to biodiversity.

We suggest that many of these benefits increase during the life of a monitoring program up to a point, and then begin to decline again. Thus, we must think about long time horizons when estimating the costs and benefits of alternative monitoring programs, and critically assess whether some particularly aged monitoring programs should be stopped. Only through developing a clear understanding of the expected benefits and costs of long-term ecological monitoring will the deployment of our limited resources have the greatest possible conservation impact.

Of these possible benefits, it is important to be explicit about which are most relevant before attempting to choose on which long-term monitoring effort to embark. Often we will expect multiple benefits; however, long-term monitoring programs should never begin without explicitly singling out at least one well-defined benefit. We briefly discuss these five benefits of monitoring with a particular emphasis on how the benefits of monitoring might be quantified so they can be compared with other options.

The ultimate gain from long-term monitoring is the sum of the five benefits – a challenging sum to calculate because, as we shall see, the benefits have different currencies.

(i) Monitoring system state to enable us to choose the best management actions

For many populations or ecological systems, the choice of which management action to implement depends on the state of the system (e.g. Johnson and Williams 1999; Chapters 4, 17). The classic example is fisheries, where there is an extensive literature on state-dependent harvesting, such as setting quotas or limiting fishing effort depending on the status of the population being harvested (Clark and Mangel 2000, Dichmont *et al.* 2010). Consequently, the best long-term monitoring data sets in Australia for populations are from managed populations, whether they be wildlife or fish (e.g. Grigg *et al.* 1999). These monitoring programs were driven by the utilitarian needs of prudent management, but have provided many other benefits, including a deeper understanding of the population ecology of the species concerned (e.g. Jonzén *et al.* 2005). Similarly, some of our better long-term monitoring data sets on vegetation state were driven by the needs of fire management (e.g. Keith *et al.* 2007).

State-dependent management demands long-term ecological monitoring, indeed continuous monitoring, so that changes in system state can be acted upon as and when they occur. As issues such as allowing appropriate flows in river systems to provide desired ecological benefits, achieving environmentally sensitive grazing, and managing fire regimes for long ecological outcomes become more important, managers will increasingly demand continuous data on system state to inform the associated management decisions.

While state-dependent management is our most successful driver of long-term ecological monitoring, rather few people have questioned exactly how precisely we need to track system state to inform management choices (but see Chapter 4). To establish how much to spend on monitoring systems where management is state-dependent, we must quantify the net benefit of accurately knowing the system state. Specifically, we need to know what cost we would pay if a poor estimate of ecological state or its current trajectory caused us to make a poor decision (Hauser *et al.* 2006; Chapter 4). For example, capturing individuals of a rare species for captive breeding when they are not in decline will increase the extinction probability of the population and waste money (Tenhumberg *et al.* 2004). On the other hand, if we intervene too late, it will be more expensive, or we will fail to secure what is left of the population. Each type of mistake has costs and benefits that need to be traded off against the investment in monitoring.

(ii) Monitoring to learn about a system: the adaptive management cycle

In many cases, managers are uncertain about the parameters that drive their system. Information that is useful to managers and can be derived from long-term monitoring data includes estimates of population parameters, identification of drivers of ecological change, and evaluation of management actions and data to build models of system dynamics. Learning is essential to the management process, both for understanding the influence of particular management actions or strategies, and for decreasing structural uncertainty about the system so as to make more informed choices among management options (Drechsler and Burgman 2004, Gerber *et al.* 2005, Nichols and Williams 2006; Chapter 4).

The key to the trade-off between learning and managing is knowing when the costs of long-term gains from additional learning are going to be compensated for by better conservation management outcomes (McCarthy and Possingham 2007, Rout *et al.* 2009). The costs and benefits of the investment in learning can be traded off with each other to find the optimal level of monitoring. This is precisely the approach used by Gerber *et al.* (2005), who discovered that the optimal monitoring period is rarely longer than 5 years when monitoring the recovery of an over-fished stock to set the fraction of that stock to protect. So when would learning-based monitoring be long-term?

If the fundamental parameters of an ecological system are constant, then monitoring for learning will rarely be long-term. This said, some fundamental characteristics inherently require a long time to learn, such as the frequency of very rare events. Furthermore, the more we learn about a system, invariably the more we discover we do not know. From the perspective of gaining such new knowledge, monitoring could go on indefinitely, although one might expect new knowledge to be gained in ever-decreasing increments. For most practical problems, sufficient knowledge to address them will usually be gained quite quickly, after which the cost of continued monitoring begins outweighing the expected practical net benefits with regard to decision making.

(iii) Monitoring to inform the public and policy makers

Long-term monitoring can inform the public, and hence decision makers, about long-term changes in our ecological systems (Chapters 3, 22). This is especially important because of the phenomenon known as "shifting baselines" whereby it is hard to perceive important long-term change because observers of a system alter their internal point of reference to reflect recent conditions (Jackson *et al.* 2001).

There are some outstanding examples of long-term monitoring programs that have fundamentally changed the way society thinks about ecological systems and the state of their environment. Long-term monitoring of changes in climate, the ozone layer, and the acidity of rain have led to huge (although not always adequate and timely) changes in policy and management (Likens *et al.* 1996). These issues have demanded long-term monitoring and dedicated, persistent scientists. Without compelling data and analyses, society would not have made the necessary and frequently expensive responses. The benefits of these long-term monitoring programs are unquestionable, and far exceed their costs. However, for some ecological problems we need to take a more precautionary approach even without sufficient data. Sometimes we need to take action in the absence of the kind of hard evidence that cannot be obtained without the expense and time of long-term monitoring (Field *et al.* 2004).

(iv) Monitoring to engage the public in ecological issues thereby leveraging effort and support

Garnering public and political support may result from publicizing the outcomes of a monitoring program (Whitelaw *et al.* 2003), as described above, or through engaging the public in the process of monitoring (Lindenmayer *et al.* 1991, Vaughan *et al.* 2003). Both of these mechanisms may influence management because policy makers often respond to publicity about detrimental environmental trends and/or pressure from the public (Cuthill 1995, Pinho 2000, Carr 2004, McNeil *et al.* 2006).

Monitoring schemes that have the objective of increasing awareness through participation include government-sponsored, frequently volunteer-driven monitoring programs mandated by some level of local, regional, or national government (e.g. community-based monitoring efforts; Vaughan *et al.* 2003, Whitelaw *et al.* 2003, Trewhella *et al.* 2005). They are useful for educating the community about local conservation issues and forming a vehicle for government agencies to provide the public with information about the status of the natural environment. These programs are often motivated by the need to involve stakeholders and citizens in planning and management processes (Cuthill 1995, Bliss *et al.* 2001) to ensure social, cultural, and economic perspectives are reflected in evidence-based decisions about conservation priorities and actions, and to encourage conflict resolution and advance social learning (Fernandez-Gimenez *et al.* 2008). However, the cost-efficiency of these monitoring programs should be considered carefully; community-based monitoring may be an expensive method of raising awareness for a small sector of society that is already committed to an issue. Managers should consider using limited resources to fund more cost-efficient forms of education that may influence more people (e.g. field days, school outings, and environmental campaigns).

Monitoring programs that are tailored to increase awareness may or may not be used to meet one or more of the other monitoring objectives. Conversely, public engagement may not be the overriding priority of many volunteer/citizen-science monitoring programs, which may be focused as much or more on obtaining meaningful data (Chapter 21). In either case, the monitoring programs may face difficult trade-offs between maximizing the scientific quality of the information produced vs. maintaining volunteer involvement and maximizing public engagement (Chapters 3, 21). In fact, these objectives may conflict because more complex or tedious monitoring programs may result in more useful data to detect trends, but may result in less engagement. The designers of community-based/citizen-science monitoring programs need to be clear about the quality and potential limits of the data collected to maintain participant confidence and avoid triggering inappropriate or unnecessary management actions based on data of poor quality or from poorly designed collection programs (Hockley *et al.* 2005, Szabo *et al.* 2007). The designers of education focused programs should be clear about the tangible benefits that are expected to arise from the investment in community education, and question whether these outweigh the benefits of investing in a more scientifically rigorous monitoring scheme or just taking action in the absence of community support. The availability of a cheap source of labor must not by itself be seen as a reason to invest resources into developing and managing a monitoring program.

(v) Monitoring to uncover novel and unpredictable patterns

Some authors are critical of untargeted monitoring (Anderson 2001, Nichols and Williams 2006, Lindenmayer and Likens 2010a; Chapters 4, 18) – monitoring that is designed to uncover the "unknown unknowns". We agree that it is hard to determine how much effort we should allocate to this sort of monitoring, especially because the chance and value of discovering something unexpected is hard to quantify. We know of no framework for optimal monitoring to detect unexpected events (Wintle *et al.* 2010). However, it is true that serendipitous discoveries, such as the impact of pesticides on North American birds (Carson 1962), have led to far-reaching conservation benefits by driving policy reform and increasing public interest (see also Chapter 3).

We perceive two kinds of serendipity in long-term ecological monitoring. First, there are the serendipitous gains simply by having people in the environment and looking, which is precisely why Lindenmayer and Likens (2010a) argue that all investigators should leave their office occasionally and actually count something. Second, there are the unexpected discoveries that emerge simply from the length of a data set – events we could not have predicted but that accumulate as a data set matures. It is conceivable that a large-scale post hoc evaluation of existing long-term monitoring programs could be used to evaluate the probabilistic benefits of serendipitous monitoring. For example, one could empirically evaluate what has been discovered accidentally and at what cost, as well as determine the action that was catalyzed by those discoveries. This sort of analysis is needed so we can determine what fraction of long-term monitoring programs should focus on serendipity vs. targeted utility (Wintle *et al.* 2010)

Given these five disparate benefits of monitoring, it is clear that a monitoring strategy well tailored to one type of benefit will not necessarily be optimal for delivering the other benefits. For example, a long-term ecological monitoring program designed to educate the public and politicians might focus on systems or species that people particularly care about (e.g. charismatic mammals or birds), and the target of monitoring may be chosen simply to encourage participation. In contrast, the long-term monitoring strategy for managing habitat may focus on species or habitat attributes that are less charismatic, but a more reliable and cost-effective way of learning and tracking system state. While a program that is optimally designed for every benefit is probably impossible, the total benefit across all the purposes of monitoring is what we need to determine for examining cost : benefit trade-offs.

Making the choices

Having outlined all the factors that we need to think about in terms of the costs and benefits of monitoring, let us speculate how such information could be used to make a decision about how much to monitor, what to monitor, and whether or when we should stop. For the simple cases of optimal harvesting (Clark and Mangel 2000) and optimal learning (Gerber *et al.* 2005, Grantham *et al.* 2009), such analyses are relatively well advanced, and we have briefly discussed them above. We will concentrate here on the more complex and hard to quantify benefits of raising awareness, engaging the public, and making serendipitous discoveries.

We speculate that the initial ability of monitoring to identify patterns of concern and inform policy or engage the public will often be low. Two or three years of data are not enough to compel even the most naïve member of the public, and certainly statistically inadequate to detect change in response to managed or unmanaged processes. Conversely, the public's enthusiasm for all things new may lead to strong engagement with volunteers. This early phase is also where most set-up costs are incurred: pilot studies must be completed and analyzed; power analysis is carried out; infrastructure and expertise are assembled. Over 5–15 years many of the most important discoveries and trends will have been uncovered. These discoveries will come from testing the original hypotheses set by the program designers, the main drivers for the study in the first place. After that, the rate of acquiring new information about long-term change will decline, although if the some of the important environmental events are rare and/or have long cycles, valuable information could continue to emerge for decades. By now costs will have stabilized and public engagement may rise again as people see the power of a long-term data set.

But how do serendipitous discoveries change the picture we have painted? We argue that there are two kinds of serendipitous discovery. First, discovery may occur simply because we are observing a system or a place that has not previously been monitored. This potential will be greatest early in the program and depend on how well-studied the system was beforehand. Then there are the serendipitous discoveries resulting from unexpected events, and many of those will only emerge after a long period of time, or when the data are amalgamated with other information in unforeseen ways. Combining these ideas together, we postulate that the cost and benefit curves of long-term ecological monitoring will generally have a shape something like those in Figure 23.1, although the absolute magnitude of each kind of cost and benefit will vary.

If we apply a short time-horizon to a cost-effectiveness analysis of monitoring, the analysis normally would not favor investing in monitoring. The set up costs are high and the short-term rewards are low. This could explain why so few scientists or agencies embark on long-term monitoring. However, the appropriate analysis is to measure the cost effectiveness of a program integrated over time. This tells us not only how efficient a proposed program is expected to be, it also gives us information about when to stop a long-term monitoring program.

Theory guiding when we should move resources from one long-term monitoring program to another is similar to the theory of optimal patch use (Charnov 1976). Switching occurs when the return on investment of the current program is outweighed by expected long-term return on investments from other programs. To understand this we need to compare the expected cost-effectiveness curves of multiple programs. Figure 23.1(c, d) shows the expected benefit and integrated cost-effectiveness of a long-term monitoring program that matures more slowly and has higher costs. We would expect to keep such a program going for much longer than the program with lower set-up costs shown in Figure 23.1(a, b), indeed we may try to keep it going for as long as possible.

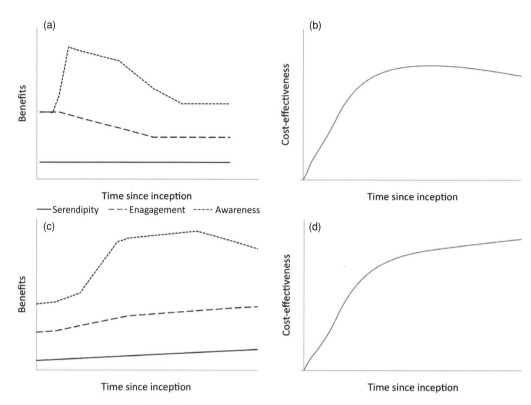

Figure 23.1 The benefits of a long-term monitoring program must be expressed over a long time horizon. Here, we show how three hypothetical benefits as well as overall cost-effectiveness of monitoring might be expected to change over time where (a) the initial set-up costs for the program are low and benefits decline after peaking, and (c) the initial set-up costs for the program are high and the benefits remain high. Cost-effectiveness peaks and then declines in what we consider to be the common situation in monitoring programs, i.e. where set-up costs are low and benefits peak early (b). However, where benefits continue to accumulate over time, cost-effectiveness is also likely to keep increasing (d).

Future research and development

How useful is this new conceptual framework? As a tool for thinking about any long-term ecological monitoring program, we think that an a priori and quantitative assessment of cost and benefits should be essential. We have never seen this attempted and it is undoubtedly difficult. However, it could be informed by some careful analysis of previous and existing programs. Such analyses are at least a rational attempt to make an informed decision and will be invaluable for choosing among new long-term monitoring programs, and eventually to determine if a given monitoring program should end (Box 23.2) – the two major questions of choosing and stopping that we posed at the beginning of this chapter.

Box 23.2 Common challenges: when to give up

Only a very small number of long-term monitoring programs have become institutionalized or endowed in perpetuity. It is usually individuals that champion most long-term monitoring programs, and people have a habit of giving up or retiring. Therefore, many monitoring programs stop by themselves, making redundant a rational decision-making process to govern this. There is a known and consistent failure rate of long-term ecological monitoring that is driven by fickle funding and the active lifespan of individual scientists. It is rare for a field ecologist to be able to commit to ecological monitoring for more than 30 years, and even rarer they can find funding to sustain monitoring over such a period. Hence, while the question of when to stop a long-term ecological monitoring program is interesting, in practice the question has only rarely arisen. However, the question will arise more frequently as monitoring programs strive to become institutionalized and remain an organizational priority as the original developers leave the program (e.g. Chapter 22). Moreover, determining which long-term project to fund among a range of choices will always be a crucial and important decision. Rather than passively letting the availability, persistence, and fate of individual scientists continue to drive the fate of many monitoring programs, we hope that organizations increasingly will base these decisions on objective cost: benefit analyses in consideration of overall organizational long-term priorities.

Some of the greatest challenges ahead will be in specifying and quantifying the expected benefits of a long-term ecological monitoring program and then integrating them into a total benefit. In this chapter, we have reviewed some of the types of benefits emerging from long-term monitoring, and we have provided examples of attempts to quantify some of these benefits, particularly for state-based decision making and learning. There is, however, an urgent need for robust techniques for predicting some of the less well understood benefits of monitoring to biodiversity of, for example, engaging the public to support and conduct conservation, informing policy makers, or uncovering novel or unpredictable threats. The last of these benefits of monitoring, serendipity, will be the most difficult to quantify, although we do not expect it to prove impossible (Wintle *et al.* 2010). Finally, there is the task of combining these five disparate benefits for any intended program.

Summary

Long-term ecological monitoring will play an increasingly important role as the threats to biodiversity strengthen and as the importance to human well-being of maintaining ecosystem services intensifies. However, we are and will continue to be faced with tough questions about which long-term ecological monitoring strategies are essential to invest in and which ones are less important to maintain: the questions of how to choose among monitoring programs, and when to stop them. To ensure that we invest in the best long-term ecological monitoring programs, it is imperative that we compare rationally the long-term costs and benefits of the options. To do this, we need to calculate the costs and predict and sum all the different potential benefits including: the ability to make state-dependent decisions, learn about a system, inform the public or policy makers, engage the public, and detect novel and unpredictable events. Crucially, all such cost-effectiveness analyses must be integrated over long time periods. Although research on quantifying these benefits is nascent, we are beginning to see approaches and case studies illustrating how to do this. As the fields of optimal monitoring and adaptive management continue to progress, it will become possible to operationalize the framework that we have presented here to identify which long-term monitoring programs will deliver the best biodiversity outcomes for an acceptable cost.

Acknowledgments

This research was largely funded by a Commonwealth Environmental Research Facility grant from the Department of Environment Water, Heritage and the Arts and Australian Research Council grants. We are grateful to Josh Millspaugh for inviting us to contribute to this volume.

References

Abelson, R. P. 1995. *Statistics as principled argument*. Third edition. Lawrence Erlbaum Associates, Hillsdale, NJ, USA.

Abhat, D. 2009. Don't hold that thought: the benefits – and challenges – of efficient data sharing. *Wildlife Professional* 3(3):24–28.

Adamowski, K., A. Prokoph, and J. Adamowski. 2009. Development of a new method of wavelet aided trend detection and estimation. *Hydrologica Processes* 23:2686–2696.

Adams, D. 1979. *The hitchhiker's guide to the galaxy*. Pan Macmillan, London, UK.

Afshinpour, B., G.-A. Houssein-Zadeh, and H. Soltanian-Zadeh. 2008. Nonparametric trend estimation in the presence of fractal noise: application of fMRI time-series analysis. *Journal of Neuroscience Methods* 171:340–348.

Afton, A. D., and M. G. Anderson. 2001. Declining scaup populations: a retrospective analysis of long-term population and harvest survey data. *Journal of Wildlife Management* 65:781–796.

Agresti, A. 1990. *Categorical data analysis*. Wiley, New York, NY, USA.

Airy, G. B. 1861. *On the algebraical and numerical theory of errors of observations and the combination of observations*. Macmillan and Co., Cambridge and London, UK.

Akaike, H. 1973. Information theory as an extension of the maximum likelihood principle. Pages 267–281 *in* B. Petrov, and F. Csaki, editors. *Second International Symposium on Information Theory*. Akademiai Kiado, Budapest, Hungary.

Allison, P. D. 2001. *Missing data*. Sage Publications, Thousand Oaks, CA, USA.

Amstrup, S. C., T. L. MacDonald, and B. F. J. Manly, editors. 2005. *Handbook of capture–recapture analysis*. Princeton University Press, Princeton, NJ, USA.

Andersen, T., J. Carstensen, E. Hernández-García, *et al.* 2008. Ecological thresholds and regime shifts: approaches to identification. *Trends in Ecology & Evolution* 24:49–57.

Anderson, D. R. 2001. The need to get the basics right in wildlife field studies. *Wildlife Society Bulletin* 29:1294–1297.

Anderson, D. R. 2003. Response to Engeman: index values rarely constitute reliable information. *Wildlife Society Bulletin* 31:288–291.

Anderson, D. R. 2008. *Model based inference in the life sciences: a primer on evidence*. Springer, New York, NY, USA.

Anteau, M. J., and A. D. Afton. 2004. Nutrient reserves of Lesser Scaup (*Aythya affinis*) during spring migration in the Mississippi Flyway: a test of the spring condition hypothesis. *Auk* 121:917–929.

Anteau, M. J., A. D. Afton, C. M. Custer, *et al.* 2007. Relationships of cadmium, mercury, and selenium with nutrient reserves of female lesser scaup (*Aythya affinis*) during winter and spring migration. *Environmental Toxicology and Chemistry* 26:515–520.

Araújo, M. B., R. G. Pearson, W. Thuiller, *et al.* 2005. Validation of species–climate impact models under climate change. *Global Change Biology* 11:1504–1513.

Arbuckle, J. L. 1996. Full information estimation in the presence of incomplete data. Pages 243–278 *in* G. A. Marcoulides and R. E. Schumacker, editors. *Advanced structural equation modeling*. Lawrence Erlbaum Associates, Hillsdale, NJ, USA.

Arnason, A. N. 1973. The estimation of population size, migration rates and survival in a stratified population. *Research in Population Ecology* 15:1–8.

Austin, J. E., M. J. Anteau, J. S. Barclay, *et al.* 2006. *Declining scaup populations: reassessment of the issues, hypotheses and research directions. Consensus report from the Second Scaup Workshop*. 17–19 January 2006, Bismarck, ND. US Department of the Interior, US Geological Survey, Northern Prairie Wildlife Research Center, Jamestown, North Dakota, USA.

Austin, J. E., C. M. Custer, and A. D. Afton. 1998. Lesser scaup (*Aythya affinis*). *In* A. Poole and F. Gill, editors. *The birds of North America*, No. 338. The Academy of Natural Sciences, Philadelphia, Pennsylvania, and The American Ornithologists' Union, Washington, DC, USA.

Azuma, D. L., J. A. Baldwin, and B. R. Noon. 1990. Estimating the occupancy of spotted owl habitat areas by sampling and adjusting for bias. *General Technical Report PSW-124*. USDA Forest Service, Berkeley, California, USA.

Bailey, L. L., J. E. Hines, J. D. Nichols, *et al.* 2007. Sampling design trade-offs in occupancy studies with imperfect detection: examples and software. *Ecological Applications* 17:281–290.

Baillie, J. E. M., C. Hilton-Taylor, and S. N. Stuart. 2004. *2004 IUCN Red List of threatened species: a global species assessment*. IUCN, Gland, Switzerland and Cambridge, UK.

Balmford, A., P. Crane, A. P. Dobson, *et al.* 2005. The 2010 challenge: data availability, information needs, and extraterrestrial insights. *Philosophical Transactions of the Royal Society of London B* 360:221–228.

Balmford, A., R. E. Green, and M. Jenkins. 2003. Measuring the changing state of nature. *Trends in Ecology & Evolution* 18:326–330.

Banerjee, M., and I. W. McKeague. 2007. Confidence sets for split points in decision trees. *Annals of Statistics* 35:543–574.

Banerjee, S., B. P. Carlin, and A. E. Gelfand. 2004. *Hierarchical modeling and analysis for spatial data*. Chapman & Hall/CRC, Boca Raton, FL, USA.

Barber, M. C. 1994. *Environmental monitoring and assessment program indicator development strategy*. Report EPA/620/R-94/022. US Environmental Protection Agency, Office of Research and Development, Environmental Research Laboratory, Athens, GA, USA.

Barbraud, C., J. D. Nichols, J. E. Hines, *et al.* 2003. Estimating rates of local extinction and colonization in colonial species and an extension to the metapopulation and community levels. *Oikos* 101:113–126.

Barker, R. J. 1997. Joint modeling of live-recapture, tag-resight, and tag-recovery data. *Biometrics* 53:666–677.

Barker, R. J. 1999. Joint analysis of mark–recapture, resighting and ring-recovery data with age-dependence and marking-effect. *Bird Study* 46(Supplement):82–91.

Bart, J., S. Brown, B. Harrington, *et al.* 2007. Population trends of North American shorebirds: population declines or shifting distributions? *Journal of Avian Biology* 38:73–82.

Bart, J., B. Collins, and R. I. G. Morrison. 2003. Estimating population trends with a linear model. *Condor* 105:367–372.

Bart, J., B. Collins, and R. I. G. Morrison. 2004. Estimating trends with a linear model: reply to Sauer *et al*. *Condor* 106:440–443.

Bart, J., M. A. Fligner, and W. I. Notz. 1998. *Sampling and statistical methods for behavioral ecologists*. Cambridge University Press, Cambridge, UK.

Bart, J., M. Hofschen, and B. G. Peterjohn. 1995. Reliability of the Breeding Bird Survey: effects of restricting surveys to roads. *Auk* 112:758–761.

Bartmann, R. M., G. C. White, L. H. Carpenter, *et al.* 1987. Aerial mark–recapture estimates of confined mule deer in pinyon–juniper woodlands. *Journal of Wildlife Management* 51:41–46.

Bas, Y., V. Devictor, J.-P. Moussus, *et al.* 2008. Accounting for weather and time-of-day parameters when analysing count data from monitoring programs. *Biodiversity and Conservation* 17:3403–3416.

Beale, C. M., J. J. Lennon, J. M. Yearsley, *et al.* 2010. Regression analysis of spatial data. *Ecology Letters* 13:246–264.

Bechtold, W. A., and P. L. Patterson, editors. 2005. The enhanced Forest Inventory and Analysis program – national sampling design and estimation procedures. *General Technical Report SRS-80*. USDA Forest Service, Asheville, NC, USA.

Beck, J. L., D. C. Dauwalter, K. G. Gerow, *et al.* 2009. Design to monitor trend in abundance and presence of American beaver (*Castor canadensis*) at the national forest scale. *Environmental Monitoring and Assessment* 164:463–479.

Bellehumeur, C., and P. Legendre. 1998. Multiscale sources of variation in ecological variables: modelling spatial dispersion, elaborating sampling designs. *Landscape Ecology* 13:15–25.

Bellhouse, D. R. 1988. Systematic sampling. Pages 125–146 *in* P. R. Krisnaiah and C. R. Rao, editors. *Handbook of statistics*, volume 6. North-Holland, Amsterdam.

Bellman, R. 1957. *Dynamic programming*. Princeton University Press, Princeton, NJ, USA.

Bellman, R., and R. Roth. 1969. Curve fitting by segmented straight lines. *Journal of the American Statistical Association* 64:1079–1084.

Benning, D. S., and D. H. Johnson. 1987. Recent improvements to sandhill crane surveys in Nebraska's central Platte valley. Pages 10–16 *in* J. C. Lewis, editor. *Proceedings of the 1985 Crane Workshop*. Platte River Whooping Crane Habitat Maintenance Trust and US Fish and Wildlife Service, 26–28 March 1985, Grand Island, Nebraska, USA.

Berliner, L. M. 1996. Hierarchical Bayesian time-series models. Pages 15–22 *in* K. Hanson and R. Silver, editors. *Maximum entropy and Bayesian methods*. Kluwer Academic Publishers, Dordrecht, The Netherlands.

Besag, J. 1974. Spatial interaction and the statistical analysis of lattice systems (with discussion). *Journal of the Royal Statistical Society, Series B* 36:192–225.

Betts, M. G., D. Mitchell, A. W. Diamond, *et al.* 2007. Uneven rates of landscape change as a source of bias in roadside wildlife surveys. *Journal of Wildlife Management* 71:2266–2273.

Beven, K. J., and M. J. Kirkby. 1979. A physically based, variable contributing area model of basin hydrology. *Hydrological Sciences Bulletin* 24:43–69.

Bickford, C. A., C. E. Mayer, and K. D. Ware. 1963. An efficient sampling design for forest inventory: the Northeast Forest Resurvey. *Journal of Forestry* 61:826–833.

Bjornstad, O. N. 2009. Package 'ncf': spatial nonparametric covariance functions. Online: <http://cran.r-project.org/web/packages/ncf/ncf.pdf>. Accessed 25 May 2011.

Bliss, J., G. Aplet, C. Hartzell, *et al.* 2001. Community-based ecosystem monitoring. *Journal of Sustainable Forestry* 12:143–167.

Bolker, B. M. 2008. *Ecological models and data in R*. Princeton University Press, Princeton, NJ, USA.

Bolker, B. M. 2009. Learning hierarchical models: advice for the rest of us (forum response). *Ecological Applications* 19:588–592.

Bolker, B. M., M. E. Brooks, C. J. Clark, *et al.* 2009. Generalized linear mixed models: a practical guide for ecology and evolution. *Trends in Ecology & Evolution* 24:127–135.

Bollen, K. A. 1989. *Structural equations with latent variables*. John Wiley & Sons, New York, NY, USA.

Bollen, K. A. 2002. Latent variables in psychology and the social sciences. *Annual Review of Psychology* 53:605–634.

Bollen, K. A., and P. J. Curran. 2004. Autoregressive latent trajectory (ALT) models: a synthesis of two traditions. *Sociological Methods and Research* 32:336–383.

Bollen, K. A., and P. J. Curran. 2006. *Latent curve models: a structural equation perspective*. John Wiley & Sons, New York, NY, USA.

Boomer, G. S., and F. A. Johnson. 2007. A proposed assessment and decision-making framework to inform scaup harvest management. Unpublished report. Available online: <http://www.fws.gov/migratorybirds/NewReportsPublications/SpecialTopics/BySpecies/ SCAUP2007Report.pdf>. Accessed 15 May 2011.

Borchers, D. L., and M. G. Efford. 2008. Spatially explicit maximum likelihood methods for capture–recapture studies. *Biometrics* 64:377–385.

Borchers, D. L., S. T. Buckland, and W. Zucchini. 2002. *Estimating animal abundance: closed populations*. Springer, London, UK.

Both, C., C. A. M. Van Turnhout, R. G. Bijlsma, *et al.* 2010. Avian population consequences of climate change are most severe for long-distance migrants in seasonal habitats. *Proceedings of the Royal Society B* 7:1259–1266.

Boulinier, T., J. D. Nichols, J. E. Hines, *et al.* 1998. Higher temporal variability of forest breeding bird communities in fragmented landscapes. *Proceedings of the National Academy of Sciences of the United States of America* 95:7497–7501.

Bowden, D. C., G. C. White, A. B. Franklin, *et al.* 2003. Estimating population size with correlated sampling unit estimates. *Journal of Wildlife Management* 67:1–10.

Box, G. E. P. 1957. Evolutionary operation: a method for increasing industrial productivity. *Journal of the Royal Statistical Society, Series C* 6:81–101.

Boyce, M. S., P. R. Vernier, S. E. Nielsen, *et al.* 2002. Evaluating resource selection functions. *Ecological Modelling* 157:281–300.

Brace, R. K., R. S. Pospahala, and R. L. Jessen. 1987. Background and objectives on stabilized duck hunting regulations: Canadian and U.S. perspectives. *Transactions of the North American Wildlife and Natural Resources Conference* 52:177–185.

Bradbury, A. 2000. Stock assessment and management of red sea urchins (*Strongylocentrotus franciscanus*) in Washington. *Journal of Shellfish Research* 19:618–619.

Bradley, N. L., A. C. Leopold, J. Ross, *et al.* 1999. Phenological changes reflect climate change in Wisconsin. *Proceedings of the National Academy of Sciences of the United States of America* 96:9701–9704.

Breidt, F. J. 1995. Markov chain designs for one-per-stratum sampling. *Survey Methodology* 21:63–70.

Breidt, F. J., and W. A. Fuller. 1999. Design of supplemented panel surveys with application to the National Resources Inventory. *Journal of Agricultural, Biological, and Environmental Statistics* 4:391–403.

Brenden, T. O., L. Wang, and Z. Su. 2008. Quantitative identification of disturbance thresholds in support of aquatic resource management. *Environmental Management* 42:821–832.

Britten, M. E., W. Schweiger, B. Flakes, *et al.* 2007. *Rocky Mountain Network Vital Signs Monitoring Plan*. Natural Resource Report NPS/ROMN/NRR-2007/010. US Department of the Interior, National Park Service, Fort Collins, CO, USA.

Brockwell, P. J., and R. A. Davis. 1996. *Introduction to time series and forecasting*. Springer-Verlag, New York, NY, USA.

Broms, K., J. R. Skalski, J. J. Millspaugh, *et al.* 2010. Using population reconstruction to estimate demographic trends in small game populations. *Journal of Wildlife Management* 74:410–417.

Brook, D. 1964. On the distinction between the conditional probability and joint probability approaches in the specification of nearest neighbor systems. *Biometrika* 51:481–483.

Brower, A. V. Z. 2006. Problems with DNA barcodes for species delimitation: "ten species" of *Astraptes fulgerator* reassessed (Lepidoptera: Hesperiidae). *Systematics and Biodiversity* 4:127–132.

Brown, J. A. 2003. Designing an efficient adaptive cluster sample. *Environmental and Ecological Statistics* 10:43–60.

Brown, J. A., M. Salehi, M. Moradi, *et al.* 2008. An adaptive two-stage sequential design for sampling rare and clustered populations. *Population Ecology* 50:239–245.

Browne, W. J., and D. Draper. 2000. Implementation and performance issues in the Bayesian and likelihood fitting of multilevel models. *Computational Statistics* 15:391–420.

Browne, W. J., and D. Draper. 2006. A comparison of Bayesian and likelihood-based methods for fitting multilevel models. *Bayesian Analysis* 1:673–514.

Browne, W. J., M. Golalizadeh, and R. M. A. Parker. 2009. A guide to sample size calculations for random effect models via simulation and the MLPowSim software package. ISBN: 0–903024-96–9. University of Bristol, UK. Online: http://seis.bris.ac.uk/~frwjb/esrc/MLPOWSIMmanual.pdf>. Accessed 19 Aug 2010.

Browne, W. J., R. H. McCleery, B. C. Sheldon, *et al.* 2007. Using cross-classified multivariate mixed response models with application to life history traits in great tits (*Parus major*). *Statistical Modelling* 7:217–238.

Browne, W. J., S. V. Subramanian, K. Jones, *et al.* 2005. Variance partitioning in multilevel models that exhibit overdispersion. *Journal of the Royal Statistical Society, Series A* 168:599–614.

Brownie, C., D. R. Anderson, K. P. Burnham, *et al.* 1985. *Statistical inference from band recovery data – a handbook*. Second edition. US Fish and Wildlife Service Resource Publication 156. US Department of the Interior, Washington, DC, USA.

Brownie, C., J. E. Hines, J. D. Nichols, *et al.* 1993. Capture–recapture studies for multiple strata including non-Markovian transitions. *Biometrics* 49:1173–1187.

Brus, D. J., and J. J. de Gruijter. 1997. Random sampling or geostatistical modeling? Choosing between design-based and model-based sampling strategies for soil. *Geoderma* 80:1–59.

Buckland, S. T. 2006. Point-transect surveys for songbirds: robust methodologies. *Auk* 123:345–357.

Buckland, S. T., D. R. Anderson, K. P. Burnham, *et al.* 1993. *Distance sampling: Estimating abundance of biological populations*. Chapman and Hall, London, UK.

Buckland, S. T., D. R. Anderson, K. P. Burnham, *et al.* 2001. *Introduction to distance sampling: estimating abundance of biological populations*. Oxford University Press, Oxford, UK.

Buckland, S. T., D. R. Anderson, K. P. Burnham, *et al.*, editors. 2004. *Advanced distance sampling*. Oxford University Press, Oxford, UK.

Buckland, S. T., K. B. Newman, L. Thomas, *et al.* 2004. State–space models for the dynamics of wild animal populations. *Ecological Modeling* 171:157–175.

Bühlmann, P., and B. Yu. 2002. Analyzing bagging. *Annals of Statistics* 30:927–961.

Bunge, J., and M. Fitzpatrick. 1993. Estimating the number of species: a review. *Journal of the American Statistical Association* 88:364–373.

Burgman, M. 2005. *Risks and decisions for conservation and environmental management.* Cambridge University Press, Cambridge, UK.

Burman, P., and R. H. Shumway. 2009. Estimation of trend in state–space models: asymptotic mean square error and rate of convergence. *Annals of Statistics* 37:3715–3742.

Burnham, K. P. 1981. Summarizing remarks: environmental influences. Pages 324–325 *in* C. J. Ralph and J. M. Scott, editors. *Estimating numbers of terrestrial birds.* Studies in Avian Biology 6. California Natural Resources Federation, Cooper Ornithological Society.

Burnham, K. P. 1993. A theory for combined analysis of ring recovery and recapture data. Pages 199–213 *in* J.-D. Lebreton and P. M. North, editors. *Marked individuals in the study of bird population.* Birkhauser Verlag, Basel, Switzerland.

Burnham, K. P., and D. R. Anderson. 2002. *Model selection and multimodel inference: a practical information-theoretic approach.* Second edition. Springer-Verlag, New York, NY, USA.

Burnham, K. P., and W. S. Overton. 1979. Robust estimation of population size when capture probabilities vary among animals. *Ecology* 60:927–936.

Busch, D. E., and J. C. Trexler. 2003. The importance of monitoring in regional ecosystem initiatives. Pages 1–2 *in* D. E. Busch and J. C. Trexler, editors. *Monitoring ecosystems: interdisciplinary approaches for evaluating ecoregional initiatives.* Island Press, Washington, DC, USA.

Butcher, G. S., and C. E. McCulloch. 1990. Influence of observer effort on the number of individual birds recorded on Christmas Bird Counts. Pages 120–129 *in* J. R. Sauer and S. Droege, editors. *Survey designs and statistical methods for the estimation of avian population trends.* Biological Report 90(1). US Department of the Interior, Fish and Wildlife Service, Washington, DC, USA.

Calabrese, J. M., and W. F. Fagan. 2004. A comparison-shopper's guide to connectivity metrics. *Frontiers in Ecology and the Environment* 2:529–536.

Callens, M., and C. Croux. 2005. Performance of likelihood-based estimation methods for multilevel binary regression models. *Journal of Statistical Computation and Simulation* 75:1003–1017.

Cam, E., J. D. Nichols, J. R. Sauer, *et al.* 2000. Relative species richness and community completeness: avian communities and urbanization in the mid-Atlantic states. *Ecological Applications* 10:1196–1210.

Cameron, A. C., and P. K. Trivedi. 1998. *Regression analysis of count data.* Cambridge University Press, Cambridge, UK.

Canadian Wildlife Service and United States Fish and Wildlife Service. 1987. Standard operation procedures (SOP) for aerial waterfowl breeding ground population and habitat surveys in North America; revised. Unpublished report.

Carle, A. C. 2009. Fitting multilevel models in complex survey data with design weights: Recommendations. *BMC Medical Research Methodology* 9:49. doi:10.1186/1471–2288-9–49.

Carlin, B. P., and T. A. Louis. 2008. *Bayes and empirical Bayes methods for data analysis.* Third edition. Chapman & Hall/CRC Press, Boca Raton, FL, USA.

Carothers, A. D. 1973. The effects of unequal catchability on Jolly–Seber estimates. *Biometrics* 29:79–100.

Carpenter, S. R. 1998. The need for large-scale experiments to assess and predict the response of ecosystems to perturbation. Pages 287–312 *in* M. L. Pace and P. M. Groffman, editors. *Successes, limitations and frontiers in ecosystem science.* Springer, New York, NY, USA.

Carr, A. J. L. 2004. Why do we all need community science? *Society & Natural Resources* 17:841–849.

Carson, R. 1962. *Silent spring.* Houghton Mifflin, Boston, MA, USA.

Casella, G., and R. L. Berger. 1990. *Statistical inference*. Duxbury Press, Belmont, CA, USA.

Chaloner, K., and I. Verdinelli. 1995. Bayesian experimental design: a review. *Statistical Science* 10:273–304.

Chamberlin, T. C. 1890. The method of multiple working hypotheses. *Science (old series)* 15:92–96.

Chance, B. L. 2002. Components of statistical thinking and implications for instruction and assessment. *Journal of Statistics Education* 10(3). Online: <www.amstat.org/publications/jse/v10n3/chance.html>.

Chandler, R. E., and E. M. Scott. 2011. *Statistical methods for trend detection and analysis in the environmental sciences*. Wiley, Chichester, UK.

Chang, Y., J. I. Miller, and J. Y. Park. 2009. Extracting a common stochastic trend: theory with some applications. *Journal of Econometrics* 150:231–247.

Chao, A., and S.-M. Lee. 1992. Estimating the number of classes via sample coverage. *Journal of the American Statistical Association* 87:210–217.

Charnov, E. L. 1976. Optimal foraging: the marginal value theorem. *Theoretical Population Biology* 9:129–136.

Chatfield, C. 1989. *The analysis of time series: an introduction*. Chapman and Hall, New York, NY, USA.

Chen, G., M. Kéry, J. Zhang, *et al.* 2009. Factors affecting detection probability in plant distribution studies. *Journal of Ecology* 97:1383–1389.

Chiu, G., R. Lockhart, and R. Routledge. 2006. Bent-cable regression theory and applications. *Journal of the American Statistical Association* 101:542–553.

Christman, M. 2004. Sequential sampling for rare and geographically clustered populations. Pages 134–145 *in* W. L. Thompson, editor. *Sampling rare or elusive species: concepts, designs, and techniques for estimating population parameters*. Island Press, Washington, DC, USA.

Clark, C. M., E. E. Cleland, J. E. Fargione, *et al.* 2007. Environmental and plant community determinants of species loss following nitrogen enrichment. *Ecology Letters* 10:596–607.

Clark, C. W., and M. Mangel. 2000. Conservation biology. Pages 173–191 *in* W. C. Clark and M. Mangel, editors. *Dynamic state variable models in ecology: methods and applications*. Oxford University Press, Oxford, UK.

Clark, J. S. 2007. *Models for ecological data: an introduction*. Princeton University Press, Princeton, NJ, USA.

Clark, J. S., and O. N. Bjørnstad. 2004. Population time series: process variability, observation errors, missing values, lags, and hidden states. *Ecology* 85:3140–3150.

Cochran, W. G. 1937. A catalogue of uniformity trial data. *Supplement to the Journal of the Royal Statistical Society* 4:233–253.

Cochran, W. G. 1977. *Sampling techniques*. Third edition. John Wiley & Sons, New York, NY, USA.

Coggins, S., N. Coops, and M. A. Wulder. 2010. Improvement of low level bark beetle damage estimates with adaptive cluster sampling. *Silva Fennica* 44:289–301.

Cohen, J. 1988. *Statistical power analysis for the behavioral sciences*. Second edition. Lawrence Erlbaum Associates, Hillsdale, NJ, USA.

Collins, J. P., A. Kinzig, N. Grimm, *et al.* 2000. A new urban ecology. *American Scientist* 88:416–425.

Comiskey, J., J. P. Schmit, S. Sanders, *et al.* 2009. Forest vegetation monitoring in eastern national parks. *ParkScience* 26:76–80.

Connelly, J. W., and M. A. Schroeder. 2007. Historical and current approaches to monitoring greater sage-grouse. Pages 3–9 *in* K. P. Reese and R. T. Bowyer, editors. *Monitoring populations of sage-grouse*. Station Bulletin 88. Idaho Forest, Wildlife and Range Experiment Station, College of Natural Resources, University of Idaho, Moscow, ID, USA.

Conover, W. J. 1980. *Practical nonparametric statistics*. Second edition. John Wiley & Sons, New York, NY, USA.

Conroy, M. J., R. J. Barker, P. W. Dillingham, *et al.* 2008a. Application of decision theory to conservation management: recovery of Hector's dolphin. *Wildlife Research* 35:93–102.

Conroy, M. J., J. P. Runge, R. J. Barker, *et al.* 2008b. Efficient estimation of abundance for patchily distributed populations via two-phase, adaptive sampling. *Ecology* 89:3362–3370.

Converse, S. J., B. G. Dickson, G. C. White, *et al.* 2004. Estimating small mammal abundance of fuels treatment units in southwestern ponderosa pine forests. Pages 113–120 *in* C. van Riper, III, and K. L. Cole, editors. *The Colorado Plateau: cultural, biological, and physical research*. University of Arizona Press, Tucson, AZ, USA.

Converse, S. J., G. C. White, and W. M. Block. 2006. Small mammal responses to thinning and wildfire in ponderosa pine-dominated forests of the southwestern United States. *Journal of Wildlife Management* 70:1711–1722.

Cooch, E. G., P. B. Conn, S. P. Ellner, *et al.* 2012. Disease dynamics in wild populations: modeling and estimation: a review. *Journal of Ornithology*: in press.

Cook, S., A. Gelman, and D. B. Rubin. 2006. Validation of software for Bayesian models using posterior quantiles. *Journal of Computational and Graphical Statistics* 15:675–692.

Cools, W., W. Van Den Noortgate, and P. Onghena. 2009. Design efficiency for imbalanced multilevel data. *Behavior Research Methods* 41:192–203.

Cooper, C. B., J. Dickinson, T. B. Phillips, *et al.* 2007. Citizen science as a tool for conservation in residential ecosystems. *Ecology and Society* 12:11.

Cooperrider, A. Y., R. J. Boyd, and H. R. Stuart, editors. 1986. *Inventory and monitoring of wildlife habitats*. US Department of the Interior, Bureau of Land Management, Denver, CO, USA.

Cordy, C. B. 1993. An extension of the Horvitz–Thompson theorem to point sampling from a continuous universe. *Statistics & Probability Letters* 18:353–362.

Cormack, R. M. 1964. Estimates of survival from the sighting of marked animals. *Biometrika* 51:429–438.

Cornfield, J., and J. W. Tukey. 1956. Average values of mean squares in factorials. *Annals of Mathematical Statistics* 27:907–949.

Corry, R. C., and J. I. Nassauer. 2004. Limitations of using landscape indices to evaluate the ecological consequences of alternative plans and designs. *Landscape and Urban Planning* 72:265–280.

Cotter, J., and J. Nealon. 1989. *Area frame design for agricultural surveys*. Area Frame Section, Research and Applications Division, National Agricultural Statistics Service, US Department of Agriculture. Washington, DC, USA.

Courbois, J. P., and N. S. Urquhart. 2004. Comparison of survey estimates of the finite population variance. *Journal of the Agricultural, Biological and Environmental Statistics* 9:236–250.

Cowell, R. G., A. P. Dawid, S. L. Lauritzen, *et al.* 2007. *Probabilistic networks and expert systems*. Springer Verlag, New York, NY, USA.

Cox, D. R., and P. J. Solomon. 2003. *Components of variance*. Chapman & Hall/CRC, Boca Raton, FL, USA.

Crawley, M. J. 2007. *The R book*. John Wiley & Sons, New York, NY, USA.

Cressie, N. A. C., C. A. Calder, J. S. Clark, *et al.* 2009. Accounting for uncertainty in ecological analysis: the strengths and limitations of hierarchical statistical modeling. *Ecological Applications* 19:553–570.

Crosbie, S. F., and B. F. J. Manly. 1985. Parsimonious modelling of capture–mark–recapture studies. *Biometrics* 41:385–398.

Cryer, J. D., and K.-S. Chan. 2010. *Time series analysis: with applications in R.* Second edition. Springer Science+Business Media, New York, NY, USA.

Curran, P. J., and K. A. Bollen. 2001. The best of both worlds: combining autoregressive and latent curve models. Pages 107–135 *in* L. M. Collins and A. G. Sayer, editors. *New methods for the analysis of change.* American Psychological Association, Washington, DC, USA.

Curtin, C. G. 2002. Integration of science and community-based conservation in the Mexico/U.S. borderlands. *Conservation Biology* 16:880–886.

Cushman, S. A., K. McGarigal, and M. Neel. 2008. Parsimony in landscape metrics: strength, universality, and consistency. *Ecological Indicators* 8:691–703.

Cuthill, M. 1995. An interpretive approach to developing volunteer-based coastal monitoring programmes. *Local Environment* 5:127–137.

Dagenais, M. G. 1969. A threshold regression model. *Econometrica* 37:193–203.

Dale, B., M. Norton, C. Downes, *et al.* 2005. Monitoring as a means to focus research and conservation – the Grassland Bird Monitoring example. Pages 485–495 *in* C. J. Ralph and T. D. Rich, editors. *Bird Conservation Implementation and Integration in the Americas: Proceedings of the Third International Partners in Flight Conference 2002.* General Technical Report PSW-GTR-191. USDA Forest Service, Albany, CA, USA.

Dale, V. H., J. Agee, J. Long, *et al.* 1999. Ecological sustainability is fundamental to managing the national forests and grasslands. *Bulletin of the Ecological Society of America* 80:207–209.

Dauwalter, D. C., and F. J. Rahel. 2009. Temporal variation in trout populations: implications for monitoring and trend detection. *Transactions of the American Fisheries Society* 138:38–51.

de Gruijter, J. J., D. J. Brus, M. F. P. Bierkens, *et al.* 2006. *Sampling for natural resource monitoring.* Springer, Berlin, Germany.

de Valpine, P. 2002. Review of methods for fitting time-series models with process and observation error, and likelihood calculations for nonlinear, non-Gaussian state–space models. *Bulletin of Marine Science* 70:455–471.

de Valpine, P. 2003. Better inferences from population-dynamics experiments using Monte Carlo state–space likelihood methods. *Ecology* 84:3064–3077.

Deming, W. E. 1950. *Some theory of sampling.* Dover Publications, Inc., New York, NY, USA.

Dennis, B., and A. Ellison. 2010. A reply to Millspaugh and Gitzen. *Frontiers in Ecology and the Environment* 8:515–516.

DeVink, J.-M., R. G. Clark, S. M. Slattery, *et al.* 2008a. Are late-spring boreal Lesser Scaup (*Aythya affinis*) in poor body condition? *Auk* 125:291–298.

DeVink, J.-M. A., R. G. Clark, S. M. Slattery, *et al.* 2008b. Is selenium affecting body condition and reproduction in boreal breeding scaup, scoters, and ring-necked ducks? *Environmental Pollution* 152:116–122.

Dhondt, A. A., S. Altizer, E. G. Cooch, *et al.* 2005. Dynamics of a novel pathogen in an avian host: mycoplasmal conjunctivitis in House Finches. *Acta Tropica* 94:77–93.

Diaz, R. 2007. Comparison of PQL and Laplace 6 estimates of hierarchical linear models when comparing groups of small incident rates in cluster randomised trials. *Computational Statistics & Data Analysis* 51:2871–2888.

Diaz-Ramos, S. M., D. L. Stevens, Jr., and A. R. Olsen. 1996. *EMAP Statistical Methods Manual.* EPA/620/R-96/XXX. US Environmental Protection Agency, Office of Research and Development, National Health Effects and Environmental Research Laboratory, Western Ecology Division, Corvallis, OR, USA.

Dichmont, C. M., S. Pascoe, T. Kompas, *et al.* 2010. On implementing maximum economic yield in commercial fisheries. *Proceedings of the National Academy of Sciences of the United States of America* 107:16–21.

Diefenbach, D. R., D. W. Brauning, and J. A. Mattice. 2003. Variability in grassland bird counts related to observer differences and species detection rates. *Auk* 120:1168–1179.

Diggle, P., and S. Lophaven. 2006. Bayesian geostatistical design. *Scandinavian Journal of Statistics* 33:53–64.

Diggle, P. J., and P. J. Ribeiro, Jr. 2007. *Model-based geostatistics.* Springer, New York, NY, USA.

Diggle, P. J., K.-Y. Liang, and S. L. Zeger. 1994. *Analysis of longitudinal data.* Oxford University Press, New York, NY, USA.

Dorazio, R. M., H. L. Jelks, and F. Jordan. 2005. Improving removal-based estimates of abundance by sampling a population of spatially distinct subpopulations. *Biometrics* 61:1093–1101.

Dorazio, R. M., J. A. Royle, B. Söderström, *et al.* 2006. Estimating species richness and accumulation by modeling species occurrence and detectability. *Ecology* 87:842–854.

Dormann, C. F., J. M. McPherson, M. B. Araujo, *et al.* 2007. Methods to account for spatial autocorrelation in the analysis of species distributional data: a review. *Ecography* 30:609–628.

Draper, D. 2008. Bayesian multilevel analysis and MCMC. Pages 77–140 *in* J. de Leeuw and E. Meijer, editors. *Handbook of multilevel analysis.* Springer, New York, NY, USA.

Draper, N., and H. Smith. 1981. *Applied regression analysis.* Second edition. Wiley, New York, NY, USA.

Drechsler, M., and M. A. Burgman. 2004. Combining population viability analysis with decision analysis. *Biodiversity and Conservation* 13:115–139.

du Toit, S. H. C., and R. Cudeck. 2001. The analysis of nonlinear random coefficient regression models with LISREL using constraints. Pages 259–278 *in* R. Cudeck, S du Toit, and D. Sorbom, editors. *Structural equation modeling: Present and future.* Scientific Software, Lincolnwood, IL, USA.

Duncan, G. J., and G. Kalton. 1987. Issues of design and analysis of surveys across time. *International Statistical Review* 55:97–117.

Duncan, T. E., S. C. Duncan, and L. A. Strycker. 2006. *An introduction to latent variable growth curve modeling.* Second edition. Lawrence Erlbaum Associates Publishers, Mahwah, NY, USA.

Dunson, D. B., J. Palomo, and K. A. Bollen. 2005. *Bayesian structural equation modeling.* SAMSI Institute. Technical Report #2005–5. Statistical and Applied Mathematical Sciences Institute, Research Triangle Park, NC, USA.

Easterling, R. G. 2010. Passion-driven statistics. *American Statistician* 64:1–5.

Easton, V. J., and J. H. McColl. 1997. Statistics glossary v1.1. Online: <http://www.stats.gla.ac.uk/steps/glossary/index.html>. Accessed 14 December 2009.

Edgar, C. B., and T. E. Burk. 2006. A simulation study to assess the sensitivity of a forest health monitoring network to outbreaks of defoliating insects. *Environmental Monitoring and Assessment* 122:289–307.

Edwards, A. W. F. 1992. *Likelihood.* Johns Hopkins University Press, Baltimore, MD, USA.

Edwards, D. 1998. Issues and themes for natural resources trend and change detection. *Ecological Applications* 8:323–325.

Edwards, T. C. Jr., D. R. Cutler, N. E. Zimmerman, *et al.* 2005. Model-based stratification for enhancing the detection of rare ecological events. *Ecology* 86:1081–1090.

Efford, M. G. 2004. Density estimation in live-trapping studies. *Oikos* 106:598–610.

Efron, B. 2003. The statistical century. Pages 27–40 *in* J. Panaretos, editor. *Stochastic musings: perspectives from the pioneers of the late twentieth century*. L. Erlbaum Associates, Inc., Mahwah, NJ, USA.

Elith, J., C. H. Graham, R. P. Anderson, *et al.* 2006. Novel methods improve prediction of species' distributions from occurrence data. *Ecography* 29:129–151.

Elith, J., J. R. Leathwick, and T. Hastie. 2008. A working guide to boosted regression trees. *Journal of Animal Ecology* 77:802–813.

Ellingson, A. R., and P. M. Lukacs. 2003. Improving methods for regional landbird monitoring: a reply to Hutto and Young. *Wildlife Society Bulletin* 31:896–902.

Elliott-Smith, E., and S. M. Haig. 2004. Piping plover (*Charadrius melodus*). In A. Poole, editor. *The Birds of North America Online*. Cornell Lab of Ornithology, Ithaca, NY, USA. Online: <http://bna.birds.cornell.edu/bna/species/002/>. Accessed 4 January 2010.

Elliott-Smith, E., S. M. Haig, and B. M. Powers. 2009. *Data from the 2006 International Piping Plover Census*. US Geological Survey Data Series 426. US Department of the Interior, US Geological Survey, Reston, VA, USA.

Ellison, A. M., and B. Dennis. 2010. Paths to statistical fluency for ecologist. *Frontiers in Ecology and the Environment* 8: 362–370.

Elmore, A. J., and S. M. Guinn. 2010. Synergistic use of Landsat Multispectral Scanner with GIRAS land-cover data to retrieve impervious surface area for the Potomac River Basin in 1975. *Remote Sensing of Environment* 114:2384–2391.

Elzinga, C. L., D. W. Salzer, and J. W. Willoughby. 1998. *Measuring and monitoring plant populations*. Technical Reference 1730–1. BLM/RS/ST-98/005+1730. US Department of the Interior, Bureau of Land Management, Denver, CO, USA.

Elzinga, C. L., D. W. Salzer, J. W. Willoughby, *et al.* 2001. *Monitoring plant and animal populations*. Blackwell Science, Malden, MA, USA.

Eng, J. 2004. Sample size estimation: a glimpse beyond simple formulas. *Radiology* 230:606–612.

Engeman, R. M. 2003. More on the need to get the basics right: population indices. *Wildlife Society Bulletin* 31:286–287.

Etterson, M. A., G. J. Niemi, and N. P. Danz. 2009. Estimating the effects of detection heterogeneity and overdispersion on trends estimated from avian point counts. *Ecological Applications* 19:2049–2066.

Fairweather, P. G. 1991. Statistical power and design requirements for environmental monitoring. *Australian Journal of Marine and Freshwater Research* 42:555–567.

Fancy, S. G., J. E. Gross, and S. L. Carter. 2009. Monitoring the condition of natural resources in US national parks. *Environmental Monitoring and Assessment* 151:161–174.

Faraway, J. J. 2006. *Extending the linear model with R: generalized linear, mixed effects and nonparametric regression models*. Chapman & Hall/CRC, Boca Raton, FL, USA.

Farnsworth, G. L., K. H. Pollock, J. D. Nichols, *et al.* 2002. A removal method for estimating detection probabilities from point-count surveys. *Auk* 119:414–425.

Fernandez-Gimenez, M. E., H. L. Ballard, and V. E. Sturtevant. 2008. Adaptive management and social learning in collaborative and community-based monitoring: a study of five community-based forestry organizations in the western USA. *Ecology and Society* 13(2):4. Online: <http://www.ecologyandsociety.org/vol13/iss2/art4/>.

Ferrari, J. R., T. R. Lookingbill, and M. C. Neel. 2007. Two measures of landscape–graph connectivity: assessment across gradients in area and configuration. *Landscape Ecology* 22:1315–1323.

Ferretti, M. 2009. Quality assurance in ecological monitoring – towards a unifying perspective. Editorial. *Journal of Environmental Monitoring* 11:726–729.

Fewster, R. M., S. T. Buckland, G. M. Siriwardena, *et al.* 2000. Analysis of population trends for farmland birds using generalized additive models. *Ecology* 81:1970–1984.

Field, S. A., P. J. O'Connor, A. J. Tyre, *et al.* 2007. Making monitoring meaningful. *Austral Ecology* 32:485–491.

Field, S. A., A. J. Tyre, and H. P. Possingham. 2005. Optimizing allocation of monitoring effort under economic and observational constraints. *Journal of Wildlife Management* 69:473–482.

Field, S. A., A. J. Tyre, N. Jonzén, *et al.* 2004. Minimizing the cost of environmental management decisions by optimizing statistical thresholds. *Ecology Letters* 7:669–675.

Fielding, A. 2003. Ordered category responses and random effects in multilevel and other complex structures. Pages 181–208 *in* S. P. Reise and N. Duan, editors. *Multilevel modeling – methodological advances, issues, and applications.* Lawrence Erlbaum, Mahwah, NJ, USA.

Fielding, A. H., and J. F. Bell. 1997. A review of methods for the assessment of prediction errors in conservation presence/absence models. *Environmental Conservation* 24:38–49.

Fink, D., W. M. Hochachka, B. Zuckerberg, *et al.* 2010. Spatiotemporal exploratory models for broad-scale survey data. *Ecological Applications* 20:2131–2147.

Fisher, R. A. 1922. On the mathematical foundations of theoretical statistics. *Philosophical Transactions of the Royal Society of London A* 222:309–368.

Fisher, R. A. 1938. Address to the Indian Statistical Congress, Sankhya. Available online: <http://www-history.mcs.st-and.ac.uk/history/Quotations2/217.html>. Accessed 15 May 2009.

Fitzpatrick, M. C., E. L. Preisser, A. M. Ellison, *et al.* 2009. Observer bias and the detection of low-density populations. *Ecological Applications* 19:1673–1679.

Ford, E. D. 2000. *Scientific method for ecological research.* Cambridge University Press, Cambridge, UK.

Forsyth, A., and L. Musacchio. 2005. *Designing small parks: a manual for addressing social and ecological concerns.* John Wiley, Hoboken, NJ, USA.

Fortin, M.-J., and M. Dale. 2005. *Spatial analysis: a guide for ecologists.* Cambridge University Press, Cambridge, UK.

Freeman, S. N., D. G. Noble, S. E. Newson, *et al.* 2007. Modelling population changes using data from different surveys: the Common Birds Census and the Breeding Bird Survey. *Bird Study* 54:61–72.

Fuentes, M., A. Chaudhuri, and D. M. Holland. 2007. Bayesian entropy for spatial sampling design of environmental data. *Environmental and Ecological Statistics* 14:323–340.

Fuller, W. A. 1999. Environmental surveys over time. *Journal of Agricultural, Biological, and Environmental Statistics* 4:331–345.

Gardner, B., J. A. Royle, and G. S. Boomer. 2007. Modeling spatial and temporal variation in scaup abundance. Unpublished report.

Gardner, B., J. A. Royle, M. T. Wegan, *et al.* 2010. Estimating black bear density using DNA data from hair snares. *Journal of Wildlife Management* 74:318–325.

Gardner, R. H., T. R. Lookingbill, P. A. Townsend, *et al.* 2008. A new approach for rescaling land cover data. *Landscape Ecology* 23:513–526.

Gardner, R. H., B. T. Milne, M. G. Turner, *et al.* 1987. Neutral models for the analysis of broad-scale landscape pattern. *Landscape Ecology* 1:19–28.

Garman, S. L., D. Witwicki, and A. Wight. 2010. Mapping ecological sites for long-term monitoring in National Parks. Pages 69–87 *in* C. Van Riper III, B. F. Wakeling, and T. D. Sisk, editors. *The Colorado Plateau IV: shaping conservation through science and management.* University of Arizona Press, Tucson, AZ, USA.

Geissler, P. H., and M. R. Fuller. 1987. Estimation of the proportion of area occupied by an animal species. *Proceedings of the Section on Survey Research Methods of the American Statistical Association* 1986:533–538.

Geissler, P. H., and B. R. Noon. 1981. Estimates of avian population trends from the North American Breeding Bird Survey. Pages 42–51 *in* C. J. Ralph and J. M. Scott, editors. *Estimating the numbers of terrestrial birds.* Studies in Avian Biology 6. California Natural Resources Federation, Cooper Ornithological Society.

Gelman, A., and J. Hill. 2007. *Data analysis using regression and multilevel/hierarchical models.* Cambridge University Press, New York, NY, USA.

Gelman, A., J. B. Carlin, H. S. Stern, *et al.* 2004. *Bayesian data analysis.* Second edition. Chapman & Hall/CRC, Boca Raton, FL, USA.

Gentle, J. E. 2007. *Matrix algebra: theory, computations, and applications in statistics.* Springer, New York, NY, USA.

Gerber, L. R., M. Beger, M. A. McCarthy, *et al.* 2005. A theory for optimal monitoring of marine reserves. *Ecology Letters* 8:829–837.

Gerrodette, T. 1987. A power analysis for detecting trends. *Ecology* 68:1364–1372.

Gerrodette, T. 1991. Models for power of detecting trends – a reply to Link and Hatfield. *Ecology* 72:1889–1892.

Gerrodette, T. 1993. Trends: software for a power analysis of linear regression. *Wildlife Society Bulletin* 21:515–516.

Gibbons, D. W., J. B. Reid, and R. A. Chapman. 1993. *The new atlas of breeding birds in Britain and Ireland: 1988–1991.* T. & A. D. Poyser, London, UK.

Gibbs, J. P., and E. Ene 2010. Program Monitor: estimating the statistical power of ecological monitoring programs. *Version* 11.0.0. Online: <www.esf.edu/efb/gibbs/monitor/>. Accessed 10 June 2011.

Gibbs, J. P., S. Droege, and P. Eagle. 1998. Monitoring populations of plants and animals. *BioScience* 48:935–940.

Gibbs, J. P., H. L. Snell, and C. E. Causton. 1999. Effective monitoring for adaptive wildlife management: lessons from the Galapagos Islands. *Journal of Wildlife Management* 63:1055–1065.

Gienapp, P., L. Hemerik, and M. E. Visser. 2005. A new statistical tool to predict phenology under climate change scenarios. *Global Change Biology* 11:600–606.

Gilbert, R. O. 1987. *Statistical methods for environmental pollution monitoring.* Van Nostrand Reinhold, NY, USA.

Gilks, W., A. Thomas, and D. Spiegelhalter. 1996. A language and program for complex Bayesian modelling. *The Statistician* 43:169–178.

Gillespie, A. J. R. 1999. Rationale for a national annual forest inventory program. *Journal of Forestry* 97:16–20.

Giudice, J., J. Fieberg, M. Zicus, *et al.* 2010. Cost and precision functions for aerial quadrat surveys: a case study of ring-necked ducks in Minnesota. *Journal of Wildlife Management* 74:342–349.

Goetz, S. J., P. Jantz, and C. A. Jantz. 2009. Connectivity of core habitat in the Northeastern United States: parks and protected areas in a landscape context. *Remote Sensing of Environment* 113:1421–1429.

Goldstein, H. 2003. *Multilevel statistical models*. Third edition. Arnold, London, UK.

Goldstein, H., and J. Rasbash. 1996. Improved approximations for multilevel models with binary responses. *Journal of the Royal Statistical Society, Series A* 159:505–513.

Goldstein, H., W. J. Browne, and J. Rasbash. 2002. Partitioning variation in multilevel models. *Understanding Statistics* 1:223–232.

Gómez, A., A. M. Kilpatrick, L. D. Kramer, *et al.* 2008. Land use and West Nile Virus seroprevalence in wild mammals. *Emerging Infectious Diseases* 14:962–965.

Goodwin, P., and G. Wright. 2004. *Decision analysis for management judgment*. Third edition. John Wiley & Sons Ltd, Chichester, UK.

Government Accountability Office [GAO]. 2006. *UDSA should improve its process for allocating funds to states for the environmental quality incentives program*. US Government Accountability Office, Washington, DC, USA.

Goward, S. N., J. G. Masek, W. B. Cohen, *et al.* 2008. Forest disturbance and North American carbon flux. *EOS Transactions* 89:105–116.

Grace, J. B. 2006. *Structural equation modeling and natural systems*. Cambridge University Press, Cambridge, UK.

Grace, J. B., and J. E. Keeley. 2006. A structural equation model analysis of postfire plant diversity in California shrublands. *Ecological Applications* 16:503–514.

Grace, J. B., T. M. Anderson, H. Olff, *et al.* 2010. On the specification of structural equation models for ecological systems. *Ecological Monographs* 80:67–87.

Grantham, H. S., K. A. Wilson, A. Moilanen, *et al.* 2009. Delaying conservation actions for improved knowledge: how long should we wait? *Ecology Letters* 12:293–301.

Gray, B. R. 2005. Selecting a distributional assumption for modelling relative abundances of benthic macroinvertebrates. *Ecological Modeling* 50:715–729.

Gray, B. R., and M. M. Burlew. 2007. Estimating trend precision and power to detect trends across grouped count data. *Ecology* 88:2364–2372.

Gray, B. R., R. J. Haro, and J. T. Rogala. 2010. Using random slopes models to estimate among-group variability in covariate effects: a case study using the association between mayfly counts and particle size. *Environmental and Ecological Statistics* 17:573–591.

Gregoire, T. G. 1998. Design-based and model-based inference in survey sampling: appreciating the difference. *Canadian Journal of Forest Research* 28:1429–1447.

Gregory, R. D., A. van Strien, P. Vorisek, *et al.* 2005. Developing indicators for European birds. *Philosophical Transactions of the Royal Society B* 360:269–288.

Gregory, R. S., and R. L. Keeney. 2002. Making smarter environmental management decisions. *Journal of the American Water Resources Association* 38:1601–1612.

Grigg, G. C., L. A. Beard, P. Alexander, *et al.* 1999. Aerial survey of kangaroos in South Australia 1978–1998: a brief report focusing on methodology. *Australian Zoology* 31:292–300.

Grimm, K. J., and N. Ram. 2009. Nonlinear growth models in Mplus and SAS. *Structural Equation Modeling* 16:676–701.

Groffman, P. M., J. S. Baron, T. Blett, *et al.* 2006. Ecological thresholds: the key to successful environmental management or an important concept with no practical application? *Ecosystems* 9:1–13.

Gross, J. E. 2003. *Developing conceptual models for monitoring programs*. Unpublished report. US Department of the Interior, National Park Service. Available online: <http://science.nature.nps.gov/im/monitor/docs/Conceptual_Modelling.pdf>. Accessed 5 Aug 2010.

Gross, J. E., L. K. Svancara, and T. Philippi. 2009. *A guide to interpreting NPScape data and analyses*. Natural Resource Report NPS/IMD/NRTR – 2009/XXX. US Department of the Interior, National Park Service, Fort Collins, CO, USA.

GUIDOS. 2008. Graphical User Interface for the Description of image Objects and their Shapes version 1.2. Online: <http://forest.jrc.ec.europa.eu/download/software/guidos>. Accessed 16 May 2011.

Guillera-Arroita, G., B. J. T. Morgan, M. S. Ridout, and M. Linkie. 2011. Species occupancy modeling for detection data collected along a transect. *Journal of Agricultural, Biological and Environmental Statistics* 16:301–317.

Guillera-Arroita, G., M. S. Ridout, and B. J. T. Morgan. 2010. Design of occupancy studies with imperfect detection. *Methods in Ecology and Evolution* 1:131–139.

Guisan, A., and N. E. Zimmermann. 2000. Predictive habitat distribution models in ecology. *Ecological Modelling* 135:147–186.

Guisan, A., O. Broennimann, R. Engler, *et al.* 2006. Using niche-based models to improve the sampling of rare species. *Conservation Biology* 20:501–511.

Guo, G., and H. Zhao. 2000. Multilevel modeling for binary data. *Annual Review of Sociology* 26:441–462.

Gurka, M. J. 2006. Selecting the best linear mixed model under REML. *American Statistician* 60:19–26.

Gurrin, L. C., K. J. Scurrah, and M. L. Hazelton. 2005. Tutorial in biostatistics: spline smoothing with linear mixed models. *Statistics in Medicine* 24:3361–3381.

Hahn, G. J., and W. Q. Meeker. 1993. Assumptions for statistical inference. *American Statistician* 47:1–11.

Hale, S. S. 1999. How to manage data badly, part 1. *Bulletin of the Ecological Society of America* 80:265–268.

Hale, S. S. 2000. How to manage data badly, part 2. *Bulletin of the Ecological Society of America* 81:101–103.

Hames, R. S., K. V. Rosenberg, J. D. Lowe, *et al.* 2001. Site reoccupation in fragmented landscapes: testing predictions of metapopulation theory. *Journal of Animal Ecology* 70:182–190.

Hammond, A., A. Adriaanse, E. Rodenburg, *et al.* 1995. *Environmental indicators: a systematic approach to measuring and reporting on environmental policy performance in the context of sustainable development*. World Resources Institute, Washington, DC, USA.

Hansen, M. H., W. N. Hurwitz, and W. G. Madow. 1953. *Sample survey methods and theory, volume I: methods and applications*. John Wiley, New York, NY, USA.

Hansen, M. H., W. G. Madow, and B. J. Tepping. 1983. An evaluation of model dependent and probability sampling inferences in sample surveys. *Journal of the American Statistical Association* 78:776–793.

Harary, F. 1969. *Graph theory*. Addison Wesley, Boston, MA, USA.

Harrell, F. E., Jr. 2001. *Regression modeling strategies: with applications to linear models, logistic regression, and survival analysis*. Springer, New York, NY, USA.

Harris, J. B. C., and D. G. Haskell. 2007. Land cover sampling biases associated with roadside bird surveys. *Avian Conservation and Ecology* 2:12.

Hastie, T., and R. Tibshirani. 1986. Generalized additive models. *Statistical Science* 1:297–318.

Hastie, T., and R. Tibshirani. 1990. *Generalized additive models*. Chapman & Hall/CRC, Boca Raton, FL, USA.

Hatheway, W. H., and E. J. Williams. 1958. Efficient estimation of the relationship between plot size and the variability of crop yields. *Biometrics* 14:207–222.

Hauser, C. E., A. R. Pople, and H. P. Possingham. 2006. Should managed populations be monitored every year? *Ecological Applications* 16:807–819.

Hawley, D. M., K. V. Dhondt, A. P. Dobson, *et al.* 2010. Common garden experiment reveals pathogen isolate but no host genetic diversity effect on the dynamics of an emerging wildlife disease. *Journal of Evolutionary Biology* 23:1680–1688.

Hayes, B. 2000a. Graph theory in practice; part I. *American Scientist* 88:9–13.

Hayes, B. 2000b. Graph theory in practice; part II. *American Scientist* 88:104–109.

Hedeker, D. 2008. Multilevel models for ordinal and nominal variables. Pages 237–274 *in* J. de Leeuw and E. Meijer, editors. *Handbook of multilevel analysis*. Springer, New York, NY, USA.

Heinz Center. 2008. *Landscape pattern indicators for the nation: a report from the Heinz Center's Landscape Pattern Task Group*, October 2008. John Heinz III Center for Science, Economics and the Environment, Washington, DC, USA.

Heisey, D. M., and E. V. Nordheim. 1995. Modelling age-specific survival in nesting studies, using a general approach for doubly-censored and truncated data. *Biometrics* 51:51–60.

Hewson, C. M., and D. G. Noble. 2009. Population trends of breeding birds in British woodlands over a 32-year period: relationships with food, habitat use and migratory behaviour. *Ibis* 151:464–486.

Heywood, J. S., and M. D. DeBacker. 2007. Optimal sampling designs for monitoring plant frequency. *Rangeland Ecology and Management* 60:426–434.

Hines, J. E., J. D. Nichols, J. A. Royle, *et al.* 2010. Tigers on trails: occupancy modeling for cluster sampling. *Ecological Applications* 20:1456–1466.

Hinkley, D. V. 1970. Inference about the change-point in a sequence of random variables. *Biometrika* 57:1–17.

Hinkley, D. V., and E. A. Hinkley. 1970. Inference about the change-point in a sequence of binomial variables. *Biometrika* 57:477–488.

Hirzel, A., and A. Guisan. 2002. Which is the optimal sampling strategy for habitat suitability modelling. *Ecological Modelling* 157:331–341.

Hochachka, W. M., and A. A. Dhondt. 2000. Density-dependent decline of host abundance resulting from a new infectious disease. *Proceedings of the National Academy of Sciences of the United States of America* 97:5303–5306.

Hochachka, W. M., R. Caruana, D. Fink, *et al.* 2007. Data-mining discovery of pattern and process in ecological systems. *Journal of Wildlife Management* 71:2427–2437.

Hochachka, W. M., M. Winter, and R. A. Charif. 2009. Sources of variation in singing probability of Florida Grasshopper Sparrows, and implications for design and analysis of auditory surveys. *Condor* 111:349–360.

Hockley, N. J., J. P. G. Jones, F. B. Andriahajaina, *et al.* 2005. When should communities and conservationists monitor exploited resources? *Biodiversity and Conservation* 14:2795–2806.

Hoerl, R. W., and R. D. Snee. 2010. Moving the statistics profession forward to the next level. *American Statistician* 64:10–14.

Hoeting, J. A. 2009. The importance of accounting for spatial and temporal correlation in analyses of ecological data. *Ecological Applications* 19:574–577.

Hoeting, J. A., R. A. Davis, A. A. Merton, *et al.* 2006. Model selection for geostatistical models. *Ecological Applications* 16:87–98.

Holland, M. D., and B. R. Gray. 2011. Multinomial mixture model with heterogeneous classification probabilities. *Environmental and Ecological Statistics* 18:257–270.

Holland, M. D., G. Meeden, and B. R. Gray. 2010. A finite population Bayes procedure for censored categorical abundance data. *Journal of the Indian Society of Agricultural Statistics* 64:171–175.

Holling, C. S. 1978. *Adaptive environmental assessment and management*. John Wiley and Sons, London, UK.

Hooke, R. 1980. Getting people to use statistics properly. *American Statistician* 34:39–42.

Hooten, M. B., C. K. Wikle, R. M. Dorazio, *et al.* 2007. Hierarchical spatio-temporal matrix models for characterizing invasions. *Biometrics* 63:558–567.

Hooten, M. B., C. K. Wikle, S. L. Sheriff, *et al.* 2009a. Optimal spatio-temporal hybrid sampling designs for ecological monitoring. *Journal of Vegetation Science* 20:639–649.

Hooten, M. B., C. K. Wikle, L. D. Carlile, *et al.* 2009b. Hierarchical population models for the red-cockaded woodpecker. Pages 354–354 *in* T. D. Rich, M. C. Arizmendi, D. Demarest, and C. Thompson, editors. *Tundra to tropics: connecting birds, habitats and people. Proceedings of the 4th International Partners in Flight Conference*, 13–16 February 2008. McAllen, TX. University of Texas-Pan American Press, Edinburg, TX, USA.

Hope, D., C. Gries, W. X. Zhu, *et al.* 2003. Socioeconomics drive urban plant diversity. *Proceedings of the National Academy of Sciences of the United States of America* 100:8788–8792.

Horvitz, D. G., and D. J. Thompson. 1952. A generalization of sampling without replacement from a finite universe. *Journal of the American Statistical Association* 47:663–685.

Hox, J. J. 2010. *Multilevel analysis – techniques and applications*. Second edition. Routledge, New York, NY, USA.

Huggett, R. J. 1993. *Modeling the human impact on nature: systems analysis of environmental problems*. Oxford University Press, New York, NY, USA.

Huggins, R. M. 1991. Some practical aspects of a conditional likelihood approach to capture experiments. *Biometrics* 47:725–732.

Hui-yong, L., L. Hui-qing, L. Qing-he, *et al.* 2008. Multi-species and multi-scale patterns and species associations of sand-fixing plantations. *Chinese Journal of Applied Ecology* 19:741–746.

Humbert, J.-Y., L. S. Mills, J. S. Horne, *et al.* 2009. A better way to estimate population trend. *Oikos* 118:1940–1946.

Hurvich, C. M., and C.-L. Tsai. 1989. Regression and time series model selection in small samples. *Biometrika* 76:297–307.

Hutto, R. L., and J. S. Young. 2002. Regional landbird monitoring: perspectives from the northern Rocky Mountains. *Wildlife Society Bulletin* 30:738–750.

Hutto, R. L., and J. S. Young. 2003. On the design of monitoring programs and the use of population indices: a reply to Ellingson and Lukacs. *Wildlife Society Bulletin* 31:903–910.

IUCN Standards and Petitions Subcommittee. 2010. Guidelines for using the IUCN Red List categories and criteria, version 8.0. Online: <http://intranet.iucn.org/webfiles/docs/SSC/RedList/RedListGuidelines.pdf>. Accessed 8 April 2010.

Jackson, J. B. C., M. X. Kirby, W. H. Berger, *et al.* 2001. Historical overfishing and the recent collapse of coastal ecosystems. *Science* 293:629–637.

Jackson, L. E., J. C. Kurtz, and W. S. Fisher, editors. 2000. *Evaluation guidelines for ecological indicators*. EPA/620/R-99/005. US Environmental Protection Agency, Office of Research and Development, Research Triangle Park, NC, USA.

Jobe, R. T., and P. S. White. 2009. A new cost–distance model for human accessibility and an evaluation of accessibility bias in permanent vegetation plots in Great Smoky Mountains National Park, USA. *Journal of Vegetation Science* 20:1099–110.

Johnson, D. H. 1995. Point counts of birds: what are we estimating? Pages 118–123 *in* C. J. Ralph, J. R. Sauer, and S. Droege, editors. *Monitoring bird populations by point counts*. General Technical Report PSW-GTR-149. USDA Forest Service, Albany, CA, USA.

Johnson, D. H. 1999. The insignificance of statistical significance testing. *Journal of Wildlife Management* 63:763–772.

Johnson, D. H. 2000. Statistical considerations in monitoring birds over large areas. Pages 115–120 *in* R. Bonney, D. N. Pashley, R. J. Cooper, and L. Niles, editors. *Strategies of bird conservation: the Partners in Flight planning process.* Proceedings RMRS-P-16. USDA Forest Service, Ogden, UT, USA.

Johnson, D. H. 2008. In defense of indices: the case of bird surveys. *Journal of Wildlife Management* 72:857–868.

Johnson, D. H., and J. W. Grier. 1988. Determinants of breeding distributions of ducks. *Wildlife Monographs* 100:3–37.

Johnson, D. H., and M. M. Rowland. 2007. The utility of lek counts for monitoring greater sage-grouse. Pages 15–23 *in* K. P. Reese and R. T. Bowyer, editors. *Monitoring populations of sage-grouse.* Station Bulletin 88. Idaho Forest, Wildlife and Range Experiment Station, College of Natural Resources, University of Idaho, Moscow, ID, USA.

Johnson, D. H., and T. L. Shaffer. 1987. Are mallards declining in North America? *Wildlife Society Bulletin* 15:340–345.

Johnson, D. H., J. P. Gibbs, M. Herzog, *et al.* 2009. A sampling design framework for monitoring secretive marshbirds. *Waterbirds* 32:203–215.

Johnson, D. H., M. J. Holloran, J. W. Connelly, *et al.* 2011. Influences of environmental and anthropogenic features on greater sage-grouse populations, 1997–2007. *Studies in Avian Biology*: in press.

Johnson, D. H., T. L. Shaffer, and W. E. Newton. 2001. Statistics for wildlifers: how much and what kind? *Wildlife Society Bulletin* 29:1055–1060.

Johnson, D. H., D. W. Sparling, and L. M. Cowardin. 1987. A model of the productivity of the mallard duck. *Ecological Modelling* 38:257–275.

Johnson, F., and K. Williams. 1999. Protocol and practice in the adaptive management of waterfowl harvests. *Conservation Ecology* 3:8.

Johnson, F. A. 2011. Learning and adaptation in the management of waterfowl harvests. *Journal of Environmental Management* 92:1385–1394.

Johnson, F. A., C. T. Moore, W. L. Kendall, *et al.* 1997. Uncertainty and the management of mallard harvests. *Journal of Wildlife Management* 61:202–216.

Johnson, N. L., and F. C. Leone. 1977. *Statistics and experimental design.* Second edition. Volume 2. John Wiley & Sons Inc., New York, NY, USA.

Jolly, G. M. 1965. Explicit estimates from capture–recapture data with both death and immigration stochastic model. *Biometrika* 52:225–247.

Jones, D. A., A. J. Hansen, K. Bly, *et al.* 2009. Monitoring land use and cover around parks: a conceptual approach. *Remote Sensing of Environment* 113:1346–1356.

Jones, K. B., H. Bogena, H. Vereecken, *et al.* 2010. Design and importance of multi-tiered ecological monitoring networks. Pages 355–374 *in* F. Müller, C. Baessler, H. Schubert, and S. Klotz, editors. *Long-term ecological research. Between theory and application.* Springer, New York, NY, USA.

Jonzén, N., A. R. Pople, G. C. Grigg, *et al.* 2005. Of sheep and rain: large-scale population dynamics of the red kangaroo. *Journal of Animal Ecology* 74:22–30.

Jöreskog, K. G. 1973. A general method for estimating a linear structural equation system. Pages 85–112 *in* A. S. Goldberger and O. D. Duncan, editors. *Structural equation models in the social sciences.* Seminar Press, New York, NY, USA.

Jöreskog, K. G. 1977. Structural equation models in the social sciences: specification, estimation and testing. Pages 265–287 *in* P. R. Krishnaiah, editor. *Applications of statistics*. North-Holland Publishing, Amsterdam, The Netherlands.

Kaiser, J. 2000. Bringing science to the National Parks. *Science* 288:34–37.

Karanth, K. U., J. D. Nichols, N. S. Kumar, *et al.* 2006. Assessing tiger population dynamics using photographic capture–recapture sampling. *Ecology* 87:2925–2937.

Karl, T. R., J. M. Melillo, and T. C. Peterson, editors. 2009. *Global climate change impacts in the United States*. Cambridge University Press, Cambridge, UK.

Kass, R. E., and A. E. Raftery. 1995. Bayes factors. *Journal of the American Statistical Association* 90:773–795.

Keeley, J. E., and C. J. Fotheringham. 2005. Plot shape effects on plant species diversity measurements. *Journal of Vegetation Science* 16:249–256.

Keeley, J. E., T. Brennan, and A. H. Pfaff. 2008. Fire severity and ecosystem responses following crown fires in California shrublands. *Ecological Applications* 18:1530–1546.

Keeley, J. E., C. J. Fotheringham, and M. Baer-Keeley. 2005. Determinants of postfire recovery and succession in mediterranean-climate shrublands of California. *Ecological Applications* 15:1515–1534.

Keith, D. A., L. Holman, S. Rodoreda, *et al.* 2007. Plant functional types can predict decade-scale changes in fire-prone vegetation. *Journal of Ecology* 95:1324–1337.

Keller, C. M. E., and J. T. Scallan. 1999. Potential roadside biases due to habitat changes along Breeding Bird Survey routes. *Condor* 101:50–57.

Kendall, M. G., and W. R. Buckland. 1971. *A dictionary of statistical terms*. Hafner Publishing Company, New York, NY, USA.

Kendall, W. L. 1999. Robustness of closed capture–recapture methods to violations of the closure assumption. *Ecology* 80:2517–2525.

Kendall, W. L. 2001. Using models to facilitate complex decisions. Pages 147–170 *in* T. M. Shenk and A. B. Franklin, editors. *Modeling in natural resource management*. Island Press, Washington, DC, USA.

Kendall, W. L., and R. Bjorkland. 2001. Using open robust design models to estimate temporary emigration from capture–recapture data. *Biometrics* 57:1113–1122.

Kendall, W. L., and J. D. Nichols. 2002. Estimating state-transition probabilities for unobservable state using capture–recapture/resighting data. *Ecology* 83:3276–3284.

Kendall, W. L., and G. C. White. 2009. A cautionary note on substituting spatial subunits for repeated temporal sampling in studies of site occupancy. *Journal of Applied Ecology* 46:1182–1188.

Kendall, W. L., S. J. Converse, P. F. Doherty, Jr., *et al.* 2009. Sampling design considerations for demographic studies: a case of colonial seabirds. *Ecological Applications* 19:55–68.

Kendall, W. L., B. C. Peterjohn, and J. R. Sauer. 1996. First-time observer effects in the North American Breeding Bird Survey. *Auk* 113:823–829.

Kendall, W. L., K. H. Pollock, and C. Brownie. 1995. A likelihood-based approach to capture–recapture estimation of demographic parameters under the robust design. *Biometrics* 51:293–308.

Kenkel, N. C., P. Juhász-Nagy, and J. Podani. 1989. On sampling procedures in population and community ecology. *Vegetatio* 83:195–207.

Kerman, J., and A. Gelman. 2007. Manipulating and summarizing posterior simulations using random variable objects. *Statistics and Computing* 17:235–244.

Kéry, M. 2010. *Introduction to WinBUGS for ecologists: a Bayesian approach to regression, ANOVA, mixed models and related analyses*. Academic Press, Burlington, MA, USA.

Kéry, M., and H. Schmid. 2006. Estimating species richness: calibrating a large avian monitoring programme. *Journal of Applied Ecology* 43:101–110.

Kéry, M., and B. R. Schmidt. 2008. Imperfect detection and its consequences for monitoring for conservation. *Community Ecology* 9:207–216.

Kéry, M., R. M. Dorazio, L. Soldaat, *et al.* 2009. Trend estimation in populations with imperfect detection. *Journal of Applied Ecology* 46:1163–1172.

Kéry, M., B. Gardner, and C. Monnerat. 2010a. Predicting species distributions from checklist data using site-occupancy models. *Journal of Biogeography* 37:1851–1862.

Kéry, M., J. A. Royle, and H. Schmid. 2005. Modeling avian abundance from replicated counts using binomial mixture models. *Ecological Applications* 15:1450–1461.

Kéry, M., J. A. Royle, H. Schmid, *et al.* 2010b. Site-occupancy distribution modeling to correct population-trend estimates derived from opportunistic observations. *Conservation Biology* 24:1388–1397.

Khodadadi, A., and M. Asgharian. 2008. Change-point problem and regression: an annotated bibliography. COBRA Preprint Series, article 44. Online: <http://biostats.bepress.com/cobra/ps/art44>.

Kincaid, T. M. 2000. Testing for differences between cumulative distribution functions from complex environmental sampling surveys. Pages 39–44 *in Proceedings of the Section on Statistics and the Environment*, American Statistical Association, Alexandria, VA, USA.

Kincaid, T. M. 2008. User guide for *spsurvey*, version 2.0. Probability survey design and analysis functions. Online: <http://www.epa.gov/NHEERL/arm/analysispages/software.htm>.

Kincaid, T. M., and A. R. Olsen. 2012. spsurvey: spatial survey design and analysis. R package version 2.3.

Kincaid, T. M., D. P. Larsen, and N. S. Urquhart. 2004. The structure of variation and its influence on the estimation of status: indicators of condition of lakes in the Northeast U.S.A. *Environmental Monitoring and Assessment* 98:1–21.

Kincaid, T., T. Olsen, D. Stevens, *et al.* 2009. spsurvey: spatial survey design and analysis. R package version 2.1. Online: <http://CRAN.R-project.org/package=spsurvey>.

Kindlmann, P., and F. Burel. 2008. Connectivity measures: a review. *Landscape Ecology* 23:879–890.

King, R., and S. P. Brooks. 2002. Model selection for integrated recovery/recapture data. *Biometrics* 58:841–851.

King, R., B. Morgan, O. Gimenez, *et al.* 2009. *Bayesian analysis for population ecology*. Chapman and Hall/CRC, Boca Raton, FL, USA.

Kish, L. 1965. *Survey sampling*. John Wiley & Sons, New York, NY, USA.

Knick, S. T., and J. W. Connelly, editors. 2011. *Greater sage grouse: ecology and conservation of a landscape species and its habitats*. Studies in Avian Biology 38. University of California Press, Berkeley, CA, USA.

Knick, S. T., D. S. Dobkin, J. T. Rotenberry, *et al.* 2003. Teetering on the edge or too late? Conservation and research issues for avifauna of sagebrush habitats. *Condor* 105:611–634.

Koss, R. S., K. Miller, G. Wescott, *et al.* 2009. An evaluation of Sea Search as a citizen science programme in Marine Protected Areas. *Pacific Conservation Biology* 15:116–127.

Krieger, A. M., and D. Pfeffermann. 1997. Testing of distribution functions from complex sample surveys. *Journal of Official Statistics* 13:123–142.

Kruse, K. L., D. E. Sharp, and J. A. Dubovsky. 2010. Status and harvests of sandhill cranes: Mid-Continent, Rocky Mountain and Lower Colorado River Valley Populations. Administrative Report. US Fish and Wildlife Service, Denver, CO, USA.

Kruskal, W., and F. Mosteller. 1979a. Representative sampling, I: non-scientific literature. *International Statistical Review* 47:13–24.

Kruskal, W., and F. Mosteller. 1979b. Representative sampling, II: scientific literature, excluding statistics. *International Statistical Review* 47:111–127.

Kruskal, W., and F. Mosteller. 1979c. Representative sampling, III: the current statistical literature. *International Statistical Review* 47:245–265.

Kruskal, W., and F. Mosteller. 1980. Representative sampling, IV: the history of the concept in statistics, 1895–1939. *International Statistical Review* 48:169–195.

La Sorte, F. A., and F. R. Thompson. 2007. Poleward shifts in winter ranges of North American birds. *Ecology* 88:1803–1812.

La Sorte, F. A., T. M. Lee, H. Wilman, *et al.* 2009. Disparities between observed and predicted impacts of climate change on winter bird assemblages. *Proceedings of the Royal Society B* 276:3167–3174.

Laird, N. M., and J. H. Ware. 1982. Random-effects models for longitudinal data. *Biometrics* 38:963–974.

Landres, P. B. 1992. Ecological indicators: panacea or liability? Pages 1295–1318 *in* D. H. McKenzie, D. E. Hyatt, and V. J. McDonald, editors. *Ecological Indicators*, Volume 2. Elsevier Applied Science, London, UK, and New York, USA.

Larsen, D. P., P. R. Kaufmann, T. M. Kincaid, *et al.* 2004. Detecting persistent change in the habitat of salmon-bearing streams in the Pacific Northwest. *Canadian Journal of Fisheries and Aquatic Sciences* 61:283–291.

Larsen, D. P., T. M. Kincaid, S. E. Jacobs, *et al.* 2001. Designs for evaluating local and regional scale trends. *BioScience* 51:1069–1078.

Larsen, D. P., N. S. Urquhart, and D. L. Kugler. 1995. Regional scale trend monitoring of indicators of trophic conditions of lakes. *Water Resources Bulletin* 31:117–140.

Larson, D. L., and J. B. Grace. 2004. Temporal dynamics of leafy spurge (*Euphorbia esula*) and two species of flea beetles (*Aphthona* spp.) used as biological control agents. *Biological Control* 29:207–214.

Lawler, J. L., and R. J. O'Connor. 2004. How well do consistently monitored Breeding Bird Survey routes represent the environments of the conterminous United States? *Condor* 106:801–814.

Le, N. D., and J. V. Zidek. 2006. *Statistical analysis of environmental space-time processes.* Springer, New York, NY, USA.

Le Lay, G., R. Engler, E. Franc, and A. Guisan. 2010. Prospective sampling based on model ensembles improves the detection of rare species. *Ecography* 33:1015–1027.

Lee, A. H., K. Wang, J. A. Scott, *et al.* 2006. Multi-level zero-inflated Poisson regression modelling of correlated count data with excess zeros. *Statistical Methods in Medical Research* 15:47–61.

Lee, S. Y. 2007. *Structural equation modeling: a Bayesian approach.* John Wiley & Sons, New York, NY, USA.

Lee, Y., and J. A. Nelder. 2001. Hierarchical generalised linear models: a synthesis of generalised linear models, random effect models and structured dispersions. *Biometrika* 88:987–1006.

Legendre, P., and L. Legendre. 1998. *Numerical ecology.* Second English edition. Elsevier Science B.V., Amsterdam, The Netherlands.

Legg, C. J., and L. Nagy. 2006. Why most conservation monitoring is, but need not be, a waste of time. *Journal of Environmental Management* 78:194–199.

Lenhard, J. 2006. Models and statistical inference: the controversy between Fisher and Neyman-Pearson. *British Journal for the Philosophy of Science* 57:69–91.

Leopold, A., and S. E. Jones. 1947. A phenological record for Sauk and Dane counties, Wisconsin, 1935–1945. *Ecological Monographs* 17:81–122.

Lepage, D., and C. M. Francis. 2002. Do feeder counts reliably indicate bird population changes? 21 years of winter bird counts in Ontario, Canada. *Condor* 104:255–270.

Lesser, V. M., and W. D. Kalsbeek. 1999. Nonsampling errors in environmental surveys. *Journal of Agricultural, Biological and Environmental Statistics* 4:473–488.

Lesser, V. M., and W. S. Overton. 1994. *EMAP status estimation: statistical procedures and algorithms*. EPA 620/R-94/008. US Environmental Protection Agency, Environmental Research Laboratory, Corvallis, OR, USA.

Lessler, J. T., and W. D. Kalsbeek. 1992. *Nonsampling error in surveys*. Wiley, New York, NY, USA.

Levin, S. A. 1992. The problem of pattern and scale in ecology. *Ecology* 73:1943–1967.

Lewis, S. 2007. The role of science in National Park Service decision-making. *George Wright Forum* 24:36–40.

Li, J., B. R. Gray, and D. M. Bates. 2008. An empirical study of statistical properties of variance partition coefficients for multi-level logistic regression models. *Communications in Statistics – Simulation and Computation* 37:2010–2026.

Likens, G. E. 1983. A priority for ecological research. *Bulletin of the Ecological Society of America* 64:234–243.

Likens, G. E., C. T. Driscoll, and D. C. Buso. 1996. Long-term effects of acid rain: response and recovery of a forest ecosystem. *Science* 272:244–246.

Lin, Y. B., Y. P. Lin, D. P. Deng, *et al.* 2008. Integrating remote sensing data with directional two-dimensional wavelet analysis and open geospatial techniques for efficient disaster monitoring and management. *Sensors* 8:1070–1089.

Lindenmayer, D. B., and G. E. Likens. 2009. Adaptive monitoring: a new paradigm for long-term research and monitoring. *Trends in Ecology & Evolution* 24:482–486.

Lindenmayer, D. B., and G. E. Likens. 2010a. *Effective ecological monitoring*. CSIRO Publishing, Collingwood, Victoria, Australia.

Lindenmayer, D. B., and G. E. Likens. 2010b. The science and application of ecological monitoring. *Biological Conservation* 143:1317–1328.

Lindenmayer, D. B., M. Tanton, T. Linga, *et al.* 1991. Public participation in stagwatching surveys of a rare mammal: applications for environmental and public education. *Australian Journal of Environmental Education* 7:63–70.

Lindley, D. V. 1985. *Making decisions*. Second edition. Wiley and Sons, New York, NY, USA.

Lindsey, J. K. 1997. *Applying generalized linear models*. Springer, New York, NY, USA.

Link, W. A. 2003. Nonidentifiability of population size from capture–recapture data with heterogeneous detection probabilities. *Biometrics* 59:1123–1130.

Link, W. A., and R. J. Barker. 2006. Model weights and the foundations of multimodel inference. *Ecology* 87:2626–2635.

Link, W. A., and R. J. Barker. 2010. *Bayesian inference; with ecological applications*. Academic Press, London, UK.

Link, W. A., and J. R. Sauer. 1997. Estimation of population trajectories from count data. *Biometrics* 53:488–497.

Link, W. A., and J. R. Sauer. 1998. Estimating population change from count data: applications to the North American Breeding Bird Survey. *Ecological Applications* 8:258–268.

Link, W. A., and J. R. Sauer. 2002. A hierarchical model of population change with application to Cerulean warblers. *Ecology* 83:2832–2840.

Link, W. A., and J. R. Sauer. 2007. Seasonal components of avian population change: joint analysis of two large-scale monitoring programs. *Ecology* 88:49–55.

Link, W. A., E. Cam, J. D. Nichols, *et al.* 2002. Of bugs and birds: Markov chain Monte Carlo for hierarchical modeling in wildlife research. *Journal of Wildlife Management* 66:277–291.

Littell, R. C., G. A. Milliken, W. W. Stroup, *et al.* 2006. *SAS for mixed models.* Second edition. SAS Institute, Cary, NC, USA.

Little, R. J. 2004. To model or not to model? Competing modes of inference for finite population sampling. *Journal of the American Statistical Association* 99:546–556.

Little, R. J. A., and D. B. Rubin. 2002. *Statistical analysis with missing data.* Second edition. John Wiley & Sons, New York, NY, USA.

Lobo, J. M., A. Jiménez-Valverde, and R. Real. 2008. AUC: a misleading measure of the performance of predictive distribution models. *Global Ecology and Biogeography* 17:145–151.

Lohr, S. L. 1999. *Sampling: design and analysis.* Duxbury Press, Pacific Grove, CA, USA.

Lohr, S. L. 2010. *Sampling: design and analysis.* Second edition. Brooks/Cole, Boston, MA, USA.

Lookingbill, T. R., S. L. Carter, B. Gorsira, *et al.* 2008. Using landscape analysis to evaluate ecological impacts of battlefield restoration. *ParkScience* 25:60–65.

Lookingbill, T. R., R. H. Gardner, J. R. Ferrari, *et al.* 2010. Combining a dispersal model with network theory to assess habitat connectivity. *Ecological Applications* 20:427–441.

Lookingbill, T. R., R. H. Gardner, P. A. Townsend, *et al.* 2007. Conceptual models as hypotheses in monitoring urban landscapes. *Environmental Management* 40:171–182.

Lovett, G. M., D. A. Burns, C. T. Driscoll, *et al.* 2007. Who needs environmental monitoring? *Frontiers in Ecology and the Environment* 5:253–260.

Lukacs, P. M., and K. P. Burnham. 2005. Estimating population size from DNA closed capture–recapture data incorporating genotyping error. *Journal of Wildlife Management* 69:396–403.

Lunn, D., D. Spiegelhalter, A. Thomas, *et al.* 2009. The BUGS project: evolution, critique, and future directions. *Statistics in Medicine* 28:3049–3067.

Lunn, D. J., A. Thomas, N. Best, *et al.* 2000. WinBUGS – a Bayesian modelling framework: concepts, structure, and extensibility. *Statistics and Computing* 10:325–337.

Lurman, J., J. H. Reynolds, and M. Robards. 2010. Recognizing when "the best scientific data available" isn't. *Stanford Environmental Law Journal* 29:247–282.

Lyons, J. E., M. C. Runge, H. P. Laskowski, *et al.* 2008. Monitoring in the context of structured decision-making and adaptive management. *Journal of Wildlife Management* 72:1683–1692.

Maas, C. J. M., and J. J. Hox. 2004. Robustness issues in multilevel regression analysis. *Statistica Neerlandica* 58:127–137.

MacKenzie, D. I. 2005a. What are the issues with presence–absence data for wildlife managers? *Journal of Wildlife Management* 69:849–860.

MacKenzie, D. I. 2005b. Was it there? Dealing with imperfect detection for species presence/absence data. *Australian and New Zealand Journal of Statistics* 47:65–74.

MacKenzie, D. I., and W. L. Kendall. 2002. How should detection probabilities be incorporated into estimates of relative abundance? *Ecology* 83:2387–2393.

MacKenzie, D. I., and J. A. Royle. 2005. Designing occupancy studies: general advice and allocating survey effort. *Journal of Applied Ecology* 42:1105–1114.

MacKenzie, D. I., L. L. Bailey, J. E. Hines, *et al.* 2011. An integrated model of habitat and species occurrence dynamics. *Methods in Ecology and Evolution* 2:612–622.

MacKenzie, D. I., L. L. Bailey, and J. D. Nichols. 2004. Investigating species co-occurrence patterns when species are detected imperfectly. *Journal of Animal Ecology* 73:546–555.

MacKenzie, D. I., J. D. Nichols, J. E. Hines, *et al.* 2003. Estimating site occupancy, colonization, and local extinction when a species is detected imperfectly. *Ecology* 84:2200–2207.

MacKenzie, D. I., J. D. Nichols, G. B. Lachman, *et al.* 2002. Estimating site occupancy rates when detection probabilities are less than one. *Ecology* 83:2248–2255.

MacKenzie, D. I., J. D. Nichols, J. A. Royle, *et al.* 2006. *Occupancy estimation and modeling: inferring patterns and dynamics of species occurrence*. Academic Press, San Diego, CA, USA.

MacKenzie, D. I., J. D. Nichols, M. E. Seamans, *et al.* 2009. Modeling species occurrence dynamics with multiple states and imperfect detection. *Ecology* 90:823–835.

MacKinnon, D. P. 2008. *Statistical mediation analysis*. Lawrence Erlbaum and Associates, New York, NY, USA.

Maddox, D., K. Poiani, and R. Unnasch. 2001. Evaluating management success: using ecological models to ask the right monitoring questions. Pages 563–584 *in* W. T. Sexton, A. J. Malk, R. C. Szaro, and N. C. Johnson, editors. *Ecological stewardship*. Elsevier Science, Oxford, UK.

Magnussen, S., W. Kurz, D. G. Leckie, *et al.* 2005. Adaptive cluster sampling for estimation of deforestation rates. *European Journal of Forest Research* 124:207–220.

Magoun, A. J., J. C. Ray, D. S. Johnson, *et al.* 2007. Modeling wolverine occurrence using aerial surveys of tracks in snow. *Journal of Wildlife Management* 71:2221–2229.

Manley, P. N., W. J. Zielinski, C. M. Stuart, *et al.* 2000. Monitoring ecosystems in the Sierra Nevada: the conceptual model foundation. *Environmental Monitoring and Assessment* 64:139–152.

Marmion, M., M. Luoto, R. K. Heikkinen, *et al.* 2009. The performance of state-of-the-art modelling techniques depends on geographical distribution of species. *Ecological Modelling* 220:3512–3520.

Marsh, D. M., and P. C. Trenham. 2008. Current trends in plant and animal population monitoring. *Conservation Biology* 22:647–655.

Martin, F. W., R. S. Pospahala, and J. D. Nichols. 1979. Assessment and population management of North American migratory birds. Pages 187–239 *in* J. Cairns, Jr., G. P. Patil, and W. E. Waters, editors. *Environmental biomonitoring, assessment, prediction, and management: certain case studies and related quantitative issues*. International Cooperative, Burtonsville, MD, USA.

Martin, J., J. D. Nichols, C. L. McIntyre, *et al.* 2009a. Perturbation analysis for patch occupancy dynamics. *Ecology* 90:10–16.

Martin, J., M. C. Runge, J. D. Nichols, *et al.* 2009b. Structured decision making as a conceptual framework to identify thresholds for conservation and management. *Ecological Applications* 19:1079–1090.

Mazzetta, C., S. Brooks, and S. N. Freeman. 2007. On smoothing trends in population index modeling. *Biometrics* 63:1007–1014.

McCarthy, M. A. 2007. *Bayesian methods for ecology*. Cambridge University Press, Cambridge, UK.

McCarthy, M. A., and H. P. Possingham. 2007. Active adaptive management for conservation. *Conservation Biology* 21:956–963.

McClintock, B. T., J. D. Nichols, L. L. Bailey, *et al.* 2010. Seeking a second opinion: uncertainty in disease ecology. *Ecology Letters* 13:659–674.

McComb, B., B. Zuckerberg, D. Vesely, *et al.* 2010. *Monitoring animal populations and their habitats: a practitioner's guide*. CRC Press/Taylor & Francis Group, Boca Raton, FL, USA.

McCullagh, P., and J. A. Nelder. 1989. *Generalized linear models*. Second edition. Chapman and Hall, London, UK.

McCulloch, C. E., and S. R. Searle. 2001. *Generalized, linear, and mixed models*. John Wiley and Sons, New York, NY, USA.

McCulloch, C. E., S. R. Searle, and J. M. Neuhaus. 2008. *Generalized, linear, and mixed models*. Second edition. John Wiley & Sons, Inc., Hoboken, NJ, USA.

McDonald, L. L. 2004. Sampling rare populations. Pages 11–42 *in* W. L. Thompson, editor. *Sampling rare or elusive species: concepts, designs, and techniques for estimating population parameters*. Island Press, Washington, DC, USA.

McDonald, T. L. 2003. Review of environmental monitoring methods: survey designs. *Environmental Monitoring and Assessment* 85:277–292.

McDonald, T. L., B. F. J. Manly, and R. M. Nielson. 2009. *Review of environmental monitoring methods: trend detection. Technical report*. WEST, Inc., Laramie, WY, USA.

McDonald, T., C. Roland, J. Fried, *et al.* 2001. Simulation of long-term monitoring sample designs in Denali National Park. Pages 407–414 *in* D. Harmon, editor. *Proceedings of the 11th Conference on Research and Resource Management in Parks and on Public Lands*. The George Wright Society, Hancock, MI, USA.

McDonald-Madden, E., P. W. J. Baxter, R. A. Fuller, *et al.* 2010. Monitoring does not always count. *Trends in Ecology & Evolution* 25:547–550.

McGarigal, K., S. A. Cushman, M. C. Neel, *et al.* 2002. FRAGSTATS: spatial pattern analysis program for categorical maps. University of Massachusetts, Amherst. Online: <http://www.umass.edu/landeco/research/fragstats/fragstats.html>. Accessed 16 May 2011.

McGarigal, K., S. Tagil, and S. A. Cushman. 2009. Surface metrics: an alternative to patch metrics for the quantification of landscape structure. *Landscape Ecology* 24:433–450.

McGee, V. E., and W. T. Carleton. 1970. Piecewise regression. *Journal of the American Statistical Association* 65:1109–1124.

McGowan, C. P., D. R. Smith, J. A. Sweka, *et al.* 2011. Multispecies modeling for adaptive management of horseshoe crabs and red knots in the Delaware Bay. *Natural Resource Modeling* 24:117–156.

McIver, J. D., R. E. J. Boerner, and S. C. Hart. 2008. The national Fire and Fire Surrogate study: ecological consequences of alternative fuel reduction methods in seasonally dry forests. *Forest Ecology and Management* 255:3075–3080.

McKelvey, K. S., and D. E. Pearson. 2001. Population estimation with sparse data: the role of estimators versus indices revisited. *Canadian Journal of Zoology* 79:1754–1765.

McNeil, B. E., E. G. Denny, and A. D. Richardson. 2008. Coordinating a Northeast Regional Phenology Network. *Bulletin of the Ecological Society of America* 89:188–190.

McNeil, T. C., F. R. Rousseau, and L. P. Hildebrand. 2006. Community-based environmental management in Atlantic Canada: the impacts and spheres of influence of the Atlantic coastal action program. *Environmental Monitoring and Assessment* 113:367–383.

McPherson, J. M., and R. A. Myers. 2009. How to infer population trends in sparse data: examples with opportunistic sighting records for great white sharks. *Diversity and Distributions* 15:880–890.

Menges, E. S., and D. R. Gordon. 1996. Three levels of monitoring intensity for rare plant species. *Natural Areas Journal* 16:227–237.

Meredith, W., and J. Tisak. 1990. Latent curve analysis. *Psychometrika* 55:107–122.

Messer, J. J., R. A. Linthurst, and W. S. Overton. 1991. An EPA program for monitoring ecological status and trends. *Environmental Monitoring and Assessment* 17:67–78.

Meyers, J. L., and S. N. Beretvas. 2006. The impact of inappropriate modeling of cross-classified data structures. *Multivariate Behavioral Research* 41:473–497.

Microsoft Corporation. 2009. Monitor (transitive verb). Encarta world English dictionary (North American edition). Online: <http://encarta.msn.com/encnet/features/dictionary/DictionaryResults.aspx?lextype=3&search=monitor>. Accessed 18 April 2011.

Millar, R. 2009. Comparison of hierarchical Bayesian models for over-dispersed count data using DIC and Bayes factors. *Biometrics* 65:962–969.

Miller, J. R., and R. J. Hobbs. 2002. Conservation where people live and work. *Conservation Biology* 16:330–337.

Miller, M. E. 2005. *The structure and functioning of dryland ecosystems – conceptual models to inform long-term ecological monitoring*. Scientific Investigations Report 2005–5197. US Department of the Interior, US Geological Survey, Reston, VA, USA.

Millspaugh, J. J., and R. A. Gitzen. 2010. Statistical danger zone. *Frontiers in Ecology and the Environment* 8:515.

Miltner, R. 2010. A method and rationale for deriving nutrient criteria for small rivers and streams in Ohio. *Environmental Management* 45:842–855.

Min, Y., and A. Agresti. 2005. Random effect models for repeated measures of zero-inflated count data. *Statistical Modelling* 5:1–19.

Minor, E. S., and D. L. Urban. 2008. A graph-theory framework for evaluating landscape connectivity and conservation planning. *Conservation Biology* 22:297–307.

Minor, E. S., S. Tessel, K. A. M. Engelhardt, *et al.* 2009. The role of landscape connectivity in assembling exotic plant communities. *Ecology* 90:1802–1809.

Moghimbeigi, A., M. R. Eshraghian, K. Mohammad, *et al.* 2008. Multilevel zero-inflated negative binomial regression modeling for over-dispersed count data with extra zeros. *Journal of Applied Statistics* 35:1193–1202.

Moilenan, A. 2002. Implication of empirical data quality to metapopulation model parameter estimation and application. *Oikos* 96:516–530.

Moineddin, R., F. I. Matheson, and R. H. Glazier. 2007. A simulation study of sample size for multilevel logistic regression models. *BMC Medical Research Methodology* 7:34.

Moore, C. T., and M. J. Conroy. 2006. Optimal regeneration planning for old-growth forest: addressing scientific uncertainty in endangered species recovery through adaptive management. *Forest Science* 52:155–172.

Moore, C. T., and W. L. Kendall. 2004. Costs of detection bias in index-based population monitoring. *Animal Biodiversity and Conservation* 27(1):287–296.

Morrison, L. W., D. R. Smith, C. C. Young, *et al.* 2008. Evaluating sampling designs by computer simulation: a case study with the Missouri bladderpod. *Population Ecology* 50:417–425.

Morrison, M. L., B. G. Marcot, and R. W. Mannan. 2006. *Wildlife-habitat relationships: concepts and applications*. Third edition. Island Press, Washington, DC, USA.

Mountford, M. D. 1982. Estimation of population fluctuations with application to the Common Bird Census. *Applied Statistics* 31:135–143.

Mouton, A. M., B. De Baets, and P. L. M. Goethals. 2010. Ecological relevance of performance criteria for species distribution models. *Ecological Modelling* 221:1995–2002.

Muggeo, V. M. R. 2003. Estimating regression models with unknown break-points. *Statistics in Medicine* 22:3055–3071.

Muggeo, V. M. R. 2008. segmented: an R package to fit regression models with broken-line relationships. *R News* 8:20–25.

Mulder, B. S., and C. J. Palmer. 1999. Introduction to effectiveness monitoring. Pages 1–19 *in* B. S. Mulder, B. R. Noon, T. A. Spies, *et al.* editors. *The strategy and design of the effectiveness monitoring program for the Northwest Forest Plan*. General Technical Report PNW-GTR-437. USDA Forest Service, Portland, OR, USA.

Muller, K. E. 2009. Analysis of variance concepts and computations. *Wiley Interdisciplinary Reviews: Computational Statistics* 1:279–282.

Müller, W. G. 2007. *Collecting spatial data*. Third edition. Springer, Berlin, Germany.

Munholland, P. L., and J. J. Borkowski. 1996. Simple Latin Square sampling + 1: a spatial design using quadrats. *Biometrics* 52:125–136.

Munson, M. A., R. Caruana, D. Fink, *et al.* 2010. A method for measuring the relative information content of data from different monitoring protocols. *Methods in Ecology & Evolution* 1:263–273.

Murphy, K. A., J. H. Reynolds, and J. M. Koltun. 2008. Evaluating the ability of the differenced Normalized Burn Ratio (dNBR) to predict ecologically significant burn severity in Alaskan boreal forests. *International Journal of Wildland Fire* 17:490–499.

Murtaugh, P. A. 2009. Performance of several variable-selection methods applied to real ecological data. *Ecology Letters* 12:1061–1068.

Muthén, L. K., and B. O. Muthén. 1998–2010. Mplus user's guide. Sixth edition. Muthén & Muthén, Los Angeles, CA, USA. Online: <http://www.statmodel.com/download/usersguide/Mplus%20Users%20Guide%20v6.pdf>. Accessed 19 Aug 2010.

National Academy of Public Administration [NAPA]. 2010. *Strengthening America's best idea: an independent review of the National Park Service's Natural Resource Stewardship and Science Directorate*. National Academy of Public Administration, Washington, DC, USA.

National Park Service [NPS]. 1999. *Natural Resource Challenge: the National Park Service's action plan for preserving natural resources*. US Department of the Interior, National Park Service, Washington, DC, USA.

National Park Service [NPS]. 2006. *Management policies 2006*. US Department of the Interior, National Park Service. ISBN 0–16-076874–8. Online: <http://www.nps.gov/policy/MP2006.pdf>. Accessed 5 Aug 2010.

National Park Service [NPS]. 2008a. *Examples of specific, measurable, monitoring objectives*. Unpublished report. US Department of the Interior, National Park Service. Available online: <http://science.nature.nps.gov/im/monitor/docs/Goals&ObjectivesGuidance.doc>. Accessed 5 Aug 2010.

National Park Service [NPS]. 2008b. *Southern Plains Network Vital Signs Monitoring Plan*. Natural Resource Report NPS/SOPN/NRR-2008/028. US Department of the Interior, National Park Service, Fort Collins, CO, USA.

National Research Council. 1995. *Review of EPA's Environmental Monitoring and Assessment Program: overall evaluation*. National Academy Press, Washington, DC, USA.

National Research Council. 2000. *Ecological indicators for the nation*. National Academy Press, Washington, DC, USA.

Nelder, J. A., and R. W. M. Wedderburn. 1972. Generalized linear models. *Journal of the Royal Statistical Society, Series A* 135:370–384.

Nelson, M., G. Moisen, M. Finco, *et al.* 2007. Forest Inventory and Analysis in the United States: remote sensing and geospatial activities. *Photogrammetric Engineering and Remote Sensing* 73:729–732.

Neter, J., W. Wasserman, and M. H. Kutner. 1990. *Applied linear statistical models*. Irwin, Homewood, IL, USA.

Newman, K. B., S. T. Buckland, S. T. Lindley, *et al.* 2006. Hidden process models for animal population dynamics. *Ecological Applications* 16:74–86.

Neyman, J. 1960. Indeterminism in science and new demands on statisticians. *Journal of the American Statistical Association* 55:625–639.

Nichols, J. D. 1991. Extensive monitoring programmes viewed as long-term population studies: the case of North American waterfowl. *Ibis* 133 (suppl. 1):89–98.

Nichols, J. D. 1992. Capture–recapture models. *BioScience* 42:94–102.

Nichols, J. D. 2000. Monitoring is not enough: on the need for a model-based approach to migratory bird management. Pages 121–123 *in* R. Bonney, D. N. Pashley, R. Cooper, and L. Niles, editors. *Strategies of bird conservation: the Partners in Flight planning process.* Proceedings RMRS-P-16. USDA Forest Service, Ogden, UT, USA.

Nichols, J. D., and J. E. Hines. 2002. Approaches for the direct estimation of λ, and demographic contributions to λ, using capture–recapture data. *Journal of Applied Statistics* 29:539–568.

Nichols, J. D., and W. L. Kendall. 1995. The use of multi-state capture–recapture models to address questions in evolutionary ecology. *Journal of Applied Statistics* 22:835–846.

Nichols, J. D., and B. K. Williams. 2006. Monitoring for conservation. *Trends in Ecology & Evolution* 21:668–673.

Nichols, J. D., T. Boulinier, J. E. Hines, *et al.* 1998. Estimating rates of local extinction, colonization and turnover in animal communities. *Ecological Applications* 8:1213–1225.

Nichols, J. D., J. E. Hines, D. I. MacKenzie, *et al.* 2007. Occupancy estimation and modeling with multiple states and state uncertainty. *Ecology* 88:1395–1400.

Nichols, J. D., J. E. Hines, J. R. Sauer, *et al.* 2000. A double-observer approach for estimating detection probability and abundance from point counts. *Auk* 117:393–408.

Nichols, J. D., R. E. Tomlinson, and G. Weggerman. 1986. Estimating nest detection probabilities for white-tailed dove nest transects in Tamaulipas, Mexico. *Auk* 103:825–828.

Nicholls, A. O. 1989. How to make biological surveys go further with generalized linear models. *Biological Conservation* 50:51–75.

Niemela, J. 1999. Ecology and urban planning. *Biodiversity and Conservation* 8:119–131.

Niles, L. J., J. Bart, H. P. Sitters, *et al.* 2009. Effects of horseshoe crab harvest in Delaware Bay on red knots: are harvest restrictions working? *BioScience* 59:153–164.

Noon, B. R. 2003. Conceptual issues in monitoring ecological resources. Pages 27–71 *in* E. D. Busch and J. C. Trexler, editors. *Monitoring ecosystems: interdisciplinary approaches for evaluating ecoregional initiatives.* Island Press, Washington, DC, USA.

Noon, B. R., T. A. Spies, and M. G. Raphael. 1999. Conceptual basis for designing an effectiveness monitoring program. Pages 21–48 *in* B. S. Mulder, B. R. Noon, T. A. Spies, *et al.*, editors. *The strategy and design of the effectiveness monitoring program for the Northwest Forest Plan.* General Technical Report PNW-GTR-437. USDA Forest Service, Portland, OR, USA.

Nowak, D. J. 2008. Assessing urban forest structure: summary and conclusions. *Arboriculture & Urban Forestry* 34:391–392.

Nusser, S. M., F. J. Breidt, and W. A. Fuller. 1998. Design and estimation for investigating the dynamics of natural resources. *Ecological Applications* 8:234–245.

Nychka, D., and N. Saltzman. 1998. Design of air-quality monitoring networks. Pages 51–76 *in* D. Nychka and W. Piegorsch, editors. *Case studies in environmental statistics.* Springer, New York, NY, USA.

O'Connor, R. J., E. Dunn, D. H. Johnson, *et al.* 2000. *A programmatic review of the North American Breeding Bird Survey: report of a peer review panel.* Unpublished report. Available online: <http://www.pwrc.usgs.gov/bbs/bbsreview/bbsfinal.pdf>. Accessed 22 December 2009.

O'Dell, T., S. Garman, A. Evenden, *et al.* 2005. *Northern Colorado Plateau Inventory and Monitoring Network, Vital Signs Monitoring Plan.* National Park Service, Moab, UT, USA.

Oakley, K. L., L. P. Thomas, and S. G. Fancy. 2003. Guidelines for long-term monitoring protocols. *Wildlife Society Bulletin* 31:1000–1003.

Ogilvie, M. A. 1967. Population changes and mortality of the mute swan in Britain. *Wildfowl* 18:64–73.

Ojiambo, P., and H. Scherm. 2010. Efficiency of adaptive cluster sampling for estimating plant disease incidence. *Phytopathology* 100:663–670.

Olea, R. A. 1984. Sampling design optimization for spatial functions. *Mathematical Geology* 16:369–392.

Olsen, A. R., and D. V. Peck. 2008. Monitoring design and extent estimates for National Wadeable Stream Assessment. *Journal of North American Benthological Society* 27:822–836.

Olsen, A. R., J. Sedransk, D. Edwards, *et al.* 1999. Statistical issues for monitoring ecological and natural resources in the United States. *Environmental Monitoring and Assessment* 54: 1–45.

Olsen, A. R., L. L. Stahl, B. D. Snyder, *et al.* 2009. Survey design for lakes and reservoirs in the United States to assess contaminants in fish tissue. *Environmental Monitoring and Assessment* 150:91–100.

Opsomer, J. D., F. J. Breidt, G. G. Moisen, *et al.* 2007. Model-assisted estimation of forest resources with generalized additive models. *Journal of the American Statistical Association* 102:400–416.

Organisation for Economic Cooperation and Development [OECD]. 2006. *Agricultural policies in OECD countries at a glance.* OECD, Paris, France.

Otis, D. L., K. P. Burnham, G. C. White, and D. R. Anderson. 1978. Statistical inference from capture data on closed animal populations. *Wildlife Monographs* 62:1–135.

Overton, W. S. 1989. *Effects of measurement and other extraneous errors on estimated distribution functions in the National Surface Water Surveys.* Technical Report 129. Department of Statistics, Oregon State University, Corvallis, OR, USA.

Overton, W. S., and S. V. Stehman. 1995. The Horvitz–Thompson theorem as a unifying perspective for probability sampling: with examples from natural resource sampling. *American Statistician* 49:261–268.

Overton, W. S., and S. V. Stehman. 1996. Desirable design characteristics for long-term monitoring of ecological variables. *Environmental and Ecological Statistics* 3:349–361.

Overton, W. S., D. White, and D. L. Stevens, Jr. 1990. *Design report for EMAP environmental monitoring and assessment program.* EPA/600/3–91/053. US Environmental Protection Agency, Washington, DC, USA.

Page, E. S. 1955. A test for a change in a parameter occurring at an unknown point. *Biometrika* 42:523–527.

Papritz, A., A. Dümig, C. Zimmermann, *et al.* 2011. Uncertainty of variance component estimates in nested sampling: a case study on the field-scale spatial variability of a restored soil. *European Journal of Soil Science* 62:479–495.

Parmenter, R. R., T. L. Yates, D. R. Anderson, *et al.* 2003. Small-mammal density estimation: a field comparison of grid-based vs. web-based density estimators. *Ecological Monographs* 73:1–26.

Pascual-Hortal, L., and S. Saura. 2006. Comparison and development of new graph-based landscape connectivity indices: towards the prioritization of habitat patches and corridors for conservation. *Landscape Ecology* 21:959–967.

Patterson, P. L., and G. A. Reams. 2005. Combining panels for forest inventory and analysis estimation. Pages 79–84 *in* P. L. Patterson and G. A. Reams, editors. *The enhanced forest inventory and analysis program – national sampling design and estimation procedures. General Technical Report SRS-80*. USDA Forest Service, Asheville, NC, USA.

Pearl, J. 1998. Graphs, causality, and structural equation models. *Sociological Methods & Research* 27:226–284.

Pearse, A. T., K. J. Reinecke, S. J. Dinsmore, *et al.* 2009. Using simulation to improve wildlife surveys: wintering mallards in Mississippi, USA. *Wildlife Research* 36:279–288.

Peterjohn, B, and K. Pardieck. 2002. *A bibliography for the North American Breeding Bird Survey*. 2002. Unpublished report. US Geological Survey, Patuxent Wildlife Research Center. Available online: <http://www.pwrc.usgs.gov/infobase/bbsbib/bbsbib.pdf>. Accessed 22 December 2009.

Peterson, S. A., N. S. Urquhart, and E. B. Welch. 1999. Sample representativeness: a must for reliable regional lake condition estimates. *Environmental Science and Technology* 33:1559–1565.

Philippi, T. 2005. Adaptive cluster sampling for estimation of abundances within local populations of low-abundance plants. *Ecology* 86:1091–1100.

Philippi, T. E., P. M. Dixon, and B. E. Taylor. 1998. Detecting trends in species composition. *Ecological Applications* 8:300–308.

Phillips, S. J., R. P. Anderson, and R. E. Schapire. 2006. Maximum entropy modeling of species geographic distributions. *Ecological Modelling* 190:231–259.

Piepho, H.-P., and J. O. Ogutu. 2002. A simple mixed model for trend analysis in wildlife populations. *Journal of Agricultural, Biological, and Environmental Statistics* 7:350–360.

Pinheiro, J. C., and D. M. Bates. 2000. *Mixed-effects models in S and S-PLUS*. Springer, New York, NY, USA.

Pinheiro, J. C., and E. C. Chao. 2006. Efficient Laplacian and adaptive Gaussian quadrature algorithms for multilevel generalized linear mixed models. *Journal of Computational and Graphical Statistics* 15:58–81.

Pinho, O. D. 2000. Community involvement in projects to reduce nonpoint source pollution. *Journal of Shellfish Research* 19:445–447.

Platt, J. R. 1964. Strong inference. *Science* 146:347–353.

Pledger, S. 2000. Unified maximum likelihood estimates for closed capture–recapture model using mixtures. *Biometrics* 56:434–442.

Plummer, M. 2003. JAGS: a program for analysis of Bayesian graphical models using Gibbs sampling. *In* K. Hornik, F. Leisch, and A. Zeileis, editors. *Proceedings of the 3rd International Workshop on Distributed Statistical Computing*, 20–22 March, Technische Universität Wien, Vienna, Austria.

Plummer, M. 2008. Penalized loss functions for Bayesian model comparison. *Biostatistics* 9:523–539.

Pollock, K. H. 1981. Capture–recapture model for age-dependent survival and capture rates. *Biometrics* 37:521–529.

Pollock, K. H. 1982. A capture–recapture design robust to unequal probability of capture. *Journal of Wildlife Management* 46:752–757.

Pollock, K. H., J. D. Nichols, T. R. Simons, *et al.* 2002. Large scale wildlife monitoring studies: statistical methods for design and analysis. *Environmetrics* 13:105–119.

Pollock, K. H., D. L. Solomon, and D. S. Robson. 1974. Tests for mortality and recruitment in a *K*-sample tag–recapture experiment. *Biometrics* 30:77–87.

Pooler, P. S., and D. R. Smith. 2005. Optimal sampling design for estimating spatial distribution and abundance of a freshwater mussel population. *Journal of the North American Benthological Society* 24:525–537.

Posthuma, L., G. W. Suter, and T. P. Traas, editors. 2002. *Species sensitivity distributions in ecotoxicology*. Lewis Publishers, Boca Raton, FL, USA.

Pradel, R. 1996. Utilization of capture–mark–recapture for the study of recruitment and population growth rate. *Biometrics* 52:703–709.

Pringle, C. M. 2000. Hydrologic connectivity and the management of biological reserves: a global perspective. *Ecological Applications* 11:981–998.

Purcell, K. L., S. R. Mori, and M. K. Chase. 2005. Design considerations for examining trends in avian abundance using point counts: examples from oak woodlands. *Condor* 107:305–320.

Qian, S. S. 2010. *Environmental and ecological statistics with R*. Chapman and Hall/CRC Press, Boca Raton, FL, USA.

Qian, S. S., and C. J. Richardson. 1997. Estimating the long-term phosphorus accretion rate in the Everglades: a Bayesian approach with risk assessment. *Water Resources Research* 33:1681–1688.

Qian, S. S., R. S. King, and C. J. Richardson. 2003. Two statistical methods for the detection of environmental thresholds. *Ecological Modelling* 166:87–97.

Qian, S. S., Y. Pan, and R. S. King. 2004. Soil total phosphorus threshold in the Everglades: a Bayesian changepoint analysis for multinomial response data. *Ecological Indicators* 4:29–37.

Quandt, R. E. 1958. The estimation of the parameter of a linear regression system obeying two separate regimes. *Journal of the American Statistical Association* 53:873–880.

Quinn, G. P., and M. J. Keough. 2002. *Experimental design and data analysis for biologists*. Cambridge University Press, Cambridge, UK.

R Development Core Team. 2011. *R: a language and environment for statistical computing*. R Foundation for Statistical Computing, Vienna, Austria. ISBN 3–900051-07–0. Online: <http://www.R-project.org>.

Rabe-Hesketh, S., and A. Skrondal. 2006. Multilevel modeling of complex survey data. *Journal of the Royal Statistical Society, Series A* 169:805–827.

Raftery, A. E., G. H. Givens, and J. E. Zeh. 1995. Inference from a deterministic population dynamics model for bowhead whales (with discussion). *Journal of the American Statistical Association* 90:402–430.

Raftovich, R. V., K. A. Wilkins, K. D. Richkus, *et al.* 2010. *Migratory bird hunting activity and harvest during the 2008 and 2009 hunting seasons*. US Fish and Wildlife Service, Laurel, MD, USA.

Ralph, C. J., and J. M. Scott, editors. 1981. Estimating numbers of terrestrial birds. *Studies in Avian Biology* 6. California Natural Resources Federation, Cooper Ornithological Society.

Ram, N., and K. J. Grimm. 2007. Using simple and complex growth models to articulate developmental change: matching method to theory. *International Journal of Behavioral Development* 31:303–316.

Ramsey, F. L., and R. Sjamsoe'oed. 1994. Habitat association studies in conjunction with adaptive cluster samples. *Environmental and Ecological Statistics* 1:121–132.

Rao, J. N. K. 2003. *Small area estimation*. Wiley, New York, NY, USA.

Rao, J. N. K., and A. J. Scott. 1981. The analysis of categorical data from complex sample surveys: chi-squared tests for goodness of fit and independence in two-way tables. *Journal of the American Statistical Association* 76:221–230.

Rao, J. N. K., and D. R. Thomas. 1988. The analysis of cross-classified categorical data from complex sample surveys. *Sociological Methodology* 18:213–269.

Rao, J. N. K., and D. R. Thomas. 1989. Chi-squared tests for contingency tables. Pages 89–114 *in* C. J. Skinner, D. Holt, and T. M. F. Smith, editors. *Analysis of data from complex surveys*. John Wiley & Sons, New York, NY, USA.

Rao, P. S. R. S. 1997. *Variance components estimation – mixed models, methodologies and applications*. Chapman & Hall/CRC, Boca Raton, FL, USA.

Rasbash, J., F. Steele, W. J. Browne, *et al.* 2009. A user's guide to MLwiN, v2.10. Centre for Multilevel Modelling, University of Bristol, UK. Online: <http://www.cmm.bristol.ac.uk/MLwiN/download/manuals.shtml>. Accessed 19 Aug 2010.

Raudenbush, S. W. 2008. Many small groups. Pages 207–236 *in* J. de Leeuw and E. Meijer, editors. *Handbook of multilevel analysis*. Springer, New York, NY, USA.

Raudenbush, S. W., and A. S. Bryk. 2002. *Hierarchical linear models: applications and data analysis methods*. Second edition. Sage, Thousand Oaks, CA, USA.

Rayfield, B., M.-J. Fortin, and A. Fall. 2011. Connectivity for conservation: a framework to classify network measures. *Ecology* 92:847–858.

Reams, G. A., W. D. Smith, M. H. Hansen, *et al.* 2005. The Forest Inventory and Analysis sampling frame. Pages 31–36 *in* P. L. Patterson and G. A. Reams, editors. *The enhanced forest inventory and analysis program – national sampling design and estimation procedures*. General Technical Report SRS-80. USDA Forest Service, Asheville, NC, USA.

Reese, G. C., K. R. Wilson, J. A. Hoeting, *et al.* 2005. Factors affecting species distribution predictions: a simulation modeling experiment. *Ecological Applications* 15:554–564.

Reid, L. M. 2001. The epidemiology of monitoring. *Journal of the American Water Resources Association* 37:815–820.

Reinhardt Adams, C., and S. M. Galatowitsch. 2008. The transition from invasive species control to native species promotion and its dependence on seed density thresholds. *Applied Vegetation Science* 11:131–138.

Renner, H., J. H. Reynolds, M. Sims, *et al.* 2011. Evaluating the power of surface attendance counts to detect long-term trends in populations of crevice-nesting auklets. *Environmental Monitoring and Assessment* 177:665–679.

Reynolds, J. H., W. L. Thompson, and B. Russell. 2011. Planning for success: identifying effective and efficient survey designs for monitoring. *Biological Conservation* 144:1278–1284.

Reynolds, J. H., C. A. Woody, N. E. Gove, *et al.* 2007. Efficiently estimating salmon escapement uncertainty using systematically sampled data. Pages 121–129 *in* C. A. Woody, editor. *Sockeye salmon ecology, evolution, and management*. American Fisheries Society Symposium No. 54. American Fisheries Society, Bethesda, MD, USA.

Reynolds, R. E., T. L. Shaffer, R. W. Renner, *et al.* 2001. Impact of the Conservation Reserve Program on duck recruitment in the U.S. Prairie Pothole Region. *Journal of Wildlife Management* 65:765–780.

Rhodes, J. R., A. J. Tyre, N. Jonzen, *et al.* 2006. Optimizing presence–absence surveys for detecting population trends. *Journal of Wildlife Management* 70:8–18.

Richardson, C. J., and S. S. Qian. 1999. Long-term phosphorus assimilative capacity in freshwater wetlands: a new paradigm for sustaining ecosystem structure and function. *Environmental Science and Technology* 33:1545–1551.

Ringold, P. L., J. Alegria, R. L. Czaplewski, *et al.* 1996. Adaptive monitoring design for ecosystem management. *Ecological Applications* 6:745–747.

Rittenhouse, C. D., A. M. Pidgeon, T. P. Albright, *et al.* 2010. Conservation of forest birds: evidence of a shifting baseline in community structure. *PLoS ONE* 5:e11938.

Robbins, C. S., D. Bystrak, and P. H. Geissler. 1986. *The Breeding Bird Survey: its first fifteen years, 1965–1979*. Resource Publication 157. US Department of the Interior, Fish and Wildlife Service, Washington, DC, USA.

Robinson, D. H., and H. Wainer. 2002. On the past and future of null hypothesis significance testing. *The Journal of Wildlife Management* 66:263–271.

Robson, D. S., and H. A. Regier. 1964. Sample size in Petersen mark–recapture experiments. *Transactions of the American Fisheries Society* 93:215–226.

Rodriguez, G., and N. Goldman. 1995. An assessment of estimation procedures for multilevel models with binary responses. *Journal of the Royal Statistical Society, Series A* 158:73–89.

Rodriguez, G., and N. Goldman. 2001. Improved estimation procedures for multilevel models with binary response: a case-study. *Journal of the Royal Statistical Society, Series A* 164:339–355.

Roesch, F. A., Jr. 1993. Adaptive cluster sampling for forest inventories. *Forest Science* 39:655–669.

Root, T. 1988. Environmental factors associated with avian distributional boundaries. *Journal of Biogeography* 15:489–505.

Rosenstock, S. S., D. R. Anderson, K. M. Giesen, *et al.* 2002. Landbird counting techniques: current practice and an alternative. *Auk* 119:46–53.

Rotenberry, J. T., and S. T. Knick. 1995. Evaluation of bias in roadside point count surveys of passerines in shrubsteppe and grassland habitats in southwestern Idaho. Pages 99–101 *in* C. J. Ralph, J. R. Sauer, and S. Droege, editors. *Monitoring bird populations by point counts*. General Technical Report PSW-GTR-149. USDA Forest Service, Albany, CA, USA.

Rout, T. M., C. E. Hauser, and H. P. Possingham. 2009. Optimal adaptive management for the translocation of a threatened species. *Ecological Applications* 19:515–526.

Royle, J. A. 2004. *N*-mixture models for estimating population size from spatially replicated counts. *Biometrics* 60:108–115.

Royle, J. A. 2006. Site occupancy models with heterogeneous detection probabilities. *Biometrics* 62:97–102.

Royle, J. A., and R. M. Dorazio. 2006. Hierarchical models of animal abundance and occurrence. *Journal of Agricultural, Biological, and Environmental Statistics* 11:249–263.

Royle, J. A., and R. M. Dorazio. 2008. *Hierarchical modeling and inference in ecology: the analysis of data from populations, metapopulations and communities*. Academic Press, San Diego, CA, USA.

Royle, J. A., and M. Kéry. 2007. A Bayesian state–space formulation of dynamic occupancy models. *Ecology* 88:1813–1823.

Royle, J. A., and W. A. Link. 2005. A general class of multinomial mixture models for anuran calling survey data. *Ecology* 86:2505–2512.

Royle, J. A., and J. D. Nichols. 2003. Estimating abundance from repeated presence–absence data or point counts. *Ecology* 84:777–790.

Royle, J. A., and K. Young. 2008. A hierarchical model for spatial capture–recapture data. *Ecology* 89:2281–2289.

Royle, J. A., S. J. Converse, and W. A. Link. In prep. Hierarchical modeling of abundance in spatially stratified populations using data augmentation.

Royle, J. A., R. M. Dorazio, and W. A. Link. 2007. Analysis of multinomial models with unknown index using data augmentation. *Journal of Computational and Graphical Statistics* 16:67–85.

Rubin, D. B. 1976. Inference and missing data. *Biometrika* 63:581–592.

Rubin, D. B. 1987. *Multiple imputation for nonresponse in surveys*. John Wiley & Sons, New York, NY, USA.

Rue, H., and L. Held. 2005. *Gaussian Markov random fields: theory and applications*. Chapman & Hall/CRC, Boca Raton, FL, USA.

Rue, H., and S. Martino. 2009. INLA: functions which allow to perform a full Bayesian analysis of structured additive models using Integrated Nested Laplace Approximaxion. R package version 0.0. Online: <http://www.r-inla.org/>.

Rue, H., S. Martino, and N. Chopin. 2009. Approximate Bayesian inference for latent Gaussian models using integrated nested Laplace approximations. *Journal of the Royal Statistical Society, Series B* 71:319–392.

Salehi, M., and J. A. Brown. 2010. Complete allocation sampling: an efficient and easily implemented adaptive sampling design. *Population Ecology* 52:451–456.

Salehi, M. M., and G. A. F. Seber. 1997. Two-stage adaptive cluster sampling. *Biometrics* 53:959–970.

Salehi, M. M., and D. R. Smith. 2005. Two-stage sequential sampling: a neighborhood-free adaptive sampling procedure. *Journal of Agricultural, Biological, and Environmental Statistics* 10:84–103.

Salehi, M., M. Moradi, J. A. Brown, *et al.* 2010. Efficient estimators for adaptive stratified sequential sampling. *Journal of Statistical Computation and Simulation* 80:1163–1179.

Salzer, D., and N. Salafsky. 2006. Allocating resources between taking action, assessing status, and measuring effectiveness of conservation actions. *Natural Areas Journal* 26:310–316.

Saracco, J. F., D. F. DeSante, M. P. Nott, *et al.* 2009. Using the MAPS and MoSI programs to monitor landbirds and inform conservation. Pages 651–658 *in* T. D. Rich, C. D. Thompson, D. Demarest, and C. Arizmendi, editors. *Proceedings of the Fourth International Partners in Flight Conference: Tundra to Topics*. University of Texas-Pan American Press, Edinburg, TX, USA.

Sargeant, G. A., M. A. Sovada, C. S. Slivinski, *et al.* 2005. Markov chain Monte Carlo estimate of species distributions: a case study of the swift fox in western Kansas. *Journal of Wildlife Management* 69:483–497.

Särndal, C.-E., B. Swensson, and J. Wretman. 1992. *Model assisted survey sampling*. Springer, New York, NY, USA.

Sarpola, M. J., R. K. Paasch, E. N. Mortensen, *et al.* 2008. An aquatic insect imaging system to automate insect classification. *Transactions of the American Society of Agricultural and Biological Engineers* 51:2217–2225.

SAS Institute. 2001. Version 8.02. SAS Institute, Cary, NC, USA.

SAS Institute. 2008. Version 9.2. SAS Institute, Cary, NC, USA.

Sauer, J. R., and W. A. Link. 2002. Hierarchical modeling of population stability and species group attributes using Markov chain Monte Carlo methods. *Ecology* 83:1743–1751.

Sauer, J. R., and W. A. Link. 2011. Analysis of the North American Breeding Bird Survey using hierarchical models. *Auk* 128:87–98.

Sauer, J. R., J. E. Hines, and J. Fallon. 2008. *The North American Breeding Bird Survey, Results and Analysis 1966–2007*. Version 5.15.2008. US Geological Survey, Patuxent Wildlife Research Center, Laurel, MD, USA.

Sauer, J. R., W. A. Link, W. L. Kendall, *et al.* 2008. A hierarchical model for estimating change in American woodcock populations. *Journal of Wildlife Management* 72:204–214.

Sauer, J. R., W. A. Link, and J. A. Royle. 2004. Estimating population trends with a linear model: technical comments. *Condor* 106:435–440.

Sauer, J. R., B. G. Peterjohn, and W. A. Link. 1994. Observer differences in the North American Breeding Bird Survey. *Auk* 111:50–62.

Schafer, J. L., and J. W. Graham. 2002. Missing data: our view of the state of the art. *Psychological Methods* 7:147–177.

Scheaffer, R. L., W. Mendenhall, and L. Ott. 1979. *Elementary survey sampling*. Second edition. Duxbury Press, North Scituate, MA, USA.

Scheaffer, R. L., W. Mendenhall, and L. Ott. 1986. *Elementary survey sampling*. Third edition. PWK-Kent Publishing Co., Boston, MA, USA.

Scheffé, H. 1956. Alternative models for the analysis of variance. *Annals of Mathematical Statistics* 27:251–271.

Schelz, C. 2002. *Vegetation long-term monitoring – Canyonlands National Park 2002 Annual report, Vols. I–III*. Southeast Utah Group, National Park Service, Moab, UT, USA.

Schmelzer, I., and P. Harris. 2009. The 2006 International Piping Plover Breeding Census in Newfoundland. Pages 153–158 *in* E. Elliott-Smith, S. M. Haig, and B. M. Powers. *Data from the 2006 International Piping Plover Census*. US Geological Survey Data Series 426. US Department of the Interior, US Geological Survey. Reston, VA, USA.

Schmit, J. P., G. Sander, M. Lehman, *et al.* 2009. *National Capital Region Network long-term forest monitoring protocol. Version 2.0*. Natural Resource Report NPS/NCRN/NRR-2009/113. US Department of the Interior, National Park Service, Fort Collins, CO, USA.

Schonewald-Cox, C. M. 1988. Boundaries in the protection of nature reserves. *BioScience* 37:480–486.

Schreuder, H. T., R. Ernst, and H. Ramirez-Maldonado. 2004. *Statistical techniques for sampling and monitoring natural resources*. General Technical Report RMRS-GTR-126. USDA Forest Service, Fort Collins, CO, USA.

Schreuder, H. T., T. G. Gregoire, and G. B. Wood. 1993. *Sampling methods for multiresource forest inventory*. Wiley, New York, NY, USA.

Schwarz, C. J., and A. N. Arnason. 1996. A general methodology for the analysis of capture–recapture experiments in open populations. *Biometrics* 52:860–873.

Schwarz, C. J., J. F. Schweigert, and A. N. Arnason. 1993. Estimating migration rate using tag recovery data. *Biometrics* 44:765–785.

Scott, A. J., and J. N. K. Rao. 1981. Chi-squared tests for contingency tables with proportions estimated from survey data. Pages 247–266 *in* D. Krewski, R. Platek, and J. N. K. Rao, editors. *Current topics in survey sampling*. Academic Press, New York, NY, USA.

Scott, C. T., W. A. Bechtold, G. A. Reams, *et al.* 2005. Sample-based estimators used by the Forest Inventory and Analysis national information management system. Pages 53–77 *in* P. L. Patterson, and G. A. Reams, editors. *The enhanced forest inventory and analysis program – national sampling design and estimation procedures*. General Technical Report SRS-80. USDA Forest Service, Asheville, NC, USA.

Scott, D. W. 1985. Averaged shifted histograms: effective nonparametric density estimators in several dimensions. *Annals of Statistics* 13:1024–1040.

Searle, S. R. 1971. *Linear models*. Wiley, New York, NY, USA.

Searle, S. R. 1987. *Linear models for unbalanced data*. Wiley & Sons, New York, NY, USA.

Searle, S. R., G. Casella, and C. E. McCulloch. 1992. *Variance components*. Wiley, New York, NY, USA.

Sebastiani, P., and H. P. Wynn. 2000. Maximum entropy sampling and optimal Bayesian experimental design. *Journal of the Royal Statistical Society, Series B* 62:145–157.

Seber, G. A. F. 1965. A note on the multiple recapture census. *Biometrika* 52:249–259.

Seber, G. A. F. 1982. *The estimation of animal abundance and related parameters*. Second edition. Macmillan, New York, NY, USA.

Seber, G. A. F., and C. J. Wild. 1989. *Nonlinear regression*. Wiley, New York, NY, USA.

Sellars, R. W. 1997. *Preserving nature in the national parks: a history*. Yale University Press, New Haven, CT, USA.

Shafer, C. L. 1995. Values and shortcomings of small reserves. *BioScience* 45:80–88.

Shaffer, T. L., and D. H. Johnson. 2008. Ways of learning: observational studies versus experiments. *Journal of Wildlife Management* 72:4–13.

Shao, Q. 2009. Nonparametric trend estimation for periodic autoregressive time series. *Communications in Statistics: Theory and Methods* 38:2418–2427.

Sher, A. A., D. L. Marshall, and J. P. Taylor. 2002. Establishment patterns of native *Populus* and *Salix* in the presence of invasive nonnative *Tamarix*. *Ecological Applications* 12:760–772.

Shewry, M. C., and H. P. Wynn. 1987. Maximum entropy sampling. *Journal of Applied Statistics* 14:165–170.

Shimotsu, K. 2009. Exact local Whittle estimation of fractional integration with unknown mean and time trend. *Econometric Theory* 26:501–540.

Shipley, B. 2000. *Cause and correlation in biology*. Cambridge University Press, Cambridge, UK.

Shipley, B. 2009. Confirmatory path analysis in a generalized multilevel context. *Ecology* 90:363–368.

Shrader-Frechette, K. 2007. *Taking action, saving lives: our duties to protect environmental and public health*. Oxford University Press, Cambridge, UK.

Shrader-Frechette, K. S., and E. D. McCoy. 1992. Statistics, costs and rationality in ecological inference. *Trends in Ecology & Evolution* 7:96–99.

Si, J. H., Q. Feng, X. Y. Zhang, *et al.* 2005. Growing season evapotransportation from *Tamarix ramosissima* stands under extreme arid conditions in northwest China. *Environmental Geology* 48:861–870.

Silvertown, J. 2009. A new dawn for citizen science. *Trends in Ecology & Evolution* 24:467–471.

Simons, T. R., M. W. Alldredge, K. H. Pollock, *et al.* 2007. Experimental analysis of the auditory detection process on avian point counts. *Auk* 124:986–999.

Sims, M., D. E. Elston, M. P. Harris, *et al.* 2007. Incorporating variance uncertainty into a power analysis of monitoring designs. *Journal of Agricultural, Biological, and Environmental Statistics* 12:236–249.

Skalski, J. R. 1992. Sample size calculations for normal variates under binomial censoring. *Biometrics* 48:877–882.

Skalski, J. R., and D. H. McKenzie. 1982. A design for aquatic monitoring programs. *Journal of Environmental Management* 14:237–251.

Skalski, J. R., and D. S. Robson. 1992. *Techniques for wildlife investigations: design and analysis of capture data*. Academic Press, San Diego, CA, USA.

Skalski, J. R., K. E. Ryding, and J. J. Millspaugh. 2005. *Wildlife demography: analysis of sex, age, and count data*. Elsevier Science, San Diego, CA, USA.

Skinner, S. P., and R. G. Clark. 2008. Relationships between duck and grassland bird relative abundance and species richness in southern Saskatchewan. *Avian Conservation and Ecology* 3(1):1. Online: <http://www.ace-eco.org/vol3/iss1/art1/>.

Smith, A. F. M. 1975. A Bayesian approach to inference about a change-point in a sequence of random variables. *Biometrika* 62:407–416.

Smith, A. F. M. 1980. Change-point problems: approaches and applications. Pages 83–98 *in* J. M. Bernardo, editor. *Bayesian Statistics. Proceedings of the First International Meeting held in Valencia (Spain)*. University Press, Valencia, Spain.

Smith, D. H. V., S. J. Converse, K. W. Gibson, *et al.* 2011. Decision analysis for conservation breeding: maximizing production for reintroduction of whooping cranes. *Journal of Wildlife Management* 75:501–508.

Smith, D. R., J. A. Brown, and N. C. H. Lo. 2004. Application of adaptive cluster sampling to biological populations. Pages 77–122 *in* W. L. Thompson, editor. *Sampling rare or elusive species: concepts, designs, and techniques for estimating population parameters.* Island Press, Washington, DC, USA.

Smith, D. R., M. J. Conroy, and D. H. Brakhage. 1995. Efficiency of adaptive cluster sampling for estimating density of wintering waterfowl. *Biometrics* 51:777–788.

Smith, D. R., B. R. Gray, T. J. Newton, *et al.* 2010. Effect of imperfect detectability on adaptive and conventional sampling: simulated sampling of freshwater mussels in the Upper Mississippi River. *Environmental Monitoring and Assessment* 170:499–507.

Smith, D. R., J. T. Rogala, B. R. Gray, *et al.* 2011. Evaluation of single and two-stage adaptive sampling designs for estimation of density and abundance of freshwater mussels in a large river. *Rivers Research and Applications* 27:122–133.

Smith, D. R., R. F. Villella, and D. P. Lemari. 2003. Application of adaptive cluster sampling to low-density populations of freshwater mussels. *Environmental and Ecological Statistics* 10:7–15.

Smith, G. W. 1995. *A critical review of the aerial and ground surveys of breeding waterfowl in North America.* Biological Science Report 5. US Department of the Interior, National Biological Service, Washington, DC, USA.

Smith, W. B. 2002. Forest Inventory and Analysis: a national inventory and monitoring program. *Environmental Pollution* 116:S233–S242.

Snedecor, G. W., and W. G. Cochran. 1967. *Statistical methods.* Sixth edition. Iowa State University Press, Ames, IA, USA.

Snee, R. 1990. Statistical thinking and its contribution to total quality. *American Statistician* 44:116–121.

Snijders, T. A. B., and J. Berkhof. 2008. Diagnostic checks for multilevel models. Pages 141–175 *in* J. de Leeuw and E. Meijer, editors. *Handbook of multilevel analysis.* Springer, New York, NY, USA.

Snijders, T. A. B., and R. J. Bosker. 1993. Standard errors and sample sizes for two-level research. *Journal of Educational Statistics* 18:237–259.

Snijders, T. A. B., and R. J. Bosker. 1999. *Multilevel analysis.* Sage Publications, London, UK.

Soldaat, L., H. Visser, M. van Roomen, and A. van Strien. 2007. Smoothing and trend detection in waterbird monitoring data using structural time-series analysis and the Kalman filter. *Journal of Ornithology* 148:S351–S357.

Spellerberg, I. F. 1991. *Monitoring ecological change.* Cambridge University Press, Cambridge, UK.

Spiegelhalter, D. J., N. G. Best, B. R. Carlin, *et al.* 2002. Bayesian measures of complexity and fit (with discussion). *Journal of the Royal Statistical Society, Series B* 64:583–639.

Spilke, J., H. P. Piepho, and X. Hu. 2005. Analysis of unbalanced data by mixed linear models using the MIXED procedure of the SAS system. *Journal of Agronomy and Crop Science* 191:47–54.

Spybrook, J., S. W. Raudenbush, R. Congdon, *et al.* 2009. Optimal design for longitudinal and multilevel research: documentation for the "Optimal Design" software. Online: <http://www.wtgrantfoundation.org/resources/overview/research_tools>. Accessed 19 Aug 2010.

StataCorp. 2009. Stata Statistical Software: Release 11. StataCorp LP, College Station, TX, USA.

Stauffer, H. B., C. J. Ralph, and S. L. Miller. 2004. Ranking habitat for marbled murrelets: a new conservation approach for species with uncertain detection. *Ecological Applications* 14:1374–1383.

Steel, R. G. D., and J. H. Torrie. 1980. *Principles and procedures of statistics, a biometrical approach*. McGraw-Hill Kogakusha, Tokyo, Japan.

Stefanski, F. A., and J. M. Bay. 1996. Simulation extrapolation deconvolution of finite population cumulative distribution function estimators. *Biometrika* 83:407–417.

Stephan, F., and J. McCarthy. 1958. *Sampling opinions: An analysis of survey procedure*. Wiley, New York, NY, USA.

Stephens, D. A. 1994. Bayesian retrospective multiple-changepoint identification. *Journal of the Royal Statistical Society, Series C* 43:159–178.

Stevens, D. L., Jr. 1997. Variable density grid-based sampling designs for continuous spatial populations. *Environmetrics* 8:167–195.

Stevens, D. L., Jr., and A. R. Olsen. 1999. Spatially restricted surveys over time for aquatic resources. *Journal of Agricultural, Biological, and Environmental Statistics* 4:415–428.

Stevens, D. L., Jr., and A. R. Olsen. 2003. Variance estimation for spatially balanced samples of environmental resources. *Environmetrics* 14:593–610.

Stevens, D. L., Jr., and A. R. Olsen. 2004. Spatially balanced sampling of natural resources. *Journal of the American Statistical Association* 99:262–278.

Stevens, D. L., Jr., and N. S. Urquhart. 2000. Response designs and support regions in sampling continuous domains. *Environmetrics* 11:13–41.

Stevens, D. L., Jr., D. P. Larsen, and A. R. Olsen. 2007. The role of sample surveys: why should practitioners consider using a statistical sampling design? Pages 11–23 *in* D. H. Johnson, B. M. Shrier, J. S. O'Neal, *et al.*, editors. *Salmonid field protocols handbook: techniques for assessing status and trends in salmon and trout populations*. American Fisheries Society, Bethesda, MD, USA.

Stoddard, J. L., A. T. Herlihy, D. V. Peck, *et al.* 2008. A process for creating multi-metric indices for large-scale aquatic surveys. *Journal of North American Benthological Society* 27:878–891.

Stokes, S. L. 1984. The Jolly–Seber method applied to age-stratified populations. *Journal of Wildlife Management* 48:1053–1059.

Stone, M. 1974. Cross-validation and multinomial prediction. *Biometrika* 61:509–515.

Strayer, D., J. S. Glitzenstein, C. G. Jones, *et al.* 1986. *Long-term ecological studies: an illustrated account of their design, operation, and importance to ecology*. Occasional Publication of the Institute of Ecosystem Studies, No. 2. Millbrook, NY, USA.

Stroud, D. 2003. Are waders world-wide in decline? Reviewing the evidence. A workshop summary. *Wader Study Group Bulletin* 101/102:8–12.

Stryhn, H., J. Sanchez, P. Morley, *et al.* 2008. Interpretation of variance parameters in multilevel Poisson regression models. *Proceedings of the 11th International Symposium on Veterinary Epidemiology and Economics*, 2006, Cairns, Australia.

Sturtz, S., U. Ligges, and A. Gelman. 2005. R2WinBUGS: a package for running WinBUGS from R. *Journal of Statistical Software* 12:1–16.

Symstad, A. J., R. A. Gitzen, C. L. Wienk, *et al.* 2011. *Plant community composition and structure monitoring protocol for the Northern Great Plains I&M Network: Version 1.00*. Natural Resource Report NPS/NRPC/NRR – 2011/291. National Park Service, Fort Collins, CO, USA.

Szabo, J. K., P. J. Davy, M. J. Hooper, *et al.* 2007. Predicting spatio-temporal distribution for eastern Australian birds using Birds Australia's atlas data: survey method, habitat and seasonal effects. *Emu* 107:89–99.

Taylor, B. L., and T. Gerrodette. 1993. The uses of statistical power in conservation biology: the vaquita and northern spotted owl. *Conservation Biology* 7:489–500.

Taylor, B. L., M. Martinez, T. Gerrodette, *et al.* 2007. Lessons from monitoring trends in abundance of marine mammals. *Marine Mammal Science* 23:157–175.

Tenhumberg, B., A. J. Tyre, K. Shea, *et al.* 2004. Linking wild and captive populations to maximize species persistence: optimal translocation strategies. *Conservation Biology* 18:1–11.

ter Braak, C. J. F., D. J. Brus, and E. J. Pebesma. 2008. Comparing sampling patterns for kriging the spatial mean temporal trend. *Journal of Agricultural, Biological, and Environmental Statistics* 13:159–176.

ter Braak, C. J. F., A. J. van Strien, R. Meijer, *et al.* 1994. Analysis of monitoring data with many missing values: which method? Pages 663–673 *in* E. J. M. Hagemeijer and T. J. Verstrael, editors. *Bird Numbers 1992. Distribution, monitoring and ecological aspects. Proceedings of the 12th International Conference of IBCC and EOAC, Noordwijkerhout, The Netherlands.* Statistics Netherlands, Voorburg/Heerlen & SOVON, Beek-Ubbergen, The Netherlands.

Theobald, D. M., D. L. Stevens, Jr., D. White, *et al.* 2007. Using GIS to generate spatially balanced random survey designs for natural resource applications. *Environmental Management* 40:134–146.

Thomas, C. D., and J. J. Lennon. 1999. Birds extend their ranges northwards. *Nature* 399:213–213.

Thomas, D. R., G. R. Roberts, and A. C. Singh. 1996. Tests of independence on two-way tables under cluster sampling. *International Statistical Review* 64:295–311.

Thomas, L. 1996. Monitoring long-term population change: why are there so many analysis methods? *Ecology* 77:49–58.

Thomas, L., and K. Martin. 1996. The importance of analysis method for Breeding Bird Survey population trend estimates. *Conservation Biology* 10:479–490.

Thomas, L., J. L. Laake, E. Rexstad, *et al.* 2009. Distance 6.0 release 'x'1. Research unit for wildlife population assessment, University of St. Andrews, UK. Online: <http://www.ruwpa.st-and.ac.uk/distance/>. Accessed 5 May 2009.

Thompson, F. R., III, and F. A. La Sorte. 2008. Comparison of methods for estimating bird abundance and trends from historical count data. *Journal of Wildlife Management* 72:1674–1682.

Thompson, S. K. 1990. Adaptive cluster sampling. *Journal of the American Statistical Association* 85:1050–1059.

Thompson, S. K. 1991. Adaptive stratified cluster sampling. *Biometrika* 78:389–397.

Thompson, S. K. 2002. *Sampling.* Second edition. Wiley, New York, NY, USA.

Thompson, S. K., and G. A. F. Seber. 1994. Detectability in conventional and adaptive sampling. *Biometrics* 50:712–724.

Thompson, S. K., and G. A. F. Seber. 1996. *Adaptive sampling.* Wiley, New York, NY, USA.

Thompson, W. L., editor. 2004. *Sampling rare or elusive species.* Island Press, Washington, DC, USA.

Thompson, W. L., G. C. White, and C. Gowan. 1998. *Monitoring vertebrate populations.* Academic Press, San Diego, CA, USA.

Thuiller, W., B. Lafourcade, R. Engler, *et al.* 2009. BIOMOD – a platform for ensemble forecasting of species distributions. *Ecography* 32:369–373.

Tierney, G. L., D. Faber-Langendoen, B. R. Mitchell, *et al.* 2009. Monitoring and evaluating the ecological integrity of forest ecosystems. *Frontiers in Ecology and the Environment* 7:308–316.

Topp-Jørgensen, E., M. K. Poulsen, J. F. Lund, *et al.* 2005. Community-based monitoring of natural resource use and forest quality in montane forests and miombo woodlands in Iringa District, Tanzania. *Biodiversity and Conservation* 14: 2653–2677.

Townsend, P. A., T. R. Lookingbill, C. C. Kingdon, *et al.* 2009. Spatial pattern analysis for monitoring protected areas. *Remote Sensing of Environment* 113:1410–1420.

Trewhella, W. J., K. M. Rodriguez-Clark, N. Corp, *et al.* 2005. Environmental education as a component of multidisciplinary conservation programs: lessons from conservation initiatives for critically endangered fruit bats in the western Indian Ocean. *Conservation Biology* 19:75–85.

Trimble, M. J., and R. J. van Aarde. 2011. Decline of birds in a human modified coastal dune forest landscape in South Africa. *PLoS ONE* 6:e16176.

Trzcinski, M. K., L. Fahrig, and G. Merriam. 1999. Independent effects of forest cover and fragmentation on the distribution of forest breeding birds. *Ecological Applications* 9:586–593.

Tukey, J. W. 1949. Memorandum on statistics in the federal government, part I, chapters I–IV. *American Statistician* 3:6–17.

Turchin, P. 2003. *Complex population dynamics: a theoretical/empirical synthesis*. Princeton University Press, Princeton, NJ, USA.

Turk, P., and J. J. Borkowski. 2005. A review of adaptive cluster sampling: 1990–2003. *Environmental and Ecological Statistics* 12:55–94.

Tversky, A., and D. Kahneman. 1982. Judgment under uncertainty: heuristics and biases. Pages 3–22 *in* D. Kahneman, P. Slovic, and A. Tversky, editors. *Judgment under uncertainty: heuristics and biases*. Cambridge University Press, New York, NY, USA.

Tyre, A. J., B. Tenhumberg, S. A. Field, *et al.* 2003. Improving precision and reducing bias in biological surveys: estimating false-negative error rates. *Ecological Applications* 13:1790–1801.

USDA Forest Service. 2007. *Forest Inventory and Analysis Strategic Plan: a history of success; a dynamic future*. FS-865. USDA Forest Service, Washington, DC, USA.

US Environmental Protection Agency [USEPA]. 1975. *National Eutrophication Survey Methods 1973–1976*. US Environmental Protection Agency, Office of Research and Development, National Environmental Research Center, Corvallis, OR, USA.

US Environmental Protection Agency [USEPA]. 1993. *Environmental Monitoring and Assessment Program Guide*, EPA/600/XX-93/XXX. US Environmental Protection Agency, Atmospheric Research and Exposure Assessment Laboratory, Research Triangle Park, NC, USA.

US Environmental Protection Agency [USEPA]. 2002. *Guidance on choosing a sampling design for environmental data collection, for use in developing a quality assurance plan*. Report EPA QA/G-5S. US Environmental Protection Agency, Washington, DC, USA.

US Environmental Protection Agency [USEPA]. 2009a. EPA's Quality System for Environmental Data and Technology. US Environmental Protection Agency. Online: <http://www.epa.gov/quality/index.html>. Accessed 12 Oct 2009.

US Environmental Protection Agency [USEPA]. 2009b. *National Lakes Assessment: a collaborative survey of the Nation's lakes*. EPA 841-R-09-001. US Environmental Protection Agency, Office of Water and Office of Research and Development, Washington, DC, USA.

US Fish and Wildlife Service [USFWS]. 2009a. *Trends in duck breeding populations, 1955–2009*. US Department of the Interior, Washington, DC, USA.

US Fish and Wildlife Service [USFWS]. 2009b. *Waterfowl population status 2009*. US Department of the Interior, Washington, DC, USA.

US Fish and Wildlife Service [USFWS]. 2010a. *Adaptive harvest management: 2010 hunting season*. US Department of the Interior, Washington, DC, USA.

US Fish and Wildlife Service [USFWS]. 2010b. *Waterfowl population status, 2010*. US Department of the Interior, Washington, DC, USA.

US Fish and Wildlife Service and US Geological Survey [USFWS and USGS]. 2010. Migratory Bird Data Center. Online: <http://migbirdapps.fws.gov/>. Accessed 5 January 2010.

US Geological Survey [USGS]. 2007. *Strategic Plan for the North American Breeding Bird Survey: 2006–2010*. Circular 1307. US Department of the Interior, US Geological Survey, Reston, VA, USA.

US North American Bird Conservation Initiative Monitoring Subcommittee (US NABCI). 2007. Opportunities for improving avian monitoring. US North American Bird Conservation Initiative Report. Online: <http://www.nabci-us.org/aboutnabci/monitoringreportfinal0307.pdf>. Accessed 4 January 2010.

Underhill, L. G., and R. P. Prys-Jones. 1994. Index numbers for waterbird populations. I: Review and methodology. *Journal of Applied Ecology* 31:463–480.

Urban, D. L. 2000. Using model analysis to design monitoring programs for landscape management and impact assessment. *Ecological Applications* 10:1820–1832.

Urban, D. L. 2002. Tactical monitoring of landscapes. Pages 294–311 *in* J. Liu and W. W. Taylor, editors. *Integrating landscape ecology into natural resource management*. Cambridge University Press, Cambridge, UK.

Urban, D., and T. Keitt. 2001. Landscape connectivity: a graph-theoretic perspective. *Ecology* 82:1205–1218.

Urquhart, N. S., and T. M. Kincaid. 1999. Designs for detecting trend from repeated surveys of ecological resources. *Journal of Agricultural, Biological, and Environmental Statistics* 4:404–414.

Urquhart, N. S., W. S. Overton, and D. S. Birkes. 1993. Comparing sampling designs for monitoring ecological status and trends: impact of temporal patterns. Pages 71–85 *in* V. Barnet and K. F. Turkman, editors. *Statistics for the environment*. John Wiley and Sons, London, UK.

Urquhart, N. S., S. G. Paulsen, and D. P. Larsen. 1998. Monitoring for policy-relevant regional trends over time. *Ecological Applications* 8:246–257.

Vallecillo, S., L. Brotons, and W. Thuiller. 2009. Dangers of predicting bird species distributions in response to land-cover changes. *Ecological Applications* 19:538–549.

Valliant, R., A. H. Dorfman, and R. M. Royall. 2000. *Finite population sampling and inference: a prediction approach*. Wiley, New York, NY, USA.

van Belle, G. 2008. *Statistical rules of thumb*. Wiley, Hoboken, NJ, USA.

Van Campenhout, B. 2006. Modeling trends in food market integration: method and an application to Tanzanian maize markets. *Food Policy* 32:112–127.

van de Pol M., Y. Vindenes, B. E. Saether, *et al.* 2010. Effects of climate change and variability on population dynamics in a long-lived shorebird. *Ecology* 91:1192–1204.

Van Grinsven, M., M. Malick, M. Moran, *et al.* 2010. *Water quality monitoring in the Northern Colorado Plateau Network, 2006–2009*. Natural Resource Technical Report NPS/NCPN/NRTR–2010/358. US Department of the Interior, National Park Service, Fort Collins, CO, USA.

van Mantgem, P. J., N. L. Stephenson, J. C. Byrne, *et al.* 2009. Widespread increase of tree mortality rates in the western United States. *Science* 323:521–524.

VanLeeuwen, D. 1993. *Completeness and sufficiency under normality in mixed model designs.* Dissertation, Oregon State University, Corvallis, OR, USA.

VanLeeuwen, D., and N. S. Urquhart 1994. Estimating components of variance for a linear model having a mixed fixed/random factor in a large data structure. Pages 1064–1068 *in* Proceedings of the 19th SAS Users Group International, 10–13 April, 1994. Dallas, TX, USA.

VanLeeuwen, D. M., L. W. Murray, and N. S. Urquhart 1996. A mixed model with both fixed and random trend components across time. *Journal of Agricultural, Biological and Environmental Statistics* 1:435–453.

Vaughan, H., G. Whitelaw, B. Craig, *et al.* 2003. Linking ecological science to decision-making: delivering environmental monitoring information as societal feedback. *Environmental Monitoring and Assessment* 88:399–408.

Vaughan, I. P., and S. J. Ormerod. 2003. Modelling the distribution of organisms for conservation: optimising the collection of field data for model development. *Conservation Biology* 17:1601–1611.

Ver Hoef, J. M., and P. L. Boveng. 2007. Quasi-Poisson vs. negative binomial regression: how should we model overdispersed count data? *Ecology* 88:2766–2772.

Ver Hoef, J. M., E. Peterson, and D. Theobald. 2006. Spatial statistical models that use flow and stream distance. *Environmental and Ecological Statistics* 13:449–464.

Viljugrein, H., N. C. Stenseth, G. W. Smith, *et al.* 2005. Density dependence in North American ducks. *Ecology* 86:245–254.

Villella, R. F., and D. R. Smith. 2005. Two-phase sampling to estimate river-wide populations of freshwater mussels. *Journal of the North American Benthological Society* 24:357–368.

Vogt, P., K. Riitters, C. Estreguil, *et al.* 2007. Mapping spatial patterns with morphological image processing. *Landscape Ecology* 22:171–177.

Wald, A. 1943. Tests of statistical hypotheses concerning several parameters when the number of observations is large. *Transactions of the American Mathematical Society* 54:426–482.

Walters, C. J. 1986. *Adaptive management of renewable resources.* MacMillan, New York, NY, USA.

Walters, C. J., and C. S. Holling. 1990. Large-scale management experiments and learning by doing. *Ecology* 71:2060–2068.

Wang, F., and A. E. Gelfand. 2002. A simulation-based approach to Bayesian sample size determination for performance under a given model and for separating models. *Statistical Science* 17:193–208.

White, D., A. J. Kimmerling, and W. S. Overton. 1992. Cartographic and geometric components of a global sampling design for environmental monitoring. *Cartography and Geographic Information Systems* 19:5–22.

White, G. C. 2005. Correcting wildlife counts using detection probabilities. *Wildlife Research* 32:211–216.

White, G. C., and K. P. Burnham. 1999. Program MARK: survival estimation from populations of marked animals. *Bird Study Supplement* 46:120–138.

White, G. C., and R. A. Garrott. 1990. *Analysis of wildlife radio-tracking data.* Academic Press, San Diego, CA, USA.

Whitelaw, G., H. Vaughan, B. Craig, *et al.* 2003. Establishing the Canadian community monitoring network. *Environmental Monitoring and Assessment* 88:409–418.

Wiens, J. A., B. van Horne, and B. R. Noon. 2002. Integrating landscape structure and scale into natural resource management. Pages 23–67 *in* J. Liu and W. W. Taylor, editors. *Integrating landscape ecology into natural resource management.* Cambridge University Press, New York, NY, USA.

Wiersma, Y. F. 2010. Birding 2.0: citizen science and effective monitoring in the Web 2.0 world. *Avian Conservation and Ecology* 5(2):13. Online: <http://www.ace-eco.org/vol5/iss2/art13/>.

Wight, J. R. 1967. The sampling unit and its effect on saltbush yield estimates. *Journal of Range Management* 20:323–325.

Wikle, C. K., and M. B. Hooten. 2010. A general science-based framework for nonlinear spatio-temporal dynamical models. *TEST* 19:417–451.

Wikle, C. K., and J. A. Royle. 2005. Dynamic design of ecological monitoring networks for non-Gaussian spatio-temporal data. *Environmetrics* 16:507–522.

Williams, B. K. 1989. Review of dynamic optimization methods in renewable natural resource management. *Natural Resource Modeling* 3:137–216.

Williams, B. K. 1996. Adaptive optimization and the harvest of biological population. *Mathematical Biosciences* 136:1–20.

Williams, B. K. 1997. Approaches to the management of waterfowl under uncertainty. *Wildlife Society Bulletin* 25:714–720.

Williams, B. K., and F. A. Johnson. 1995. Adaptive management and the regulation of waterfowl harvests. *Wildlife Society Bulletin* 23:430–436.

Williams, B. K., J. D. Nichols, and M. J. Conroy. 2002. *Analysis and management of animal populations: modeling, estimation and decision making.* Academic Press, San Diego, CA, USA.

Williams, B. K., R. C. Szaro, and C. D. Shapiro. 2007. *Adaptive management: the US Department of the Interior technical guide.* Adaptive Management Working Group, US Department of the Interior, Washington, DC, USA.

Williams, S. J. 2006. A comparison variance estimates of stream network resources. Masters project report. Department of Statistics, Colorado State University, Fort Collins, CO, USA. Online: <http://www.stat.colostate.edu/starmap/williams.csu.masters.report.2006.pdf>.

Willmott, C. J., and J. J. Feddema. 1992. A more rational climatic moisture index. *Professional Geographer* 44:84–87.

Wilson, D. E., F. Russell, J. D. Nichols, *et al.*, editors. 1996. *Measuring and monitoring biological diversity: standard methods for mammals.* Smithsonian Institution Press, Washington, DC, USA.

Wilson, K. R., and D. R. Anderson. 1985. Evaluation of two density estimators of small mammal population size. *Journal of Mammalogy* 66:13–21.

Wintle, B. A., M. A. McCarthy, K. M. Parris, *et al.* 2004. Precision and bias of methods for estimate point survey detection probabilities. *Ecological Applications* 14:703–712.

Wintle, B. A., M. C. Runge, and S. A. Bekessey. 2010. Allocating monitoring effort in the face of unknown unknowns. *Ecology Letters* 13:1325–1337.

Witczuk, J., S. Pagacz, and L. S. Mills. 2008. Optimizing methods for monitoring programs: Olympic marmots as a case study. *Wildlife Research* 35:788–797.

Witherington, B., P. Kubilis, B. Brost, *et al.* 2009. Decreasing annual nest counts in a globally important loggerhead sea turtle population. *Ecological Applications* 19:30–54.

Witwicki, D. 2010. *Variance and power for trend assessment of pilot data: integrated upland monitoring, Canyonlands National Park.* Natural Resource Report NPS/NCPN/NRR-2010/171. US Department of the Interior, National Park Service, Fort Collins, CO, USA.

Wood, S. N. 2006. *Generalized additive models: an introduction with R.* Chapman & Hall/CRC, Boca Raton, FL, USA.

Woodall, C. W., B. L. Conkling, M. C. Amacher, *et al.* 2010. *The Forest Inventory and Analysis Database Version 4.0: Database Description and Users Manual for Phase 3.* General Technical Report NRS-61. USDA Forest Service, Newtown Square, PA, USA.

Woodby, D. 1998. Adaptive cluster sampling: efficiency, fixed sample sizes, and an application to red sea urchins (*Strongylocentrotus franciscanus*) in southeast Alaska. Pages 15–20 *in* G. S. Jamieson and A. Campbell, editors. *Proceedings of the North Pacific Symposium on Invertebrate Stock Assessment and Management.* Canadian Special Publication of Fisheries and Aquatic Sciences No. 125, National Research Council of Canada, Ontario, Canada.

Wright, J. A., R. J. Barker, M. R. Schofield, *et al.* 2009. Incorporating genotype uncertainty into mark–recapture-type models for estimating abundance using DNA samples. *Biometrics* 65:833–840.

Wright, S. 1918. On the nature of size factors. *Genetics* 3:367–374.

Wright, S. 1921. Correlation and causation. *Journal of Agricultural Research* 10:557–585.

Yates, F. 1935. Some examples of biased sampling. *Annals of Eugenics* 6:202–213.

Yoccoz, N. G., J. D. Nichols, and T. Boulinier. 2001. Monitoring of biological diversity in space and time. *Trends in Ecology & Evolution* 16:446–453.

Yoshizaki, J., K. H. Pollock, C. Brownie, *et al.* 2009. Modeling misidentification errors in capture–recapture studies using photographic identification of evolving marks. *Ecology* 90:3–9.

Zhang, X., F. W. Zwiers, G. C. Hegerl, *et al.* 2007. Detection of human influence on twentieth-century precipitation trends. *Nature* 448:461–465.

Zucchini, W., D. L. Borchers, M. Erdelmeier, *et al.* 2007. *WiSP 1.2.4. Institut fur Statistik und Okonometrie, Geror-August-Universitat Gottingen*, Gottingen, Germany.

Zuckerberg, B., D. N. Bonter, W. M. Hochachka, *et al.* 2011. Climatic constraints on wintering bird distributions are modified by urbanization and weather. *Journal of Animal Ecology* 80:403–413.

Zuckerberg, B., A. M. Woods, and W. F. Porter. 2009. Poleward shifts in breeding bird distributions in New York State. *Global Change Biology* 15:1866–1883.

Index